Interpreting Quantum Theories

Interpreting Quantum Theories

The Art of the Possible

Laura Ruetsche
Department of Philosophy
University of Michigan

OXFORD
UNIVERSITY PRESS

Great Clarendon Street, Oxford OX2 6DP

Oxford University Press is a department of the University of Oxford.
It furthers the University's objective of excellence in research, scholarship,
and education by publishing worldwide in

Oxford New York

Auckland Cape Town Dar es Salaam Hong Kong Karachi
Kuala Lumpur Madrid Melbourne Mexico City Nairobi
New Delhi Shanghai Taipei Toronto

With offices in

Argentina Austria Brazil Chile Czech Republic France Greece
Guatemala Hungary Italy Japan Poland Portugal Singapore
South Korea Switzerland Thailand Turkey Ukraine Vietnam

Oxford is a registered trade mark of Oxford University Press
in the UK and in certain other countries

Published in the United States
by Oxford University Press Inc., New York

© Laura Ruetsche 2011

The moral rights of the author have been asserted
Database right Oxford University Press (maker)

First published 2011

All rights reserved. No part of this publication may be reproduced,
stored in a retrieval system, or transmitted, in any form or by any means,
without the prior permission in writing of Oxford University Press,
or as expressly permitted by law, or under terms agreed with the appropriate
reprographics rights organization. Enquiries concerning reproduction
outside the scope of the above should be sent to the Rights Department,
Oxford University Press, at the address above

You must not circulate this book in any other binding or cover
and you must impose the same condition on any acquirer

British Library Cataloguing in Publication Data
Data available

Library of Congress Cataloging in Publication Data
Data available

Typeset by SPI Publisher Services, Pondicherry, India
Printed in Great Britain
on acid-free paper by
MPG Books Group, Bodmin and King's Lynn

ISBN 978–0–19–953540–8

10 9 8 7 6 5 4 3 2 1

Contents

Preface	xi
List of Abbreviations	xv
List of Symbols	xvi
1. Exegesis Saves: Interpreting Physical Theories	1
1.1. Why be so particular?	1
1.2. Interpreting physical theories	5
1.3. In praise, and in defense, of the standard account	10
1.4. Criteria of adequacy for interpretations	10
1.5. Realism and pristine interpretation	12
1.6. Interpreting QM_∞	14
2. Quantizing	19
2.1. Rival quantum theories?	20
2.2. Physical equivalence and unitary equivalence	24
2.3. The Stone–von Neumann uniqueness theorem	30
2.3.1. Mechanics: classical to quantum	30
2.3.2. Classical theories as symplectic vector spaces	33
2.3.3. Hamiltonian quantization	35
2.3.4. The Stone–von Neumann theorem	37
2.3.5. Uniqueness illustrated	42
2.4. Pause	45
3. Beyond the Stone–von Neumann Theorem	46
3.1. Suspending weak continuity	46
3.1.1. The position and momentum representations	46
3.1.2. Are they "physical"?	50
3.2. Non-vanilla configuration spaces	53
3.2.1. Mr. Heisenberg's group	54
3.2.2. Phase spaces of other topologies	57
3.3. Infinitely many degrees of freedom	59
3.3.1. The thermodynamic limit: the infinite spin chain	59
3.3.2. QFT: the Klein–Gordon field	65
3.4. Conclusion	72
4. Representation Without Taxation: An Unstrenuous Tour of Algebraic Notions	73
4.1. What an algebra is	73
4.2. C^* algebras, abstractly	75

4.3.	C^* algebras, concretely	77
4.4.	The representation of a C^* algebra	82
	4.4.1. Faith and reducibility	84
	4.4.2. Unitary equivalence of representations	85
4.5.	Von Neumann algebras	86
4.6.	States on C^* algebra	89
	4.6.1. The GNS representation	89
	4.6.2. Foliums	94
4.7.	Von Neumann factors	98
4.8.	Conclusion and précis	99

5. Axioms for QM_∞ — 102
- 5.1. A declaration of sentiment — 102
- 5.2. Axioms for QFT — 104
- 5.3. Entanglement and locality in QFT — 109
  - 5.3.1. Motivating the Microcausality axiom — 109
  - 5.3.2. Entanglement in QFT — 113
- 5.4. Axioms for QSM — 114
- 5.5. Conclusion — 116

6. Interpreting QM_∞: Some Options — 117
- 6.1. What are we interpreting? — 119
- 6.2. Hilbert Space Conservatism — 122
  - 6.2.1. Some privileging strategies — 123
  - 6.2.2. Underprivileged? — 124
  - 6.2.3. Overprivileged? — 125
- 6.3. DHR selection theory — 127
  - 6.3.1. Motivation and set up — 127
  - 6.3.2. Reservations — 131
- 6.4. Algebraic Imperialism — 132
  - 6.4.1. Apologetic Imperialism — 133
  - 6.4.2. Bold Imperialism — 136
  - 6.4.3. Instantiation and necessity — 139
  - 6.4.4. Imperialism: excess and deficiency — 142
- 6.5. Mixed strategies — 143
- 6.6. Universalism — 145
- 6.7. Unpristine approaches — 146

7. Extraordinary QM — 148
- 7.1. Typing von Neumann factors — 149
  - 7.1.1. Dimension functions and subtypes — 154
- 7.2. Atomlessness and normality — 155
- 7.3. KMS states — 159
- 7.4. A modicum of modular theory — 162
- 7.5. Extraordinarily physical — 166

	7.5.1. Non-atomic von Neumann algebras in QFT	166
	7.5.2. Non-atomic von Neumann algebras in QSM	167
7.6.	Interlude	168
8. Interpreting Extraordinary QM		169
8.1.	Preparation	170
8.2.	The MBA and ordinary QM	173
	8.2.1. Characterizing the MBA	174
	8.2.2. Lattice entertain you	177
	8.2.3. A scheme and some instances	179
8.3.	The MBA and QM_∞	181
	8.3.1. Hope	182
	8.3.2. Tempered	183
	8.3.3. Interpretation unbound?	186
8.4.	Conclusion	188
9. Is Particle Physics Particle Physics?		190
9.1.	Particle physics	191
9.2.	Particle interpretations	193
9.3.	Pro particles	195
	9.3.1. Fock space, heuristically	195
	9.3.2. The particle notion as fundamental	199
9.4.	Anti particles: an argument, and a loophole	204
9.5.	Closing the loophole: incommensurable particle notions	206
	9.5.1. Fock space without the heuristic	206
	9.5.2. Unitary inequivalence and incommensurability	211
9.6.	The incommensurability of Jack's and Jill's particle notions	214
	9.6.1. Jill's spacetime	214
	9.6.2. Jack's notion, and Jill's	215
9.7.	Operationalizing the particle notion?	216
9.8.	The Unruh effect without particles	218
9.9.	Conclusion: the case against, restated and extended	219
10. Particles and the Void		221
10.1.	Extended particle notions	222
	10.1.1. An appeal to exfoliated predictions	222
	10.1.2. An appeal to Fell's theorem	225
	10.1.3. Universalizing	226
10.2.	Spacetime matters	229
	10.2.1. Killing particles and μ-born particles	229
	10.2.2. Adulteration?	232
10.3.	Matter matters	234
	10.3.1. For Jack	235
	10.3.2. More adulteration?	237
10.4.	Coherent states	239
	10.4.1. One degree of freedom	240
	10.4.2. n degrees of freedom	241

viii CONTENTS

		10.4.3. QFT	242
		10.4.4. Some properties of coherent representations	243
		10.4.5. Coherence: what is it good for?	244
	10.5.	Conclusion	246
11.	**Phenomenological Particle Notions**		247
	11.1.	Particle physics, redux	249
		11.1.1. The interaction picture and Haag's theorem	250
		11.1.2. Hope: Haag–Ruelle?	253
		11.1.3. The $(\phi^4)_2$ theory	255
	11.2.	Cosmological particle creation	256
	11.3.	Conclusion	260
12.	**A Matter of Degree: Making Sense of Phase Structure**		261
	12.1.	The thermodynamic limit: why go there?	262
		12.1.1. Ergodicity	263
		12.1.2. Phase structure	268
		12.1.3. Broken symmetry	270
	12.2.	Phase structure: a closer look	272
		12.2.1. Pure phases	272
		12.2.2. The set of equilibrium states	277
	12.3.	Extremist obstructions to explanation	281
		12.3.1. The phase argument	281
		12.3.2. The W^* argument	284
	12.4.	Complicating content	288
13.	**Interlude: Symmetry Breaking in QSM**		291
	13.1.	Introduction: Coalesced Structure	291
	13.2.	Symmetry and the Sharp Distinction	292
	13.3.	Broken symmetry in QM_∞	300
		13.3.1. The individual sense	300
		13.3.2. The decompositional sense	301
		13.3.3. Comparing the accounts	303
	13.4.	The decompositional account illustrated	305
		13.4.1. The ferromagnet	305
		13.4.2. Superconductivity	306
	13.5.	Coalesced structures in broken symmetry	309
14.	**Broken Symmetry and Physicists' QFT**		312
	14.1.	Introduction	312
	14.2.	Broken symmetry in QFT	314
		14.2.1. Overview	315
		14.2.2. An example: the massless Klein–Gordon field	317
	14.3.	Goldstone bosons	320
	14.4.	The Higgs mechanism	324
		14.4.1. The saga of the electroweak theory	325
		14.4.2. The abelian Higgs model	328

14.5. Coalesced structures in QFT?	330
14.5.1. Promissory notes	330
14.5.2. Unphased	332
14.6. A Sounder Principle?	336
15. Re: Interpreting Physical Theories	**340**
15.1. An anatomy of scandal	340
15.2. Realism sophisticated	344
15.2.1. The global argument	344
15.2.2. Structural Realism	346
15.3. QM_∞ and the morals	349
15.4. Law and possibility	351
15.5. Virtue reconceived	354
15.6. Fundamental physics	355
References	**357**
Index	**371**

Preface

Around the turn of the century, in an unkempt faux-gothic skyscraper in Pittsburgh, I got interested in quantum field theory (QFT). Reflecting a hope that philosophical engagement with quantum theories could address something other than the measurement problem and/or the Bell Inequalities, the interest was stoked by outstanding work on QFT going on all around me: Aristidis Arageorgis's monumental dissertation on QFT in curved spacetime (Arageorgis 1995); Hans Halvorson and Rob Clifton's beautiful papers on entanglement, Bell correlations, and complementarity (Clifton and Halvorson 2001a; 2001b; 2002; Clifton et al. 1998; Halvorson and Clifton 2000); John Earman's cryptic mutterings in the free weight room at Club One. I wanted to do philosophy of QFT too. Alas, there was a significant difference between me and the sages just mentioned. I lacked the technical wherewithal to plunge into the existing literature. I knew enough about Hilbert space to readily follow (and occasionally write) journal articles about the philosophy of ordinary quantum mechanics (QM), but I hadn't the foggiest notion what a C^* algebra was. Fighting back a chilling suspicion that this was a mystery too subtle for me, I set off in search of a published work (a brief article, I hoped) that would tell me what's what.

This being a preface, it should come as no surprise that, not finding the work I was after, I've tried to write it.[1] I think I've mostly failed: the work I was after would have been significantly shorter, much clearer, and immensely more rigorous than what I seem to have produced. But I did my best, and I like to think the following pages have some redeeming features. I'll try in this preface to disclose what the reader needs to know to make the most of them.

In the course of my efforts, I became convinced that some of QFT's most philosophically provocative features are ones it shares with the thermodynamic limit of quantum statistical mechanics (QSM). Adding the thermodynamic limit to my portfolio, I've labelled the collective of theories it contains "QM_∞." In spite of tending to be the sort of person who either neglects to have views or refrains from advertising them when she does, I managed while working on this project to acquire a variety of views, both about the interpretive issues I aspire to explicate and about their relevance to more general philosophical debates. Some of the material that follows attempts to sketch these views. Thus the book has three, somewhat entangled, aspirations. First, to serve

---

[1] Teller's 1995 *An Interpretive Introduction to Quantum Field Theory* does an excellent job of delivering what its title promises. It is not only accessible but also wide-ranging and adroit in the interpretive agendas it charts. Laudably, and unlike the volume before you, its purview includes physicists' QFTs. But just because of some of these virtues, it steers clear of the language of operator algebras, the language that served as the vernacular for the literature I was hoping to penetrate.

as an introduction to the philosophy of QM_∞. Second, to contribute to that endeavor. Third, to persuade those fortunate enough to not themselves be philosophers of physics that the endeavor in question might have something to do with them.

Sorted according to these aspirations, the large-scale structure of the book is something like this. Chapters 1 and 15 sketch the general issues and what I take QM_∞ to offer them. Chapters 2–5 describe how and why questions of interpretation *peculiar to* QM_∞ arise, and develop tools for addressing those questions. Chapter 6 catalogs a range of strategies for interpreting QM_∞. The groups of chapters 7–8, 9–11, and 12–14 each sustain discussions that, although interacting with other issues broached in the book, aim to be reasonably self-contained contributions to the philosophy of QM_∞. Chapters 7–8 address the adequacy to the demands of QM_∞ of strategies for interpreting ordinary QM. Chapters 9–11 tackle the problem of particles in QFT. Chapters 12–14 discuss the significance of phase structure and broken symmetry in QSM and QFT.

Throughout, my expositions of technical notions are adamantly informal, and meant to be accessible to readers who encountered QM through its philosophy, for instance through such exemplary works as Redhead (1989), Hughes (1989), or Albert (1994). Thus I will assume readers to have a working acquaintance with vector spaces, operators, and the so-called axioms of ordinary non-relativistic QM, but I won't presuppose systematic mathematical knowledge beyond that. QM's most provocative features—entanglement, with its whiff of 'non-locality', and the measurement problem—can be and typically are illustrated by means of finite dimensional Hilbert spaces.[2] Much of the standard philosophical literature is accordingly insulated from the niceties of convergent sequences, measure theory, and continuous functions. Not presupposing acquaintance with such notions, I resolve to offer brief "gistifications" of them when they appear in the text. Indeed, my governing principle will be that if you can be familiar with standard philosophy of QM texts without knowing about X, I should at least gistify X for you.

This policy, and the widely dispersed gistifications it spawns, is liable to drive those brought up as mathematicians, and so conditioned to regard such notions as elementary, crazy. I've taken several steps meant to minimize tedium and annoyance for readers with more extensive mathematical backgrounds. First, I'll here warn them when chunks of the text that follows are largely expository. Chapter 4 presents the basics of operator algebra theory; §7.1 sketches type theory for von Neumann algebras; §7.4 gives a satellite-level view of modular theory; §8.2.2 reviews some lattice theory. (I told you the gistifications were dispersed.)

Second, where possible, I employ a convention to distinguish between notes that gloss mathematical jargon that may be unfamiliar to the uninitiated and notes that discharge standard note functions of issuing asides, apologies, recommendations, and

---

[2] This isn't to say QM_∞ is irrelevant to these problems. See e.g. Barrett (2002).

so on. Counters for the former will be unadorned arabic numerals (like ^{7}); counters for the latter will be arabic numerals subscripted by a "g" (for "glossing"; like $^7{}_g$). Gistifications too long and involved to happily fit in notes will appear in Scholia, which the mathematically sophisticated have leave to skip. Some elementary notions will be gistified by way of examples, examples whose features will sometimes matter to later discussion. The mathematically sophisticated also have leave to skip examples, but the leave comes with a warning that they may occasionally be compelled by later developments to retrace their steps.

To make things easy to find, each chapter deploys two parallel numbering systems. The first applies to equations, which are numbered consecutively: "equation 4.12" will be the next equation after "equation 4.11," and so on. The second applies to the barbarous category of "everything else"—facts, examples, definitions, and so on. These "non-equations" are also numbered consecutively, so that "Fact 7.2" may separate "Example 7.1" from "Example 7.3." (A note on nomenclature. Results that would, according to something like their "depth", appear in proper mathematical expositions under the various headings "Theorem," "Corollary," "Proposition," etc., will be indiscriminately labeled here as "facts." My possibly idiosyncratic reason for this ugly terminology is that I have a scruple against a labeling system that sorts results according to their depth, when I myself lack the taste requisite to execute such a sort.)

In spite of its dearth of technical sophistication, the text occasionally goes into mathematical or physical detail that can be omitted without blurring the big picture. Sometimes passages or sections developing these details are labeled "digressions" or equipped with "short versions," meant to serve those whose appetite for details is flagging as surrogates for the section itself. The following lists of abbreviations and symbols, as well as what I hope is a decent index, should also direct the reader to expository passages.

So much for my mathematical apologies. Now for my physics apologies. This is neither a QFT textbook nor a guide to calculation. I try to offer a picture of where theories come from and how they fit together, not an account, reliable in precise detail, about genealogical relations between different significant formulas. The physics formulas I display are usually in "natural units" which set the physical constants h (Planck's constant), c (the speed of light), and G (the gravitational constant) to 1. Although, I've made every effort to be consistent, I've drawn upon a variety of texts, deploying a variety of different notations, sign, and normalization conventions. So there may be places where it's charitable to take me to set not only these physical constants but also 2π, i, and -1 to 1. While we're on the subject, be warned that in what follows, both the overbar and "*" denote the conjugate of a complex number; the latter also sometimes denotes the involution of an element of an algebra. When the involution is furnished by the adjoint operation on Hilbert space, it is also sometimes denoted †. To complete the circle, sometimes the overbar indicates the closure of a set of operators. When I think disambiguation is needed, I try to supply it.

It is my pleasure to close with a brief catalog of characters who, in one way or another, helped me through this. The usual disclaimers apply: I and I alone am to blame for the errors that remain in this work; I'm especially sorry if those errors include omitting to thank those who deserve thanks. For their patience, good humor, and competence, people at or affiliated with Oxford University Press: editor Peter Momtchiloff, assistant editor Eleanor Collins, production editor Elmandi du Toit, cover designer Laura O'Brien, copy-editor Sarah Barrett, and proofreader Lesley Rhodes. For his wizardry with diagrams: Dmitri Gallow. For partially funding leaves during which I worked on the project: The American Council of Learned Societies (Ryskamp Fellowship, 2001–2002), and the Center for Advanced Study in the Behavior Sciences (CASBS Fellowship, 2006–2007). (At CASBS I was assigned the office in which Thomas Kuhn is reputed to have drafted *Structure*; he evidently succeeded much better than I in resisting what struck me as the imperative to repeatedly ride a bicycle up Old la Honda Road.[3]) For donating its heart valve to John Earman, an anonymous cow. For remaining mad, bad, and dangerous to know, John Earman. For reading and commenting on all or part of the manuscript, in some cases, repeatedly: Dave Baker, Jeremy Butterfield, John Earman, Doreen Fraser, Hans Halvorson, Richard Healey, Stephen Leeds, Naomi Luce, Tracy Lupher, John Manchak, Brad Skow, Paul Teller, and an anonymous reviewer for Oxford University Press. (Hans, Paul, and the Johns were particularly heroic.) For aiding and abetting the project in other ways: Frank Arntzenius, Bob Batterman, Jemina Ben-Menahem, Caro Brighouse, Harvey Brown, Jeff Bub, Rob Clifton,†2002 Cian Dorr, Meir Hemmo, Nick Huggett, Chuang Liu, David Malament, Sandy Mitchell, Wayne Myrvold, the Pittsburgh Center for Philosophy of Science, Itamar Pitowsky,†2010 David Sipfle, Chris Smeenk, Alexander Wilce, Andrea Woody. For participating, participants in reading groups and seminars at Pitt, and other audiences in other venues. For providing mammalian companionship during all or part of my efforts, such as they were: Grisbi, Stargell, and Honus.†2009 For most of the above, and then some, Gordon Belot.

---

[3] The last trip I timed: 23:50. Many have done it faster, but Thomas Kuhn is not among them.

List of Abbreviations

The arabic numerals in the rightmost column give the page number (where appropriate) of the first occurrence of the abbreviation in question.

BCS	Bardeen–Cooper–Schrieffer (model of superconductivity)	167
CARs	canonical anti-commutation relations	14
CCCRs	circular canonical commutation relations	58
CCRs	canonical commutation relations	14
DHR	Doplicher–Haag–Roberts (selection theory)	127
ETCCRs	equal time canonical commutation relations	195
GNS	Gelf'and–Neimark–Segal (construction)	92
GTR	general theory of relativity	7
HEP	high energy physics	191
iff	if and only if	...
KMS	Kubo–Martin–Schwinger (criterion of equlibrium)	125
l.h.s.	left-hand side (of an equation)	...
MBA	Maximal Beable Approach	169
QFT	quantum field theory	xi
QM	quantum mechanics	xi
QM_∞	quantum mechanics for "infinite" systems	xi
QSM	quantum statistical mechanics	xi
r.h.s.	right-hand side (of an equation)	...
STR	special theory of relativity	109
w.r.t.	with respect to	...

List of Symbols

In quasi-alphabetical order. The arabic numerals in the rightmost column give the page number of the first significant occurrence of the symbol in question.

Symbol	Description	Page
$\hat{A}$	Hilbert space operator	...
$\hat{A}^\dagger$	adjoint of the Hilbert space operator $\hat{A}$	...
$\hat{a}_k, \hat{a}_k$	annihilation operator for the mode k	68
$\hat{a}_k^\dagger, \hat{a}_k^\dagger$	creation operator for the mode k	68
$\mathfrak{A}$	C^* algebra	75
$\mathfrak{A}'$	commutant of the algebra $\mathfrak{A}$	86
$\mathfrak{A}_{CAR}$	canonical anti-commutation algebra	60
$\mathfrak{B}(\mathcal{H})$	algebra of bounded operators on $\mathcal{H}$	21
$\mathbb{C}, \mathbb{C}^n$	(n-tuples of) complex numbers	...
$C(X)$	algebra of complex-valued functions on X	74
$\mathfrak{C}$	algebra of compact operators on $\mathcal{H}$	265
$\mathcal{F}, \mathcal{F}_\omega, \mathcal{F}_\pi$	folium, of state ω, of a representation π	94
$\mathfrak{F}(\mathcal{H})$	Fock space over $\mathcal{H}$	68
$\mathfrak{F}$	DHR field algebra	127
$\mathcal{H}$	Hilbert space	21
$\ell^2(n)$	Hilbert space of square summable n-tuples	23
$\ell^2(\infty)$	Hilbert space of square summable infinite sequences	23
$L^2(X)$	Hilbert space of square integrable functions of X	23
$\mathcal{L}$	Lagrangian	...
$\mathfrak{M}$	von Neumann algebra	86
$\mathfrak{M}(2)$	von Neumann algebra of 2×2 matrices	61
N_k, N_k^π	variety k number operator (for representation π)	69, 212
N, N^π	total number operator (for representation π)	69, 216
π	representation	83
$(\mathcal{Q}, \mathcal{S})$	kinematic pair	25
$\mathbb{R}, \mathbb{R}^n$	(n-tuples of) real numbers	...
$\mathfrak{R}$	von Neumann factor	98
(S, Ω)	symplectic vector space	33
$\mathcal{S}_\mathfrak{A}$	States on $\mathfrak{A}$	132

$\sigma(x), \sigma(y), \sigma(z)$	Pauli spins	59
σ_t^ω	Modular group for the state ω	169
$\mathfrak{T}^+(\mathcal{H})$	Trace-class operators on $\mathcal{H}$	21
$|\psi\rangle$	Hilbert space vector	22
$\mathfrak{Z}\mathfrak{M}$	center (of $\mathfrak{M}$)	99
$\mathbb{Z}$	positive and negative integers	...
$\emptyset$	null set	...

1

Exegesis Saves
Interpreting Physical Theories

1.1 Why be so particular?

If, like Plato's Eleatic Stranger, we had a mania for dichotomous division, we might sort philosophers into two types: the generalists, distinguished by their sweeping and lofty concerns with (for instance) the natures of truth, meaning, virtue, and knowledge; and the particularists, distinguished by their obsessive attention to earthly details, such as the allocation of resources in an industrial society, or the physics of quantum non-locality. In his 1974 presidential address to the Eastern Division of the American Philosophical Association,[1] John Rawls considers his particular philosophy, moral theory, which he understands as

> the study of substantive moral conceptions, that is, how the basic notions of the right, the good, and moral worth may be arranged to form different moral structures. Moral theory tries to identify the chief similarities and differences between these structures and to characterize the way in which they are related to our moral sensibilities and natural attitudes, and to determine the conditions they must satisfy if they are to play their expected role in human life. (1974, 5)

Philosophers of physics may recognize their own labors as analogously particularized. Studying substantive physical theories, philosophers of physics seek to articulate the structures by which those theories represent, predict, and explain. They try to identify the chief similarities and differences between prominent theoretical structures, and to characterize the way those structures are related to empirical data as well as to metaphysical home truths. They seek to determine the grounds for, and the meaning of, acceptance of theoretical structures as successful science.

In his presidential address, Rawls takes on the prejudice that moral theorists can't get on with their jobs until generalists—epistemologists, philosophers of language, philosophers of mind—have finished theirs. "It is thought," he observes, "first that other philosophical questions cannot be satisfactorily resolved until the problems of

---

[1] Brought to my attention by Jon Mandle, to whom I am grateful.

epistemology, or nowadays the theory of meaning, are already settled; and second that these prior questions can be investigated independently: their answers neither rest on nor require any conclusions from other parts of philosophy" (1974, 6). The prejudice is that generalist philosophy can be conducted in isolation from particularist philosophy, and indeed that the former's deliverances serve the latter in some sort of legislative role. Brought home to the philosophy of physics, one expression of the prejudice is the thought that one must consult a generalist's account of explanation to tell whether quantum mechanics (QM) explains the distant correlations it predicts. Another expression is the expectation that the scientific realism debate can be conducted (and maybe even settled) in abstraction from the details of particular scientific theories.

Against the expression of prejudice in the sphere of moral philosophy, Rawls contends not only that "preoccupation with the problems that define [generalists'] subjects may get in the way and block the path to advance" (1974, 6) but also that ground-level engagement with the structures of substantive moral conceptions may clear that path. Contending that "the analysis of moral concepts, the existence of objective moral truths, and the nature of persons and personal identity, depend on an understanding of these structures," Rawls concludes that "[t]he problems of moral philosophy that tie in with the theory of meaning and epistemology, metaphysics, and the philosophy of mind, must call upon moral theory" (1974, 6). His conclusion liberates the particularist, and gives moral theory a voice in moral philosophy more generally cast.

I mean this book as an exercise in liberatory particularism. Its primary focus is the interpretation of quantum field theory (QFT) and the thermodynamic limit of quantum statistical mechanics (QSM)—theories I gather under the heading "QM_∞." (The subscript will be explained presently.) The book's primary aim is to suggest that Rawls was right about more than moral theory. I will contend that questions generalists in the philosophy of science typically address in lofty abstraction—questions such as, "What are laws of nature?" and "What epistemic attitude should we strike toward successful physical theories?"—are transformed when brought into contact with QM_∞. For example: it is generally supposed by those who consider the matter in lofty abstraction that to interpret a physical theory is to identify the set of worlds possible according to that theory. The idea is that a theory says what it says by allowing some states of affairs (the worlds possible according to the theory) and forbidding others (the worlds by its lights physically impossible). It is also usually supposed that it's the job of a theory's laws to circumscribe its collection of possible worlds: its possible worlds are exactly those consistent with its laws. While which (if any!) of a theory's possible worlds is actual may be a contingent matter (the historical accident of what the initial conditions were, say), what the theory's possible worlds are depends only on its laws.[2] Thus loftily considered, physical possibility is what one might call

---

[2] See e.g. Suppe (1977, 226–8) and van Fraassen (1989, 222–3) on "laws of coexistence" and "laws of succession": the former determine which instantaneous configurations are "physically possible" (Suppe 1977, 226); the latter "select the physically possible trajectories in the phase space" (ibid).

"unimodal." Everything that's physically possible is physically possible in the same way, and interpretation is the *pristine* business of identifying a theory's physical possibilities by bringing one's own metaphysical scruples and technical sophistication to bear on its laws.[3]

Underlying this methodological *ideal of pristine interpretation* is a distinction between what holds at each world of which a theory T is true and what varies—as John Earman put it in his 2002 presidential address to the Philosophy of Science Association, "a distinction between what holds as a consequence of the laws of physics and what is compatible with but does not follow directly from those laws" (2004a, 1229).[4] The class of what applies in all settings where T applies includes T's laws, as well as metaphysical, methodological, and mathematical truths; the class of what changes from setting to setting includes initial/boundary conditions, as well as practical considerations parochial to the settings which give rise to them. To adhere to the ideal of pristine interpretation is to invoke only considerations from the former class when circumscribing the collection of worlds that are possible according to T. It is to aim at a commodity aptly described as *the* interpretation of QM, all told, rather than a set of commentaries on the theory's subspecies and/or piecemeal applications. A virtue Everett famously claims for his own relative state formulation of QM aptly expresses the limiting case of ideal interpretation: all extra-theoretical considerations drop away, and "the theory itself sets the framework for its interpretation" (Everett 1957, 462).

The "Samurai TV repairman" portrayed by John Belushi on *Saturday Night Live* in the late 70s is an apt caricature of the pristine interpreter and his commitment to unimodality. No matter what ailing appliance was brought forth for his attention, Belushi's TV repairman would respond in the same way. Without pausing to examine the workings of the ailing appliance, he'd unsheath an enormous sword conveniently hung from his belt, and with deranged vigor cut the TV set in two. Just so, T's pristine interpreter, confronted with the space of logically possible worlds, splits off the worlds possible according to T in one fell swoop. Just as Belushi's TV repairman defiantly ignores the interior details of the appliance before him, the pristine interpreter consigns differences *between* circumstances that fall under T to "the astronomer, geographer, geologist, etc." (as Wigner, Houtappel, and van Dam (1965, 596) once memorably put it). For the pristine interpreter, variations between worlds to which T applies make no difference to what's possible according to T.

The ideal of pristine interpretation is not often directly voiced.[5] But it pervades the practice of philosophy of physics over the last half-century. Witness competing interpretations of spacetime theories, which by muscular application of differential geometry and the like, tie their banners to competing accounts of the metaphysics of

---

[3] I refer those who are wondering whether an uninterpreted theory even *has* laws to §1.2's contention that the theories we care about are already partially interpreted.

[4] Ch. 13 addresses Earman and Wigner's contention that this "sharp distinction" is *presupposed* by notions of symmetry and invariance.

[5] An exception may be Fine (1986, 148), whose NOA parts ways with the ideal.

substance and identity.[6] And observe how Redhead's classic *Incompleteness, Nonlocality, and Realism* characterizes the task of interpreting QM:

> [An interpretation of QM] is simply some account of the nature of the external worlds and/or our epistemological relation to it that serves to *explain* how it is that the statistical regularities predicted by the formalism with the minimal statistical interpretation come out the way they do.... (1989, 44)

By suggesting that an interpretation of QM should emanate from general principles of metaphysics and epistemology, Redhead evokes the ideal of pristine interpretation. He moreover implicates the *generality* of the principles informing an interpretation in its capacity (constitutive, he reckons, of its status *as* an interpretation) to explain. Contending that "theories that lack an interpretation...simply do not contribute to our *understanding* of the natural world" (p. 45), Redhead attributes the explanatory oomph of an interpretation X of a theory Y to "the 'unifying' effect of X. A few general principles about the nature of reality expressed in X comprehend a wide variety of seemingly unconnected observational regularities, including Y" (p. 45).

There is something *extremist* about the ideal of pristine interpretation. The pristine interpreter of quantum theories announces principles for their interpretation, and holds to them come what may. This policy ensures that each theory receives at most one interpretation, the one determined for it by the antecedent principles, an interpretation whose content is independent of "geographical" considerations that arise along with particular applications of the theory. We can imagine, in contrast, a less principled and more pragmatic approach to interpreting physical theories, one which allows "geographical" considerations to influence theoretical content, and so allows the same theory to receive different interpretations in different contexts. I will try in this book not only to imagine such an approach to interpreting physical theories, but also to defend it against the ideal of pristine interpretation. I will argue that the best way to support QM_∞'s explanatory aspirations is to give up on the idea that a theory's laws on their own—or even in concert with "a few general principles" in the form of duly abstract mathematical, metaphysical, methodological, etc., considerations—delimit what worlds are possible according to that theory. Instead, I will contend, those worlds should be characterized in different ways for different extranomic (factual or material or explanatory or maybe even practical) circumstances. Far from unimodal, physical possibility fractures on these circumstances into a kaleidoscope of varieties, varieties indexed the extranomic contingencies that condition them. The details of particular settings matter in something like the way laws are generally supposed to matter: they matter to what generalizations hold (and perhaps also to what properties are appropriate for involvement in generalizations) and thus to what explanations those generalizations support. On this view, contingency adulterates the delimitation of physical possibility.

---

[6] The vast literature on the dreaded hole argument (catalyzed by Earman and Norton 1987) is a good place to see this in action.

The philosopher's task is accordingly adulterated. It's to notice which contingencies matter, to understand how they matter, and to try to decide whether their mattering makes a difference to characteristically philosophical concerns.

My argument supposes that a theory ought to be interpreted in a way that enables its explanatory capacity. The nature and significance of a theory's explanatory capacity is a central bone of contention in the scientific realism debate, and an interpretation of a theory is what a realist about that theory believes. So it shouldn't be surprising that my discussion of QM_∞ has a collateral consequence for the realism debate. It is that there often isn't a single interpretation under which a theory enjoys the full range of virtues realists are wont to cite as reasons for believing that theory. The theory manifests different virtues under different ways to characterize its set of possible worlds. One might be tempted to count this as a strike against realism—evidence of an "underdetermination of interpretation by theory" exacerbating whatever underdetermination of theory by evidence may obtain. But I think it points to a mistaken attitude toward interpretation, an attitude with which most parties to the realism debate operate.

Reflecting the prejudice Rawls bemoans, the attitude is that interpretation is an afterthought. Once the lofty generalist decides whether or not we should believe successful scientific theory T, the narrow particularist tells us what we believe (or not) when T is set equal to (say) QFT. One problem with this attitude is that arguments which look compelling when conducted about T in lofty abstraction can disintegrate when T is instantiated as a genuine, interpreted theory. Another problem is that accounts of theoretical virtues (such as empirical adequacy, explanatory reach, internal and external coherence, continuity with future science, and so on) that adhere cleanly to T considered in abstraction may disintegrate when T is instantiated and interpreted. The lofty debate is conducted in terms that obscure how real theories possess the virtues they possess. It is often a theory *under an interpretation* that predicts, explains, and promotes understanding. To the disappointment of the realist, there may not be a single interpretation under which a given theory accomplishes all those things. But it is to misunderstand this circumstance to treat it as grounds *against* epistemic commitment to a theory. A theory that underdetermines its own interpretation can be capable of a sort of semantic indecision that's a scientific resource, a ground for commitment to that theory. Or so I will suggest.

The balance of this chapter reviews a lofty account—the best lofty account I can muster—of the interpretation of physical theories. It also sketches QM_∞ in sufficient depth to suggest how its exploration could refine and redirect this account, and with it, our thought about the natures of theories and theoretical virtue.

1.2 Interpreting physical theories

This section reviews an account of what it is to interpret a physical theory that I take to be widespread, although often implicit. Explicit variations on the account are most often found in the course of discussions of the model-theoretic "semantic view" of theories. I greet this circumstance with trepidation. There is a vast literature evaluating the semantic view, and another vast literature examining models in the sciences. I fear that my invocation here of the notion of model will be taken for a stand on issues debated in those literatures. I don't mean to take a stand; I mean only to help myself to the two uncomplicated and appealing (if vague) ideas that mathematical physics is formulated in terms of mathematical structures, and that mathematical structures can have physical instantiations.

The account of interpretation starts with the idea, shared in some form by philosophers as various as Ludwig Wittgenstein and David Lewis, that the content of a proposition is given by the set of possible worlds of which that proposition is true. The idea is widely embraced because there is much to recommend it. After all, someone who understands a proposition—who grasps its content—is able, at least in principle, to recognize when it obtains, as well as to distinguish the states of affairs that make it true from those that falsify it. Extended in a straightforward way to theories entire, the idea becomes

> [The Standard Account] The content of a theory is given by the set of worlds of which that theory is true.

This idea is sufficiently deep-seated and widespread that I've called it the *standard account*.

A testament to the deepseatedness of the standard account is how squarely it frames projects central to metaphysics and philosophy of physics. A basic metaphysical concern is how to zone the space of *logically possible* worlds. Take a logically possible world to be a maximal set of consistent propositions.[7] A significant zoning restriction is to consider only those possible worlds consistent with propositions stating the laws of nature. The zone so defined is the space of *nomologically possible* worlds.

There are sweeping and homely ways for philosophers of science to concern themselves with nomologically possible worlds. The sweeping way addresses the principles behind the zoning restriction, by articulating and evaluating general accounts of laws of nature. Van Fraassen describes the homely way: "When we come to a specific theory, there is an immediate philosophical question, which concerns the content alone: *how can the world possibly be the way this theory says it is?* This is for me the foundational question *par excellence*" (van Fraassen 1989, 193). Take some physical theory T, and someone who can't say what laws of nature (in general) are. This someone might nevertheless have enough of a handle on T to embark on the project of characterizing and individuating the worlds of which T would be true—the task (van Fraassen remarks) of equipping T with content.

To characterize the worlds of which a theory T would be true is to supply as good an account of possibility-according-to-T as anyone could demand. It is to explicate the

---

[7] I'll soon admit that the propositional calculus is an inappropriate vehicle for the content of physical theories.

genus of nomological possibility that is physical possibility, with T acting as the species. It is also to answer on behalf of T what van Fraassen calls "*the question of interpretation*: Under what conditions is the theory true? What does it say the world is like?" (1991, 242). To interpret a physical theory is not only to equip it with content but also to explicate the notion of physical possibility allied with it.

Interpreting a theory articulates its truth conditions, but it doesn't follow that one has to be committed to the truth of theories to care about interpreting them. Van Fraassen is agnostic, and he cares. So might you, if you think interpretation promotes understanding of the theory (presuming that understanding is afforded by the grasp of truth conditions), or articulates its explanatory potential (presuming that a theory explains a phenomenon by situating it in the space of possibilities its laws allow), or serves what Sellars identifies as the philosopher's characteristic task of "understand[ing] how things in the broadest possible sense of the term hang together in the broadest possible sense of the term" (1963, 37) (presuming, as Sellars does, that among the hanging things are our best scientific theories). Describing what the world would be like if the theory were true hardly commits one to believing that it *is* true; characterizing a theory's possible worlds or models hardly commits one to claiming that the actual world is among them. The old-fashioned realist takes on these extra commitments: believing that the theory is true, she numbers the actual world among its models.

On the standard account, then, to interpret a theory is to characterize the worlds possible according to it. These possible worlds are (i) models (in something like the logician's sense) of the theory, and (ii) characterized as physical. I'm not going to try to say exactly what (ii) amounts to, although I hope it's clear that models characterized in terms of configurations of matter in spacetime count, and that models characterized in terms of sequences of natural numbers don't (unless, of course, they are accompanied by an informative account of why we should regard sequences of natural numbers as physical).

It is essential to understand that, in the present sense of interpretation, the vast majority of the theories philosophers talk about *are already partially interpreted*. Otherwise they wouldn't be theories of *physics*. These theories typically come under philosophical scutiny already having been equipped, by tradition and by lore, with an interpretive core almost universally acknowledged as uncontroversial. In the case of the general theory of relativity (GTR), this received core includes Einstein's field equations, along with the idea that their solutions describe space and time, energy and matter; in the case of QM, this received core incorporates canonical [anti]commutation relations, along with the quantization and statistical algorithms. There may also be, by tradition and by lore, theoretical foci of longstanding disagreement: the interpretation of the Born rule, say. And sustained philosophical attention to a theory is likely to trigger additional interpretive disputes that arise along with the attempt to extend and refine its received core: disputes about how exactly a substantivalist should individuate models of the Einstein field equations, say.

That the theories philosophers care about arrive already partially interpreted explains how the pristine interpreter can take a theory's laws to constrain its interpretation: the principles we call its laws can be part of what's left in the received core of a theory when you strip away what different interpreters disagree about. I think that partially interpreted theories can explain, predict, and so on; but I also think (and will try to argue in what follows) that sometimes a theoretical explanation's *bona fides* can be secured *only* by saying more about its interpretation than the received core does.

We can distinguish two phases in the interpretation of a physical theory (see van Fraassen 1989, ch. 9, or Suppe 1977, 221–30). One phase is *structure-specifying*. In this phase, the interpreter characterizes the structures by which the theory would represent physical reality. For example, the interpreter of general relativity might nominate structures of the form $(\mathcal{M}, G_{ab}, T_{ab})$, where the first entry is a manifold and the second and third entries are tensor fields defined on that manifold and satisfying Einstein's Field Equations. (Notice that the presentation proceeds with less formality than is generally demanded of axiom systems admitting models in the logician's sense. This is entirely appropriate. The philosopher of physics interprets physics, not logic. Notice also that the structure-specification is an interpretive enterprise in which physicists, *qua* physicists, engage. This underscores the observation of the last paragraph, that physical theories come to philosophers' attention already partially interpreted.) In the other, *semantic*, phase of interpretation, the interpreter identifies the physical worlds which model the theory so structured. So our sample interpreter might declare that a collection of spacetime points conceived as substances and bearing properties encoded (in a way the interpreter is beholden to elaborate) by the stress energy and curvature tensors instantiates the structure $(\mathcal{M}, G_{ab}, T_{ab})$. Moving from the structure-specifying through the semantic phase, the interpreter characterizes, and characterizes *as physical*, the worlds possible according to the theory. We will see that, like thermodynamic phases, the semantic and structure-specifying phases of interpretation can intermingle and coexist.

For the myriad physical theories admitting a Hamiltonian formulation, the structure-specifying phase of interpretation decomposes neatly into three elements (see, again, van Fraassen 1989 or Suppe 1977). The first element is a specification of the theory's *state space*. In the mercifully simple case of the classical theory of a particle of mass m moving in one dimension, the state space comprises ordered pairs (q, p) of position and momentum values. The second element is a specification of the set of physical magnitudes or *observables* the theory recognizes. In the same simple case, observables are functions from elements of state space to the real numbers $\mathbb{R}$; for example, the physical magnitude *kinetic energy* is given by the function $f_E(q, p) = p^2/2m$. Together, these two elements constitute a *kinematics* for the theory.

The third element of structure specification is the theory's *dynamics*, that is, its account of the time development of states and observables. In classical mechanics, dynamics takes the form of trajectories through state space determined by Hamilton's equations, once a Hamiltonian function $H(q, p)$ is provided. According to this structure-specification for classical mechanics, the structures by which the theory would represent physical reality are state spaces with dynamical trajectories imposed. (Each trajectory corresponds to what I've been calling a "world" possible according to the theory.) Insofar as these trajectories are all instances of a dynamical scheme furnished by Hamilton's equations, the structures form "a family of state space types" (van Fraassen 1989, 223), with the Hamiltonian scheme supplying the kinship relation. "Thus theories are structures, these structures being phase spaces with configurations imposed on them in accordance with the laws of the theory" (Suppe 1977, 227).

In the *semantic* phase of interpretation, the project is to characterize models of the theory—instantiations of the structure specified—*in physical terms*. Straightforwardly construed, a semantics can take the form of an account of which propositions attributing determinate values to magnitudes recognized by the theory are true of a system represented by a state of the theory. The characterization qualifies as physical if we have grounds for commitment to the status of (some of) the magnitudes in question as physical. A straightforward semantics for the simple classical theory is: *"Observable O has value x in state (q, p)" is true just in case* $f_o(q,p) = x$. But semantics needn't be so straightforward. It can, for instance, be conducted in terms exogenous to those used to present the theory structure. Terms supplied by antecedent ontology are an example, as when an interpreter of QFT takes Fock space structures to describe possible *particle* configurations.

Let us summarize the standard account: to interpret a physical theory is to characterize the worlds possible according to that theory. Two phases of this characterization can be distinguished. One phase identifies the theory's structures: its states, observables, and dynamics. The other characterizes the physical situations that count as models of the theory so structured. Interpretation is an exercise in nomic articulation: a theory's laws guide the characterization of its possible worlds; the interpretation of a theory is at the same time an explication of the notion of nomological possibility allied with the theory.

Understanding interpretation as an exercise in nomic articulation enables us to notice one way the ideal of pristine interpretation is reflected in the laws of nature literature (see e.g. Carroll 2004). An interpretation identifies the worlds possible according to a theory. So do that theory's laws. According to most accounts, laws of nature are (or imply) generalizations that aren't accidental. The intuition is that "All unsupported bodies in the vicinity of the earth fall toward it" is, or has something to do with the, nomic; "all the coins in my pocket are dimes" does not. John Carroll offers a diagnosis of what's un-nomic about propositions like the latter: "What blocks it from being a law is that something in nature, or really a certain sort of initial condition of the universe, an absence of something in nature, explains the regularity" (Carroll 2007, 74). Refracted through this concern with distinguishing real laws from pretender generalizations, the idea that a theory's laws determine what's possible according to that theory assumes a different form. It becomes the idea that what's possible according to a theory shouldn't depend on accidental particulars, like actual initial or boundary conditions. But such non-dependence is exactly what the commitment to unimodality requires, and the ideal of pristine interpretation pursues. By allowing only considerations of metaphysics and theoretical structure to guide the identification of the worlds possible according to a theory, the ideal insulates that identification from the influence of contingencies and accidents.

1.3 In praise, and in defense, of the standard account

The practice of philosophers of physics speaks in favor of the standard account of interpretation. Most things philosophers of physics count as interpretive questions are interpretive questions in its sense. A question about a physical theory's state space: does the state of a quantum system include, à la Bohm, a determinate configuration? A question about the observables appropriate to a physical theory: must the observables of general relativity be generally covariant, and if so, in what sense? A question about the dynamics of a physical theory: can a full Newtonian dynamics be formulated on a reduced phase space encoding information only about *relative* positions and velocities? A question about a physical theory's semantics: do all propositions attributing determinate values to observables pertaining to quantum systems have truth values, or just some of them? If the latter, which ones? The list could be extended. The standard account of interpretation enables us to regard its entries as different expressions of a single enterprise.

The words of some philosophers of physics speak against the standard account. Matthias Frisch (2005), among others (see e.g. the essays in Morgan and Morrison 1999), has recently challenged the idea that theories represent by picking out sets of possible worlds as models of their fundamental mathematical structures. But I'm not ready to give up on that idea yet. I take Frisch's consideration of classical electrodynamics to establish that the theory is not simple but manifold, that conjoining its different variations can induce inconsistency, and that physicists blithely ignore this as they opportunistically leap from one formulation to another as the problems they confront demand. I am not hostile to a single one of those points. Indeed, I will try to establish their QM_∞ analogs in the balance of this book. But I do not conclude from them that classical electrodynamics is inconsistent, and so unsusceptible to interpretation in the standard sense. Rather, I conclude that electrodynamics admits multiple interpretations in the standard sense, and celebrate the opportunity this affords philosophers to articulate and adjudicate those interpretations, as well as to reflect on what their presence might be telling us about physics.

1.4 Criteria of adequacy for interpretations

This section announces some criteria of adequacy for interpretation as it is standardly understood. These criteria are provisional; they're meant to help initiate, not terminate, the interrogation of interpretive options. The exigencies of interpreting QM_∞ may induce us to revise our sense of what it takes to comprehend the physical world by means of an empirical theory. To my mind, the capacity of interpretive projects to inspire such revision is part of what makes them philosophical.

In "The Theoretician's Dilemma," Hempel (1965) considers the case for taking the content of a scientific theory to be its "Craigified" part—that is, the connections it draws between observable phenomena, connections presented without the intermediary of (nonlogical) theoretical apparatus. Hempel resists the Craigification of theoretical content because he attributes theories a function which he considers their Craigified parts ill-suited to fulfill. That function is inductive systematization. What promotes it, Hempel suggests, is the systematic unification between observable phenomena effected by the theoretical apparatus.

Without going deeper into the details of Hempel's position, I claim it as an inspiration for one criterion of interpretive adequacy. Supposing that explanation is a central function of physical theories, the criterion is: *An interpretation should enable the theory it interprets to discharge its explanatory duties.* I don't need—I'm not sure there exists—a categorical imperative for identifying these duties. In applying the criterion, we can start from the explanatory aspirations of physicists using a theory, and aim for a reflective equilibrium between the assessment of interpretations for sustaining those aspirations and the assessment of those aspirations as appropriate.

"Explanatory duties" is not only vague but also synechdochical. There are a host of things a theory ought to do or be, and an interpretation of a theory is adequate insofar as it enables the theory to do or be these things. An interpreted theory ought to be logically consistent, empirically adequate (if we think the theory is), and maybe even simple and mathematically elegant. What's more, an interpreted theory ought to make sense of experimental practice. Huge tracts of the literature on the interpretation of QM concern whether it is possible to give an interpretation of quantum mechanics consistent with the empirical predictions of the minimally interpreted quantum statistical algorithm. Still more of this literature tackles the measurement problem, which constrains interpretation to secure the fundamental presupposition of laboratory practice: that experiments have outcomes.

I've been proceeding as though each theory were an island, complete unto itself. In this mode of address, adequacy of the sort under discussion is a purely internal affair: an interpretation ought to equip a theory with content sufficient to that theory's own aims. But externally oriented questions of sufficiency arise as well. We may, for instance, want to interpret a theory in a way that makes sense of its fit with environing theories and their interpretations. Thus in the case of the thermodynamic limit of QSM, we might hope for an interpretation that brings the theory into a substantial explanatory relationship with thermodynamics. Again, existing theories are conceptual resources from which new theories are forged. Thus we might want an an interpreted theory to serve as such a resource. Semiclassical quantum gravity gives a good example of how this desire plays out in the interpretation of QFT.

Semiclassical quantum gravity considers a quantum field on spacetime manifold subject to Einstein's field equations, with a quantum commodity (the expectation value $\langle T_{ab}\rangle$ of the stress energy tensor T_{ab}) substituted for a classical one (plain old T_{ab})[8]:

$$G_{ab} = 8\pi \langle T_{ab}\rangle \qquad (1.1)$$

One point of this hybridization is to glean from it what a purebred quantum theory of gravity might look like. An interpretation of QFT that withheld physical significance from a stress energy observable (or something like it) would blunt that point. It would fail to recognize enough observables to sustain QFT in this developmental task.

If the foregoing criteria of adequacy had a bumper sticker on their car, that bumper sticker would say:

Physically significant is significant for physics.

To interpret a physical theory is to delimit what that theory recognizes as physically significant. The criteria demand the delimitation to respect how the theory matters to the present and future of physics.

The criteria discussed so far all fall roughly under the heading of "sufficiency." They demand that an interpretation of a theory say *enough* about the world of which that theory is true to make sense of the theory as successful science. Should we also demand that interpretations not say *too much*? Although intuitively appealing, the demand is difficult to motivate ecumenically. It is preaching to converted Ockhamists to declare parsimony desirable in itself. Perhaps it helps to observe that parsimony can serve recognized epistemological and metaphysical goals. For instance, there is a failure of parsimony that frustrates the venerable goal of *determinism*. For a theory to have any shot at determinism, its dynamics must specify how the values of some set of magnitudes at one time depend on the values of some set of magnitudes at another. Unparsimoniously recognizing more magnitudes than are dynamically tractable, or more states than can be put into one-to-one correspondence with valuations of dynamically tractable magnitudes, the structure-specifying phase of an interpretation can threaten determinism. Of course, its semantic phase could work to defuse this threat, as some defenses of substantivalism against the hole argument do (see e.g. Brighouse 1995).

1.5 Realism and pristine interpretation

There is a loftier way to motivate the 'sufficiency' criteria just announced. It is to adopt the desideratum:

---

[8] Good sources to consult for a review of the basics of differential geometry and its use in spacetime theories include the appendix to Friedman (1983) and Wald (1984). Instead of a comprehensive formal review, I will issue informal sketches of key ideas as they come up, and trust to context and prose explanations to orient the uninitiated reader through the rest.

An interpretation of T should, as far as possible, attribute T the virtues realists are wont to cite as reasons for believing that T is true or approximately true.

The desideratum motivates the "sufficiency" criteria because these virtues include inductive systematization and other varieties of empirical success, invoked by the realist's Miracles Argument, as well as continuity with successor theories, invoked by standard realist replies to the pessimistic meta-induction (for an anatomy of these debates, see Psillos 1999). Adopting the desideratum above preserves a bridge between generalists and specialists in the form of the idea that an interpretation of T is what a realist about T believes. But it preserves that bridge in a way that neither prejudges the outcome of the generalists' debate nor implies the insulation of that debate from particularist inquiries. Even supposing T admits an interpretation outfitting it with a complete wardrobe of virtues, antirealists can refrain from regarding those virtues as reasons for belief. And particularists' close examination of an instantiation of T may reveal that "as far as possible" isn't very far at all: T may exhibit different virtues only under different interpretations.

We are in a position to exhibit a connection between realism and pristine interpretation. Suppose that in different settings, T exhibited different virtues salient to the realist. Applied to theoretical high energy physics, it afforded the inductive systematization implicit in the notion of "particle" and exhibited consistency with another regnant theory; applied to quantum optics, it sacrificed some of that consistency for the sake of joining *other* regnant physical theories in the task of modeling collider phenomena (later chapters try to put some meat on the bones of these suppositions). And suppose further that the interpretation explicating the virtues T exhibits in one setting is different from and rival to the interpretation explicating the virtues T exhibits in another. Then there isn't a single interpretation of T under which it exhibits all the virtues realists regard as reasons for believing T. The impediment this presents to realism is: to be wholly virtuous, T requires different interpretations in different settings. Reasons to believe T invoking T's successes in these settings thus apply to different, competing interpretations of T. There isn't an interpretation of T in which they collectively and coherently license belief.

Pristine interpretation is a strategy for negotiating this impediment. To adhere to the ideal of pristine interpretation is to ignore details of T's particular applications—details that are merely consistent with but not compelled by T's laws—when identifying the worlds possible according to T. To adhere to the ideal is to equip T with a content that's invariant under changes in T's applications. Pristinely interpreted, T won't admit different interpretations under different settings.

There is no guarantee that the strategy will succeed. It may be that under no pristine interpretation does T emerge as wholly virtuous. But despairing of pristine interpretation is frightfully close to despairing of warranted realism. If T only exhibits *some* of its virtue in each setting to which it applies, unpristinely keying T's contents to its circumstances of application guarantees that there is no single interpretation under which T possesses all its virtues.

1.6 Interpreting QM$_\infty$

Here's one reason to think QM$_\infty$ could use some interpretation. What I'll call theories of *ordinary QM* concern systems of finitely many particles in Euclidean space. There is a standard notion of what a quantum theory of such a system requires. A quantum theory requires a *Hilbert space representation*, that is, a Hilbert space, on which act symmetric operators obeying relations characteristic of the system quantized—in general, canonical commutation relations, a.k.a. CCRs (for mechanical systems) or canonical anticommutation relations, a.k.a. CARs (for spin ones). Possible states for the quantum system are density matrices on the representing Hilbert space; observables pertaining to the system are self-adjoint operators on the representing Hilbert space. Most interpreters of ordinary QM take quantum kinematics, in the form determined by a Hilbert space representation, as their point of departure.

After that, they typically diverge. They disagree about whether to supplement these kinematics' bare quantum states with hidden variables. They disagree about whether quantum *dynamics* (i.e. the time evolution of quantum states and observables) is collapse-ridden. They disagree about whether a quantum system can exhibit a determinate observable value its state cannot predict with certainty. They disagree about whether quantum reality comprises one world or many, and about the role in its constitution of minds, the environment, and the biorthogonal decomposition theorem. But amid all these disagreements, it doesn't occur to interpreters of ordinary QM to worry that they're not even talking about the same Hilbert space structure to begin with.

Interpreters of ordinary QM don't worry about this because the Stone–von Neumann theorem suggests that they needn't. The theorem states that all Hilbert space representations of the CCRs for a particular classical Hamiltonian theory of finitely many particles in Euclidean space stand to one another in a mathematical relation called *unitary equivalence*. Unitary equivalence is widely accepted as a standard of physical equivalence for Hilbert space representations. It follows that all these representations are simply and unalarmingly different ways of expressing the same quantum kinematics. For finitely many spin systems subject to CARs, the Jordan–Wigner theorem likewise guarantees uniqueness. For systems of finitely many particles, the directive "quantize!" has a unique outcome. That is why many philosophical discussions of QM have proceeded on the assumption that the basic kinematics of a quantum theory takes the form of a Hilbert space representation.

One of my missions here is to chronicle the disintegration of that assumption in the context of QM$_\infty$, and the consequent disruption to an interpretive landscape familiar from philosophers' discussions of ordinary QM. The uniqueness theorems do not extend to QFTs, which one obtains not by quantizing a system of finitely many particles, but by quantizing a field, an entity defined at every point of space(time). Neither do they apply to the thermodynamic limit of quantum statistical mechanics, where the number of systems one considers and the volume they occupy are taken to be infinite. This brings us to the provocative feature of QM_∞, not shared by ordinary QM. *According to very same criterion of physical equivalence by whose lights Hilbert space representations of the CCRs/CARs for ordinary quantum theories are reassuringly unique, the CCRs/CARs of a theory of QM_∞ can admit infinitely many physically inequivalent Hilbert space representations.*

Something has gone terribly wrong—but what? Are we mistaken about what a quantum theory is, or about when Hilbert space representations are physically equivalent? Has some crucial aspect of what it is to be *physical*—a consideration that would alleviate the apparent non-uniqueness—escaped our notice? Was it fundamentally misguided to expect "the theory of the mass m free boson field" (and other terms presuming to denote particular theories of QM_∞) to have a single referent?

These are among the questions I aim to explore here. My means of exploration will be the articulation and evaluation of a variety of accounts of the content of theories of QM_∞ in a variety of circumstances where unitarily inequivalent representations arise. I will try to argue that no overarching account—no single interpretation of QM_∞—is adequate to all of those circumstances. I think that this should attenuate our commitment to the ideal of pristine interpretation, as well as complicate our notion of physical possibility and our conduct of the scientific realism debate, and I will try to suggest how.

The plot of the book is as follows. Chapter 2, "Quantizing," reviews the procedure of Hamiltonian quantization, explicates the Stone–von Neumann theorem as a theorem about the uniqueness, up to unitary equivalence, of quantizations thus obtained, and identifies the presuppositions underlying the reception of this result as a proof of the *physical equivalence* of those quantizations. These are assumptions about the shape of quantum theories and the nature of physical equivalence. Chapter 3 shows how theories of QM_∞ and other sorts of extraordinary QM escape the clutches of the Stone–von Neumann and Jordan–Wigner theorems to admit unitarily inequivalent Hilbert space representations—sometimes continuously many of them.

Suggestively, there is a level of abstraction at which even unitarily inequivalent Hilbert space representations of a theory of QM_∞ share a common structure, the structure of an abstract algebra. The aim of Chapter 4, "Representation without Taxation," is to introduce the technicalia needed to ascend to this level of abstraction. The introduction is informal, and meant to be accessible to readers who encountered QM through its philosophy. (Please see the Preface for various devices I employ in an attempt to keep this from annoying those with more extensive mathematical backgrounds.) Thus I will assume readers to have a working acquaintance with vector spaces, operators, and the so-called axioms of ordinary non-relativistic QM, but I won't presuppose systematic mathematical knowledge beyond that.

Once the basic notions of algebraic quantum theory have been introduced, they can be used to characterize theories of QM_∞, as well as interpretive engagement with those theories. After Chapter 5 reviews some axiom systems for QM_∞, Chapter 6 presents a preliminary overview of candidate interpretive approaches. These range from "Hilbert Space Conservatism," which continues to insist, in the face of QM_∞'s superabundance of unitarily inequivalent representations, that theories of QM_∞ take the same form as theories of ordinary QM, and are physically equivalent only when unitarily equivalent; to "Algebraic Imperialism" (so labeled by Arageorgis 1995), which identifies the content of a theory of QM_∞ in terms of the abstract algebraic structure common to all its concrete Hilbert space representations. Both the Conservative and the Imperialist promise to adhere to the ideal of pristine interpretation: their identification of the worlds possible according to a theory of QM_∞ rests on considerations so general that the adulterating option, of contouring the theory's content in response to non-"nomic" details of its applications, is suppressed. Chapter 5 unkindly labels such pristine interpretive positions "extremist," due to their resolution to adhere to overarching interpretive principles in the face of the contingent particularities that frame applications of the theories they interpret.

The algebraic apparatus developed in Chapter 4 makes visible a startling mathematical possibility. The backdrop against which it emerges as startling is provided by a family of pristine approaches to ordinary quantum semantics, a family which includes most familiar interpretations of QM. Advocates of these *maximal beable strategies* take the task of quantum semantics to be the specification of the largest set of propositions concerning a quantum system that can be attributed simultaneously determinate truth values *obedient to classical truth tables*. Gripped by different metaphysical scruples and technical insights, different versions of the maximal beable strategy identify this "maximal beable algebra" in different ways. But for each, the identification hinges on *a privileged set of pure states* of the system under scrutiny (e.g. possible endpoints of collapse, states of determinate position, "value states"). The startling mathematical possibility which emerges from Chapter 4 is the possibility of algebras of quantum observables which admit no pure normal (that is, countably additive) states. Chapters 7 and 8 continue Chapter 4's unpunishing exposition of algebraic notions by explaining how such algebras of observables come to be instantiated in QM_∞, and how they stymie the very maximal beable strategies which apply so well to ordinary QM. The upshot is that QM_∞ rewrites what serve ordinary QM as the usual rules for quantum semantics.

Chapters 9–14 confront candidate interpretations of QM_∞ with phenomena they might be expected to save. Chapter 9 asks, "Is Particle Physics Particle Physics?" That is, should we understand QFT as a theory about particles? A notorious case against rests on the Unruh effect, according to which an observer accelerating through Minkowski spacetime in its QFT vacuum state—a state distinguished by the absence of particles—finds herself bombarded by . . . particles. Chapter 9 dissects this case against, to reveal the working therein of unitarily inequivalent representations of the quantum field theoretic commutation relations. Chapter 10 characterizes the circumstances conducive to a particle interpretation, and argues that many QFTs meriting interpretation fall outside those circumstances.

1.6 INTERPRETING QM$_\infty$ 17

This is already a refutation of a species of extremism that would subject *every* theory of QM$_\infty$ to a particle interpretation. Chapter 10 also targets the extremism of Hilbert Space Conservatism by producing examples of theories of QM$_\infty$ with the following feature: different applications of the theory are explicated by different (and ergo competing) Conservative interpretations of it. Pristinely pursued, Conservatism undermines the empirical reach of such a theory.

Chapter 11 is one of the few places in the book that makes contact with the QFTs high energy particle physicists test with scattering experiments. (Chapter 14's treatment of broken symmetry in QFT is another place.) It investigates a route from the physicists' techniques of calculation and detection—techniques which employ eerily particulate Feynman diagrams and cloud chamber tracks—to a particle interpretation of QFT. It travels that route only long enough to conclude that however such a "phenomenological particle interpretation" might be executed, it will differ in fundamental respects from particle interpretations of the sort at issue in Chapters 9 and 10.

This extended discussion of the particle notion concludes that there's a particular extremist, the Hilbert Space Conservative, who can't interpret a particular theory of QM$_\infty$ in a way that explicates all its empirical virtues. Chapters 12–14, which treat phase structure and spontaneous symmetry breaking in QM$_\infty$, argue for a more radical conclusion: there occur in QM$_\infty$ *individual explanations* that *no* extremist can make sense of. Chapter 12, "A Matter of Degree," presents phase structure—understood as the existence, at a single critical temperature, of distinct equilibrium states, corresponding to distinct pure thermodynamic phases available to a system in equilibrium at that temperature—as an *explanandum* that might require an *explanans* no extremist interpretation can sustain. To construct an interpretation that makes sense of phase structure, I suggest, we should suspend an assumption, shared by all extremisms, about how to specify the content of physical theories. This is the unimodality assumption that physical theories sort states of affairs into the unqualifiedly possible and the unqualifiedly impossible, the assumption motivating the ideal of pristine interpretation. Using examples from QSM, I develop principles for *adulterated* interpretation, principles meant to take some of the mystery out of the suggestion that what's possible according to a theory could vary along with its circumstances of application.

Chapter 12's case, that no extremism can sustain the explanations countenanced in the thermodynamic limit, is dubbed "the Coalesced Structures Argument." This argument is liable to be met with an immediate, and apparently devastating, objection. The objection is that the thermodynamic limit, which inspires the Coalesced Structures Argument, is a rank idealization from which no serious foundational conclusions can be drawn. Bluntly put, the objection is that the Coalesced Structures Argument is irrelevant because steaming cups of coffee are finite. Chapters 13 and 14, which treat broken symmetry in QSM and QFT respectively, start out as an end-run around this objection. Broken symmetry in QM$_\infty$ shares with phase structure many of the features on which the Coalesced Structures Argument turns. Moreover, broken symmetry is reputed to characterize theories of QM$_\infty$ that come by their ∞ honestly, for instance, the QFTs allied into the Standard Model. The end-run Chapters 13 and 14 attempt is to devise a QFT analog of the Coalesced Structures Argument. The attempt does not fail so much as evaporate into speculation, for the simple reason that the QFTs that are reputed to break symmetry aren't formulated explicitly enough to tell whether or how the concepts at play in the Coalesced Structures Argument apply. But this observation prompts another about the apparently devastating objection to the phase argument: it rests on a *guess* about the future of physics. So too does my willingness to rest foundational conclusions on the Coalesced Structures Argument. Chapter 14 closes by explaining why I like my guess better.

Although I claim in this book that "extremist" interpretations of QM_∞ fail to make sense of the full range of empirical successes those theories enjoy, that claim is not the point of this book. The point of this book is how different the debate over scientific realism looks in the wake of the claim. Chapter 15, "Re: Interpreting Physical Theories," develops the point. It brings the particularist labors of the previous dozen chapters home to roost in the generalists' debating halls. It tries to persuade the generalist that he should care about the presence and role in QM_∞ of unitarily inequivalent representations, by describing the lessons the present particularist sees in QM_∞, lessons about the natures of interpretation, physical possibility, theoretical virtue, and grounds for warranted belief in our best scientific theories.

2
Quantizing

Note: This chapter introduces a convention, described more completely in the Preface, to distinguish between notes that gloss mathematical jargon and notes that discharge other standard note functions.

Proponents of Heisenberg's (b. 1925) matrix mechanics and Schrödinger's (b. 1926) wave mechanics did not immediately form a mutual admiration society. "I am convinced that you have made a decisive advance with your formulation of the quantum condition, just as I am equally convinced that the Heisenberg–Born route is off the track," Einstein wrote Schrödinger in 1926 (Przibaum 1986, 28). In a note to a paper published that same year, Schrödinger divulged that the Heisenberg–Born route left him "discouraged, if not repelled, by what appeared... a rather difficult method of transcendental algebra, defying visualization" (as quoted in Jammer 1966, 272). Heisenberg, for his part, told Pauli, "the more I think of the physical part of the Schrödinger theory, the more detestable I find it. What Schrödinger writes about visualization makes scarcely any sense, in other words I think it is shit" (Letter of 8 June 1926, quoted in Moore 1989, 221).

Wave and matrix mechanics were initially considered, by their progenitors and others, to be rival quantum theories. In the 1930s, von Neumann put this rivalry definitively to rest, by producing what was received as a demonstration of the physical equivalence of the theories. Both the demonstration and its reception have presuppositions, which this chapter aims to uncover. Chapter 3 will aim to exhibit quantum theories operating outside the scope of those presuppositions. Subsequent chapters examine such theories, in search of philosophical lessons.

2.1 Rival quantum theories?

In the Heisenberg theory,[1] complex Hermitian matrices Q and P, representing position and momentum respectively, were postulated to obey a rule geared toward the recovery of the Bohr–Sommerfeld quantization condition. From these assumptions, and a requirement on the form of the Hamiltonian matrix, follow the *canonical commutation relations*, which for a quantum system with n degrees of freedom are

$$[Q_i, Q_j] = [P_i, P_j] = 0, \quad [P_i, Q_j] = -i\hbar \mathbb{I}\delta_{ij} \qquad (2.1)$$

Here Q_i and P_i are the position and momentum matrices for the ith degree of freedom, $[A, B] := AB - BA$, $\mathbb{I}$ is the identity matrix, and $\hbar$ is Planck's constant h (4.136 × 10^{-15} eV-sec) divided by 2π. Factors of h and $\hbar$ will typically be set to 1 in what follows.

By apparent contrast, Schrödinger's wave mechanics rested on two differential equations, the time-independent and time-dependent Schrödinger equations, for a complex-valued wave function $\psi(x)$, where x ranges over the system's configuration space (e.g., for a system moving in one Euclidean dimension, the real line $\mathbb{R}$). To solve these equations, Schrödinger introduced a differential operator H whose eigenfunctions satisfy the time-independent equation. H is moreover the infinitesimal generator of solutions to the time-dependent equation, in the sense that those solutions (expressed in terms of the time parameter t) take the form:

$$\psi(x, t) = e^{-iHt}\psi(x, 0) \qquad (2.2)$$

In 1926, Schrödinger put a damper on the matrix-wave rivalry. Consider a system with one degree of freedom. Schrödinger noted that, provided its position and momentum are identified with the differential operators:

$$q : q\psi(x) = x\psi(x) \qquad p : p\psi(x) = -i\frac{d\psi(x)}{dx} \qquad (2.3)$$

respectively, they satisfy the Heisenberg commutation relations. Moreover, the differential operator H of the Schrödinger theory has the same functional dependence on position and momentum as does the Hamiltonian of the Heisenberg theory. These reassuring results generalize to systems of any finite number of degrees of freedom, and suggest that the Heisenberg theory is at least a special case of the Schrödinger theory.

Von Neumann's work of the early 1930s consolidates the suggestion. Von Neumann recognized that wave and matrix mechanics are each instances of what I'll call a theory of *ordinary quantum mechanics*, that is, a theory according to which[2]

---

[1] My expositions of matrix and wave mechanics are broad-brush. One place to find more details is Emch (1972, §8.1).

[2] For an introduction, aimed at philosophers, to ordinary quantum mechanics, see Hughes (1989).

OQM Physical magnitudes (or observables) pertaining to a physical system correspond to the self-adjoint elements of the set $\mathfrak{B}(\mathcal{H})$ of bounded operators acting on a *separable*[3,g] Hilbert space $\mathcal{H}$;[4]

OQS The possible states of the system stand in one-to-one correspondence with the set $\mathfrak{T}^+(\mathcal{H})$ of density operators (that is, positive trace-class operators of trace 1) on $\mathcal{H}$. The expectation value of an observable $\hat{A}$ pertaining to a system in the state $\hat{W}$ is $Tr(\hat{A}\hat{W})$;

OQD Where $\hat{H}$ is the Hamiltonian of an isolated system with initial state $\hat{W}(0)$, its state at time t is given by $e^{-i\hat{H}t}\hat{W}(0)e^{i\hat{H}t}$.

I will say more about dynamics (OQD) presently. As for kinematics (OQM and OQS): a density operator $\hat{\rho}$ defines a map from $\mathfrak{B}(\mathcal{H})$ to the complex numbers $\mathbb{C}$ via the trace prescription: for each $\hat{A}$ in $\mathfrak{B}(\mathcal{H})$, $\rho(\hat{A}) = Tr(\hat{\rho}\hat{A})$. Such a map qualifies as a state because of how sanely it assigns expectation values to observables. Expectation values are operationalized by long run experimental averages, which are real numbers. So are the values $\rho(\hat{A})$, when $\hat{A}$ is self-adjoint. What's more, ρ is linear ($\rho(\hat{A} + \hat{B}) = \rho(\hat{A}) + \rho(\hat{B})$), positive ($\rho(\hat{A}) \geq 1$ when $\hat{A}$ is positive), normed (where $\hat{I}$ is the identity operator on $\mathcal{H}$, $\rho(\hat{I}) = 1$), and countably additive (when $\{\hat{E}_i\}$ is a countable set of mutually orthogonal projection operators acting on $\mathcal{H}$, $\rho(\sum_i \hat{E}_i) = \sum_i \rho(\hat{E}_i)$). Interpreting $\rho(\hat{E})$ as the probability assigned to the eventuality associated with the projection operator $\hat{E}$ (and speaking a bit roughly in order to cover what is probably familiar ground quickly): positivity ensures no eventuality is assigned negative probability; normalization ensures that the "fail-safe" eventuality associated with the identity operator is assigned probability 1; and countable additivity ensures that the probability assigned to a conjunction of pairwise exclusive eventualities is the sum of the probabilities assigned to those eventualities individually.

Gleason's theorem tells us that, for a Hilbert space $\mathcal{H}$ of dimension greater than 2, the trace prescription puts density operators in one-to-one correspondence with states, conceived as normed, positive, countably additive assignments of probabilities to projection operators in $\mathfrak{B}(\mathcal{H})$. Call this the *normal conception of states*. For systems associated with Hilbert spaces of dimension greater than 2, OQS follows from the conjunction of OQM and the normal conception of states.

The convexity of the set of states on $\mathfrak{B}(\mathcal{H})$ grounds a distinction between *pure* and *mixed* states.

Definition 2.1 (pure and mixed states on $\mathfrak{B}(\mathcal{H})$). A state ρ on $\mathfrak{B}(\mathcal{H})$ is *mixed* if and only if it can be written as a non-trivial convex combination of other states ρ_1 and ρ_2 on $\mathfrak{B}(\mathcal{H})$; i.e.

---

[3] Recall that a Hilbert space is separable if and only if it has a countable basis. Some non-separable Hilbert spaces make an appearance in §3.1.

[4] I will often use a notation in which Hilbert space operators wear party hats $[\hat{A}]$, and I will commit the usual solecism of using (e.g.) "$\hat{A}$" to denote both the observable and the operator.

$$\rho(\hat{A}) = \lambda \rho_1(\hat{A}) + (1-\lambda)\rho_2(\hat{A}) \text{ for all } \hat{A} \in \mathfrak{B}(\mathcal{H}) \tag{2.4}$$

where $0 < \lambda < 1$ and $\rho_1 \neq \rho_1$. If ρ admits no such decomposition, it is an extremal element of the set of states on $\mathfrak{B}(\mathcal{H})$, that is, a *pure* state.

In ordinary QM, the density operator implementing a pure state is a projection operator for a one-dimensional subspace of $\mathcal{H}$. A pure state on $\mathfrak{B}(\mathcal{H})$ is also a *vector state*.

Scholium: "Bra-ket" notation. The definition below introduces Dirac's barbarous "bra-ket" notation, which encloses elements of a Hilbert space $\mathcal{H}$ in "kets" (like so: $|\psi\rangle$), and encloses elements of the dual space of linear functionals $\phi : \mathcal{H} \to \mathbb{C}$ in "bras" (like so: $\langle \phi |$). For each $|\phi\rangle \in \mathcal{H}$, the Hilbert space inner product determines a linear functional $\langle \phi |$ that maps an arbitrary element of $\mathcal{H}$ to its inner product with $|\phi\rangle$. The very hilarious upshot is that in Dirac notation, the "bra-ket" $\langle \phi | \psi \rangle$ gives the inner product between $|\phi\rangle$ and $|\psi\rangle$. For $\hat{A} \in \mathfrak{B}(\mathcal{H})$, the inner product of the vectors $|\phi\rangle$ and $\hat{A}|\psi\rangle$ is often written $\langle \phi | \hat{A} | \psi \rangle$; when $\hat{A}$ is self-adjoint and $|\psi\rangle$ a vector state, $\langle \psi | \hat{A} | \psi \rangle$ gives the expectation value of the observable implemented by $\hat{A}$ in the state $|\psi\rangle$. See Redhead (1989, §1.1) for a crash course.

From Graham Farmelo's recent Dirac biography:

...after an evening meal in St John's, Dirac was listening to dons reflecting on the pleasures of coining a new word, and, during a lull in the conversation, piped up with four words: "I invented the bra." There was not a flicker of a smile on his face. The dons looked at one another anxiously, only just managing to suppress a fit of giggling, and one of them asked him to elaborate. But he shook his head and returned to his habitual silence, leaving his colleagues mystified. (Farmelo 2009, 326–7) ♠

Definition 2.2 (vector state on $\mathfrak{B}(\mathcal{H})$). A state ρ on $\mathfrak{B}(\mathcal{H})$ is a *vector state* if and only if there is some vector $|\psi\rangle \in \mathcal{H}$ such that $\rho(\hat{A}) = \langle \psi | \hat{A} | \psi \rangle$ for all $\hat{A} \in \mathfrak{B}(\mathcal{H})$.

The probabilities assigned by the vector state coded by $|\psi\rangle$ are duplicated by the probabilities assigned, via the trace prescription, by the one-dimensional projection operator for the ray containing $|\psi\rangle$. Conversely, an extremal element $\hat{E}$ of the set of density operator states can be encoded as a vector state, simply by selecting a normed vector from the ray onto which $\hat{E}$ projects. Thus, in ordinary QM *pure states are vector states and vice versa*.

It is utterly standard for physics texts (e.g. Messiah 1961; Shankar 1994) and for philosophical treatments (e.g. Redhead 1989; Hughes 1989) to characterize quantum theories in the foregoing terms. Both Heisenberg's and Schrödinger's theories can be so characterized: Heisenberg's theory realizes the structure OQM and OQS in Hilbert spaces $\ell^2(\infty)$ of square summable infinite sequences of complex numbers; Schrödinger's theory realizes this structure in Hilbert spaces $L^2(\mathbb{R}^n)$ of square integrable complex-valued functions of $\mathbb{R}^n$. The next two scholia introduce these to those who haven't made their acquaintance.

Scholium: Measurable functions. A *measure* μ on a space is a map from subsets of that space to the positive real numbers s.t. $\mu(\emptyset) = 0$ and $\mu(\cup_{i=1}^{\infty} A_i) = \sum_{i=1}^{\infty} \mu(A_i)$ if $A_i \cap A_j = \emptyset$ for all $i \neq j$. (The subsets in question must form a σ-ring; see Reed and Simon 1980a, I.4 for details.) With respect to a measure μ on the real line $\mathbb{R}$, a real-valued function $f : \mathbb{R} \to \mathbb{R}$ is *measurable* if and only if for every open interval $\Delta \subset \mathbb{R}$, μ assigns $f^{-1}(\Delta)$ a measure. A complex-valued function $f : \mathbb{R} \to \mathbb{C}$ is measurable if and only if both its real and complex parts are. ♠

Scholium: The Hilbert spaces $\ell^2(n)$, $\ell^2(\infty)$ and $L^2(X, \mu)$. A Hilbert space is a linear vector space over the complex numbers equipped with an inner product and complete in the norm determined by that inner product. ("Complete" here means that every sequence of elements in the space that converges according to the inner product norm converges to a limit that's also an element of the space.) The Hilbert space $\ell^2(n)$ consists of sequences x_i, $i = 1$ to n, of complex numbers, which sequences are *square summable* in the sense that $\sum_{i=1}^{n} |x_i|^2 < \infty$. The inner product is given by $\langle \{x_i\} | \{y_i\} \rangle := \sum_{i=1}^{n} \bar{x}_i y_i$ (where the overbar denotes complex conjugation). The Hilbert space $\ell^2(\infty)$, whose elements are *infinite* sequences, is constructed analogously. The Hilbert space $L^2(\mathbb{R})$ consists of complex-valued functions $f : \mathbb{R}^n \to \mathbb{C}$, which functions are *square integrable* in the sense that $\int_{\mathbb{R}} |f(x)|^2 dx < \infty$. The inner product is given by $\langle f(x) | g(x) \rangle := \int_{\mathbb{R}} \overline{f(x)} g(x) dx$. The definitions of square integrability and the inner product appeal to the Lebesgue measure dx on $\mathbb{R}$, which (roughly speaking) provides a notion of the "size" of arbitrary subsets of $\mathbb{R}$, and thereby tells us how to integrate. The Lebesgue measure dx turns $\mathbb{R}$ into the *measure space* $(\mathbb{R}, dx)$. (For a less impressionistic treatment of measures, integration, and measure spaces, see Reed and Simon 1980a, I.3 and I.4.) Now consider an arbitrary measure space (X, μ). The Hilbert space $L^2(X, \mu)$ consists of complex valued functions $X \to \mathbb{C}$ that are square integrable with respect to the measure μ. ♠

OQM and OQS furnish a *kinematics* for quantum theory, that is, an account of the physical magnitudes and the instantaneous states it recognizes. To say how those states and magnitudes change over time is to give a *dynamics* for quantum theory. With $\psi(x)$ understood as an element of $L^2(\mathbb{R})$, Schrödinger's equation (2.2) does that for pure states of a quantum particle moving in one dimension. OQD generalizes Schrödinger's equation to cover not just vector but also density operator states. We could just as well implement dynamics by keeping states static and letting observables evolve. The result is the *Heisenberg picture*, according to which $\hat{A}(t) := e^{-i\hat{H}t} \hat{A} e^{i\hat{H}t}$ is the "evolute" through time t of the observable $\hat{A}$ pertaining to a system governed by a Hamiltonian $\hat{H}$ and in enduring state $\hat{W}$.

Because $\hat{H}$ is self-adjoint, the operators $\hat{U}(t) := e^{-i\hat{H}t}, t \in \mathbb{R}$, implementing ordinary quantum dynamics form a one (real) parameter *group*[5,g] of unitary operators on $\mathcal{H}$.

---

[5] A *group* is just a set equipped with an associative binary operation, an inverse operation, and an identity element. The real numbers $\mathbb{R}$ are an example: they form a group under the binary operation of addition, whose inverse is subtraction and whose identity is 0. Because addition is commutative, this group is *abelian*.

That is to say, dynamics are implemented unitarily in ordinary quantum mechanics. The converse of the inference from $\hat{H}$'s self-adjointness to the unitarity of the group $\hat{U}(t)$ is Stone's theorem. Since it matters later (e.g. §3.1), I'll state it now.

Fact 2.3 (Stone's Theorem). Suppose that $\hat{V}(x)$, a one (real) parameter group of unitary operators on some Hilbert space $\mathcal{H}$, is *continuous* in the sense that $\left|\left(\hat{V}(x+\delta) - \hat{V}(x)\right)|\psi\rangle\right| \to 0$ as $\delta \to 0$ for all $|\psi\rangle \in \mathcal{H}$ and for all x. (Chapter 4 will divulge that this is continuity in the strong operator topology.) Then $\hat{V}(x)$ has a unique self-adjoint generator $\hat{G}$ such that $\hat{V}(x) = e^{-i\hat{G}x}$. (For a proof see e.g. Kadison and Ringrose 1997a, thm. 5.6.36 (367–70).)

Viewed through the prism of Stone's theorem, the time-dependent Schrödinger equation casts the system Hamiltonian in the role of self-adjoint generator of a one-parameter group $\hat{U}(t)$ of unitary time evolution operators.

Having attributed quantum theories a Hilbert space structure, von Neumann demonstrated in 1931 what had been conjectured the previous year by Stone: the *unitary equivalence* (up to multiplicity) of any pair of Hilbert space quantizations of a classical system with configuration space $\mathbb{R}^n$. Under some natural assumptions about what constitutes the content of a physical theory, it follows from this theorem that wave mechanics and matrix mechanics are *physically equivalent*. The next section exposes these assumptions. The Stone–von Neumann uniqueness theorem itself occupies the balance of the chapter.

2.2 Physical equivalence and unitary equivalence

On what Chapter 1 called "the standard account," the content of a physical theory consists in the worlds possible according to that theory. On the wholly reasonable supposition that theories are equivalent only if their contents coincide, the standard account implies that *two theories are physically equivalent only if the set of worlds possible according to the first is the same as the set of worlds possible according to the second*. My mission here is to develop a case that unitarily equivalent theories of ordinary QM satisfy this necessary condition for physically equivalence.

Such a mission is obviously doomed. A theory of ordinary QM conforms to the template provided by OQM, OQS, and OQD. Theories so specified can, we shall see, stand to one another in a relation of unitary equivalence. *But they can't be established to be physically equivalent in the content-coincidence sense.* The reason is simple. A theory conforming to the template of ordinary QM is not fully interpreted. It's interpreted, and only provisionally so, up to what Chapter 1 called "structure-specification." There is a thriving philosophy of ordinary QM because interpreters who agree about the Hilbert space structure of a quantum theory can disagree, even vehemently, about what Chapter 1 called its "semantics." They disagree, that is, about what it is about a world that enables a Hilbert space theory to save its phenomena. Their disagreements can moreover develop in ways that induce them to revise the template provided by by OQM, OQS, and OQD. It follows that any criterion of physical equivalence applicable—as the criterion of unitary equivalence is—to theories characterized only up to structure-specification is provisional. The presumptive physical equivalence of theoretical structures satisfying such a criterion can be undone during the semantic stage of interpretation.

With these caveats in mind, we can justify unitary equivalence as a *presumptive* criterion of physical equivalence for theories of ordinary QM as follows. To adapt the content coincidence criterion of physical equivalence to incompletely interpreted theories, identify a physical possibility with a *state*, understood as a map $\omega : \mathcal{Q} \to \mathbb{R}$, from a set $\mathcal{Q}$ of physical quantities (a.k.a. magnitudes or observables) to their expectation values.[6] By sweeping questions about the physical and metaphysical underpinnings of these expectation value assignments under the rug, the identification brings the content coincidence criterion of physical equivalence to bear on incompletely interpreted theories. Under the identification, the kinematic content of a theory may be expressed by the pair $(\mathcal{Q}, \mathcal{S})$, where $\mathcal{S}$ is its set of states and $\mathcal{Q}$ is the set of observables in the domain of those states. Even though to assign a theory a kinematic pair is not to interpret it completely, the question of how to assign a theory a kinematic pair is hardly interpretively inert. This might be best illustrated by the case of gauge theories (see e.g. Healey 2007 or Belot 2007, §6.2), whose hotly debated issues include: should $\mathcal{Q}$ contain only gauge-independent quantities? Should gauge-connected states be distinct or identical members of $\mathcal{S}$? Here's a first try at a criterion of content coincidence for theories specified up to a kinematic pair $(\mathcal{Q}, \mathcal{S})$:

Definition 2.4 (Weak Kinematic Equivalence (ke)). $(\mathcal{Q}, \mathcal{S})$ and $(\mathcal{Q}', \mathcal{S}')$ are weakly kinematically equivalent if and only if there exist bijections $i_s : \mathcal{S} \to \mathcal{S}'$ and $i_q : \mathcal{Q} \to \mathcal{Q}'$ such that for all $\omega \in \mathcal{S}$ and for all $A \in \mathcal{Q}$, $\omega(A) = [i_s(\omega)](i_q(A))$.

More prosaically, (ke) demands a one-to-one correspondence between instantaneous states (articulated as expectation value assignments) possible according to the first theory and instantaneous states (likewise articulated) possible according to the second theory.

Typically, a theory of ordinary QM identifies $\mathcal{Q}$ with some $\mathfrak{B}(\mathcal{H})$ by starting with a collection $\{\hat{P}_i\}$ of symmetric operators on $\mathcal{H}$ *satisfying the [anti]commutation relations canonical for the theory in question*, then adding products (e.g. $\hat{p}\hat{q}$) and linear combinations (e.g. $\hat{p}\hat{q} - \hat{p}\hat{q}$), and limits of sequences of products and linear combinations, of these operators. As Chapter 4 elaborates, this amounts to building an *algebra* using the canonical observables $\{\hat{P}_i\}$ as a basis. Chapter 4 will formally introduce notions for describing and distinguishing between algebraic structures; for now we'll proceed informally, and postpone precise statements and proofs of some of the claims that follow until we've taken on board the apparatus required for constructing them.

---

[6] In the interest of generality, this notion of state is laxer than the normal conception, which subjects the expectation value assignments to additional criteria of adequacy.

26 QUANTIZING

Thinking of $\mathfrak{B}(\mathcal{H})$ as algebraically structured reveals a sense in which the criterion (ke) is weaker than we might like. Part of the content of a theory is the functional relationships it posits between the physical magnitudes it recognizes. After all, these functional relationships are entangled in its laws: $\hat{H} = \hat{p}^2/2m$ makes the energy of a free system of mass m a function of its momentum; the Schrödinger equation uses the energy of an isolated system to build a family $\hat{U}(t) = e^{-i\hat{H}t}$ of operators describing that system's time evolution, which implies that the operator $\hat{H}$ is a limit of a sequence of functions of the operators $\hat{U}(t)$. Confining attention to relations between magnitudes that are uncontentiously kinematic, the *algebraic structure* of a collection $\mathcal{Q}$ of magnitudes—e.g. the genealogy of their descent from a collection of canonical magnitudes generating them—is part of the kinematic content of the pair $(\mathcal{Q}, \mathcal{S})$. So let us strengthen (ke) to further require that the bijection i_q preserve the "algebraic structure," including the generator structure, of kinematically equivalent theories.[7]

Definition 2.5 (Kinematic Equivalence (KE)). $(\mathcal{Q}, \mathcal{S})$ and $(\mathcal{Q}', \mathcal{S}')$ are kinematically equivalent if and only if there exist a bijection $i_s : \mathcal{S} \to \mathcal{S}'$ and an *algebraic-structure-preserving* bijection $i_q : \mathcal{Q} \to \mathcal{Q}'$ such that for all $\omega \in \mathcal{S}$ and for all $A \in \mathcal{Q}$, $\omega(A) = [i_s(\omega)](i_q(A))$. Where $\{P_i\}$ and $\{P'_i\}$ are the canonical observables generating $\mathcal{Q}$ and $\mathcal{Q}'$ respectively, i_o must be such that $i_o(P_i) = P'_i$.

Unitary equivalence is one sort of isomorphism that can obtain between theories whose magnitudes take the form of collections of Hilbert space operators.

Definition 2.6 (Unitary Equivalence). A Hilbert space $\mathcal{H}$ and a collection of operators $\{\hat{O}_i\}$ is *unitarily equivalent* to $(\mathcal{H}', \{\hat{O}'_i\})$ if and only if there exists a one-to-one, linear, invertible, norm-preserving transformation ("unitary map") $U : \mathcal{H} \to \mathcal{H}'$ such that $U^{-1}\hat{O}'_i U = \hat{O}_i$ for all i.

While falling far short of a formal proof, the following considerations suggest that the bijection established by U between collections of magnitudes moreover preserves significant aspects of their structure. U's linearity ensures that when an unprimed magnitude $\hat{A}$ is a linear function of magnitudes $\hat{O}_i$ in the unprimed theory, the primed magnitude $U^{-1}\hat{A}U$ is the corresponding linear function of the primed magnitudes $U^{-1}\hat{O}_i U$. U's unitarity ensures that when $\hat{A}$ is a product $\hat{B}\hat{C}$ of magnitudes in the unprimed theory, the primed magnitude $U^{-1}\hat{B}\hat{C}U$ is the corresponding product of the primed magnitudes $U^{-1}\hat{B}U$ and $U^{-1}\hat{C}U$ (since $U^{-1}\hat{B}UU^{-1}\hat{C}U = U^{-1}\hat{B}\hat{C}U$). Finally, U's norm-preservation ensures that when a sequence of unprimed operators $\hat{A}_i$ converge to an unprimed operator $\hat{A}$, their primed counterparts under the mapping induced by U converge to $U^{-1}\hat{A}U$. To get a sense of why this is, consider the ultraweak topology, which will be formally introduced in §4.3. A sequence of operators $\hat{A}_i$ on $\mathcal{H}$ converges to an operator $\hat{A}$ on $\mathcal{H}$ in this topology if and only if $|Tr(\hat{W}(\hat{A} - \hat{A}_i)|$

---

[7] A more formal argument that this *is* an additional requirement begins on page 97. Clifton and Halvorson (2001b) develop and deploy a criterion similar to (KE), which they attribute to Glymour (1970), as a criterion of physical equivalence.

goes to 0 as $i \to \infty$ for each density operator $\hat{W}$ on $\mathcal{H}$. Now, if U is unitary, it will establish a one-to-one correspondence between density operators $\hat{W}$ on $\mathcal{H}$ and density operators $\hat{W}' := U^{-1}\hat{W}U$ on $\mathcal{H}'$ such that $|Tr(\hat{W}(\hat{A} - \hat{A}_i))| = |Tr(\hat{W}'(\hat{A} - \hat{A}_i))|$. But thanks to the invariance of the trace under cyclic permutation, $|Tr(\hat{W}'(\hat{A} - \hat{A}_i))| = |Tr(\hat{W}U(\hat{A} - \hat{A}_i)U^{-1})|$. It follows that $\hat{A}_i$ converges to $\hat{A}$ in the ultraweak topology of the unprimed representation if and only if $U^{-1}\hat{A}_i U$ converges to $U^{-1}\hat{A}U$ in the ultraweak topology of the primed representation.[8]

We can now say what it is for ordinary quantum theories to be unitarily equivalent to one another. Key to the definition is the idea that the magnitudes recognized by an ordinary quantum theory descend from a set of operators satisfying the [anti]commutation relations constitutive of that theory.

Definition 2.7 (Unitary Equivalence of Ordinary Quantum Theories). Where the canonical operators $\hat{P}_i$ on $\mathcal{H}$ generate the algebra $\mathfrak{B}(\mathcal{H})$, and the canonical operators $\hat{P}'_i$ on $\mathcal{H}'$ generate the algebra $\mathfrak{B}(\mathcal{H}')$, the ordinary quantum theories $(\mathfrak{B}(\mathcal{H}), \mathfrak{T}^+(\mathcal{H}))$ and $(\mathfrak{B}(\mathcal{H}'), \mathfrak{T}^+(\mathcal{H}'))$ are unitarily equivalent just in case $(\mathcal{H}, \{\hat{P}_i\})$ and $(\mathcal{H}', \{\hat{P}'_i\})$ are unitarily equivalent in the sense of Definition 2.6.

Now the result we've been looking for:

Fact 2.8 (KE ↔ UE). $(\mathfrak{B}(\mathcal{H}), \mathfrak{T}^+(\mathcal{H}))$ and $(\mathfrak{B}(\mathcal{H}'), \mathfrak{T}^+(\mathcal{H}'))$ are unitarily equivalent if and only if they satisfy (KE).

The U effecting their unitary equivalence furnishes both the bijection i_s from the first theory's state space to the second's and the bijection i_q from the first theory's observable set to the second's.[9] We've just suggested that such a U preserves relevant algebraic structure. (Chapter 4 will make this suggestion precise.) So if ordinary quantum theories are unitarily equivalent, then they're kinematically equivalent. The converse follows simply from the facts that the observable algebras $\mathfrak{B}(\mathcal{H})$ and $\mathfrak{B}(\mathcal{H}')$ of kinematically equivalent ordinary quantum theories are isomorphic, and that the only isomorphisms which obtain between such observable algebras are implemented unitarily (Kadison and Ringrose 1997b, thm. 9.3.4 (664)).

Notice that the unitary map establishing the kinematic equivalence of two ordinary quantum theories does more than identify for each state and each observable in one theory its counterpart in the other. This inner product is also arguably itself part of the content of an ordinary quantum theory. Understanding $\langle\phi|\psi\rangle$ as determining the probability for a transition from a vector state $|\phi\rangle$ to a vector state $|\psi\rangle$—the sort of transition constituted by a measurement collapse, say—and taking the empirical content of an ordinary quantum theory set in a Hilbert space $\mathcal{H}$ to reside in its complete

---

[8] For a formal explication and proof of this claim, see Bratteli and Robinson (1987, corr. 2.3.21, (58)).
[9] For each $\hat{A} \in \mathfrak{B}(\mathcal{H})$, $i_q(\hat{A}) = U\hat{A}U^{-1}$; i_s is just i_q restricted to those elements of $\mathfrak{B}(\mathcal{H})$ that are density operators. It follows from the unitarity of U and the invariance of the trace under cyclic permutation that for all $\hat{\rho} \in \mathfrak{T}^+(\mathcal{H})$ and for $\hat{A} \in \mathfrak{B}(\mathcal{H})$, $\rho(A) = [i_s(\hat{\rho})](i_q(\hat{A}))$.

set of transition probabilities, Wigner famously argued that any symmetry of such a theory is implemented by a unitary or anti-unitary operator (Wigner 1959, 233–6). These are the operators, he showed, under which the theory's transition probabilities, as determined by its inner product structure, are invariant.

The Hilbert space inner product can be taken to supply a metric for pure (i.e. vector) states: where $|\psi\rangle$ and $|\phi\rangle$ are unit vectors, the closer the absolute value of their inner product $\langle\phi|\psi\rangle$ is to 1, the closer the states implemented by those vectors are to one another.[10] Such a metric could have implications for what counterfactuals the theory supports, for what explanations it offers, what notion of error (and so confirmation within experimental error) is appropriate to it, for how and whether it accommodates chaotic behavior, and so on.[11] One might want implications like these to be preserved under physical equivalence.

Let us take these considerations to heart, and announce a criterion for the equivalence of theories which incorporate metrics on their pure state spaces. Model such a theory as a triple $(\mathcal{Q}, \mathcal{S}, \mu)$ consisting of a kinematic pair $(\mathcal{Q}, \mathcal{S})$ plus a "possibility metric" $\mu : \mathcal{S} \times \mathcal{S} \to [0, 1]$ imposed on its pure states. Then we could amplify the criterion [KE] to require physically equivalent theories to recognize isomorphic possibility metrics:

Definition 2.9 (Kinematic★ Equivalence (K★E)). $(\mathcal{Q}, \mathcal{S}, \mu)$ and $(\mathcal{Q}', \mathcal{S}', \mu')$ are kinematically★ equivalent if and only if $(\mathcal{Q}, \mathcal{S})$ and $(\mathcal{Q}', \mathcal{S}')$ satisfy (KE) and for all pure $\omega, \omega' \in \mathcal{S}$, $\mu(\omega, \omega') = \mu'(i_s(\omega), i_s(\omega'))$.

Now, unitarily equivalent Hilbert space theories are equivalent in even this more demanding sense. The U implementing their unitary equivalence will identify vector states $|\psi\rangle, |\phi\rangle$ in one theory with vector states $U|\psi\rangle, U|\phi\rangle$ in the other. Then U's unitary ensures that $|\langle\phi|\psi\rangle| = |\langle\phi|U^{-1}U\psi\rangle|$, in conformity with (K★E).

(K★E) is a *modal* amplification of the content coincidence criterion expressed by (KE). Let us consider next a dynamical amplification. Kinematically equivalent theories $(\mathcal{Q}, \mathcal{S})$ and $(\mathcal{Q}', \mathcal{S}')$ are hardly physically equivalent if the time developments of states instantaneously identified under i_s diverge. Suppose that a theory's dynamics takes the form of a set $\mathcal{D} = \{d_t\}$ of one (real) parameter flows on state space, where the action of $d_t \in \mathcal{D}$ on a $\omega \in \mathcal{S}$ is understood as follows: $d_t(\omega)$ is the state into which ω evolves under the influence of d_t in a time t. So, for example, an ordinary quantum theory recognizes an element of $\mathcal{D}$ for each candidate Hamiltonian in $\mathfrak{B}(\mathcal{H})$. For the element d_t corresponding to the Hamiltonian $\hat{H}$, the state $\hat{W}$ evolves as follows:

$$d_t(\hat{W}) = e^{-i\hat{H}t}\hat{W}e^{i\hat{H}t}$$

---

[10] The restriction to pure states is meant to finesse questions about how to understand mixed states and the transition probabilities between them.

[11] This talk of possibility metrics shouldn't be read as implying that David Lewis would adopt the inner product as a metric for the space of possible worlds, or that its adoption would cohere with antecedently established intuitions about which worlds are most like which others (see eg. Lewis 1979).

2.2 PHYSICAL EQUIVALENCE AND UNITARY EQUIVALENCE

Modeling a dynamical theory as a triple $(\mathcal{Q}, \mathcal{S}, \mathcal{D})$, we can announce a criterion of dynamical equivalence:

Definition 2.10 (Dynamical Equivalence (DE)). $(\mathcal{Q}, \mathcal{S}, \mathcal{D})$ and $(\mathcal{Q}', \mathcal{S}', \mathcal{D}')$ are dynamically equivalent if and only if $(\mathcal{Q}, \mathcal{S})$ and $(\mathcal{Q}', \mathcal{S}')$ satisfy (KE) and there exists a bijection $i_d : \mathcal{D} \to \mathcal{D}'$ such that for all $\omega \in \mathcal{S}$, all $A \in \mathcal{Q}$, all $d_t \in \mathcal{D}$ and all t, $d_t(\omega)(A) = i_d(d_t)([i_s(\omega)])(i_q(A))$.

Unitarily equivalent theories of ordinary QM are also dynamically equivalent. Where U implements their unitary equivalence, i_d identifies the flow generated by $\hat{H}$ in one theory with the flow generated by $U\hat{H}U^{-1}$ in the other. Dynamical equivalence follows from properties of the trace and of unitarity already harped upon.

In the course of an "informal" explication of theoretical equivalence, Putnam reckons equivalent theories should not only be "completely mutually interpretable" (1989, 39), but also such that "the 'translation' of each into the other preserves the relation of *explanation*". Implying (KE) for ordinary quantum theories, unitarily equivalence ensures complete mutual translation respecting functional relations—including those (e.g. $F = ma$) implicated in theory's nomic structure. Insofar as the relation of explanation rests on a theory's dynamical, nomic, and modal structures, unitary equivalence, by implying (KE★) and (DE), promises to preserve explanation as well. Unitarily equivalent ordinary quantum theories agree about which observables are physical, about which states on those observables are possible, about how those states can change in time, about how to equip those states with a possibility metric, about which data confirm or falsify the statistical predictions they're capable of making. Their agreement is stalwart enough to suggest that *unitarily equivalent ordinary quantum theories are presumptively physically equivalent*.

They're only "presumptively" equivalent because unitarily equivalent theories subject to different semantics could come out physically inequivalent. Consider a Bohm-type semantics for ordinary quantum theory declaring position always determinate, and another Bohm-type semantics according that privilege to momentum. Applied to unitarily equivalent theories, the fraternal Bohmians identify very different possible worlds. This underscores the remark with which this section opened: that physical equivalence is properly understood as a relation between *fully interpreted physical theories*. (It can even happen that unitarily inequivalent theories of ordinary quantum mechanics subject to different, and slyly compensating, semantics turned out to be physically equivalent—although I myself would be very suspicious of the motives of the interpreters if that came to pass.) Still, the case is strong that unitarily equivalent ordinary quantum theories are as physically equivalent as they can be, given that they're not fully interpreted!

These considerations elevate the Stone–von Neumann theorem—which states (roughly) that any two quantizations of a (suitable) classical theory T are unitarily equivalent—to a heartening uniqueness theorem: there is only one ordinary quantum theory that quantizes T! To that theorem we now turn.

2.3 The Stone–von Neumann uniqueness theorem

2.3.1 Mechanics: classical to quantum

Ideologies of quantization and their sustaining techniques are richly deserving of a foundational study on their own.[12] It is a decided limitation of this book that almost all the quantizations it discusses are instances of a single species: Hamiltonian (a.k.a. canonical) quantization, the straightforward and powerful conception of what it is to quantize a classical theory that frames the Stone–von Neumann uniqueness theorem. The Hamiltonian quantization procedure applies to a classical theory cast in Hamiltonian form. And so we will start with a very brief sketch of Hamiltonian mechanics.

Lagrangian to Hamiltonian mechanics One often reaches the Hamiltonian formulation of a classical theory from its Lagrangian formulation.[13] Consider a very simple mechanical system: a particle of mass m moving in a three-dimensional Euclidean space. The system's *configuration space*, the space of its possible configurations (a.k.a. positions), is $\mathbb{R}^3$. Let $\mathbf{q} = (q_1, q_2, q_3)$ denote an arbitrary element of this space. The first time derivative of the system's configuration, $\dot{\mathbf{q}} := \frac{d\mathbf{q}}{dt}$, is just the system's velocity. The pair $(\mathbf{q}, \dot{\mathbf{q}})$ specifies the state of the system: the values of all significant quantities pertaining to the system are fixed by its $(\mathbf{q}, \dot{\mathbf{q}})$ values. The moral generalizes: a mechanical system freed from the confines of a configuration space assumed to be $\mathbb{R}^3$ still has a state of the form $(\mathbf{q}, \dot{\mathbf{q}})$, where q_i are *generalized coordinates* appropriate to the system's configuration space; observables pertaining to the system are functions from its state space to $\mathbb{R}$.

Noteworthy among these is the system's *Lagrangian*, a function $\mathcal{L}(\mathbf{q}, \dot{\mathbf{q}})$ of the system's configuration and its first derivative. When it comes time to consider field theories, we'll often take $\mathcal{L}$ to arise from a spatial Lagrangian density $L(\mathbf{x})$ via $\mathcal{L} = \int L(\mathbf{x}) d\mathbf{x}$. For a mechanical system subject only to forces that depend only on its configuration, $\mathcal{L}(\mathbf{q}, \dot{\mathbf{q}})$ is just kinetic minus potential energy. For our simple mechanical system, for instance,

$$\mathcal{L} = \frac{1}{2} m \dot{\mathbf{q}}^2 - V(\mathbf{q}) \quad (2.5)$$

where $V(\mathbf{q})$ is a configuration-dependent potential function. Applying a variational principle to the Lagrangian results in the Euler–Lagrange equations, which govern how a system's state $(\mathbf{q}, \dot{\mathbf{q}})$ changes over time and are equivalent to equations of motion determined by Newton's second law.

To obtain the Hamiltonian formulation of classical mechanics, introduce for each configuration variable q_i a *conjugate momentum* p_i, defined by:[14]

---

[12] For a taste, see §4 of Landsman (2004).
[13] For a more rigorous and systematic introduction, see Marsden and Ratiu (1994).
[14] For a constrained Hamiltonian system, there fail to be as many independent such equations as there are dimensions of configuration space. Gauge freedom is present; so too are a host of difficulties attending

2.3 THE STONE–VON NEUMANN UNIQUENESS THEOREM

$$p_i := \frac{\partial \mathcal{L}}{\partial \dot{q}_i} \tag{2.6}$$

With $\mathcal{L}$ as given above, p_i is (reassuringly enough) just $m\dot{q}_i$. Next, introduce a Hamiltonian observable $H(q_i, p_i)$, defined by:

$$H(q_i, p_i) := \sum_{j=1}^{n} p_j \dot{q}_i - \mathcal{L}(q_i, \dot{q}_i) \tag{2.7}$$

In Hamiltonian mechanics, the state of a simple mechanical system is determined by its position and momentum. The position and momentum variables q_i and p_j in whose terms the Hamiltonian is given therefore serve as coordinates for the phase space M of possible states of the system. They're known as *canonical coordinates*. System observables are functions from M to $\mathbb{R}$. The position and momentum *observables*, for example, map points in phase space to their q_i and p_j coordinate values, respectively. All other observables can be expressed as functions of these observables. In this sense, q_i and p_j are *fundamental*: to attribute a system a configuration and a momentum is thereby to determine the value of every other magnitude pertaining to that system. This variety of fundamentality, which I will call "fundamental in the physicist's sense," will reappear in connection with Chapter 9's discussion of the particle notion in QFT.

In Hamiltonian mechanics, dynamics flow from the Hamiltonian observable H, which usually coincides with the sum of the system's kinetic and potential energies. The Hamiltonian helps identify dynamically possible trajectories $\mathbf{q}(t), \mathbf{p}(t)$ through phase space as those obedient to Hamilton's equations of motion, also equivalent to Newton's second law:

$$\frac{dq_i}{dt} = \frac{\partial H}{\partial p_i}, \quad \frac{dp_i}{dt} = -\frac{\partial H}{\partial q_i} \tag{2.8}$$

Here, and for the sake of mathematical tractability, we'll confine our attention to linear dynamical systems, that is those whose phase spaces admit linear canonical coordinates and whose equations of motion are linear in those coordinates.[15]

Let's review the dossier we've compiled on Hamiltonian mechanics. Possible states of a Hamiltonian system are elements of the phase space M appropriate to that system; observables pertaining to the system are functions $M \to \mathbb{R}$; system dynamics are given by the flow imposed by the system's Hamiltonian observable on M via Hamilton's equations. To prepare the ground for Hamiltonian quantization, we ought to say a little more. We ought to say something about the *algebraic* structure of classical observables.

---

the system's quantization (see Henneaux and Teitelboim 1992). For the time being, I will ignore these. For a sustained philosophical treatment of gauge theories, see Healey (2007).

[15] For a precise characterization, see Wald (1994, 14 ff.); Marsden and Ratiu (1994, §2.7 ff.).

As smooth functions on phase space, classical observables form a set that is also a *vector space* over the real numbers.[16] To first approximation, an algebra is a linear vector space V endowed with a (not necessarily associative) *multiplicative structure*.[17] More specialized algebras have more specialized multiplicative structures. The vector space of classical observables becomes a *Lie algebra* upon being equipped with a multiplicative structure supplied by the Poisson *bracket*.[18] The Poisson bracket $\{f,g\}$ of classical observables $f : M \to \mathbb{R}$ and $g : M \to \mathbb{R}$ is:

$$\{f,g\} := \sum_i \left(\frac{\partial f}{\partial q_i} \frac{\partial g}{\partial p_i} - \frac{\partial f}{\partial p_i} \frac{\partial g}{\partial q_i} \right) \tag{2.9}$$

The Poisson bracket provides another way of formulating Hamilton's equations (2.8).

$$\frac{dq_i}{dt} = \{q_i, H\}, \quad \frac{dp_i}{dt} = \{p_i, H\} \tag{2.10}$$

It follows that for a general observable $f : M \to \mathbb{R}$

$$\frac{df}{dt} = \{f, H\} \tag{2.11}$$

The Poisson brackets between observables p_i and q_j are particularly nifty.

$$\{p_i, p_j\} = \{q_i, q_j\} = 0, \quad \{p_i, q_j\} = -\delta_{ij} \tag{2.12}$$

Considered both as coordinates and as observables, p_j and q_i are canonical *because* they satisfy (2.12).

Here is the payoff. With the fundamental Poisson brackets (2.12), we have identified the structure of a classical theory essential to its quantization. *According to the Hamiltonian quantization scheme, the fundamental Poisson brackets of a classical theory T encapsulate the algebraic structure any quantization of T must realize.* As §2.3.3 elaborates, the realization takes the form of symmetric Hilbert space operators obeying *canonical commutation relations* mimicking the classical Poisson brackets.

Hamiltonian mechanics generalizes straightforwardly to more complicated systems. The configuration of a composite mechanical system consists in the positions of its constituents; thus, a system of two (non-identical) particles moving in a three-dimensional Euclidean space has configuration space $\mathbb{R}^6$ (where the first three coordinates represent the position of the first particle and the second three coordinates represent the position of the second particle). These examples notwithstanding, a configuration space

---

[16] Where f and g are elements of this vector space, vector addition is defined by $(f+g)(x) = f(x) + g(x)$ for all $x \in M$ and scalar multiplication defined by $(af)(x) = af(x)$ for all $a \in \mathbb{R}$ and all $x \in M$.

[17] Which is to say a function $V \times V \to V$, linear in each argument, that maps each pair of vectors u, v to a vector $u \cdot v$ which is their product.

[18] A *bracket* is a map from pairs of vectors u, v to a vector $[u, v]$, which map is linear, antisymmetric, and satisfies the Jacobi identity: $[u, [v, w]] + [v, [w, u]] + [w, [u, v]] = 0$ for all $u, v, w \in V$. A vector space equipped with a bracket structure is a Lie algebra. Lie algebras have numerous astonishing features. For instance, they can be "exponentiated" to obtain Lie groups. For more on Lie groups and Lie algebras, see Bryant (1995).

needn't have the topology $\mathbb{R}^n$. A single system moving on the circle S^1 has S^1 as its configuration space. And one way to treat a system consisting of n identical particles moving in a 2-dimensional Euclidean space is to start with a configuration space that is the n-fold Cartesian product of $\mathbb{R}^2$ with itself, then identify any two configurations that differ only in the exchange of two particles (see Morandi 1992, 124 ff., for further complications).

2.3.2 Classical theories as symplectic vector spaces

The symplectic vector space structure... of the manifold of solutions for a linear dynamical system is the fundamental structure that underlies the construction of the quantum theory of a linear field. (Wald 1994, 17)

The foregoing presentation of Hamiltonian mechanics is probably old hat. Let's restyle it. Details of the restyling are not essential to what follows. For the sakes of readers already familiar with them, as well as readers liable to regard them as unpleasant distractions, herewith:

The Short Version. The aim of this section is to disclose the phase space of a classical Hamiltonian theory as a variety of vector space called *a symplectic vector space*. The point of the disclosure is to sow the idea that an inner product vector space (or Hilbert space), in which the *quantization* of a classical theory is set, may descend from that classical theory's phase space, understood as symplectic vector space.

A classical theory can be modeled as a pair (M, Ω) where M is its phase space and the symplectic product Ω is an antisymmetric, linear, non-degenerate map from pairs of points of M to the real numbers. The Poisson brackets between canonical classical observables can be expressed in terms of Ω; ergo, so too can be expressed the algebraic structure the quantization of the classical theory should mirror. (M, Ω) is "like" a Hilbert space because M is a linear vector space on which $\Omega : M \times M \to \mathbb{R}$ defines a product. (M, Ω) is not like a Hilbert space because the product Ω defines is not an inner product. ♠

Consider Hamilton's equations for a system with configuration space $\mathbb{R}$:

$$\frac{dq}{dt} = \frac{\partial H}{\partial p}, \quad \frac{dp}{dt} = -\frac{\partial H}{\partial q}$$

Letting $y = (q, p) = (y_1, y_2)$ stand for an arbitrary element of the system's phase space, we can rewrite this in one fell swoop by means of a 2×2 matrix Ω:

$$\Omega = \begin{pmatrix} 0 & 1 \\ -1 & 0 \end{pmatrix} \tag{2.13}$$

Hamilton's equations formulated in terms of Ω become:

$$\sum_{ij} \Omega_{ij} \frac{dy_i}{dt} = \frac{\partial H}{\partial y_j} \tag{2.14}$$

34 QUANTIZING

This trick generalizes to Hamiltonian systems with configuration space $\mathbb{R}^n$ (and beyond): an antisymmetric, linear, non-degenerate $2n \times 2n$ matrix Ω enables us to write Hamilton's equations in (2.14)'s compact form.

What makes this conservation of ink possible is the fact that the phase space M of a Hamiltonian system is a *symplectic vector space*, that is, a vector space equipped with a symplectic product.

Definition 2.11 (Symplectic Product). A *symplectic product* on a vector space V over the field $\mathcal{F}$ is a map $\Omega : V \times V \to \mathcal{F}$ which is antisymmetric (that is $\Omega(f,g) = -\Omega(g,f)$, which implies that $\Omega(f,f) = 0$ for all $f \in V$); and linear in each argument. If $\Omega(f,g) = 0$ for all g only if $f = 0$, Ω is *non-degenerate*.

A Hamiltonian system with configuration space $\mathbb{R}^n$ has a phase space $\mathbb{R}^{2n}$ with a natural vector space structure.[19] Let $\gamma = (q_1, \ldots, q_n, p_1, \ldots, p_n)$ denote an arbitrary element of that space. The non-degenerate symplectic product

$$\Omega(\gamma, \gamma') = \sum_{j=1}^{n} (p_i q'_j - p'_i q_j) \tag{2.15}$$

turns it into a symplectic vector space. Reverting to the case of one dimension, we see that $\Omega(\gamma, \gamma') = \sum_{ij} \Omega_{ij} \gamma'_i \gamma_j$, with Ω given by (2.13), is an instance of (2.15). The lesson generalizes: the Ω that enables the expeditious expression of Hamilton's equations for a system with phase space $\mathbb{R}^{2n}$ is a non-degenerate symplectic product on $\mathbb{R}^{2n}$.

Considered from loftier heights than we'll typically attain here, Hamiltonian mechanics has symplectic structure built into its very *geometry*. Viewed from the vantage of differential geometry, the phase space of a Hamiltonian theory is a manifold M equipped with a closed non-degenerate two-form ω, also known as a *symplectic form*. Given a Hamiltonian function $h : M \to \mathbb{R}$, ω identifies a smooth vector field on M, the integral curves of which are solutions to Hamilton's equations.[20]

We'll nevertheless stick with regarding the phase space of a classical Hamiltonian theory as a low-brow symplectic vector space (even though it's also a sophisticated symplectic manifold). For a fixed point $\gamma_0 = (q_0, p_0)$ of $\mathbb{R}^{2n}$, $\Omega(\gamma_0, \cdot)$ is a function $\mathbb{R}^{2n} \to \mathbb{R}$ that's linear, because Ω is bilinear. By a judicious choice of γ_0, we can induce $\Omega(\gamma_0, \cdot)$ to be the function whose value at each point of $\mathbb{R}^{2n}$ is just the q_1 coordinate of that point.[21] We can do likewise for each component of the canonical observables. It follows that we can fix γ_0 so that $\Omega(\gamma_0, \cdot)$ is any linear function we like of the canonical observables.[22] In particular, we can express the Poisson brackets for the canonical observables (2.12) *in their entirety* as follows:

---

[19] With addition defined by $(p_i, q_i) + (p'_i, q'_i) = (p_i + p'_i, q_i + q'_i)$, and so on.

[20] For elaboration and explication of geometrical approaches to mechanics, and their fecundity for foundational questions, see Butterfield (2007); Belot (2007). The digressive §3.2 is one place that will make use of symplectic manifolds.

[21] For those of you keeping score at home, the choice is $\gamma_0 = (0, \ldots, 0; 1, 0, \ldots)$.

[22] E.g. $\gamma_0 = (2, 0, \ldots; 3, 0, \ldots)$ reproduces the function $f(q_i, p_i) = 3q_1 - 2p_1$.

2.3 THE STONE–VON NEUMANN UNIQUENESS THEOREM

$$\{\Omega(\gamma_1,\cdot),\Omega(\gamma_2,\cdot)\} = -\Omega(\gamma_1,\gamma_2) \qquad (2.16)$$

Wald explains why we should bother with all this compact re-expression.

> It may seem to the reader that [2.16] does little more than express the perfectly satisfactory and familiar relations [2.12] in a much more obscure form. However, even in the finite dimensional case, [2.16] has a notable advantage over [2.12] in that it allows us to write down the fundamental Poisson bracket relations without requiring us to make a particular choice of linear canonical coordinates on M. More importantly, when we treat fields in curved spacetime (for which M is infinite dimensional), it is far from clear how to generalize the relations [2.12] in such a way as to define the Poisson bracket on a natural class of linear functions on M. On the other hand, the Poisson bracket relation in the form [2.16] generalizes straightforwardly to the infinite dimensional case. (Wald 1994, 15)

We should bother because the compact re-expression enables the extension of techniques of Hamiltonian quantization to systems of the sort treated by QM_∞.

Where (M,Ω) is the phase space, considered as a symplectic vector space, of a classical Hamiltonian theory, and $\gamma_1, \gamma_2 \in M$, $\Omega(\gamma_1,\gamma_2)$ is their *symplectic product*, Hamilton's equations define dynamically allowed trajectories $\gamma_1(t)$ and $\gamma_2(t)$ through γ_1 and γ_2. Hamiltonian time evolution conserves symplectic products: the quantity $\Omega(\gamma_1(t),\gamma_2(t))$ is independent of the choice of t. This enables us to perform the following trick. We have the phase space M of the classical theory, along with its symplectic structure. Down the road, we will be interested in the space S of duly well-behaved solutions to equations of motion for a system of phase space M. For instance, we'll build the Hilbert space in which the quantization of the Klein–Gordon field lives from the space of solutions to the classical Klein–Gordon equation. The trick is to identify elements of S with elements of M by picking some time $t=0$ and treating each element of M as the initial data for an element of S. Because the symplectic product is preserved by time evolution, the symplectic product between γ_1 and γ_2, understood as points of the phase space M, is also unambiguously the symplectic product between their Hamiltonian evolutes $\gamma_1(t), \gamma_2(t)$, for all t. But that makes it the symplectic product between the pair of elements of the solution space S for which γ_1 and γ_2 are initial data! Thus the symplectic structure Ω on M can be transferred to a symplectic structure Ω on S. Often in what follows, symplectic vector spaces such as (M,Ω)—equivalently, (S,Ω)—will constitute the scaffolding of a classical theory from which its quantization is constructed.

2.3.3 Hamiltonian quantization

> ...most theoretical physicists, following Schwinger, regard the *action principle* as fundamental. Theories are defined by Lagrangians...There is, however, another point of view, more consistent with the spirit of axiomatic field theory. *Any* commutation or anticommutation relations consistent with the dynamics can define a possible model... In this approach a formal theory consists of equations of motion plus commutation rules (or some more general algebraic relations). (Fulling 1989, 126)

In the Hamiltonian quantization scheme, a classical theory cast in Hamiltonian form is quantized by promoting its canonical observables to symmetric operators $\hat{q}_i, \hat{p}_i$ acting on some Hilbert space $\mathcal{H}$ and obeying commutation relations corresponding to the fundamental Poisson brackets of the classical theory. The familiar Heisenberg form of the canonical commutation relations (CCRs), representing a quantization of our mercifully simple classical theory with phase space $\mathbb{R}^{2n}$ and canonical observables q_i and p_i, is (cf. (2.1) and (2.12)):

$$[\hat{q}_i, \hat{q}_j] = [\hat{p}_i, \hat{p}_j] = 0, \quad [\hat{p}_i, \hat{q}_j] = -i\hat{I}\delta_{ij} \tag{2.17}$$

where $\hat{I}$ is the identity operator.

Should the aspiring quantum mechanic construe her classical theory in terms of a symplectic vector space (S, Ω), she'd apply the same recipe to the CCRs of that theory, expressed in terms of Ω (cf. (2.16)). That is, she'd assign to each $y \in S$ a symmetric operator $\hat{\Omega}(y, \cdot)$ acting on Hilbert space $\mathcal{H}$ and obeying:

$$[\hat{\Omega}(y_1, \cdot), \hat{\Omega}(y_2, \cdot)] = -i\hat{\Omega}(y_1, y_2)\hat{I} \tag{2.18}$$

which is equivalent to (2.17).

In a slight abuse of terminology, I'll call a set of operators $\{\hat{p}_i, \hat{q}_j\}$ [$\{\hat{\Omega}(y, \cdot)\}$] acting on a Hilbert space $\mathcal{H}$ to satisfy (2.17) [(2.18)] a *representation* of the Heisenberg form of the CCRs on that Hilbert space. By a "quantization," I have meant and will continue to mean a representation of appropriate commutation relations. The epigram from Fulling conveys a conviction common among mathematical physicists: the commutation relations form the physical core of a quantum theory. Section 1.2 contended that, coming to philosophers' attention already endowed with such a core, physical theories come to philosophers' attention already partially interpreted.

Ordinary quantum mechanics is one way to build a theory around that core. For, having found a representation of the CCRs, one can prosecute ordinary quantum mechanics as usual, that is, in the terms set by §2.1's [OQM], [OQS], and [OQD]. Density matrices on the Hilbert space $\mathcal{H}$ carrying the representation correspond to quantum states; bounded self-adjoint operators on $\mathcal{H}$—we'll see in Chapter 4 that these are the ones we can obtain as polynomials, or limits (in an appropriate sense) of sequences of polynomials, of the canonical observables $\{\hat{q}_i, \hat{p}_j\}$—correspond to quantum observables. The Schrödinger equation gives a template for dynamical evolution. Thus a quantization plus the precepts of ordinary quantum mechanics specify the structure of a quantum theory.

The quantum theory so specified embraces in one kinematic structure a family of dynamical models—as many as there are candidate Hamiltonian observables in $\mathfrak{B}(\mathcal{H})$. To precipitate a particular dynamical model out of the family, that is, to apply the quantum theory to a particular physical system, one needs to assign that system a Hamiltonian observable $\hat{H}$. Via the Schrödinger equation, $\hat{H}$ generates dynamics for the system. These dynamics imply that the time rate of change of an observable $\hat{f}$ pertaining to the system is given by:

$$\frac{d\hat{f}}{dt} = [\hat{f}, \hat{H}] \tag{2.19}$$

Comparison with the expression (2.11) for the time evolution of a classical observable motivates the "Poisson bracket goes to commutator" heuristic: following the heuristic establishes a correspondence between classical and quantum dynamics.

The usual realization of the Heisenberg form of the CCRs for a system with configuration space $\mathbb{R}^n$ is the one Schrödinger found, called the *Schrödinger Representation*. It uses the Hilbert space $L^2(\mathbb{R}^n)$ of square-integrable complex-valued functions $\psi(x_1, ..., x_n)$ of $\mathbb{R}^n$, with:

$$\hat{x}_i \psi(x_1, ..., x_n) = x_i \psi(x_1, ..., x_n)$$
$$\hat{p}_j \psi(x_1, ..., x_n) = -i \frac{d\psi(x_1, ..., x_n)}{dx_j} \tag{2.20}$$

(compare. Eq. (2.3)). According to the Stone–von Neumann theorem, any other representation of the CCRs is essentially equivalent to this one. The next section presents that theorem.

2.3.4 The Stone–von Neumann theorem

The Weyl relations The Heisenberg form of the CCRs is mathematically infelicitous (for elaboration, see Summers 1999). The one-dimensional case suffices to exhibit the infelicity. The unboundedness of the position and momentum operators prevents them from being jointly defined on all of $L^2(\mathbb{R})$, so that (2.17) fails to be an operator equation for that Hilbert space. As §3.1 explains, this can sometimes be addressed by painstaking attention to domains of definition for canonical operators. But there is a more convenient fix. It is to exponentiate $\hat{q}$ and $\hat{p}$ to obtain unitary operators, bounded and everywhere defined on a Hilbert space $\mathcal{H}$, and to re-express (2.17) in terms of those unitaries. In particular, where $\hat{U}(a) = \exp(-ia\hat{q})$ and $\hat{V}(a) = \exp(-ib\hat{p})$, if (2.17) holds, then for all $a, b \in \mathbb{R}$,

$$\hat{U}(a)\hat{V}(b) = \exp(-iab)\hat{V}(b)\hat{U}(a) \tag{2.21}$$

Call (2.21) the *Weyl form* of the CCRs, or *Weyl relations* for short. Their generalization for a classical theory with phase space $\mathbb{R}^{2n}$ is straightforward. Where **a** and **b** are both n-tuples of real numbers, one defines:

$$\hat{U}(\mathbf{a}) = exp(-i\sum_{i=1}^{n} a_i \hat{q}_i) \text{ and } \hat{V}(\mathbf{b}) = exp(-i\sum_{i=1}^{n} b_i \hat{p}_i).$$

Then one requires that for all $\mathbf{a}, \mathbf{b} \in \mathbb{R}^n$,

$$\hat{U}(\mathbf{a})\hat{V}(\mathbf{b}) = \exp(-i\mathbf{a}\cdot\mathbf{b})\hat{V}(\mathbf{b})\hat{U}(\mathbf{a}) \tag{2.22}$$

where $\mathbf{a}\cdot\mathbf{b}$ is the dot product of those vectors in $\mathbb{R}^n$.

The Weyl relations encapsulating the quantization of a classical theory can also be expressed in terms of the symplectic product Ω on that classical theory's phase space M or solution space S. Suppose for simplicity that $M = \mathbb{R}^2$. Then each pair (a, b) of real numbers corresponds to an element $\gamma = (a, b) \in M$. We can use the unitaries $U(a)$ and $V(b)$ satisfying the Weyl Relations (2.21) to define a map from elements of M to unitary operators acting on $\mathcal{H}$. For each $\gamma = (a, b) \in \mathbb{R}$ set:

$$\hat{W}(\gamma) = \exp(i\frac{1}{2}ab)U(a)V(b) \tag{2.23}$$

Then it follows from the unitarity of $U(a)$, $V(b)$ and the standard rules for taking adjoints that $\hat{W}^\dagger(\gamma) = \hat{W}(-\gamma)$. Rewriting the Weyl relations (2.21) in terms of $\hat{W}(\gamma)$s, we obtain:

$$\hat{W}(\gamma_1)\hat{W}(\gamma_2) = \exp[\frac{i}{2}(a_2 b_1 - a_1 b_2)]\hat{W}(\gamma_1 + \gamma_2) \tag{2.24}$$

But the quantity $(a_2 b_1 - a_1 b_2)$ in the exponent, Eq. (2.15) tells us, is just the standard symplectic product $\Omega(\gamma_1, \gamma_2)$ between elements $\gamma_1 = (a_1, b_1)$ and $\gamma_2 = (a_2, b_2)$ of $M = \mathbb{R}^2$. This generalizes. To quantize the classical theory associated with the symplectic vector space (S, Ω), the aspiring quantum mechanic must find for each $\gamma \in S$ an operator $\hat{W}(\gamma)$ which is unitary, and which is an element of a set of such operators satisfying the relations:

$$\hat{W}(\gamma_1)\hat{W}(\gamma_2) = \exp[\frac{i}{2}\Omega(\gamma_1, \gamma_2)]\hat{W}(\gamma_1 + \gamma_2)$$
$$\hat{W}^\dagger(\gamma) = \hat{W}(-\gamma) \tag{2.25}$$

The relations (2.25) are equivalent to (2.22), and will also be called *the Weyl relations*. A *representation* of the Weyl relations is a map from M or S to a collection of unitary operators on a Hilbert space $\mathcal{H}$ satisfying (2.25).

Returning to the case that $M = \mathbb{R}^2$, the Weyl unitaries $\hat{U}(a)$ and $\hat{V}(b)$ may be familiar under a different description. Recall that in the familiar Schrödinger representation, momentum is the infinitesimal generator of position translations and vice versa. Letting $\exp(-ib\hat{p})$ act on a normed vector $|\psi\rangle \in L^2(\mathbb{R})$ (i.e. a pure state) translates its position expectation value by an amount b; letting $\exp(-ia\hat{q})$ act on $|\psi\rangle$ translates its momentum expectation value by an amount a. But these position and momentum translation operators are just the Weyl unitaries $\hat{U}(a)$ and $\hat{V}(b)$! Indeed, in the Schrödinger representation of the Weyl relations, $\hat{U}(a)$ and $\hat{V}(b)$ act on an arbitrary $\psi \in L^2(\mathbb{R}^n)$ as follows:

$$\hat{U}(a)\psi(x) = e^{ia\cdot x}\psi(x) \tag{2.26}$$

$$\hat{V}(b)\psi(x) = \psi(x+b) \tag{2.27}$$

2.3 THE STONE–VON NEUMANN UNIQUENESS THEOREM

It's easy to see from Eq. (2.27) that the Weyl unitary $\hat{V}(b)$ shifts the expectation value $\psi \in L^2(\mathbb{R})$ assigns to the position observable by b. To appreciate Eq. (2.26) as describing a shift in momentum, apply the Schrödinger representation momentum operator $-i\frac{d}{dx}$ to the r.h.s., and compare the result to $-i\frac{d}{dx}\psi(x)$.

Stone's theorem assures us that, provided the families $\{\hat{U}(a)\}$ and $\{\hat{V}(b)\}$ are continuous with respect to the real parameters a and b, the canonical operators $\hat{q}$ and $\hat{p}$ can be recovered as the unique infinitesimal generators of those families. In this case, the Heisenberg CCRs (2.17) are essentially the infinitesimal form of the Weyl relations (2.22). Supposing continuity (a supposition suspended presently), the former hold if the latter do. The converse is not true. For example, on the Hilbert space $L^2(0, 1)$ of square integrable functions of the real interval $[0, 1]$, $\hat{q}$ and $\hat{p}$ as defined by (2.3) satisfy the Heisenberg CCRs but not the Weyl Relations—naively, because once $\hat{V}(b)$'s action is patched up to keep it from translating functions out of $L^2(0, 1)$, the Weyl Relations break down (see Segal 1967, §1).

Some definitions will clear the way for a statement of the Stone–von Neumann theorem.

Definition 2.12 (invariant subspace of a Hilbert space). A subspace $\mathcal{K}$ of $\mathcal{H}$ is *invariant* under a collection $\mathfrak{Q}$ of operators on $\mathcal{H}$ if and only if the result of acting on any vector in $\mathcal{K}$ with an operator in $\mathfrak{Q}$ is a vector in $\mathcal{K}$.

Invariant subspaces are like islands you can't get off using tools in $\mathfrak{Q}$. Informally, irreducible representations are ones without interesting islands. Officially:

Definition 2.13 ((ir)reducible representations). A representation $(\hat{U}(a), \hat{V}(b))$ $[\hat{W}(\gamma)]$ of the Weyl relations acting on a Hilbert space $\mathcal{H}$ is *irreducible* if and only if no non-trivial subspace of $\mathcal{H}$ is invariant under all $\hat{U}(a), \hat{V}(b)$ $[\hat{W}(\gamma)]$. Other representations are *reducible*.

Reducible representations flout parsimony: each invariant subspace of a representation of the Weyl relations itself carries a representation—called a *subrepresentation*—of the Weyl relations, obtained by restricting the reducible representation to that subspace. (Subrepresentations will be dealt with more systematically in Chapter 4.) An irreducible representation has itself as its only subrepresentation; a reducible representation is a direct sum of its subrepresentations, where this is understood as follows:

Scholium: Direct sum of Hilbert spaces. Where $\mathcal{H}$ and $\mathcal{H}'$ are Hilbert spaces, their direct sum $\mathcal{H} \oplus \mathcal{H}'$ is a vector space consisting of elements of the form $|\alpha\rangle \oplus |\beta\rangle$, $|\alpha\rangle \in \mathcal{H}$ and $|\beta\rangle \in \mathcal{H}'$. The rule for vector addition is $|\alpha_1\rangle \oplus |\beta_1\rangle + |\alpha_2\rangle \oplus |\beta_2\rangle = (|\alpha_1\rangle + |\alpha_2\rangle) \oplus (|\beta_1\rangle + |\beta_2\rangle)$. Inner products work like so: $((\langle\alpha_1| \oplus \langle\alpha_2|)(|\beta_1\rangle + |\beta_2\rangle)) = \langle\alpha_1|\alpha_2\rangle + \langle\beta_1|\beta_2\rangle$.) $\mathcal{H} \oplus \mathcal{H}'$ contains *only* those elements: thanks to the rule for vector addition, and in contrast to the entanglement endemic to the tensor product of two Hilbert spaces, $\mathcal{H} \oplus \mathcal{H}'$ contains only vectors that "factorize" into direct products of vectors from each summand Hilbert space. ♠

Superselection Reducible representations are lurking in the notion of *superselection*, which may be familiar from discussions of ordinary QM (e.g. Hughes 1989, 285). My present, modest aim is to exploit this posited familiarity to illustrate the notion of a reducible representation.

Consider a quantum theory set in a Hilbert space $\mathcal{H}$. A superselection rule is in force if it is not the case that every normed vector in $\mathcal{H}$ corresponds to a pure state of the theory.[23] A normed non-trivial superposition of vectors $|\chi\rangle = a|\phi\rangle + b|\psi\rangle$ is not a pure state if the expectation value assignments it generates coincide with those arising from the convex combination of distinct states $\hat{P}_{|\phi\rangle}$ and $\hat{P}_{|\psi\rangle}$ implemented by the non-extremal density operator $\hat{W} = |a|^2 \hat{P}_{|\phi\rangle} + |b|^2 \hat{P}_{|\psi\rangle}$. (Here $\hat{P}_{|\psi\rangle}$ is the projection operator for the subspace spanned by $|\psi\rangle$.) Let $\mathfrak{Q}$ denote the full set of physical observables, which we are for the purposes of this discussion identifying with observables providing a representation of the Weyl relations. In order for $|\chi\rangle$ to be empirically distinguishable from $\hat{W}$, there must be some observable $\hat{A} \in \mathfrak{Q}$ susceptible to interference between $|\psi\rangle$ and $|\phi\rangle$, in the sense that $\langle\phi|\hat{A}|\psi\rangle \neq 0$. Thus, a superselection rule is in force only if there is some non-colinear pair of vectors $|\psi\rangle$ and $|\phi\rangle$ such that $\mathfrak{Q}$ lacks an observable susceptible to their interference. But if there is such a pair of vectors, they will belong to distinct subspaces of $\mathcal{H}$ invariant under the action of $\mathfrak{Q}$, contrary to the assumption that $\mathfrak{Q}$ is an irreducible representation. Thus no superselection rule applies to an irreducible representation.

In the presence of a superselection rule, $\mathcal{H}$ resolves itself into a set of orthogonal subspaces $\{\mathcal{K}_i\}$—called variously *superselection subspaces*, *superselection sectors*, or *coherent subspaces*—that collectively span $\mathcal{H}$. That is, $\mathcal{H} = \oplus_i \mathcal{K}_i$. If $|\phi\rangle$ and $|\psi\rangle$ come from different superselection sectors, then for each $\hat{A} \in \mathfrak{Q}$, $\langle\phi|\hat{A}|\psi\rangle = 0$. Interference between vectors from different superselection sectors is obliterated because each superselection sector is invariant under $\mathfrak{Q}$. It follows that in the presence of superselection, a quantum theory set in $\mathcal{H}$ doesn't recognize every self-adjoint element of $\mathfrak{B}(\mathcal{H})$ as an observable. Otherwise, the projection operator $\hat{E}_\chi$, a self-adjoint element of $\mathfrak{B}(\mathcal{H})$ corresponding to a normed superposition of vectors from different superselection sectors, would render $|\chi\rangle$ empirically distinguishable from $\hat{W}$. In what we've been calling ordinary QM, observables correspond to self adjoint elements of $\mathfrak{B}(\mathcal{H})$. So superselection moves us away from ordinary QM.

The impoverishment of the ordinary quantum observable set makes possible a new kind of quantum observable, a *superselection observable*, whose defining feature is that it's a nontrivial element of the set $\mathfrak{Q}$ of observables that commutes with every other element of $\mathfrak{Q}$.[24] Likewise, when $\mathfrak{Q}$ acts reducibly on $\mathcal{H}$, there's an element of $\mathfrak{B}(\mathcal{H})$, different from the identity, that commutes with every element of $\mathfrak{Q}$ (Bratteli and Robinson 1987, prop. 2.3.8 (47)).

---

[23] For a more nuanced discussion of superselection, see Earman (2008).

[24] The definition of superselection does not guarantee the existence of such an observable. The definition implies that $\mathfrak{Q} \subset \mathfrak{B}(\oplus_i \mathcal{K}_i)$. Observables of the form $\oplus_i a_i \hat{I}_i$, where $\hat{I}_i$ is the identity operator for $\mathcal{K}_i$, commute with every element of $\mathfrak{B}(\oplus_i \mathcal{K}_i)$, and so with every element of $\mathfrak{Q}$. But the definition doesn't guarantee that observables of the specified form belong to $\mathfrak{Q}$.

2.3 THE STONE–VON NEUMANN UNIQUENESS THEOREM

The theorem we've all been waiting for! The Stone–von Neumann theorem concerns weakly continuous representations of the Weyl relations. These should be understood as representations in which the collection $\{\hat{W}(\gamma)\}$ of unitary operators acting on $\mathcal{H}$ to satisfy the Weyl relations over (M, Ω) varies continuously (in the weak topology) with γ. (NB: Restricted to unitary operators, the weak topology and the strong topology, introduced in the statement of Stone's theorem (Fact 2.3), coincide.)

Fact 2.14 (Stone–von Neumann theorem). Let (M, Ω) be a finite dimensional symplectic vector space, with $M = \mathbb{R}^{2n}$. Every weakly continuous irreducible representation of the Weyl Relations over (M, Ω) is unitarily equivalent to every other. In particular, each is unitarily equivalent to the Schrödinger representation, which is irreducible. (Proofs can be found in Simon 1972, thm. 7.5, or Prugovečki 1971, 342 ff.)

In the case that a representation $(\mathcal{H}, \hat{W}(\gamma))$ of the Weyl relations is weakly continuous but reducible, each of its irreducible subrepresentations is unitarily equivalent to the Schrödinger representation. That is, $(\mathcal{H}, \hat{W}(\gamma))$ is unitarily equivalent to a direct sum of Schrödinger representations ("equivalent to the Schrödinger representation up to multiplicity").

The Stone–von Neumann theorem assuages an anxiety a worrywart might have had about the Hamiltonian quantization recipe. The anxiety is that different physicists, each starting with the same classical theory and each competently following the recipe, could produce *different quantum theories*. We're accustomed, for instance, to understand "the quantum theory of a particle moving in one dimension"—the theory which, when equipped with a suitable Hamiltonian, describes what everyone always calls "*the* one-dimensional quantum oscillator"—as *the* theory determined by finding a representation of the Weyl relations over $\mathbb{R}^2$. The anxiety is that there's no *single* such theory, that different skillful applications of the Hamiltonian quantization scheme to the same classical theory result in *physically distinct*, maybe even mutually contradictory, quantum theories. The anxiety is that even having fixed a classical theory T, "T's quantization" is ambiguous at best and inconsistent at worst.

Section 2.2 catalogued inducements to accept that unitarily equivalent quantizations are presumptively physically equivalent. Let us accept that. Then the Stone–von Neumann theorem assures us that, when we set off to quantize a classical theory in its scope, the quantization we seek is unique.

In the version of the Stone–von Neumann theorem just presented, it is irreducible continuous representations of the Weyl relations that are unique up to unitary equivalence. The theorem can also be couched (after somewhat lengthier mathematical preliminaries) as a theorem about irreducible representations of the Heisenberg group, which is the group of position and momentum translations, characteristically entangled, for a phase space $\mathbb{R}^{2n}$, and which will make an appearance in §3.1. The uniqueness of the Schrödinger representation of the CCRs follows as well from Mackey's imprimitivity theorem (see Mackey 1998 for an exposition). These other versions of the uniqueness result (see Rosenberg 2004 for a history) incorporate more information about the structure of the classical theory—by, for instance, treating $\mathbb{R}^n$ not only as a configuration space but also as a translational symmetry group—than does the present exposition of the Stone–von Neumann theorem stated in terms of the Weyl relations. (For a denunciation of the latter on this score, see Landsman 2006, fn. 9.) My apology for focusing on an impoverished version of the uniqueness result is that (perhaps just because it is impoverished) it is the most accessible.

2.3.5 Uniqueness illustrated

The quantization of a classical point system moving in one Euclidean dimension, and its application to the one-dimensional simple harmonic oscillator, illustrates the uniqueness vouchsafed by the Stone–von Neumann theorem. It also provides a useful prologue to the constitution within QFT of a particle notion, a topic later chapters treat in more detail. Here I sketch the quantization, as well as different attitudes one might adopt toward different ways of accomplishing it.

We are looking for a representation of the CCRs arising from classical Hamiltonian mechanics in one Euclidean dimension. Once we have that representation, the particular classical dynamical model we'll attempt to render in quantum form is the simple harmonic oscillator. In terms of canonically conjugate position and momentum observables q and p, the classical Lagrangian is:

$$\mathcal{L} = \frac{1}{2}(\dot{q}^2 - \omega^2 q^2) \tag{2.28}$$

The classical Hamiltonian is:

$$H = \frac{1}{2}p^2 + \frac{1}{2}\omega^2 q^2 \tag{2.29}$$

The standard quantization of the theory is set in the Schrödinger representation. Symmetric operators $\hat{p}$ and $\hat{q}$ defined on the model of Eq. (2.20) act on the Hilbert space $L^2(\mathbb{R})$ to satisfy the CCRs (2.17) for $i = 1$. "Exponentiating" $\hat{p}$ and $\hat{q}$, we obtain a weakly continuous, irreducible representation of the Weyl relations.

"Argument" 2.15 (irreducibility of the Schrödinger representation). Equations (2.26) and (2.27) describe how the unitaries conspiring in the Schrödinger representation of the Weyl relations act on arbitrary elements of $L^2(\mathbb{R})$. To make the irreducibility of this representation plausible, focus on the unitaries $\hat{V}(b)$, which take an arbitrary $\psi(x) \in L^2(\mathbb{R})$ and shift it b units to left (yielding $\psi(x+b)$). Intuitively and pictorially, given any pair ψ, ϕ of wave functions in $L^2(\mathbb{R})$, it's going to be possible to use unitaries in $\hat{V}(b)$ to shift one so that its support—the set of points in $\mathbb{R}$ where the amplitude of the wave function is non-zero—overlaps with the other's. Of course, even with $\psi(x)$ shifted by some $\hat{V}(b)$ to overlap with $\phi(x)$, the quantity $\langle \phi(x)|\hat{V}(b)|\psi(x)\rangle = \int \phi^*(x)\hat{V}\psi(x)dx$ could still be 0, due to phase cancellation. This is where the unitaries $\hat{U}(a)$ that join the unitaries $\hat{V}(b)$ to represent the Weyl relation come in: they alter the phase of the wavefunctions on which they act, and so can act on the shifted wave function $\hat{V}(b)|\psi(x)\rangle$ to undo its phase cancellation with $\phi(x)$. Thus, for any ψ, ϕ there will be elements $\hat{V}(b), \hat{U}(a)$ of the Schrödinger representation such that $\langle\phi|\hat{U}(a)\hat{V}(b)|\psi\rangle \neq 0$. No non-trivial subspace of $L^2(\mathbb{R})$ is left invariant by the collection of operators representing the Weyl relations, which is just to say that the representation is irreducible. As for weak continuity, consider the family $\hat{V}(b), b \in \mathbb{R}$, that must be a weakly continuous function of b. Eq. (2.27) tells us how members of this family act on an arbitrary element of the Hilbert space. Focus on $\hat{V}(b)\psi(x) = \psi(x + b)$. It should be easy to persuade yourself that for each b, $\hat{V}(b + \delta)$ converges weakly to $\hat{V}(b)$ as δ approaches 0. Similar considerations establish that the family $\hat{U}(a)$ is weakly continuous. This makes the representation weakly continuous. ♠

We can use $\hat{p}$ and $\hat{q}$ to introduce operators $\hat{a}$ and $\hat{a}^\dagger$:

$$\hat{a} = \sqrt{\frac{\omega}{2}}\hat{q} + i\sqrt{\frac{1}{2\omega}}\hat{p} \quad \hat{a}^\dagger = \sqrt{\frac{\omega}{2}}\hat{q} - i\sqrt{\frac{1}{2\omega}}\hat{p} \quad (2.30)$$

The CCRs for $\hat{p}$ and $\hat{q}$ imply CCRs for these operators:

$$[\hat{a}, \hat{a}^\dagger] = \hat{I} \quad (2.31)$$

Introducing the quantities $\hat{a}^\dagger$ and $\hat{a}$ serves the cause of applying the representation we've constructed to that most venerable of systems, the harmonic oscillator. Start by supposing that the dynamics for the *quantum* oscillator is governed by a Hamiltonian obtained by replacing the canonical classical observables p and q in the *classical* Hamiltonian (2.29) with their quantum counterparts $\hat{p}$ and $\hat{q}$. Re-expressed in terms of $\hat{a}$ and $\hat{a}^\dagger$, the quantum Hamiltonian becomes:

$$\hat{H} = \omega(\hat{a}^\dagger\hat{a} + \frac{\hat{I}}{2}) \quad (2.32)$$

So expressed, its eigenstates and eigenvalues are readily forthcoming. The lowest energy eigenstate is a state, call it $|0\rangle$, that $\hat{a}$ "annihilates"; that is, $\hat{a}|0\rangle = 0$. Ergo $\hat{H}|0\rangle = \frac{\omega}{2}|0\rangle$ and this ground state $|0\rangle$ takes energy eigenvalue $\frac{\omega}{2}$. An eigenstate of $\hat{H}$, $|0\rangle$ is invariant under the evolution $\hat{H}$ generates.

To obtain the system's next lowest energy eigenstate (a.k.a. its first excited state), let $\hat{a}^\dagger$ act once on the ground state to yield an eigenstate $|\phi_1\rangle = \hat{a}^\dagger|0\rangle$ of the Hamiltonian with eigenvalue ω. To obtain the system's n^{th} excited state, let $\hat{a}^\dagger$ act n times on the ground state and normalize. The result $(|\phi_n\rangle = \sqrt{\frac{1}{n!}}(\hat{a}^\dagger)^n|0\rangle)$ will be an eigenstate of the Hamiltonian with eigenvalue $\frac{(n+1)\omega}{2}$. The set of energy eigenstates $\{(\hat{a}^\dagger)^n\}|0\rangle$ spans the Hilbert space for the quantum oscillator.

What $\hat{a}^\dagger$ "giveth," $\hat{a}$ "taketh away": applying $\hat{a}$ to the system's n^{th} excited state yields its $(n-1)^{st}$ excited state. Thus $\hat{a}^\dagger$ and $\hat{a}$ have been tagged "raising" and "lowering" operators. The former can be understood to add quanta of excitation to the states on which they act; the latter, to remove them. The operator $\hat{N} = \hat{a}^\dagger\hat{a}$ takes non-negative

44 QUANTIZING

integers as eigenvalues, and eigenstates of the Hamiltonian as eigenvectors. With each eigenstate of the Hamiltonian is associated an $\hat{N}$ eigenvalue equal to the number of quanta of excitation that state contains. Because $\hat{N}$ can be understood to count quanta of excitation, it is called a "number operator". The quanta it counts bear properties in the sense that (for example) in an n-quanta state, each quantum contributes $\frac{\omega}{2}$ to the total energy. Thus the raising and lowering operators $\hat{a}^\dagger$, $\hat{a}$, the ground (no quantum) state $|0\rangle$, and the number operator $\hat{N}$ foster an understanding of the quantum oscillator in terms of quanta of excitation admitting principled ascriptions of energy and momentum values.[25]

Now suppose that some renegade theorists undertook to quantize the classical theory of a point particle on the real line and apply the resulting quantization to the simple harmonic oscillator. These mavericks construct a representation of the CCRs not from $\hat{p}$ and $\hat{q}$ but from:

$$\hat{p}' = \frac{1}{b}\hat{p}, \quad \hat{q}' = b\hat{q} \qquad (2.33)$$

acting on this same Hilbert space of square integrable functions. These operators satisfy the CCRs, and they can be exponentiated into families of unitaries giving a weakly continuous, irreducible representation of the Weyl Relations. From $\hat{p}'$ and $\hat{q}'$, the maverick theorists next use (2.30) as a model to construct primed raising and lowering operators $\hat{a}'^\dagger$ and $\hat{a}'$, which satisfy the CCRs (2.31); they express the quantum Hamiltonian for the oscillator in terms of these primed operators. The primed raising and lowering operators can be used to define a primed ground state ($|0'\rangle$ such that $\hat{a}'|0'\rangle = 0$) and a primed number operator ($\hat{N}' = \hat{a}'^\dagger \hat{a}'$). The primed commodities $\hat{a}'^\dagger$, $\hat{a}'$, $|0'\rangle$, and $\hat{N}'$ also circumscribe a "quanta of excitation" concept for the oscillator.

The primed raising and lowering operators are linear combinations of the original unprimed ones: e.g. the primed lowering operator $\hat{a}' = \frac{(b^2+1)}{2b}\hat{a} + \frac{(b^2-1)}{2b}\hat{a}^\dagger$. Owing to the action of the $\hat{a}^\dagger$ term, the unprimed ground state $|0\rangle$ will not be annihilated by the primed lowering operator. Unprimed states $|\phi_n\rangle$, containing (by the lights of the unprimed number operator) exactly n quanta of primed excitation, will not be eigenstates of the primed number operator $\hat{N}'$. And so on. Thus we can imagine belligerent primed and unprimed theorists arguing about, say, how many quanta of excitation are represented by a given normed element of $L^2(\mathbb{R})$.

What are we to make of such disputes? One option is to construe them as disputes about *rival theories*. The disputants do *appear* to be disagreeing about matters of physical significance, such as the presence, absence, and distribution of momentum and energy. But, because both theorists have constructed the Weyl relations for the

---

[25] As Teller (1995, 40–44) emphasizes (and Chs. 9–11 below discuss), in the QFT context, this understanding is a matter of *interpretation*, rendering the nomenclature which frames it problematic. I evoke the nomenclature here, without endorsing it, in order to convey a sense of the structure of the standard quantization of the simple harmonic oscillator.

phase space $\mathbb{R}^2$, the Stone–von Neumann theorem assures us that their representations are unitarily equivalent. The equivalence is implemented by $\hat{U}: \psi(x) \to \frac{1}{\sqrt{b}}\psi(x)$, a unitary operator on the Schrödinger representation.

Glymour reckons that "the natural necessary condition for the equivalence of two theories framed in different languages is that they be intertranslatable" (1970, 279). $\hat{U}$ provides the translation manual he's after. Armed with this manual, the disputants can see that the primed theorist attributes the system the state $|\psi\rangle$ exactly when the unprimed theorist attributes it the state $\hat{U}^{-1}|\psi\rangle$, and so on, and so on. $\hat{U}$ establishes an isomorphism between the algebraic structures canonical for each theory. The translation manual reveals not only that any state the primed theorist thinks possible for the system has a counterpart the unprimed theorist thinks possible, but also that transition probabilities between unprimed states coincide with those between their primed counterparts. Any data which elements of the primed theory accommodate, counterpart elements of the unprimed theory accommodate as well—and *mutatis mutandis* for falsifying data. The theories stand, fall, and metrize their possibility spaces together.

This suggests that disputes between primed and unprimed theorists are disputes about *different formulations of the same theory*. Reflecting that (2.33) amounts to a rescaling—a tractable deformation of our measuring stick—we might have expected this. No physical considerations privilege one formulation over another. Like centimeters and inches, they are options between which there is nothing to choose but habit and taste.

2.4 Pause

The next chapter is a continuation of this one. This chapter made a case for unitary equivalence as a presumptive criterion of physical equivalence for ordinary quantum theories specified up to kinematic pairs. It described the Hamiltonian quantization recipe, and raised an anxiety about that recipe: what if different test kitchens, using the same ingredient classical theory T and scrupulously following the Hamiltonian recipe, could obtain distastefully different, even physically inequivalent, quantum theories? To what (if anything) would "the quantization of T" refer? The chapter closed by explicating the Stone–von Neumann theorem, which secures the unitary equivalence of Hamiltonian quantizations in its scope. This assuages the anxiety only insofar as the quantum theories we care about fall under the scope of the Stone–von Neumann theorem. The next chapter catalogs quantum theories that don't.

3

Beyond the Stone–von Neumann Theorem

The Stone–von Neumann theorem has assumptions: it assumes representations of the Weyl Relations to be weakly continuous; it assumes the phase space of the classical theory to be $\mathbb{R}^{2n}$; it assumes that n is finite. Suspending any one of the assumptions cancels the theorem's guarantee of uniqueness. This chapter considers the suspension of each in turn. QM_∞ emerges from these considerations as a provocation to reconsider received views about the nature and content of quantum theories.

3.1 Suspending weak continuity

Take the simple case of a particle moving in one Euclidean dimension. The Weyl relations (2.21) over $M = \mathbb{R}^2$ encapsulate the quantum theory of such a particle. Representations in which the families of unitaries $\hat{U}(a)$ and $\hat{V}(b)$ satisfying the Weyl Relations fail to be weakly continuous—I'll call these *non-regular representations*—fall outside the scope of the Stone–von Neumann theorem. Even in this most straightforward case, the Weyl relations admit unitarily inequivalent non-regular representations, corresponding to presumptively physically inequivalent quantum theories. This section introduces a few simple non-regular representations, in order to ask whether they should be taken seriously as *physical* theories rival to the one furnished by the Schrödinger representation.

3.1.1 The position and momentum representations

A disorienting fact about ordinary QM is that it incorporates no *exact position eigenstates*, that is, eigenstates of the position operator associated with punctal eigenvalues $\lambda \in \mathbb{R}$ (for a discussion, see Teller 1979). This is plain from the standard spectral decomposition of self-adjoint Hilbert space operators with continuous spectra.

"Spectral theory": a gistification. (Note the scare quotes. For a proper introduction, see Prugovečki 1971, III.5–6, IV.2.2.) Let $\hat{Q}$ on $\mathcal{H}$ be a self-adjoint operator. We're going to develop a scheme for talking about functions of $\hat{Q}$ (like $f(\hat{Q}) = e^{ia\hat{Q}}$). The

3.1 SUSPENDING WEAK CONTINUITY

scheme exploits two facts. First, where λ is a number, we know how to talk about functions $f(\lambda)$. Second every self-adjoint operator is associated with a set of numbers: its *spectrum*. In the easiest case $\hat{Q}$'s spectrum $\{q_i\}$ is discrete. Then:

$$\hat{Q} = \sum_i q_i \hat{E}_i \tag{3.1}$$

where $\hat{E}_i$ is the projection operator for the eigenspace associated with the eigenvalue q_i. The *spectral decomposition* (3.1) of $\hat{Q}$ enables us to define functions of $\hat{Q}$ like so:

$$f(\hat{Q}) := \sum_i f(q_i) \hat{E}_i \tag{3.2}$$

Note that $f(\hat{Q})$ will be bounded if f is bounded on $\hat{Q}$'s spectrum.

The harder case is when $\hat{Q}$'s spectrum, like that of the standard position and momentum operators, is continuous. We can still think of $\hat{Q}$ as allied with a function E^Q from subsets of $\mathbb{R}$ to projection operators on $\mathcal{H}$. Let the projection operator $\hat{E}^Q(\Delta)$ be the image under this map for a subset $\Delta \subset \mathbb{R}$. According to the quantum statistical algorithm, the probability that a $\hat{Q}$ measurement performed on a system in the state $|\psi\rangle$ yields an outcome in Δ is $\langle \psi | \hat{E}^Q(\Delta) | \psi \rangle$.

E^Q is a *spectral measure*. Its features include

(a) $E^Q(\mathbb{R}) = \hat{I}$ [normalization]; and
(b) $E^Q(\cup_{i=1}^\infty S_i) = \sum_{i=1}^\infty E^Q(S_i)$ if $S_i \cap S_j = \emptyset$ when $i \neq j$ [countable additivity]

In feature (b), $\sum_{i=1}^\infty E^Q(S_i)$ should be understood as the $n \to \infty$ limit, in the strong operator topology, of $\sum_{i=1}^n E^Q(S_i)$ (see the measurable function Scholium on p. 23).

To such a spectral measure there corresponds a unique *spectral function*, which I won't try to characterize directly. But by its works shall ye know it. The spectral function E^Q_λ is a map from $\mathbb{R}$ to projection operators on $\mathcal{H}$ which is related to the spectral measure E^Q like so: when Δ is the semi-open interval $(a, b]$, with $a < b$, $E^Q(\Delta) = E^Q_b - E^Q_a$. The foregoing are informal introductions to, not definitions of, the notions of spectral measure and spectral function. They're meant to position you to get the gist of the Spectral Theorem, which alerts us that even operators with continuous spectra can be expressed on the model of (3.1):

Fact 3.1 (Spectral Theorem). To each self-adjoint operator $\hat{A}$ on $\mathcal{H}$, there corresponds a unique spectral function E^A_λ such that the domain D_A of $\hat{A}$ consists of all vectors ψ such that:

$$\int_\mathbb{R} \lambda^2 d\langle \psi | E^A_\lambda | \psi \rangle < +\infty$$

and for any $\psi \in D_A$ and $\phi \in \mathcal{H}$:

$$\langle \phi | A \psi \rangle = \int_\mathbb{R} \lambda d\langle \phi | E^A_\lambda | \psi \rangle$$

(see Prugovečki 1971, thm. 6.3 (250)).

(Don't let expressions like "$d\langle\phi|E_\lambda^A|\psi\rangle$" disturb you. For each ϕ and ψ, $\langle\phi|E_\lambda^A|\psi\rangle$ will be some scalar-valued function of λ, call it $\mu(\lambda)$. And there's nothing mysterious about $d\mu(\lambda)$: it's just $\frac{d\mu(\lambda)}{d\lambda}d\lambda$.) Note that E_λ^A has been rigged so that for any state ψ:

$$\int_\Delta d\langle\psi|E_\lambda^A|\psi\rangle = \langle\psi|E_\Delta^A|\psi\rangle$$

The payoff, of course, is that we can use E_λ^Q to say what $f(\hat{Q})$ is.

Fact 3.2 (defining $f(\hat{Q})$). Where $f(\lambda)$ is a bounded measurable function on $\mathbb{R}$ and E_λ^Q is the spectral measure for $\hat{Q}$, there is a unique, bounded linear operator $\hat{A}$ such that for all $\psi, \phi \in \mathcal{H}$:

$$\langle\phi|\hat{A}\psi\rangle = \int_\mathbb{R} f(\lambda)d\langle\phi|E_\lambda^Q\psi\rangle$$

The operator $\hat{A}$ is none other than the function $f(\hat{Q})$ we've been after (see Prugovečki 1971, thm. 2.1 (270)). ♠

The spectral measure for the operator $\hat{A}$ acting on $\mathcal{H}$ is a map from measurable subsets Δ of $\mathbb{R}$ to projection operators on $\mathcal{H}$. Let $\hat{P}_\Delta^A$ be the image of this map for the set Δ. $\hat{P}_\Delta^A$ serves as a device for assigning probabilities: in a density operator state $\hat{\rho}$, the probability that a measurement of $\hat{A}$ yields an outcome in the set Δ is given by $Tr(\hat{\rho}\hat{P}_\Delta^A)$. The standard spectral measure for the position operator maps degenerate intervals $\Delta = [a, a]$ to the zero operator 0. Thus in any density operator state $\hat{\rho}$, the probability that position assumes a value in $[a, a]$—a point value—is $Tr(\hat{\rho}0) = 0$. In other words, no density operator state is an exact position eigenstate. In ordinary QM, every state is a density operator state. Ergo in ordinary QM, no state is an exact position eigenstate. The continuous momentum observable likewise lacks exact eigenstates.

Ordinary QM is conducted in a separable Hilbert space (i.e. one with a countable basis). But a non-separable Hilbert space is exactly what we need to sustain a full set of exact position eigenstates. For should an eigenstate of position exist for every possible position $\lambda \in \mathbb{R}$, there would be as many pairwise orthogonal position eigenvectors in the Hilbert space as there are real numbers (and *mutatis mutandis* for momentum). Lacking a countable basis, the Hilbert space would fail to be separable.

Non-regular representations are typically non-separable (Cavallaro, Morchio, and Strocchi 1999). Those who hanker after exact position or momentum eigenstates might therefore hope to find them in non-regular representations. And find them they do. The non-regular *position representation*,[1] for example, is set in the non-separable vector space $\ell_2(\mathbb{R})$ of square summable functions $f : \mathbb{R} \to \mathbb{C}$. These are functions supported on a countable subset S_f of $\mathbb{R}$ and such that $\sum_{x \in S_f} |f(x)|^2 < \infty$. The functions

---

[1] Halvorson (2004), which I follow here, gives a lucid exposition of position and momentum representations. Clifton and Halvorson (2001b; 2002) deploy them in the service of a rigorous formulation of Bohrian complementarity.

3.1 SUSPENDING WEAK CONTINUITY

$$\varphi_\lambda(x) = 1 \text{ if } \lambda = x \tag{3.3}$$

$$\varphi_\lambda(x) = 0 \text{ if } \lambda \neq x$$

for $\lambda \in \mathbb{R}$ furnish an (uncountable!) orthonormal basis for $\ell_2(\mathbb{R})$. Introduce unitary operators $\hat{U}(a)$ and $\hat{V}(b)$ that act as follows on elements of this basis:

$$\hat{U}(a)\varphi_\lambda(x) = e^{ia\lambda}\varphi_\lambda(x) \tag{3.4}$$

$$\hat{V}(b)\varphi_\lambda(x) = \varphi_{\lambda-b}(x) \tag{3.5}$$

$\hat{U}(a)$ and $\hat{V}(b)$ not only satisfy the Weyl Relations, but do so irreducibly: any basis element φ_λ of $\ell_2(\mathbb{R})$ can be obtained from any other by acting on it with the appropriate element of $\{\hat{V}(b)\}$. So no non-trivial subspace of $\ell_2(\mathbb{R})$ is invariant under the action of $\{\hat{V}(b), \hat{U}(a)\}$.

The family of unitaries $\hat{U}(a)$ is weakly continuous.[2] By Stone's theorem, there is some self-adjoint $\hat{Q}$ on $\ell_2(\mathbb{R})$ such that $\hat{U}(a) = e^{ia\hat{Q}}$. From (3.4) and a Taylor series expansion of $\hat{U}(a)$ in terms of $\hat{Q}$, it follows that:

$$\hat{Q}\varphi_\lambda(x) = \lambda\varphi_\lambda(x) \tag{3.6}$$

In other words, $\varphi_\lambda(x)$ is an eigenvector of $\hat{Q}$ associated with eigenvalue λ. Interpreting $\hat{Q}$ as the position operator, $\varphi_\lambda(x)$ is an exact position eigenstate.

If *both* families $\hat{U}(a)$ and $\hat{V}(b)$ giving an irreducible representation of the Weyl relations are weakly continuous, then that representation is set in a separable Hilbert space (Summers 1999). As the position representation is non-separable and the family $\hat{U}(a)$ is weakly continuous, the family $\hat{V}(b)$ cannot be.[3] With the family $\hat{V}(b)$ disappointing a presupposition of Stone's theorem (Fact 2.3), no self-adjoint operator $\hat{P}$ on $\ell_2(\mathbb{R})$ can be recovered as the infinitesimal generator of that family. Insofar as the momentum observable gets its identity as a physical magnitude from the functional relations the commutation relations establish between it and other physical magnitudes, part of what it is to be the momentum observable is to be the infinitesimal generator of position translations. Thus in the position representation, no operator plays the traditional functional role of a momentum observable.

In the Schrödinger representation, members of the family $\hat{U}(a)$ translate the momentum of the states on which they act by an amount a. Note that in the position representation, $\hat{U}(a)\varphi_\lambda(x)$ lies in the same ray—and so assigns the same quantum probabilities—as $\varphi_\lambda(x)$. In at least a Pickwickian sense, the states φ_λ are invariant under

---

[2] To appreciate the continuity of the map $a \mapsto \hat{U}(a)$, consider the behavior of $\lim_{\delta \to 0} \langle \varphi_\lambda | \hat{U}(a+\delta) | \varphi_\lambda \rangle$ for arbitrary a. It's just $\lim_{\delta \to 0} e^{i\delta\lambda} \langle \varphi_\lambda | \hat{U}(a) | \varphi_\lambda \rangle$, which smoothly approaches $1 \times \langle \varphi_\lambda | \hat{U}(a) | \varphi_\lambda \rangle$ as δ decreases.

[3] To see that it isn't, notice that $\langle \varphi_\lambda | \hat{V}(b) | \varphi_\lambda \rangle = 0$ unless $b = 0$, in which case $\langle \varphi_\lambda | \hat{V}(b) | \varphi_\lambda \rangle = 1$. This exposes the map $b \mapsto \hat{V}(b)$ as discontinuous at that point.

translations in momentum. Of course, the scrupulous might complain that where there is no momentum, there can be no genuine momentum translations.

There is a completely analogous non-regular representation, the *momentum representation*, with exact eigenstates of momentum. The uncountable orthonormal basis of this space is furnished by plane waves e^{ikx}, eigenstates of the momentum operator for each real number k. Owing to the family $\hat{U}(a)$'s failure to be weakly continuous, the momentum representation lacks a position operator—a fact for which Emch offers the quasi-physical explanation that it makes no sense to speak of the position of a plane wave (1972, 231–2). $\hat{V}(b)$, which serves the Schrödinger representation as a translation operator for position, acts trivially on the exact momentum eigenstates, considered as quantum states. These states are, in at least a Pickwickian sense, translation-invariant (Beaume et al. 1974 introduces them in this light).

With the assumption of weak continuity suspended, the Stone–von Neumann theorem ceases to apply. The position and momentum representations have uncountable bases; the Schrödinger representation has a countable one. Each of the former is unitarily inequivalent to the latter. Less obviously, each of the former is unitarily inequivalent to the other (Halvorson 2004, 51).

3.1.2 Are they "physical"?

The Stone–von Neumann theorem reassured us that if we found any way to quantize a classical theory in its scope, we found the only way. Although mathematically consistent, the position and momentum representations should cancel this reassurance only if those representations constitute genuine physical theories. As a matter of fact, in (what as far as I know was) the debut of the momentum representation, Segal branded it pathological. In order to exhibit "the connection between undesirable states and 'discontinuous Weyl systems'" (1967, 120), he constructed a non-regular representation of the Weyl relations. Displaying one of its basis elements, he declared it "mathematically well defined but physically ... rather suspect, because it is a state with an exact value ... for the momentum."

Why should Segal be suspicious? Here's one reconstruction. Systems whose wave functions are narrowly and strictly confined in configuration space are systems which should, according to the Heisenberg Uncertainty Principle, exhibit spreads in momenta more dramatic than we're accustomed to witnessing. Non-regular representations incorporating (e.g.) exact position eigenstates are unphysical in the sense of empirically pointless. There aren't, or so our evidence indicates, natural phenomena to which they might apply.

A proponent of the physical significance of non-regular representations needn't be moved. For one thing, what the evidence shows is that exact position eigenstates don't occur *very often* in nature—and surely it's the worst sort of chauvinism to brand something unphysical just because it's rare. For another thing, on the standard picture, the content of a physical theory consists of the worlds *possible* according to that theory—at most one of which could be actual. So even if they're never actualized, non-regular representations could furnish part of a physical theory's content. Some merely possible worlds foster understanding of the theory, the proponent might continue, even if they're not the actual world, provided that they perspicuously reveal key theoretical structures, particularly ones cloaked by obfuscating actuality. Thus a proponent of non-regular representations could admit that they're unnatural, but insist that they're unnatural in this content-illuminating sense.

I am not going to take sides in the dispute between the fan and the foe of non-regular representations. Not only aren't there straightforward rules for the application of terms of empirical endearment such as "physical" and "natural," there shouldn't be. Considerations of many sorts can count for or against reckoning a bit of candidate theory "physical," and more insight into the theory is to be gained by exploring this space of considerations than can be gained by setting out to settle the question once and for all. To further exploration let us now turn.

An ideology of quantization The aim of understanding quantum theories, lately invoked by the fan of non-regular representations in their defense, can also be invoked by their foe to their detriment. Non-regular representations, this foe might claim, are an affront to an ideology of quantization that has guided both how physicists build and how they think about quantum theories. Here's a typical statement of the creed:

By the "quantization problem" we shall mean the problem of setting up a correspondence between classical observables and quantum observables, i.e. between functions on $\mathbb{R}^{2n}$ and self-adjoint operators on $L^2(\mathbb{R}^n)$, such that the properties of the classical observables are reflected as much as possible in their quantum counterparts in a way consistent with the probabilistic interpretation of quantum mechanics. (Folland 1989, 15–16)[4]

The ideology is: *the properties of classical observables should be reflected as much as possible*[5] *in their quantum counterparts*. It provides at once a powerful heuristic for the construction of quantum theories and a powerful inducement to philosophy.

To see the heuristic in action, let us revisit the Poisson bracket goes to commutator rule. A further vindication of the rule is that it serves the aim of reflecting the structure of the classical theory in its quantization. The multiplication of classical observables (functions on $\mathbb{R}^{2n}$) is commutative. The multiplication of quantum observables

---

[4] Notice that we can make progress on the quantization problem for a *theory*, and so find ourselves with a collection of Hilbert space states and Hilbert space observables, without thereby learning how to model the dynamics for each *system* in a theory's scope. Questions such as how to construct a Hamiltonian for a system, which may be bedevilled by operator ordering problems, can remain. Thanks to Paul Teller for emphasizing this.

[5] About the hedge: a complete quantization of a classical theory (M, Ω) would be a map, from the *entire* set of classical observables (smooth functions on M) to symmetric Hilbert space operators, under which the commutator brackets of the latter mirror the Poisson brackets of the former. It has been known since the 1940s that any such map is in one way or another deficient as a quantization (see Gotay 2000 for a recent review). The usual defect, the one at issue in (for instance) the notorious "van Hove obstruction," is reducibility.

(self-adjoint operators on $L^2(\mathbb{R}^n)$) is not. Enforcing the rule endows both sets of observables with a common multiplicative structure, that provided by the Poisson [classical]/commutator [quantum] bracket.

To appreciate the inducement to philosophy latent in the ideology, note that the heuristic is guided by an account of what the properties of the classical observables are. To generate such an account is to investigate the foundations of classical physics.

In cases where the classical theory to be quantized has a set of canonical observables (i.e. observables satisfying Eq. 2.12), the aim of reflecting the properties of the classical observables in their quantum counterparts is best promoted by the usual procedure of incorporating canonical observables in the theory's quantization, then building other quantum observables from these by analogy to the classical case. Consider for example the Hamiltonian for the quantum oscillator, whose functional dependence on position and momentum operators mirrors the functional dependence of the classical Hamiltonian on those magnitudes (see §2.3.5). Failing to identify correlates to the full set of fundamental classical observables, the position and momentum representations frustrate this procedure. Their foe might complain that they reflect too little of the structure of the classical theory to qualify as physically significant.

Underlying this complaint is the conviction that the physical core of the quantum theory lies in the Heisenberg form of the CCRs, so that representations of the Weyl relations ought be taken seriously only when they correspond to representations of the Heisenberg CCRs. The conviction would be tenuous if it were impossible to give rigorous and useful expression to the Heisenberg CCRs: if the mathematical niceties motivating their exponentiation into the Weyl relations could be pursued only by that expedient. But, for the particle moving in one Euclidean dimension, and for other topologically uncomplicated classical theories, there are other ways to address the failure of position and momentum operators to have a dense domain in common. One is to set the quantization in Schwartz space, the dense linear manifold of all continuously differentiable functions $f : \mathbb{R} \to \mathbb{C}$ that, along with all their derivatives, go to 0 faster as $|x| \to \infty$ than does any inverse power of x. A subspace of $L^2(\mathbb{R})$, Schwartz space lies in the common domain of $\hat{p}$ and $\hat{q}$. The Heisenberg form of the CCRs *is* well-defined as an operator equation on Schwartz space, and rigor restored without moving to the Weyl relations (see Emch 1972, 226 ff., for a sketch; Jørgensen and Moore 1984 or Schmüdgeon 1990 for full-blown theories of the Heisenberg CCRs).

Another cause for concern derives from our expectations about what a *quantum* theory should be like. As encapsulated by (OQD), those expectations include the expectation that quantum dynamics are implemented by a *weakly continuous* one parameter family of unitary operators—the one generated by the system Hamiltonian, in conventional treatments. David Malament and John Manchak have shown that there is a precise sense in which the position representation disappoints that expectation (Manchak 2008). If what it is to be physical is to obey the laws of physics, and if the Schrödinger equation is a law of physics, this reveals a sense in which non-regular representations aren't physical: they break the connection Schrödinger dynamics would establish between Hamiltonian operators and continuous families of evolution operators. (§ 8.3.2 returns to this point.)

Applications of non-regular representations Of course, this is all very dainty. A muscular proponent of non-regular representations could remark, on their behalf, that they have, after all, found physical application (see Acerbi, Morchio, and Strocchi 1993a; 1993b). Non-regular representations play a role in the quantization of systems subject to periodic potentials or to external fields that make spatial translations non-commutative. They are instrumental to one method of "fermion bosonization," the project of exhibiting fermion fields as an appropriate limit of bosonic structures. They've been suggested as framework for QED without an indefinite metric (Thirring and Narnhofer 1992), and share formal properties with "polymer representations" of Loop Quantum Gravity (see Ashtekar, Lewandowski, and Sahlmann 2003). And over the last decade, non-regular representations have played an increasing role in quantum cosmology. (For a sampling, see Ashtekar 2007; Ashtekar, Fairhurst, and Willis 2003; Ashtekar, Pawlowski, and Singh 2006; Husain and Winkler 2004; Modesto 2006.)

Non-regular representations even suggest themselves as appropriate models for a simple mechanical system moving in one Euclidean dimension. In the "free particle" limit as $\omega \to 0$, the equilibrium state of the one-dimensional quantum oscillator at zero temperature resides in a non-regular representation of the Weyl relations. Intuitively, this is because the equilibrium state of a free particle should define a uniform probability distribution over possible positions in $\mathbb{R}^n$. That means it should be translationally invariant, and so an inhabitant of (something like) the non-regular momentum representation.

Given these potentials for physical application, and the principle that metaphysical or ideological scruples shouldn't block fruitful avenues of theory development, Segal's suspicion of non-regular representations should not be codified into a ban against them. But neither should we indiscriminately accept non-regular representations without physical application as full-fledged bearers of empirical content, when features constitutive of a classical theory's quantization threaten to leak out through their discontinuities.

3.2 Non-vanilla configuration spaces

This section is both impressionistic and digressive. For those who wish to skip ahead, herewith:

The Short Version. Not all systems are aptly attributed phase spaces of the form $\mathbb{R}^{2n}$. Systems with phase spaces of different topologies fall outside the scope of the Stone–von Neumann theorem. Treatments of the theorem that link representations of the Weyl relations to the symmetries of $\mathbb{R}^{2n}$ indicate why this should be so. Possessing different symmetries than does $\mathbb{R}^{2n}$, phase spaces of different topologies fall outside the jurisdiction of uniqueness theorems about representations of $\mathbb{R}^{2n}$'s symmetries. An example of a system whose phase space is not $\mathbb{R}^{2n}$ is a bead confined to move on a circle. Its phase space is the cylinder $S_1 \times \mathbb{R}$ (with x, an element of the unit circle S_1, giving the configuration of the system and $l \in \mathbb{R}$ giving its angular momentum). Hilbert space representations of relations that stand to the symmetries of the cylinder as the Weyl relations stand to the symmetries of $\mathbb{R}^{2n}$ are not unique up to unitary equivalence. ♠

The quantization of a classical theory whose phase space M isn't $\mathbb{R}^{2n}$ escapes the clutches of the Stone–von Neumann theorem. One way to see how the assumption that $M = \mathbb{R}^{2n}$ figures in the Stone–von Neumann theorem, is to re-express that theorem in terms of *the Heisenberg group*. The Heisenberg group is a $2n + 1$ dimensional Lie group[6g] that, in a sense, encodes the global symmetries of the classical phase space $\mathbb{R}^{2n}$. We will see that the "extra" dimension reflects a freedom to specify *what* constant appears as a coefficient of $\delta_{ij}\hat{I}$ in the CCRs. Expressed as a theorem about the Heisenberg group, the Stone–von Neumann theorem asserts that, once a constant whose value is set to Planck's constant for empirical reasons is specified, all irreducible representations of the Heisenberg group are unitarily equivalent to the Schrödinger representation.

Expositions of the Stone–von Neumann theorem often embark from the Heisenberg form of the CCRs, which lurks in this version of the theorem as the Lie algebra of which the Heisenberg group is the universal covering group. Understood as so lurking, these CCRs are not principles of quantization posited *ab initio*, and in the same way for every quantization, but a reflection of the structure of a particular (and significant) class of classical phase space. We shouldn't expect the Stone–von Neumann theorem to apply to the quantization of classical theories whose phase spaces lack that structure. The first part of this section sketches the Stone–von Neumann theorem as a theorem about the Heisenberg group; the second part gives some examples of classical theories whose phase spaces place them outside the ambit of the theorem, so understood.

3.2.1 Mr. Heisenberg's group

The Heisenberg form of the CCRs made their appearance in something like the following manner. The variables q_i, p_j were introduced as coordinates of the classical phase space $\mathbb{R}^{2n}$. The standard symplectic product Ω as defined by (2.15) on $\mathbb{R}^{2n}$ endowed q_i, p_j (considered now as classical observables) with the canonical Poisson bracket structure (2.12). Following the Poisson bracket goes to commutator rule yielded the CCRs. Let us now regard them in another light.

---

[6] A Lie group G is a group that is also a manifold, and whose product $(G \times G \to G)$ and inverse $(G \to G)$ operations are differentiable. The local structure of a Lie group G is determined by a Lie algebra $\mathfrak{g}$ (see note 18 of Ch. 2): *very* roughly speaking, you get the group by exponentiating the algebra. For a good introduction to Lie groups and Lie algebras in mechanics, see Bryant (1995).

3.2 NON-VANILLA CONFIGURATION SPACES

Letting $n = 1$ for the sake of simplicity, consider the symplectic *manifold* (see §2.3.2) $(\mathbb{R}^2, \Omega)$, with Ω the standard symplectic form $dp \wedge dq$. $\mathbb{R}^2$ has basis elements q, p. Translations in position q, as well as translations in momentum p, are *symplectic symmetries*; that is, they are transformations under which the symplectic form Ω is invariant.[7] Each set of translations corresponds to the abelian (i.e. commutative) additive group $\mathbb{R}$. Together, the groups of translations have the power, given any element (q, p) of the classical phase space $\mathbb{R}^2$, to extract any other element $(q + a, p + b)$, $a, b \in \mathbb{R}$, from it. That is, together the group of translations acts *transitively* on the phase space.

Now for the general case. Consider a phase space $\mathbb{R}^{2n} = \mathbb{R}^n \times \mathbb{R}^n$, with coordinates $q_1, ..., q_n; p_1, ..., p_n$, equipped with the standard symplectic form Ω. Acting on the first n coordinates of the phase space, spatial translations form an abelian additive group $\mathbb{R}^n$ of symplectic symmetries. Acting on the second n coordinates, momentum translations do likewise. Together, the groups act transitively on the phase space. Now let us suppose that we're after a *unitary representation* of each of these additive groups on some Hilbert space $\mathcal{H}$. Such a representation maps each $\mathbf{a} = \{a_1, a_2 ... a_n\} \in \mathbb{R}^n$ to a unitary $\hat{U}(\mathbf{a})$ acting on $\mathcal{H}$, and $\mathbf{b} = \{b_1, b_2 ... b_n\} \in \mathbb{R}^n$ to a unitary $\hat{V}(\mathbf{b})$ acting on $\mathcal{H}$. Looking down the road to a quantization of $(\mathbb{R}^{2n}, \Omega)$, if we expect that quantization to be set in a Hilbert space and to inherit the symplectic symmetries of its classical antecedent, then we have reason to look for this unitary representation. In quantum theories, Wigner's theorem is supposed to tell us, symmetries are implemented unitarily. Ergo in the quantization we pursue, the inherited symplectic symmetries should be implemented unitarily.

In the Schrödinger representation of the CCRs for $M = \mathbb{R}^{2n}$, one can realize $U(\mathbf{a})$ and $V(\mathbf{b})$ as operators that act on $L^2(\mathbb{R}^n)$. $U(\mathbf{a})$ varies continuously with each a_i; the corresponding family of unitaries has infinitesmal generator $\{\hat{q}_i\}$; and analogously for $V(\mathbf{b})$ and $\{\hat{p}_i\}$. Weyl remarked that one can use these unitaries $U(\mathbf{a})$ and $V(\mathbf{b})$ to define a map W from $\mathbb{R}^n \times \mathbb{R}^n$ to unitaries on $\mathcal{B}(\mathcal{H})$:

$$(\mathbf{a}, \mathbf{b}) \mapsto \hat{W}(\mathbf{a}, \mathbf{b}) := \exp(\frac{i\mathbf{a} \cdot \mathbf{b}}{2})\hat{U}(\mathbf{a})\hat{V}(\mathbf{b}) \tag{3.7}$$

The map $(\mathbf{a}, \mathbf{b}) \mapsto \hat{W}(\mathbf{a}, \mathbf{b})$ affords a representation of the abelian additive group $\mathbb{R}^{2n}$—the group of position and momentum translations that acts transitively on $\mathbb{R}^{2n}$—as follows. Use $\hat{W}(\gamma)$ to abbreviate $\hat{W}(\mathbf{a}, \mathbf{b})$ as defined above. Then the unitaries $\hat{W}(\gamma)$ obey the group law:

$$\hat{W}(\gamma)\hat{W}(\gamma') = \exp(\frac{i}{2}[\mathbf{a}' \cdot \mathbf{b} - \mathbf{a} \cdot \mathbf{b}'])\hat{W}(\gamma + \gamma') \tag{3.8}$$

(The expressions (3.7) and (3.8) should look familiar from §2.3.4's introduction of the Weyl relations over $(\mathbb{R}^{2n}, \Omega)$. We're now exhibiting the connection between those

---

[7] Such transformations deserve to be called symmetries, the thought is, because the symplectic form determines which trajectories through phase space satisfy Hamilton's equations. Transformations leaving Ω invariant are transformations under which this solution space is unchanged.

relations and $(\mathbb{R}^{2n}, \Omega)$'s symplectic symmetries.) The map W is a *projective unitary representation* of the abelian additive group $\mathbb{R}^{2n}$. Although W maps elements of $\mathbb{R}^{2n}$ to unitary operators, it's not a straight-up unitary representation, because its group law (3.8) departs from the simple form:

$$U(a)U(b) = U(a+b) \qquad (3.9)$$

by the incorporation of a complex number of square modulus that varies with the group elements being composed.

So let's next consider the factor $\exp(\frac{i}{2}\left[\mathbf{a}' \cdot \mathbf{b} - \mathbf{a} \cdot \mathbf{b}'\right])$ that keeps W from being a straight-up unitary representation. By means of this complex phase factor, W determines a bilinear, non-degenerate map from pairs $((\mathbf{a},\mathbf{b}), (\mathbf{a}',\mathbf{b}'))$ of elements of $\mathbb{R}^{2n}$ to the quantity $\left[\mathbf{a}' \cdot \mathbf{b} - \mathbf{a} \cdot \mathbf{b}'\right]$ in the exponent of this phase factor. Hence the group law defines a map $\Omega: M \times M \to \mathbb{R}$ given by

$$\Omega((\mathbf{a},\mathbf{b}),(\mathbf{a}',\mathbf{b}')) = (\mathbf{a}' \cdot \mathbf{b} - \mathbf{a} \cdot \mathbf{b}') \qquad (3.10)$$

Ω is not only bilinear and non-degenerate, but also skew-symmetric. Replacing the variables $\mathbf{a}$ and $\mathbf{b}$ with the more suggestive $\mathbf{q}$ and $\mathbf{p}$, we recognize that Ω is none other than the standard symplectic product (2.15) on the phase space $\mathbb{R}^{2n}$!

This is no accident. Weyl showed that an affine space, such as $\mathbb{R}^{2n}$, has a faithful irreducible projective unitary representation only when it is symplectic, in which case the phase factors of the projective representation are determined, as in (3.8), by the symplectic product. So if $\mathbb{R}^{2n}$ has a projective unitary representation, it is of (3.8)'s form.

(An idea Mackey (1998) attributes to Weyl is that the *kinematics of a quantization of a classical theory with phase space $\mathbb{R}^{2n}$ is expressed by a projective unitary representation of the abelian additive group $\mathbb{R}^{2n}$ with phase factors given by the symplectic structure on $\mathbb{R}^{2n}$.* So Mackey believes Weyl believes (3.8) constitutes quantum kinematics!)

Any such projective unitary representation is at the same time a representation of the Heisenberg group, which I won't attempt to describe rigorously (see Folland 1989, 17–20). Returning to the simplest case of $M = \mathbb{R}^2$, the rough idea is that just as the family $V(a) = \exp(-ia\hat{p}), a \in \mathbb{R}$, implements the group of position translations by exponentiating the momentum operator $\hat{p}$, and the family $U(a) = \exp(-ia\hat{q}), a \in \mathbb{R}$, implements the group of momentum translations by exponentiating the position operator $\hat{q}$, "exponentiating" the CCRs for $\mathbb{R}^{2n}$ yields unitaries implementing all position and momentum translations, and governed by the group law (3.8). That is, "exponentiating" the CCRs for $\mathbb{R}^{2n}$ yields the Heisenberg group H_n.

The *interesting* algebraic structure of the CCRs is the bracket relation they impose on canonical observables. But to fully specify the CCRs, we need to supply Planck's constant as the coefficient in the r.h.s. of expressions like $[\hat{p}_i, \hat{q}_j] = -i\delta_{ij}\hat{I}$. The Stone–von Neumann theorem states that, once this constant is specified, every irreducible representation of the Heisenberg group H_n is unitarily equivalent to every other. The

Schrödinger representation is an irreducible unitary representation of the Heisenberg group H_n for Planck's constant. The Stone–von Neumann theorem implies that every other irreducible representation of H_n for Planck's constant is unitarily equivalent to the Schrödinger representation.

This incidentally suggests another complaint the foe of non-regular representations can level against them: in non-regular representations, the structure of the Heisenberg group—in particular, the structure of the CCR algebra that generates it—goes missing. Although the non-regular representations discussed above concern systems with classical phase space $\mathbb{R}^{2n}$, they reflect the symmetries and topology of that phase space less thoroughly than regular representations do. This circumstance also suggests that the Stone–von Neumann theorem won't guarantee uniqueness when theories whose phase spaces *aren't* $\mathbb{R}^{2n}$ (or are $\mathbb{R}^{2n}$, but equipped with a different group structure) are quantized.

3.2.2 *Phase spaces of other topologies*

Consider, for instance, the extraordinarily simple system consisting of a single particle constrained to move on the unit circle S_1. The canonical variables of a classical Hamiltonian treatment of this system are its position, given by an angular variable $\phi \in [0, 2\pi]$, and its angular momentum $\ell \in \mathbb{R}$. Its configuration space is the circle S_1 and its phase space is the cylinder $S_1 \times \mathbb{R}$.

Before quantizing,[8] a change of variables from the standard cylindrical coordinates (ϕ, ℓ) is in order. We're trying to keep track of how the symmetries and topological structure of the classical phase space—as encoded in the Lie algebra of canonical observables—is reflected in the ensuing quantization. The variable ϕ is ill-suited to this project, because it does not correspond to an observable, that is, a continuous function on phase space. The discontinuity occurs as ϕ approaches 2π, and occurs because the configuration space of the system is a circle. We'll work instead with the variables:

$$x = \cos\phi, \quad y = \sin\phi, \quad z = \ell \tag{3.11}$$

which are continuous on the cylinder.[9]

Among these variables, the standard symplectic form for the cylinder defines the Poisson brackets:

$$\{x, y\} = 0, \quad \{z, x\} = y, \quad \{z, y\} = -x \tag{3.12}$$

---

[8] A project in which my impressionistic exposition tries to follow the careful treatments of Isham (1983) and Gotay (2000). Thanks are owed to Gordon Belot for help with this. Blame for persisting misunderstandings is not.

[9] Another approach is to let the configuration variable be unitary instead of self-adjoint. See Lévy-Leblond (1976).

Subject to the bracket (3.12), x, y, z are the generators of a Lie algebra that plays the role for the phase space $S_1 \times \mathbb{R}$ that the CCR algebra played for the phase space $\mathbb{R}^{2n}$: the role of generating a group of symplectic symmetry transformations that act transitively on the phase space. So let us suppose that to build a quantum theory of the particle on the circle, we must find a Hilbert space representation of the Circular Canonical Commutation Relations (CCCRs):

$$[x, y] = 0, \quad [z, x] = iy, \quad [z, y] = -ix \qquad (3.13)$$

Although these follow from applying the Poisson bracket goes to commutator rule to (3.12), we've approached them via the symmetries of the classical phase space instead.

The standard representation of the CCCRs (3.13) is given by:

$$\hat{x} = \cos\phi, \quad \hat{y} = \sin\phi, \quad \hat{z} = i\frac{d}{d\phi} \qquad (3.14)$$

acting on $L^2(S_1)$, the space of functions $\psi(\phi) : S_1 \to \mathbb{C}$ that are square integrable with respect to the measure $\frac{d\phi}{2\pi}$. Notice that the spectrum of $\hat{x}$ and $\hat{y}$ is $[-1, 1]$—as befits the configuration space S_1—and the spectrum of $\hat{z}$ is $2\pi n$, $n \in \{0, 1, 2, ...\}$—which is the angular momentum spectrum for a particle on a circle suggested by boundary conditions and the de Broglie relations.

Now, the CCCRs are not the CCRs, and the group of symmetries they generate is not the Heisenberg group. About the uniqueness, or lack thereof, of irreducible representations for the bead on the circle, the Stone–von Neumann theorem is silent. It cannot vouch that the standard quantization realizes the CCCRs uniquely. And indeed, a family of quantizations labelled by $\theta \in [0, 1]$ also satisfy the CCCRs (see Isham 1983, 1270–72). These quantizations are related to standard one (3.14) as follows:

$$\hat{x}_\theta = \hat{x}, \quad \hat{y}_\theta = \hat{y}, \quad \hat{z}_\theta = i\frac{d}{d\phi} + \theta \qquad (3.15)$$

The $\theta = 0$ member of this family is just the standard quantization Q_0. For $\theta \neq 0$, the spectra of the angular momentum operators z and z_θ bear witness to Q_θ's unitary inequivalence to Q_0: as already noted, the former has spectrum $\{2\pi n\}$ for integer n; the spectrum of the latter is $\{2\pi(n - \theta)\}$. Worlds possible according to the first quantum theory include worlds where the system has no angular momentum; no such world is possible according to the second quantum theory.[10]

The family Q_θ of unitarily inequivalent quantizations of the particle confined to a circle are not idle mathematical curiosities. They are related to the θ angles of Yang–Mills theory, and, mildly adapted, they find physical application in the Bohm–Aharonov effect, which (extremely briefly stated) is that an electron translated around an (infinitely extended) solenoid experiences a phase rotation determined by

---

[10] Dürr et al. argue that Bohmian mechanics "provides a sharp mathematical justification" (2006, 791) of the capacity of classical theories with non-vanilla configuration spaces to admit competing quantizations.

the flux through the solenoid.[11] Banishing the electron from the region of space occupied by the solenoid, we attribute it a configuration space that is $\mathbb{R}^3$ with a cylinder removed. Like the circle, this topologically non-$\mathbb{R}^n$ configuration space frames a Hamiltonian theory that admits a family of inequivalent quantizations. Different members of this family correspond to different fluxes through the solenoid and hence to different phase shifts for the transported electron (see Landsman 1990 for details).

Thinking big, observe that mechanics set in a spatially compact universe (such as, presumably, our own) will have a topologically non-$\mathbb{R}^{2n}$ phase space, placing its quantization outside the scope of the Stone–von Neumann theorem.

3.3 Infinitely many degrees of freedom

The Stone–von Neumann theorem concerns the quantization of a classical theory with phase space $\mathbb{R}^{2n}$, where n is finite. However, many significant quantum theories arise by quantizing classical theories whose phase spaces are infinite-dimensional. I've grouped quantum theories so obtained under the heading "QM_∞." Examples include classical fields, which associate a field magnitude (scalar, spinor, vector, or what have you) with every point of space(time), and the idealized systems considered at the thermodynamic limit of statistical mechanics, obtained by letting the number of systems considered and the volume they occupy become infinite, while keeping their density finite. Stupendously successful, quantum field theories are keepers of the Standard Model's exotic particle zoo, and uniters of fundamental forces. The thermodynamic limit of statistical mechanics makes available a rigorous treatment of phenomena as mundane as phase transitions. Quantized, it accounts for phenomena as otherworldly as superconductivity.

Theories of QM_∞ are conceptually, empirically, and practically fundamental. But the Stone–von Neumann theorem cannot assure us that any one of them has a unique Hilbert space realization. Indeed, unitarily—and so presumptively physically—inequivalent quantizations abound in QM_∞: an alarming circumstance that will command our attention for the rest of this book. This chapter closes with some simple illustrations.

3.3.1 The thermodynamic limit: the infinite spin chain

Anyone even minimally acquainted with the simple quantum mechanical systems much discussed by philosophers has on hand the resources to describe a quantum system admitting unitarily inequivalent representations: a chain of infinitely many spin $\frac{1}{2}$ systems.

An ordinary quantum theory of a single spin system features symmetric operators $\{\hat{\sigma}(x), \hat{\sigma}(y), \hat{\sigma}(z)\}$, the *Pauli spin observables*, acting on a Hilbert space $\mathcal{H}$ to satisfy the *Pauli relations*.

---

[11] For why this might be interesting, see Belot (1998) or Healey (2007). Dürr et al. (2006) catalog other topologically non-$\mathbb{R}^{2n}$ phase spaces with physical applications.

$$[\hat{\sigma}(x), \hat{\sigma}(y)] = i\hat{\sigma}(z), \quad [\hat{\sigma}(y), \hat{\sigma}_z] = i\hat{\sigma}(x), \quad [\hat{\sigma}(z), \hat{\sigma}(x)] = i\hat{\sigma}(y) \qquad (3.16)$$

$$\hat{\sigma} \cdot \hat{\sigma} = (\hat{\sigma}(x))^2 + (\hat{\sigma}(y))^2 + (\hat{\sigma}(z))^2 = 3I$$

Next consider a finite number of spin $\frac{1}{2}$ systems, arranged in a one-dimensional lattice. A quantum theory for this composite system equips each location k with a Pauli spin $\hat{\sigma}_k = (\hat{\sigma}(x)_k, \hat{\sigma}(y)_k, \hat{\sigma}(z)_k)$ satisfying the Pauli Relations, expanded to include the requirement that spin observables for different systems commute:

$$[\hat{\sigma}(x)_k, \hat{\sigma}(y)_{k'}] = i\delta_{kk'}\hat{\sigma}(z)_k, \text{ etc., } \hat{\sigma}_k \cdot \hat{\sigma}_k = 3I \qquad (3.17)$$

Consider the Pauli Relations (3.17) for a fixed n. A collection of operators satisfying those relations generates the *CAR algebra* for n: this algebra consists of anything and everything you can obtain by forming polynomials of Pauli spins. Do not be distressed that "CAR" stands for "canonical *anti*commutation relations," while the Pauli relations are commutation relations. For:

Fact 3.3. A representation of the CARs for n uniquely determines a representation of the CAR algebra for n, and conversely (see Emch 1972, 271–2).

The validity of this fact extends beyond representations where n is finite. It extends, for instance, to representations whose index set ranges over the positive and negative integers $\mathbb{Z}$. The upshot is that for each n, a representation of the CARs is a representation of the Pauli spin relations and the CAR algebra they generate, and vice versa.

The Pauli relations and the CARs: a gistification. (For a proper introduction, see Emch 1972, 269–75.) Suppose $\{\hat{p}_i, \hat{q}_i\}$ act on some Hilbert space to satisfy the CCRs. Then the "creation" and "annihilation" operators $\hat{a}_i$ and $\hat{a}_i^\dagger$, defined by:

$$\hat{a}_i = \frac{1}{\sqrt{2}}(\hat{p}_i - i\hat{q}_i) \qquad (3.18)$$

$$\hat{a}_i^\dagger = \frac{1}{\sqrt{2}}(\hat{p}_i + i\hat{q}_i) \qquad (3.19)$$

satisfy:

$$[\hat{a}_i, \hat{a}_j] = 0 = [\hat{a}_i^\dagger, \hat{a}_j^\dagger], \quad [\hat{a}_i, \hat{a}_j^\dagger] = -i\delta_{ij}I \qquad (3.20)$$

which are also CCRs. CCRs describe bosons, typically particles with integer spin.

Fermions, typically particles with half-integer spin, obey canonical *anticommutation* relations.[12] To obtain the CARs, replace the commutators $[A, B] := AB - BA$ in (3.20) with *anticommutators*[13] $\{A, B\} := AB + BA$, and lose an i:

---

[12] Another topic given short shrift in this book is the connection between spin and statistics. For the classic account, see Streater and Wightman (1964). Axioms of coventional QFT mediate the connection.

[13] Notational infelicity: the brackets that here denote the anticommutator serve in the rest of the book to denote Poisson or Lie brackets.

3.3 INFINITELY MANY DEGREES OF FREEDOM

$$\{\hat{a}_i, \hat{a}_j\} = 0 = \{\hat{a}_i^\dagger, \hat{a}_j^\dagger\}, \quad \{\hat{a}_i, \hat{a}_j^\dagger\} = -\delta_{ij}I \quad (3.21)$$

You can think of the indices as labeling sites on a lattice. Thus $\hat{a}_i^\dagger, \hat{a}_i$ are "creation" and "annihilation" operators for a fermion at the site i. Suppose the index in (3.21) ranges from 1 to n. The resulting set of equations gives the CARs for n degrees of freedom.

Where $\{\hat{\sigma}(x), \hat{\sigma}(y), \hat{\sigma}(z)\}$ are Pauli spin operators for $n = 1$, they determine a representation of the CARs for one degree of freedom via:

$$\hat{a} = \frac{1}{2}(\hat{\sigma}(x) + i\hat{\sigma}(y)) \quad (3.22)$$

$$\hat{a}^+ = \frac{1}{2}(\hat{\sigma}(x) - i\hat{\sigma}(y))$$

The *CAR algebra for* $n = 1$ is defined to be the algebra $\mathfrak{M}(2)$ of all complex 2×2 matrices. This algebra is generated by the Pauli spins in the sense that every element of $\mathfrak{M}(2)$ can be written as a linear combination of Pauli spins and the identity operator (which is the square of any Pauli spin). To extend the Pauli relations to n (finite) systems, supplement the obvious generalization of (3.17) with the requirement that components of spin pertaining to different sites commute. A representation of these extended relations generates the CAR algebra for n, which is the tensor product of n copies of $\mathfrak{M}(2)$.[14] ♠

Suppose we set out to build an ordinary quantum theory for a collection of spin $\frac{1}{2}$ systems. Our first step will be to find a representation of the appropriate Pauli relations on a Hilbert space $\mathcal{H}$. The Pauli spins furnishing this representation are our canonical observables. Other observables pertaining to the system correspond to self-adjoint elements of $\mathfrak{B}(\mathcal{H})$ which can be obtained from the Pauli spins by forming polynomials. Ordinary quantum states are given by density operators on $\mathcal{H}$. (A complication we can presently ignore is that, because $\mathcal{H}$ for a single spin system is two-dimensional, there are non-density matrix states conforming to the normal conception of states.) We've all seen ordinary quantum theories of the single spin system.

One way to build a quantization fitting these specifications for a chain of n spin systems is to use a vector space $\mathcal{H}$ spanned by a basis consisting of sequences s_k, where each entry takes one of the values ± 1, and k ranges from 1 to n. (NB There are finitely many distinct such sequences—finitely many ways to map a set of finite cardinality into the set $\{+1, -1\}$.) Operators $\hat{\sigma}(z)_j, j = 1$ to n, are introduced in such a way that sequences s_k whose j^{th} entry is ± 1 correspond to eigenvectors associated with the eigenvalue ± 1. Operators $\hat{\sigma}(y)_j, \hat{\sigma}(x)_j$ conspiring with these to satisfy the Pauli Relations (3.17) are introduced by analogy to their single electron counterparts (see Sewell 2002, §2.3).

---

[14] If you were worried about equations (3.22) giving $\hat{\sigma}(z)$ nothing to do, take heart. He'll be employed in representations of the CARs for $n \geq 2$. See Emch (1972, 269–75) for how.

Because we are dealing with canonical *anti*commutation relations, not canonical commutation relations, we are once more beyond the reach of the Stone–von Neumann theorem. But a theorem due to Jordan and Wigner (1928) fills the breach:

Fact 3.4 (Jordan–Wigner Uniqueness Theorem). *For each finite n, every irreducible representation of the CARs is unitarily equivalent to every other.*

Once n is fixed, the CARs/Pauli Relations have an irreducible representation which is unique up to unitary equivalence.

At the risk of pedantry, let's spell out what that means. (NB For the duration of this explication, spin operators will doff their hats to minimize notational clutter.) Suppose Werner and Erwin each find an irreducible representation of the Pauli Relations (3.17) for a finite spin chain. Let $\sigma(i)_k^W$ be the operator on $\mathcal{H}_W$ by which Werner represents the ith component of spin for the kth particle; let $\sigma(i)_k^E$ be the operator on $\mathcal{H}_E$ by which Erwin represents the ith component of spin for the kth particle. According to Def. 2.7, Werner's representation and Erwin's are unitarily equivalent if and only if there exists a unitary map $U : \mathcal{H}_E \to \mathcal{H}_W$ such that:

$$U\sigma(i)_k^E U^{-1} = \sigma(i)_k^W \quad \text{for all } i \in \{x, y, z\}, k \in \{1, 2, \ldots, n\} \quad (3.23)$$

Let $f(\{\sigma(i)_k\})$ be a linear function of Pauli spins. Because unitary maps are linear, the U relating Werner's Pauli spins to Erwin's also induces the following relation between linear functions of those spins:

$$f(\{\sigma(i)_k^W\}) = f(U\{\sigma(i)_k^E\}U^{-1}) = Uf(\{\sigma(i)_k^W\})U^{-1} \quad (3.24)$$

$f(\{\sigma(i)_k^W\})$ and $f(\{\sigma(i)_k^E\})$ are, respectively, elements of $\mathfrak{B}(\mathcal{H}_W)$ and $\mathfrak{B}(\mathcal{H}_E)$.

Now let $f_i(\{\sigma(i)_k^E\})$ be a sequence of functions of Erwin's Pauli spins that converges in $\mathcal{H}_E$'s weak topology to an operator F_E, which will thereby be a member of $\mathfrak{B}(\mathcal{H}_E)$. (Every element of $\mathfrak{B}(\mathcal{H}_E)$ can be obtained in this way.) Because unitary maps are norm-preserving, $Uf_i(\{\sigma(i)_k^E\})U^{-1}$ is a sequence of functions of Werner's Pauli spins that converges in $\mathcal{H}_W$'s weak topology to an operator $F_W \in \mathfrak{B}(\mathcal{H}_W)$. (Every element of $\mathfrak{B}(\mathcal{H}_W)$ can be obtained in this way.) Using lim^W and lim^E to denote, respectively, limits in Werner's and Erwin's weak topologies, it follows that:

$$UF_E U^{-1} = U(lim^E f_i(\{\sigma(i)_k^E\})U^{-1} = lim^W Uf_i(\{\sigma(i)_k^E\})U^{-1} = F_W \quad (3.25)$$

Here's what's crucial about the unitary map establishing a correspondence between the Pauli spin operators figuring in Erwin's representation and the Pauli spin operators figuring in Werner's representation. That unitary map doesn't merely establish that Erwin and Werner use the same number of canonical observables. The map establishing the correspondence between Erwin's and Werner's Pauli spins *extends in a way that preserves that correspondence* to the full set of bounded operators on each quantizer's Hilbert space. Supposing that each is prosecuting an ordinary quantum theory, the ordinary quantum theories they prosecute are (presumptively) physically equivalent. In general, the Jordan–Wigner theorem implies that any (irreducible) representation of the Pauli relations for a finite

3.3 INFINITELY MANY DEGREES OF FREEDOM   63

spin chain we concoct is unitarily—and so, considered as the basis for an ordinary quantum theory, presumptively, physically—equivalent to any other.

The *polarization* of a spin chain is described by a vector whose magnitude ($\in [0, 1]$) gives the strength and whose orientation gives the direction of the system's magnetization. On a single electron, it is represented by an observable $\hat{\mathbf{m}}$ whose three components correspond to three orthogonal components of quantum spin. For example, in the $+1$ eigenstate $|+\rangle$ of $\hat{\sigma}_z$ (understood as the z-component of spin), $\langle \hat{\mathbf{m}} \rangle = +1$ along the z axis.

For the finite spin chain, the polarization observable has components $\hat{m}_i := \frac{1}{n} \sum_{k=1}^{n} \hat{\sigma}(i)_k$. $\hat{\mathbf{m}}$ belongs to $\mathfrak{B}(\mathcal{H})$ because its components are polynomials of Pauli spins. Let $[s_k]_j \in \{\pm 1\}$ stand for the jth entry of the sequence s_k. In the basis sequence s_k, the zth-component of polarization $\hat{m}_z$ takes an expectation value of magnitude $\frac{1}{n} \sum_{j=1}^{n} [s_k]_j$. This quantity attains extreme values (of ± 1) for those sequences every term of which is the same.

Let $\hat{W}$ be a state in the present representation assigning $\hat{m}_z$ the expectation value $+1$. The Jordan–Wigner theorem ensures that any other representation of the Pauli Relations for n will be unitarily equivalent to the present one. Any other representation of those relations is thus guaranteed to contain a state $\hat{W}'$ (the image of $\hat{W}$ under the unitary map implementing the equivalence of the representations) and an observable $\hat{m}'_z$ (the image of $\hat{m}_z$ under that map) such that the expectation value of $\hat{m}'_z$ in the state $\hat{W}'$ is $+1$.

Finally, consider a doubly infinite chain, labeled by the positive and negative integers $\mathbb{Z} = \{..., -2, -1, 0, 1, 2, ...\}$, of spin $\frac{1}{2}$ systems. As before, to construct a quantum theory of this system is to associate with each site k a Pauli spin $\hat{\sigma}_k = (\hat{\sigma}(x)_k, \hat{\sigma}(y)_k, \hat{\sigma}(z)_k)$ satisfying the Pauli Relations (3.17). But if we follow the strategy adopted for the finite spin chain, and attempt to construct our Hilbert space from a basis consisting all possible maps from $\mathbb{Z}$ to $\{\pm 1\}$, we are foiled. Because the set of such maps is non-denumerable, the Hilbert space we'd construct would be non-separable, breaking the tradition of using separable Hilbert spaces for physics. (A typical expression of traditionalism is the one Simon issues at the outset of an exposition of those aspects of functional analysis he deems most important for physics. "Throughout, all our Hilbert spaces will be separable unless otherwise indicated. Many of the results extend to non-separable spaces, but we cannot be bothered with such obscurities" (1972, 18).[15])

Here's one way to build a separable Hilbert space representation of the Pauli relations for an infinite chain of spins. Start with the sequence $[s_k]_j = +1$ for $j \in \mathbb{Z}$, and add all sequences obtainable therefrom by finitely many local modifications. The resulting basis consists of all sequences for which all but a finite number of sites take the value $+1$. Continue to follow the model of the finite spin chain to introduce operators $\hat{\sigma}(i)_k^+$ satisfying the Pauli Relations (see Sewell 2002, §2.3). I will call this

---

[15] Halvorson (2001; 2004) contains reasons to bother.

the $\mathcal{H}^+$ *representation*—but please keep in mind that it matters to the algebraic structure of this representation *which* elements of $\mathfrak{B}(\mathcal{H}^+)$ play the role of *which* Pauli spins.

Before considering how the polarization observable $\hat{\mathbf{m}}^+$ behaves on the $\mathcal{H}^+$ representation, let us assure ourselves that there is such an observable. To be an element of $\mathfrak{B}(\mathcal{H}^+)$, the polarization observable must be a polynomial of Pauli spins, or the limit (in the appropriate sense) of a sequence of such polynomials. The sequence $\hat{m}_i^N := \frac{1}{2N+1} \sum_{k=-N}^{N} \hat{\sigma}(i)_k^+$ has a limit as $N \to \infty$ in $\mathcal{H}^+$'s weak topology, which provides the appropriate sense of convergence. A sequence $\hat{A}_i$ of operators on $\mathcal{H}$ converges to an operator $\hat{A}$ in $\mathcal{H}$'s weak operatory topology iff for all $|\psi\rangle, |\phi\rangle \in \mathcal{H}$, $|\langle\psi|(\hat{A} - \hat{A}_i)|\phi\rangle|$ goes to 0 as i goes to ∞. Confining attention to the sequence $\hat{m}_z^N$, observe that $\langle s_k | \hat{m}_z^N | s_k' \rangle = \frac{1}{2N+1} \sum_{j=-N}^{N} (\frac{[s_k]_j + [s_k']_j}{2})$ for basis sequences s_k and s_k'. Because -1 occurs only finitely many times in each basis sequence, $\langle s_k | \hat{m}_z^N | s_k' \rangle$ will converge to 1 as $N \to \infty$, no matter what s_k and s_k' are. This shows that the sequence $\hat{m}_z^N$ converges weakly to the identity operator.

We can imagine realizations of the Pauli Relations, modeled on the $\mathcal{H}^+$ representation, for which this convergence is not to be had. One example is a representation whose "ground state" s_k is a sequence for which $\lim_{N \to \infty} \langle s_k | \hat{m}_i^N | s_k \rangle$ does not converge (e.g. the sequence $[s_k]_j = +1$ for $2^n < |j| \leq 2^{n+1}$ and n odd; $[s_k]_j = -1$ otherwise).

So it is noteworthy that $\mathcal{H}^+$'s weak topology underwrites the convergence of the sequence $\hat{m}_z^N$, the convergence in virtue of which the z component of $\hat{\mathbf{m}}^+$ is well-defined. Other components of global polarization are also captured as weak limits of polynomials of Pauli spins. Each component of polarization is thus an element of $\mathfrak{B}(\mathcal{H}^+)$, and so an observable in ordinary QM's sense.

$\mathcal{H}^+$ is spanned by a basis whose elements correspond to doubly infinite sequences s_k only finitely many of whose entries are -1. For every member of this basis, $\langle \hat{\mathbf{m}}^+ \rangle$ will be oriented along the z axis and take the value $\lim_{n \to \infty} \frac{1}{2n+1} \sum_{j=-n}^{n} [s_k]_j$. Because for each basis element, all but a finite number of its entries take the value $+1$, this limit will be 1. Every ordinary quantum state on the $\mathcal{H}^+$ representation will inherit this feature from the basis vectors in terms of which it is expressed: every state in the representation will have unit polarization in the positive z direction.

Because the chain is infinite, the Jordan–Wigner theorem does not imply that the representation just constructed is unique up to unitary equivalence. And it is not. Consider, for contrast, a representation set in a Hilbert space whose basis elements correspond to the sequence $[s_k]_j = -1$ for $j \in \mathbb{Z}$, along with all sequences obtainable from this one by finitely many local modifications. The basis consists, then, of sequences for which all but a finite number of sites take the value -1. Operators $\hat{\sigma}(i)_k^-$ satisfying the Pauli relations are introduced in such a way that $[s_k]_j$, the jth entry in the basis sequence s_k, gives the expectation value of $\hat{\sigma}(z)_j$ (see Sewell 2002, §2.3). Call this *the $\mathcal{H}^-$ representation*. By parity of reasoning, $\hat{\mathbf{m}}^-$ is an observable in this quantization, one assigned the expectation value -1 by every ordinary quantum state in the $\mathcal{H}^-$ representation.

We can now see that the $\mathcal{H}^-$ and $\mathcal{H}^+$ representations are not unitarily equivalent. Following Sewell (2002, §2.3.3), suppose, for contradiction, that they were. Then there'd be a unitary $U: \mathcal{H}^+ \to \mathcal{H}^-$ such that $U\hat{\sigma}(z)_j^+ U^{-1} = \hat{\sigma}(z)_j^-$ for all $j \in \mathbb{Z}$. Where $\hat{m}(z)_N^\pm := \frac{1}{2N+1}\sum_{k=-N}^{N} \hat{\sigma}(z)_k^\pm$, this implies that $\hat{m}(z)_N^- = U\hat{m}(z)_N^+ U^{-1}$. For $|\psi^+\rangle$ and $|\psi^-\rangle$, unit vectors in $\mathcal{H}^+$ and $\mathcal{H}^-$ related by $|\psi^-\rangle = U|\psi^+\rangle$, it follows that:

$$\langle \psi^+ | \hat{m}(z)_N^+ | \psi^+ \rangle = \langle \psi^- | \hat{m}(z)_N^- | \psi^- \rangle \tag{3.26}$$

But in the limit $N \to \infty$, (3.26) breaks down: the r.h.s. (which gives the expectation value the state $|\psi^-\rangle$ assigns the z-component of $\hat{m}^-$) and the l.h.s. (which gives the expectation value the state $|\psi^+\rangle$ assigns the z-component of $\hat{m}^+$) go to -1 and $+1$ respectively. The expectation values $|\psi^+\rangle$ assigns observables in $\mathfrak{B}(\mathcal{H}^+)$ do not replicate the expectation values $|\psi^-\rangle$ assigns the images of those observables (in $\mathfrak{B}(\mathcal{H}^-)$) under the map U. This exposes the failure of the $\mathcal{H}^+$ and $\mathcal{H}^-$ representations of the infinite spin chain to be unitarily equivalent.

Variations on the foregoing example will appear repeatedly in subsequent chapters.

3.3.2 QFT: the Klein–Gordon field

The theory of the mass m free boson field, obtained by quantizing the classical mass m Klein–Gordon field in Minkowski spacetime, may be the simplest QFT there is. The quantization of Klein-Gordon theory will be reviewed in detail in Chapter 9. The aim of this section's cursory treatment is to place this theory of QM_∞ outside the scope of the Stone–von Neumann theorem, and in the sights of philosophers interested in foundational questions.

Spacetime crash course. (For a proper introduction, consult Wald 1994.) Minkowski spacetime $\mathcal{M}$ is a four-dimensional real vector space on which is defined a non-degenerate, symmetric, bilinear map $\eta: \mathcal{M} \times \mathcal{M} \to \mathbb{R}$ (a.k.a. a *pseudo-Reimannian metric*) of signature $(+++-)$. To see what is meant by signature, notice that η is an inner product (although not necessarily a positive definite one) on $\mathcal{M}$ and as such furnishes a notion of an orthonormal basis for that vector space: a vector v is normal if $\eta(v, v) = \pm 1$; vectors v, w are orthogonal if $\eta(v, w) = 0$; thus a set of vectors $\{x_\mu, \mu \in \{1, 2, 3, 4\}\}$ is an orthonormal basis if they're each normal and they're pairwise orthogonal. η's signature describes its action on an arbitrary orthonormal basis of $\mathcal{M}$: for one element x_ν of that basis, $\eta(x_\nu, x_\nu) = -1$; for the other three, $\eta(x_{\nu'}, x_{\nu'}) = +1$, and this is true no matter what orthonormal basis we consider. Bases can always be re-indexed so that x_4 is the element mapped to -1; thus the signature $(+++-)$ characterizes η's treatment of $\mathcal{M}$'s orthonormal bases.

$\mathcal{M}$'s elements are spacetime "events," that is (speaking naively) ordered pairs of instants of time and points of space. An orthonormal basis $\{x_\mu\}$ for $\mathcal{M}$ determines a way to coordinate events: an event $v \in \mathcal{M}$ corresponds to the quadruple v^μ of real numbers that are v's components in this basis. That is, $v = \sum_{\mu=1}^{4} v^\mu x_\mu := v^\mu x_\mu$ (where the last step implicitly defines *the Einstein summation convention,* which is to sum over all values

66 BEYOND THE STONE-VON NEUMANN THEOREM

of an index that appears as both a subscript and a superscript in an expression). Think of v^4 as the event's temporal coordinate, and $v^i, i \in \{1,2,3\}$, as its spatial coordinates, in the coordinate system with basis vectors $\{x_\mu\}$.

Given an event $v \in \mathcal{M}$, η sorts every other event $w \in \mathcal{M}$ into one of three classes: those that are *lightlike separated* from v (w such that $\eta(v,w) = 0$); those that are *timelike separated* from v (w s.t. $\eta(v,w) < 0$); and those that are *spacelike separated* from v (w s.t. $\eta(v,w) > 0$). Thus η imposes a *lightcone structure* (see Fig. 3.1[16]) on Minkowski spacetime that is the same everywhere. Taking $\mathcal{M}$ to be a manifold, we can use η to equip that manifold with a *metric tensor* $\eta_{\mu\nu}$, a map, defined at each point of the manifold, from pairs of vectors tangent to that point to $\mathbb{R}$. Because the space of vectors tangent to a point $p \in \mathcal{M}$ is isomorphic to $\mathcal{M}$, $\eta_{\mu\nu}(p)$ just acts on that space as η acts on $\mathcal{M}$. (This is called a metric because you can use it to define the differential displacement ds in the near neighborhood of an arbitrary point of the manifold, via $ds^2 = \eta_{\mu\nu} dx^\mu dx^\nu$.) The Minkowski metric $\eta_{\mu\nu}$ is invariant. Indeed, $\mathcal{M}$ admits global coordinates (e.g. inertial coordinates) in which at each point $\eta_{\mu\nu}$ assumes the form:

$$\begin{pmatrix} 1 & 0 & 0 & 0 \\ 0 & 1 & 0 & 0 \\ 0 & 0 & 1 & 0 \\ 0 & 0 & 0 & -1 \end{pmatrix}$$

Symmetries of Minkowski spacetime are maps $\Lambda : \mathcal{M} \to \mathcal{M}$ which preserve the Minkowski metric in the sense that $\eta_{\mu\nu}(p) = \eta_{\mu\nu}(\Lambda(p))$ for all $p \in \mathcal{M}$. For obvious reasons, they're also called *isometries* of Minkowski spacetime. Isometries of Minkowski spacetime include translations, rotations, and Lorentz boosts (which can be thought of as translating between inertial coordinate systems adapted to observers moving at constant velocity with respect to one another). These isometries form a group called the *Poincaré* group.

Minkowski spacetime is the spacetime of Special Relativity. General Relativity traffics in more general spacetimes $(\mathcal{M}, g_{\mu\nu})$, where $\mathcal{M}$ is a differential manifold and $g_{\mu\nu}$ a metric tensor defined at each point of $\mathcal{M}$ but *allowed to vary from point to point*. The Einstein Field equations are 256 coupled partial differential equations which can be abbreviated

$$G_{\mu\nu} = 8\pi T_{\mu\nu}$$

They make the metric tensor (which is hiding in the Einstein curvature tensor $G_{\mu\nu}$) at a point depend on the stress–energy tensor $T_{\mu\nu}$ describing the distribution of energy and matter near that point. A solution $(\mathcal{M}, g_{\mu\nu}, T_{\mu\nu})$ to the Einstein field equations imposes a light cone structure at each point of M, but not necessarily the *same* light cone structure. In a coordinate system in which the light cone at some reference point

---

[16] Cribbed, with gratitude, from John Norton's "Einstein for Everyone" website: www.pitt.edu/jdnorton/teaching/HPS_0410/chapters/spacetime/index.html#Light.

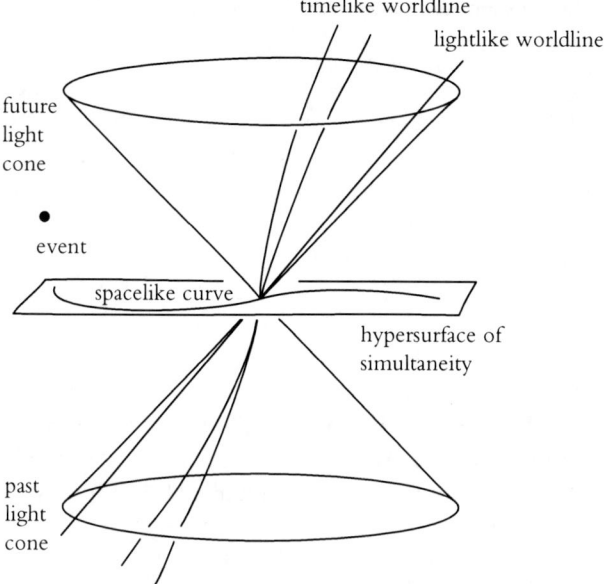

Figure 3.1. The lightcone structure of Minkowski spacetime

p assumes the familiar 45° shape of Fig. 3.1—for any point, there is such a coordinate system—light cones at other points can be narrower, wider, or tipped with respect to the reference light cone. Among the cool things this makes possible according to general relativity are black holes and time travel (for introductions, see Geroch 1978 and Malament 1984). ♠

Where (x_1, x_2, x_3, t) are inertial coordinates for Minkowski spacetime and x ranges over points in that spacetime, the classical Klein–Gordon equation for mass m is:

$$\left(\frac{\partial^2}{\partial x_1^2} + \frac{\partial^2}{\partial x_2^2} + \frac{\partial^2}{\partial x_3^2} - \frac{\partial^2}{\partial t^2} - m^2\right)\varphi(x) = 0 \tag{3.27}$$

Scholium: Notational variations. You'll also see this equation written in terms of the D'Alembertian operator $\Box := \left(\frac{\partial^2}{\partial x_1^2} + \frac{\partial^2}{\partial x_2^2} + \frac{\partial^2}{\partial x_3^2} - \frac{\partial^2}{\partial t^2}\right)$ as:

$$\Box \phi(x) = m\phi(x)$$

or in terms of the Laplacian operator $\Delta := \left(\frac{\partial^2}{\partial x_1^2} + \frac{\partial^2}{\partial x_2^2} + \frac{\partial^2}{\partial x_3^2}\right)$ (itself expressed in terms of the vector differential del operator $\nabla := \left(\frac{\partial}{\partial x_1} x_1 + \frac{\partial}{\partial x_2} x_2 + \frac{\partial}{\partial x_3} x_3\right)$ as $\Delta := \nabla \cdot \nabla = \nabla^2$), as:

$$\left(\Delta - \frac{\partial^2}{\partial t^2}\right)\phi(x) = m\phi(x) \quad \text{or} \quad \left(\nabla^2 - \frac{\partial^2}{\partial t^2}\right)\phi(x) = m\phi(x)$$

or in terms of the Einstein summation convention[17] as:

$$(\partial^\mu \partial_\mu - m)\phi(x) = 0 \quad \spadesuit$$

We'll focus on solutions $\phi(x)$ to (3.27) which are maps from points $x := (x_1, x_2, x_3, t) := (\vec{x}, t)$ of Minkowski spacetime to $\mathbb{R}$.

The Klein–Gordon equation results from extremizing the action $S = \int \mathcal{L} dt$ determined by the Lagrangian density:

$$\mathcal{L} = \int \left(\frac{1}{2} [\dot{\varphi}(\vec{x})^2 - (\nabla \varphi(\vec{x}))^2 - m^2 \varphi(\vec{x})^2] \right) d^3\vec{x} \tag{3.28}$$

(see Wald 1994, §3.1). In a globally hyperbolic spacetime, such as Minkowski spacetime, a Lagrangian density such as (3.28) imposes a symplectic product on the space of solutions to (3.27) (Woodhouse 1992, thm. 7.2.2). Let (S, Ω) denote the symplectic vector space determined by (3.27) and (3.28). To follow the Hamiltonian recipe for quantizing classical Klein–Gordon theory is to find a Hilbert space representation of the Weyl relations over (S, Ω).

In Minkowski spacetime, this can be done by following a textbook "frequency-splitting" heuristic for quantizing Klein–Gordon theory (see Kaku 1993, ch. 3.2; Wald 1994, ch. 4.3). Solutions to the Klein–Gordon equation can be Fourier-decomposed into uncoupled normal modes with angular frequency ω_k. This suggests an infinite collection of independent oscillators as a model of the classical field. Exploiting the analogy with the simple harmonic oscillator, the textbook approach represents the (infinitesimal version of) the Weyl Relations by means of "creation" and "annihilation" operators $\hat{a}_k^\dagger$ and $\hat{a}_k$ for positive frequency (with respect to inertial time t) field modes ω_k:

$$[\hat{a}_k, \hat{a}_{k'}] = 0 = \left[\hat{a}_k^\dagger, \hat{a}_{k'}^\dagger\right], \quad \left[\hat{a}_k, \hat{a}_{k'}^\dagger\right] = \delta_{kk'} \hat{I}. \tag{3.29}$$

These operators act on the symmetric Fock space over a "single particle Hilbert space" $\mathcal{H}$.

Scholium: Fock space. (See Wald 1994, A.2.) The *symmetric Fock space* $\mathfrak{F}_s(\mathcal{H})$ over $\mathcal{H}$ is the completed direct sum $\bigoplus_{i=0}^\infty [\otimes_i \mathcal{H}]_s$. Here $\otimes_i \mathcal{H}$ is the i-fold tensor product of $\mathcal{H}$ with itself. $[\otimes_i \mathcal{H}]_s$ is the symmetric subspace of that i-fold tensor product, that is, the subspace spanned by vectors $\otimes_i |\xi_i\rangle$ symmetric in the labels i. (This means the vectors don't change when their direct summands swap places: for $i = 2$, the vector $|\phi\rangle \otimes |\psi\rangle + |\psi\rangle \otimes |\phi\rangle$ is an example.) $\otimes_0 \mathcal{H}$ is stipulated to be $\mathbb{C}$. The *antisymmetric Fock space* $\mathfrak{F}_a(\mathcal{H})$ over $\mathcal{H}$ is defined analogously, with $[\otimes_i \mathcal{H}]_a$ being the subspace of the i-fold tensor product spanned by vectors antisymmetric in the labels i. (This means a

---

[17] Understanding ∂_μ as $\frac{\partial}{\partial x_\mu}$ and using the Minkowski metric $\eta^{\mu\nu} = \eta_{\mu\nu}$ to raise the index μ: $\partial^\mu = \eta^{\mu\mu} \partial_\mu$.

vector's direct summands' swapping places multiplies the vector by -1: for $i = 2$, the vector $|\phi\rangle \otimes |\psi\rangle - |\psi\rangle \otimes |\phi\rangle$ is an example.) ♠

The textbook quantization constructs the single particle Hilbert space $\mathcal{H}$ by decomposing the space S of solutions to (3.27) into a space S^+ of solutions that oscillate with positive frequency with respect to inertial time and a space S^- of negative frequency solutions. The symplectic product Ω defines a positive-definite inner product on S^+; $\mathcal{H}$ is the completion of S^+ with respect to that inner product. Chapter 9 will review this construction in more detail.

The action of the creation and annihilation operators furnishing a representation of the Weyl relations on the Fock space $\mathfrak{F}_s(\mathcal{H})$ facilitates the interpretation of quantized Klein–Gordon theory as particle physics. $\mathfrak{F}_s(\mathcal{H})$ features a state $|0\rangle$ such that $\hat{a}_k |0\rangle = 0$ for all k. $|0\rangle$ is the lowest energy eigenstate of the quantum Hamiltonian for the free boson field (also handled by analogy with the simple harmonic oscillator). The state $\hat{a}_k^\dagger |0\rangle$ is an eigenstate of the Hamiltonian with the same total energy a particle with frequency ω_k would have—provided the momentum and rest mass of this particle are given by standard relativistic expressions. The pattern repeats. The state $|n_k\rangle = \left(\hat{a}_k^\dagger\right)^n |0\rangle$ has the energy appropriate to n particles of frequency ω_k; the state $\hat{a}_k |n_k\rangle$ has the energy appropriate to $(n-1)$ such particles. So it is natural to speak as if $\hat{a}_k^\dagger$ adds a particle of frequency ω_k to a state on which it acts, and as if $\hat{a}_k$ removes a particle of this type. $|0\rangle$, the state from which no further particles can be removed, is the vacuum state. (Please keep in mind that the natural *façon de parler*, which I adopt here, suggests ontological commitments—to particles—and an interpretation of the action of the operators $\hat{a}_k^\dagger$ and $\hat{a}_k$ that should not be accepted uncritically.) The operator $\hat{N}_k = \hat{a}_k^\dagger \hat{a}_k$ serves as the *number operator* for particles of variety k. Its eigenvalues are natural numbers; the state obtained by acting on the vacuum n times with the creation operator $\hat{a}_k^\dagger$ is an $\hat{N}_k$ eigenstate with eigenvalue n. Thus $\hat{N}_k$ is understood to count particles of variety k: its expectation value in a state gives the mean number of particles of variety k one would find in that state. Hence, $\hat{N} = \sum_j \hat{a}_j^\dagger \hat{a}_j$ is the *total number operator*. That for all k, $\langle 0|\hat{N}_k|0\rangle = \langle 0|\hat{N}|0\rangle = 0$ reinforces $|0\rangle$'s claim to be the vacuum or no-particle state.

States that admit analysis in terms of their particle contents are not isolated. Every state in the Fock space can be approximated to arbitrary precision by acting on $|0\rangle$ with linear combinations of creation and annihilation operators. That makes the vacuum state *cyclic* with respect to the set of operators (3.29) furnishing a representation of the Weyl relations. It also suggests that the cluster of operators framing the particle notion has some claim to be physically fundamental—a suggestion Chapters 9–11 will examine in some detail.

The frequency-splitting heuristic builds a representation of the Weyl relations for the free boson field from classical solutions to the Klein–Gordon equation which are positive frequency with respect to Minkowski time. Because the solution space S is infinite in dimension, the Stone–von Neumann theorem cannot vouch that this representation is unique up to unitary equivalence. Indeed, the Weyl Relations for the Klein–Gordon field admit *continuously many* inequivalent representations: even analogs of the scale transformation (2.33) visited upon the simple harmonic oscillator generate inequivalent representations (Segal 1967, 124 ff.). We can draw on the example of the infinite spin chain to see why Hilbert space quantizations of classical Klein–Gordon theory are so dramatically non-unique.

We will construct the Fock space carrying a representation of the CCRs (3.29) for the Klein–Gordon field as follows.[18] Its basis elements will consist of infinite sequences n_i of non-negative integers. We'll eventually think of each sequence as a simultaneous eigenstate of all the variety number operators $\hat{N}_k$, with the mth entry of the sequence specifying the associated $\hat{N}_m$ eigenvalue. But first we must negotiate a hurdle reminiscent of one we encountered when quantizing the infinite spin chain. It's that infinite sequences of non-negative integers are uncountable in number, but the basis of this quantization's Fock space had better be countable, if that space is to be (as decency demands) separable. One way to clear this hurdle and ensure separability is to consider as elements of our basis only the sequence n_i^0, whose m^{th} entry is 0 for all m, and sequences that differ from n_i^0 in only finitely many places.

Ignoring normalization factors, we can introduce creation and annihilation operators satisfying (3.29) that act as follows on elements of this basis. The creation operator $\hat{a}_k^\dagger$ adds 1 to the kth entry of each basis sequence n_i. It leaves other entries unchanged. The annihilation operator $\hat{a}_k$ subtracts 1 from the kth entry of each basis sequence n_i, unless that entry is 0, in which case it maps the sequence to the zero vector. Variety number operators $\hat{N}_k$ and the total number operator $\hat{N}$ can be constructed from the creation and annihilation operators so introduced. The sequence n_i^0 emerges as the vacuum state of this representation. It is annihilated by each annihilation operator; an eigenstate, with eigenvalue 0, of every number operator; and such that every state in the Hilbert space can be approximated to arbitrary precision by acting on it with polynomials of creation and annihilation operators. Let's call this the n^0 *quantization* of the Klein–Gordon field.[19]

Next let's, for the sake of variety, quantize the Klein–Gordon field again, this time negotiating the separability hurdle by constructing our Fock space from the basis element n_i^j, whose m^{th} entry is 0 for $m < j$ and 1 for $m \geq j \neq 0$, and sequences that differ from n_i^j in only finitely many places. (NB n_i^0, and every other basis element of the n^0 quantization, differs from n_i^j in *infinitely* many places.) As before, we can introduce creation and annihilation operators acting on this space and satisfying (3.29). The kth creation operator will act just like $\hat{a}_k^\dagger$; the kth annihilation operator will act a bit more intricately, annihilating sequences whose k^{th} entry is 0 for $k < j$ and sequences whose

---

[18] See Gårding and Wightman (1954); Arageorgis (1995, Appendix C) for more rigorous presentations.
[19] We've informally lapsed into the *occupation number formalism* for QFT. See Teller (1995, ch. 3) for an accessible introduction.

k^{th} entry is 1 for $k \geq j$. Variety and total number operators are constructed from these creation and annihilation operators. With respect to these operators, the base sequence n_i^j emerges as a vacuum with credentials completely parallel to those of n_i^0. Let's call this the n^j *quantization* of the Klein–Gordon field.

Are the quantizations we've just sketched unitarily equivalent? Because the phase space of classical Klein–Gordon theory is infinite dimensional, the Stone–von Neumann theorem cannot assure us that they are, and the following informal consideration suggests that they are not. No state in the n^j quantization can be expressed as a convergent superposition of states in the n^0 quantization, and vice versa. More formally, a theorem due to Fulling (1989, 145) states that Fock space representations $(|0\rangle, \hat{a}_k^\dagger, \hat{a}_k)$ and $(|0'\rangle, \hat{a}'^\dagger_k, \hat{a}'_k)$ are unitarily equivalent iff $\langle 0'|\hat{N}|0'\rangle$ is finite. But the expectation value of n^0's total number operator in the n^j vacuum state is $\sum_{i \geq j} 1_i$, which is infinite. (So is the expectation value of n^0's total number operator in any state in the n^j representation.) Therefore the n^0 and n^j representations of the Klein–Gordon field are unitarily inequivalent. Since there are continuously many infinite sequences of non-negative integers that differ from one another in infinitely many entries,[20] there are continuously many analogously inequivalent quantizations of the Klein–Gordon field.

The principles (OQS), (OQM), and (OQD) of ordinary quantum mechanics, in concert with the standard picture of theoretical content, urges that quantizations must be unitarily equivalent to be physically equivalent. Given this criterion, *there are continuously many physically inequivalent quantizations of the free boson field.* The vacuum state of one of a pair of unitarily inequivalent representations can't be realized as a density operator state on the other representation. The number operator of one of a pair of unitarily inequivalent representations is undefined on every state in the other representation. Theorists wielding inequivalent representations don't simply disagree about which state is the ground state of the free boson field. Each denies that the other's ground state is a state *at all*. That is, they're not disagreeing about how to describe worlds they both agree are possible. A translation manual of the sort set up by a unitary map between their theories could resolve (or at least drastically temper) that disagreement. *They're disagreeing about which worlds are possible to begin with.* Because the sets of worlds each deems possible are (in a sense the next chapter will make precise) disjoint, theorists wielding unitarily inequivalent representations can't both be even partially right. Regarded as theories of ordinary QM, unitarily inequivalent representations of the Weyl relations quantizing classical Klein–Gordon theory are out-and-out rivals. This leaves us with continuum-many starkly distinct theories vying for the title of free boson field—a veritable embarrassment of riches for particle physics.

---

[20] Here's an argument sketch: there are continuously many infinite sequences of non-negative integers; each equivalence class of sequences that differ from one another in only finitely many entries is countable; so there are continuously many equivalence classes.

3.4 Conclusion

The standard account of theoretical content and the principles of ordinary quantum mechanics form a powerful alliance—provided they operate within the scope of the Stone–von Neumann and Jordan–Wigner theorems. Philosophers seeking to interpret a theory in the ambit of the alliance can share at least (and often do share at most) an unambiguous concrete Hilbert space structure as a point of departure. Others have chronicled the ensuing foundational odysseys. What I want to explore here are repercussions for interpretive projects that emanate from the fact that many significant quantum theories fall outside the scope of the uniqueness theorems. In QM_∞, many unitarily inequivalent representations of the physics are available. Adherents to the standard account and to the principles of ordinary quantum mechanics have to say which we take seriously, and why. These questions are particularly daunting because there are contexts where multiple inequivalent quantizations are required to save the phenomena: consider, for example, an iron bar, susceptible both to spontaneous magnetization in the $+z$ direction and to spontaneous magnetization in the $-z$ direction. If there is no single Hilbert space representation of the physics in which both states of magnetization appear (as the foregoing section accurately suggests), then the conjunction of the standard account and the principles of ordinary quantum mechanics imply that it's not the case that both magnetizations are possible for the iron bar. But it's hard to see why we would want to talk about physical possibility at all, if we had to say things like that.

One way out of the interpretive quagmire created from unitarily inequivalent representations is to deny that they correspond to distinct physical theories after all. The simplest form of the denial embraces the standard account of theoretical content but jilts the principles of ordinary quantum mechanics, preferring instead to identify the worlds possible according to (say) the quantization of a classical theory (S, Ω) in terms of *structures all Hilbert space representations of the Weyl relations over (S, Ω) share*. The nature and existence of such structures, as well as the interpretive options they underwrite, are the topic of the next three chapters.

4

Representation Without Taxation
An Unstrenuous Tour of Algebraic Notions

> All discourse owes its existence to the interweaving of forms.
> (Plato, *Sophist* 259e)

The central aim of this chapter is to offer a minimally punishing introduction to algebraic quantum theory. Because familiar theories of ordinary QM are instances of algebraic quantum theories, the reader may discover that she's better acquainted with algebras and their physical application than she might have expected. Because theories of ordinary QM are *special cases* of algebraic quantum theories, the reader may nevertheless be in for some surprises.

First, a disclaimer: I cleave to the heretical (to philosophers) opinion that absolute precision can hinder communication. Absolute precision sometimes demands heaping so many details and complications onto an exposition as to make its dialectical core impenetrable to the uninitiated. Although in what follows I tell no outright lies (of which I am aware), I also consign complications to notes and offer somewhat impressionistic examples or argument sketches in lieu of exhaustive derivations. For readers hungry for more rigor, I provide references to works where they'll find it.[1] Readers already conversant with the notions here introduced might be happier skipping ahead.

4.1 What an algebra is

Intuitively, an algebra is just a collection of elements along with a way of taking their products and linear combinations. Insofar as it's the business of natural science to weave physical magnitudes into functional relationships, organizing physical magnitudes into an algebra underwrites that business. Officially,

Definition 4.1 (algebra). An algebra $\mathfrak{A}$ over the field $\mathbb{C}$ of complex numbers is a set of elements $(A, B, \ldots)$ with the following features:

---

[1] Kadison and Ringrose (1997a; 1997b) and Bratteli and Robinson (1987; 1997) are standard general references.

i. First, $\mathfrak{A}$ is closed under a commutative, associative operation $+$ of binary addition, with respect to which $\mathfrak{A}$ forms a group. The zero element 0 of $\mathfrak{A}$ is identity element of the group: i.e. $A + 0 = A$ for all $A \in \mathfrak{A}$.
ii. Second, $\mathfrak{A}$ is closed with respect to *scalar multiplication by complex numbers*.[2g]
iii. Finally, $\mathfrak{A}$ is closed with respect to a binary multiplication operation $\cdot$, which is associative and distributive with respect to addition, but not necessarily commutative. (In what follows, $A \cdot B$ will often be abbreviated AB.) If the algebra has a multiplicative identity—i.e. an element I such that $AI = IA = A$ for all $A \in \mathfrak{A}$—it is a *unital* algebra. Unless otherwise announced, all the algebras discussed here are unital.

Features (i) and (ii) reveal an algebra to be something well known to those acquainted with ordinary QM—a linear vector space over the complex numbers—which feature (iii) equips with a binary multiplication operation.

Example 4.2. The space $C(X)$ of all continuous, complex-valued functions f on the closed interval $X \in \mathbb{R}$ forms a unital algebra. Each function in $C(X)$ is bounded, because a continuous function f on a compact space attains a maximum value. On this collection of functions, addition and multiplication are defined *pointwise*. That is, $(f + g)(x) = f(x) + g(x)$ and $(fg)(x) = f(x)g(x)$. The additive and multiplicative identities are the constant functions $f(x) = 0$ and $f(x) = 1$, respectively. ♠

Example 4.3 ($\mathfrak{B}(\mathcal{H})$). Although a linear vector space, a Hilbert space isn't on its own an algebra. Its problem is that it lacks a suitable multiplication operation, a vector-valued function that takes ordered pairs of vectors as arguments. Scalar-valued, a Hilbert space inner product is *not* a multiplication operation of the relevant sort. But we don't have to look far to find examples of algebras in ordinary QM. Consider the set $\mathfrak{B}(\mathcal{H})$ of bounded operators on a Hilbert space $\mathcal{H}$. If $\hat{A}$ and $\hat{B}$ are elements of $\mathfrak{B}(\mathcal{H})$, so are $\hat{A} + \hat{B}$, $\hat{A}\hat{B}$, and $c\hat{A}$ for all $c \in \mathbb{C}$. $\mathfrak{B}(\mathcal{H})$ is closed under binary addition, binary multiplication, and multiplication by complex numbers, which operations have the features required by (i)–(iii). (The closure desideratum explains the restriction to *bounded* operators on $\mathcal{H}$: as we saw with the Heisenberg form of the CCRs, the products and sums of unbounded operators can fail to be well-defined on all of $\mathcal{H}$.) The projection operators associated with the null space and with $\mathcal{H}$ entire are the additive and multiplicative identities respectively. $\mathfrak{B}(\mathcal{H})$ is an algebra, a unital one. ♠

$\mathfrak{B}(\mathcal{H})$ is, moreover, a $*$ *algebra*.

---

[2] where this operation satisfies the usual requirements: $c_1(A + B) = c_1 A + c_1 B$, $(c_1 + c_2)A = c_1 A + c_2 A$, $c_1(c_2 A) = (c_1 c_2)A$, and $(c_1 A)B = A(c_1 B) = c_1(AB)$ for all $A, B \in \mathfrak{A}$ and $c_1, c_2 \in \mathbb{C}$.

Definition 4.4 (* algebra). A * algebra is an algebra $\mathfrak{A}$ closed under an *involution* $* : \mathfrak{A} \to \mathfrak{A}$ satisfying:

$$(A^*)^* = A, \quad (A+B)^* = A^* + B^*, \quad (cA)^* = \bar{c}A^*, \quad (AB)^* = B^*A^*$$

for all $A, B \in \mathfrak{A}$ and all complex c (where the overbar denotes complex conjugation).[3,g]

The adjoint operation † defines an involution for $\mathfrak{B}(\mathcal{H})$. In ordinary QM, an $\hat{A} \in \mathfrak{B}(\mathcal{H})$ such that $\hat{A}^\dagger = \hat{A}$ is *self-adjoint*, and represents an observable magnitude. This is generalized by algebraic QM, which deploys * algebras (without prejudice to whether they coincide with some $\mathfrak{B}(\mathcal{H})$ or not), and identifies their self-adjoint elements with observables.

The genus *algebra* encompasses many species individuated by conditions added to (i)–(iii). We've already encountered the species *unital*, which will be endemic to this work. A species evolved in classical environments is the *abelian* algebra, whose binary multiplication operation is commutative. More adapted to quantum environments is the species *C* algebra*, and its subspecies *von Neumann* (a.k.a. *W**) *algebra*. C^* and von Neumann algebras can be characterized either abstractly, as a kind of Banach space, or concretely, as duly structured sets of bounded Hilbert space operators. We will consider each characterization in turn.

4.2 C^* algebras, abstractly

The algebraic approach to quantum theories associates the physical magnitudes of a quantum theory with the (self-adjoint) elements of an algebra characteristic to that theory. A standard recipe for generating such an algebra is to start with the set of magnitudes satisfying the theory's CCRs (or CARs), form polynomials of those magnitudes, then close the resulting set, by adding to it the limit points of convergent sequences of elements it already contains.[4] This procedure yields an algebra that has the canonical magnitudes as its *generators*. The closure step reflects the fact that one sort of interweaving of physical magnitudes constitutes one magnitude—the global polarization of an infinite spin chain, say—as the limit of a sequence of other magnitudes. The closure step thereby presupposes a criterion of convergence for sequences, which the unadorned algebraic structure thus far presented fails to supply. To remedy this defect, we'll supplement the structure to obtain a type of algebra of extraordinarily wide application.

To turn an unadorned * algebra into a C^* algebra, we need to add a suitable *norm*, which is a function that assigns a non-negative real number to each element of the

---

[3] An involution on a space is generally a function that is its own inverse.

[4] Life isn't always that simple. Some varieties of closure—e.g. closure in the weak and strong topologies, discussed below—add limit points of generalized sequences called nets (see Kadison and Ringrose 1997a, 113–16, 304–9). I ignore this subtlety in the text.

algebra.[5,g] Familiar examples of norms include the *Hilbert space vector norm*, which maps each vector $|\psi\rangle$ in a Hilbert space to $\sqrt{\langle\psi|\psi\rangle}$, and the Euclidean length function, which maps a vector (a, b) in Euclidean two-space to $\sqrt{(a^2 + b^2)}$.

A space equipped with a norm is thereby equipped with the criterion of convergence (a.k.a., speaking somewhat loosely, a *topology*) induced by that norm: a sequence v_i of elements of the space converges to the element v in the norm induced by $\|\ \|$ iff the sequence of real numbers $\|v_n - v\|$ converges to 0 in the usual sense. Similarly, v_i is a *norm-wise Cauchy sequence* iff, for any $\epsilon > 0$, there is a natural number N_ϵ such that $\|v_i - v_j\| < \epsilon$ for all $i, j > N_\epsilon$. A space is *complete with respect to a norm* iff the limit of every norm-wise Cauchy sequence of elements of the space is also an element of the space.

Now we can say what a C^* algebra is.

Definition 4.5 (C^* algebra (abstract)). A C^* algebra $\mathfrak{A}$ is a $*$ algebra over $\mathbb{C}$, equipped with a norm, satisfying $\|A^*A\| = \|A\|^2$ and $\|AB\| \leq \|A\|\|B\|$ for all $A, B \in \mathfrak{A}$, with respect to which norm it is complete.

A normed vector space that is complete with respect to its norm is a *Banach space*. An algebra is a vector space, so a C^* algebra is a Banach space whose norm has the features just stated.

Construing physical magnitudes as elements of a C^* algebra, we have the wherewithal to understand some as polynomials and/or limits of sequences of others. We can, for instance, take the energy H of a free particle to be proportional to the *square* of its momentum p; or take an evolution operator e^{-iHt} to be the *limit* of a sequence of polynomials in H generated by a Taylor series expansion. Construing physical magnitudes as elements of a C^* algebra, we have the wherewithal to conduct the business of physics.

Example 4.6 (the C^* algebra $C(X)$). All the algebra $C(X)$ from Example 4.2 needs to be a C^* algebra are a properly behaved norm and an involution. The involution $f^* = \bar{f}$, and the norm $\|f\| = \sup_{x \in X} |f(x)|$, i.e. the maximum value $|f(x)|$ takes in the interval X, will do the trick. ♠

Example 4.7 (the C^* algebra $\mathfrak{B}(\mathcal{H})$). Perhaps a more familiar example of a C^* algebra is the set $\mathfrak{B}(\mathcal{H})$ of bounded operators on a separable Hilbert space $\mathcal{H}$. The adjoint operation on $\mathfrak{B}(\mathcal{H})$ provides the involution, and the Hilbert space operator norm,

$$\|A\| := \sup_{\sqrt{\langle\psi|\psi\rangle}=1} |A|\psi\rangle| \qquad (4.1)$$

---

[5] In more detail, a norm on a vector space V is map $\|\ \|$ from V into the non-negative reals satisfying $\|v\| = 0$ only if $v = 0$, $\|cv\| = |c|\|v\|$, $\|v\| + \|w\| \geq \|v + w\|$ for all $v, w \in V$ and all $c \in \mathbb{C}$.

(where $|\psi\rangle$ ranges over $\mathcal{H}$ and where $|\circ|$ is the vector norm on $\mathcal{H}$) provides the C^*-algebraic norm. Rendered somewhat loosely into prose, Eq. (4.1) says that the operator norm of an element $A \in \mathfrak{B}(\mathcal{H})$ is the "length" of the longest vector that results from having A act on unit (a.k.a. *normed*) vectors in $\mathcal{H}$. ♠

In a sense, these are the only examples of C^* algebras there are: every unital abelian C^* algebra is isomorphic to the algebra of continuous complex valued functions on a compact Hausdorff space such as X,[6] and every C^* algebra, abelian or not, is isomorphic to a norm-closed self-adjoint subalgebra of $\mathfrak{B}(\mathcal{H})$ for some $\mathcal{H}$ (Kadison and Ringrose 1997a, thms. 4.4.3 (270) and 4.5.6 (281)). Here "isomorphism" means a 1:1 map that preserves algebraic structure. More precisely:

Definition 4.8 (∗-morphism). A ∗-morphism from a C^* algebra $\mathfrak{A}$ to a C^* algebra $\mathfrak{D}$ is a map π such that $\pi(c_1 A + c_2 B) = c_1 \pi(A) + c_2 \pi(B)$, $\pi(AB) = \pi(A)\pi(B)$, and $\pi(A^*) = \pi(A)^*$ for all $A, B \in \mathfrak{A}$ and all $c_1, c_2 \in \mathbb{C}$. A ∗-*isomorphism* is 1:1 and onto.

A ∗-morphism from $\mathfrak{A}$ to $\mathfrak{D}$ takes positive elements of the former to positive elements of the latter. It is also continuous: if a sequence $A_i \in \mathfrak{A}$ converges in $\mathfrak{A}$'s C^* norm, then the sequence $\pi(A_i) \in \mathfrak{D}$ converges in $\mathfrak{D}$'s C^* norm (Bratteli and Robinson 1987, thm. 2.3.1 (42)). Since most of the algebras at hand are ∗ algebras, "∗-morphism" (and cognates) will often be abbreviated to "morphism" (and cognates) in the sequel.

A C^* algebra needn't be *presented* as an algebra of Hilbert space operators. In Example (4.6)'s presentation of the C^* algebra $C(X)$, no mention whatsoever was made of a Hilbert space. In what follows, I'll call a C^* algebra conceived in abstraction from its possible Hilbert space realizations *an abstract C^* algebra*.

4.3 C^* algebras, concretely

WARNING: To minimize notational clutter, we will hereupon suspend the behatting of Hilbert space operators when the context makes clear that the objects under discussion are Hilbert space operators.

Concretely characterized,

Definition 4.9 (C^* algebra (concrete)). A C^* algebra $\mathfrak{A}$ is a self-adjoint subalgebra of $\mathfrak{B}(\mathcal{H})$, for some $\mathcal{H}$, that is closed in $\mathcal{H}$'s uniform topology.

An algebra $\mathfrak{A}$ of concrete Hilbert space operators is self-adjoint iff $A \in \mathfrak{A}$ implies $A^\dagger \in \mathfrak{A}$. It's uniformly (also known as "norm") closed iff it contains the limit points of its norm-wise Cauchy sequences. These limit points are understood as follows:

---

[6] If, noting that there exist unital abelian C^* algebras that have projection operators as members, you're worried about how a space of continuous functions could ever include such projection operators, take heart: Hausdorff spaces can be *disconnected* and for X sufficiently disconnected, $C(X)$ can include projection operators. (Thanks to Mike Tamir for getting to the bottom of this.)

Definition 4.10 (uniform (a.k.a. norm) convergence). A sequence A_n of operators on $\mathcal{H}$ converges to an operator A on $\mathcal{H}$ in the *uniform* (also called the *norm*) *operator topology* iff:

$$\|(A_n - A)\| \to 0 \text{ as } n \to \infty \tag{4.2}$$

where $\|\circ\|$ is the Hilbert space operator norm (4.1).

While every $\mathfrak{B}(\mathcal{H})$ is a C^* algebra, the converse is not the case: a C^* algebra need only be (up to isomorphism) a *subalgebra* of $\mathfrak{B}(\mathcal{H})$. To see how some elements of $\mathfrak{B}(\mathcal{H})$ could be left out of the uniform closure of others, reflect that the uniform topology is relatively strict: very roughly speaking, uniform convergence demands that $|(A_n - A)|\psi\rangle|$ converge to 0 at "the same speed" for every $\psi \in \mathcal{H}$. It isn't easy to be a convergent sequence by the lights of the uniform topology. A Hilbert space sustains other operator topologies with looser admissions criteria. One is the *strong operator topology*:

Definition 4.11 (strong convergence). A sequence A_n of operators on $\mathcal{H}$ converges to an operator A on $\mathcal{H}$ in the *strong operator topology* iff for each $|\psi\rangle \in \mathcal{H}$:

$$|(A_n - A)|\psi\rangle| \to 0 \text{ as } n \to \infty \tag{4.3}$$

where $|\circ|$ is the vector norm on $\mathcal{H}$.

Notice the different roles "for all $|\psi\rangle \in \mathcal{H}$" plays in the criteria for uniform (4.2) and strong (4.3) convergence. In the former, the clause hides in the definition (4.1) of the operator norm $\|\circ\|$: for each n, the quantity $\|(A_n - A)\|$ is determined by considering all unit vectors $|\psi\rangle \in \mathcal{H}$. Thus *the* operator norm $\|\circ\|$ is a *single* map from operators to real numbers in which every vector in $\mathcal{H}$ conspires. It's the quantity $\|(A_n - A)\|$ that must approach 0 as n grows for uniform convergence to obtain. Strong convergence, by contrast, is a *point-wise*, or vector-by-vector, criterion of convergence. Each vector $|\psi\rangle$ in $\mathcal{H}$ determines (via the vector norm) a map, $|\circ|_\psi$ such that $|O|_\psi = |O|\psi\rangle|$ for all $O \in \mathfrak{B}(\mathcal{H})$, from operators to real numbers. Strong convergence obtains only if for each such map—that is, for each vector $|\psi\rangle$ in $\mathcal{H}$—$|(A_n - A)|_\psi$ goes to zero as n grows.

Like the strong topology, the weak operator topology employs a pointwise criterion of convergence. This time, a pair of vectors $|\psi\rangle, |\phi\rangle$ defines via $\mathcal{H}$'s inner product a map, call it $|\circ|_{\psi,\phi}$, from $\mathfrak{B}(\mathcal{H})$ to real numbers: $|O|_{\psi,\phi} = |\langle\psi|O|\phi\rangle|$ for each $O \in \mathfrak{B}(\mathcal{H})$. For each such map—that is, for each pair of vectors $|\psi\rangle, |\phi\rangle$—$|(A_n - A)|_{\psi,\phi}$ must go to zero as n grows for weak convergence to obtain.[7]

---

[7] As anticipated in fn. 4, the weak topology cares about the differences between sequences and nets, in the sense that the weak closure of a set can contain limits of nets in the set that aren't limits of sequences in the set. But I'm suppressing that complication, as well as a mare's nest of other topologies. See Bratteli and Robinson (1987, §2.4.1) for a disentanglement.

Definition 4.12 (weak convergence). A sequence A_n of operators on $\mathcal{H}$ converges to an operator A on $\mathcal{H}$ in the *weak operator topology* iff for all $|\psi\rangle, |\varphi\rangle \in \mathcal{H}$:

$$|\langle\psi|A_n - A|\varphi\rangle| \to 0 \text{ as } n \to \infty \tag{4.4}$$

We'll also have occasion to refer to the σ-weak (also known as the ultraweak) topology:

Definition 4.13 (ultraweak convergence). A sequence A_n of operators on $\mathcal{H}$ converges to an operator A on $\mathcal{H}$ in the *ultraweak operator topology* iff for each density operator ρ on $\mathcal{H}$

$$|Tr(\rho(A_n - A))| \to 0 \text{ as } n \to \infty \tag{4.5}$$

If a sequence converges uniformly, then it converges weakly, but not vice versa; if a sequence converges uniformly, then it converges strongly, but not vice versa. (Be warned that one encounters other forms of words that express the same claim. Thinking of a topology as a criterion of convergence, one might regard topologies as "stronger" the fewer sequences meet the criterion. Hence: the uniform topology is stronger than the strong topology, which in turn is stronger than the weak topology. Stronger topologies are also said to be "finer," and weaker ones "coarser.")

Here are some examples of sequences that converge in the strong, but not the uniform, topology. It's no accident that they are drawn from infinite-dimensional Hilbert spaces. Provided $\mathcal{H}$ is finite-dimensional, all the topologies just painstakingly distinguished coincide.[8]

Example 4.14 (countable additivity). Let $\mathcal{H}$ be the Hilbert space $\ell^2(\infty)$ of square summable countable sequences of complex numbers. And let $\{P_i\}$ be a complete orthonormal set of projection operators in $\mathcal{H}$. Then the sequence of partial sums $S_n = (P_1 + P_2 + \cdots + P_n)$ converges in the strong, but not the uniform, operator topology.

Strong convergence. To establish strong convergence, we'll show that S_n is a Cauchy sequence in the strong topology. According to Def. (4.11), this means that for each $|\psi\rangle \in \mathcal{H}$,

$$|(S_n - S_{n-1})|\psi\rangle| \to 0 \text{ as } n \to \infty \tag{4.6}$$

Now $S_n - S_{n-1}$ is just P_n. And $|\psi\rangle$ is just $\sum_i P_i|\psi\rangle$. In order to belong to $\ell^2(\infty)$, $|\psi\rangle$ must be square summable, that is, such that $\sum_i |P_i|\psi\rangle|^2 < \infty$. The sum will converge only if $|P_n|\psi\rangle|^2 \to 0$ as $n \to \infty$, and hence only if $|P_n\psi\rangle| \to 0$ as $n \to \infty$. But this is just the criterion for S_n to be a Cauchy sequence in the strong topology.

---

[8] If I may be permitted a parenthetical conjecture about the something like the sociology of our discipline: the reason philosophical treatments of quantum theory have (generally) ignored these operator topologies and the differences between them is because those treatments (generally) address foundational questions that can be raised in the context of finite-dimensional Hilbert spaces, where the differences between topologies don't matter. Indeed, for a while, the main game in philosophy of QM was to find the *smallest* Hilbert space in which a certain foundational point—a "no go" result—could be made.

But not uniform convergence. Strong convergence is a battle fought vector by vector: for each $|\psi\rangle \in \mathcal{H}$, we fix $|\psi\rangle$ and consider the behavior of $|(S_n - S_{n-1})|\psi\rangle|$ as $n \to \infty$. But we cannot fight a vector-by-vector battle to establish uniform convergence. The quantity $\|S_n - S_{n-1}\| = \|P_n\|$ that must approach 0 as $n \to \infty$ for S_n to be a Cauchy sequence in the uniform topology is a quantity that *for each n* is a supremum *over every normed vector in* $\mathcal{H}$ (see Def. (4.10)). This makes the battle a losing one. For each P_n, there's some unit vector $|\phi\rangle$ that lies in the subspace onto which P_n projects. (As n grows, this $|\phi\rangle$ can change, but there's always going to be such a $|\phi\rangle$, or $\mathcal{H}$ wouldn't be infinite-dimensional.) That is, there's some unit vector $|\phi\rangle$ such that $|P_n|\phi\rangle| = 1$. Thus, no matter what n is, the supremum over unit vectors of $|P_n|\phi\rangle|$ will be at least 1. It will also be at most 1, because P_n is a projection operator. No matter how much n grows, $\|S_n - S_{n-1}\|$ stays stubbornly fixed at 1, and spoils S_n's claim to be a Cauchy sequence in the uniform topology. ♠

The example illustrates a general truth: countable sequences of pairwise orthogonal projections converge strongly, but not uniformly. Recall that a state $\omega : \mathfrak{B}(\mathcal{H}) \to \mathbb{C}$ is defined to be countably additive just in case for any countable set $\{E_i\}$ of pairwise orthogonal projections in $\mathfrak{B}(\mathcal{H})$:

$$\omega(\sum_{i=1}^{\infty} E_i) = \sum_{i=1}^{\infty} \omega(E_i) \tag{4.7}$$

This condition adverts to the strong operator topology: in (4.7), $\sum_{i=1}^{\infty} E_i$ should be understood as the $n \to \infty$ limit, in the strong operator topology, of $\sum_{i=1}^{n} E_i$.

Example 4.15 (the polarization of an infinite spin chain). The discussion of the infinite spin chain in §3.3.1 harbors a sequence that converges in the weak topology of some Hilbert space, but not its uniform topology: the sequence that defines the z-component of the global polarization of the system. Consider the representation of the CARs on the Hilbert space $\mathcal{H}^+$, spanned by a base sequence s_k all of whose entries are 1 and all sequences differing from s_k in only finitely many entries. Section 3.3.1 showed that the sequence $\hat{m}(z)_N = \frac{1}{2N+1} \sum_{k=-N}^{N} \hat{\sigma}(z)_k$ has a limit as $N \to \infty$ in $\mathcal{H}^+$'s weak topology. That weak limit, an operator that had every unit vector in the Hilbert space as an eigenvalue one eigenvector, was none other than the identity operator $\hat{I}$ for $\mathcal{H}^+$. But the same sequence lacks a uniform limit. In order for $\hat{m}(z)_N$ to converge as $N \to \infty$ to $\hat{I}$ in $\mathcal{H}^+$'s uniform topology, it must be the case that (see Def. (4.10)):

$$\|\hat{m}(z)_N - \hat{I}\| := \sup_{\sqrt{\langle\psi|\psi\rangle}=1} |(\hat{m}(z)_N - \hat{I})|\psi\rangle|$$

approaches 0 as $N \to \infty$. But this won't happen. For each N, one can find a basis sequence s_k (one split as evenly as possible between $+1$'s and -1's) such that

$|\frac{1}{2n+1} \sum_{k=-N}^{N} \hat{m}(z)_k |s_k\rangle| \leq \frac{1}{N}$. This implies that $|(\hat{m}(z)_N - \hat{I})|s_k\rangle| \geq 1 - \frac{1}{N}$, and so that $\sup_{\sqrt{\langle\psi|\psi\rangle}=1} |(\hat{m}(z)_N - \hat{I})|\psi\rangle| \geq 1 - \frac{1}{N}$. This spoils uniform convergence. ♠

Example 4.16 (characteristic functions on $L^2(X)$). $L^2(X)$ is the Hilbert space of square integrable (with respect to the Lebesgue measure) complex-valued functions on the compact space X (Example (2.1)). The abelian C^* algebra $C(X)$ of continuous functions $f : X \to \mathbb{C}$ acts multiplicatively on $L^2(X)$: for $f \in C(X)$ and $|\psi\rangle = \psi(x) \in L^2(X)$, $f|\psi\rangle = f(x)\psi(x)$. The characteristic function $\chi_\Delta : X \to \mathbb{C}$ for the interval $\Delta \subset X$ is defined as follows:

$$\chi_\Delta(x) = 1 \text{ if } x \in \Delta$$

$$\chi_\Delta(x) = 0 \text{ otherwise} \qquad (4.8)$$

Notice that χ_Δ is a projection operator: $\chi_\Delta \chi_\Delta = \chi_\Delta = \chi_\Delta^\dagger$ (for a formal argument, see Kadison and Ringrose 1997a, ex. 2.5.12 (117)). Consider a family of continuous real-valued functions $\chi_\Delta^i \in C(X)$ that, intuitively, progressively approximate χ_Δ. Each of these χ_Δ^i's takes the value 0 outside (but not everywhere outside) Δ; each takes the value 1 inside (but not everywhere inside) Δ, and as n grows, the regions where the continuous function χ_Δ^n differs from χ_Δ shrink. Thus for each $x \in X$,

$$\lim_{n \to \infty} \chi_\Delta^n(x) = \chi_\Delta(x) \qquad (4.9)$$

Implying that the vector norms $|(\chi_\Delta^n - \chi_\Delta)|\psi\rangle|$ go to 0 as $n \to \infty$ for each $|\psi\rangle$ in the Hilbert space, this secures the convergence of χ_Δ^i to χ_Δ in $C(X)$'s strong operator topology.

However, these χ_Δ^i do not converge to χ_Δ in $C(X)$'s uniform operator topology. What's required for uniform convergence is:

$$\sup_{\sqrt{\langle\psi|\psi\rangle}=1} |(\chi_\Delta^n - \chi_\Delta)|\psi\rangle| \to 0 \text{ as } n \to \infty \qquad (4.10)$$

But for any n and any $\epsilon > 0$, we can find a unit vector ϕ in $L^2(X)$ such that $|(\chi_\Delta^n - \chi_\Delta)|\phi\rangle| > \epsilon$. Here's how. For simplicity, suppose $0 < \epsilon < 1$. For each n, there's going to be a subset Γ of X such that $\Gamma \cap \Delta = \emptyset$ and $\chi_\Delta^n(x) \geq \epsilon$ for all $x \in \Gamma$. If there weren't, $\chi_\Delta^n(x)$ wouldn't be continuous. It follows that $(\chi_\Delta^n(x) - \chi_\Delta(x)) \geq \epsilon$ for all $x \in \Gamma$. What ruins the uniform convergence of χ_Δ^i to χ_Δ is the supply of unit vectors $\phi \in L^2(X)$ such that:

$$\int_\Gamma \phi^* \phi \, dx = 1 \qquad (4.11)$$

For any such ϕ,

$$|(\chi_\Delta^n - \chi_\Delta)|\phi\rangle|^2 = \int_X (\chi_\Delta^n - \chi_\Delta)^2 \phi^* \phi \, dx \qquad (4.12)$$

Because $\phi(x) = 0$ outside of Γ, we can rewrite this as:

$$|(\chi_\Delta^n - \chi_\Delta)|\phi\rangle|^2 = \int_\Gamma (\chi_\Delta^n - \chi_\Delta)^2 \phi^* \phi \, dx \tag{4.13}$$

Because within Γ, $(\chi_\Delta^n - \chi_\Delta) > \epsilon$, and because ϕ is a unit vector,

$$|(\chi_\Delta^n - \chi_\Delta)|\phi\rangle|^2 > \int_\Gamma \epsilon^2 \phi^* \phi \, dx = \epsilon^2 \tag{4.14}$$

But this implies that $|(\chi_\Delta^n - \chi_\Delta)|\phi\rangle| > \epsilon$, which is what we needed to spoil uniform convergence. ♠

This example also illustrates a significant truth. The C^* algebra $C(X)$, acting on the Hilbert space $L^2(X)$ and closed in the uniform topology, excludes elements of $\mathfrak{B}(L^2(X))$: the projection operators (characteristic functions $\chi(\Delta)$). The closure of $C(X)$ in the weak topology, by contrast, captures those operators (along with every other measurable function; see Kadison and Ringrose 1997a, ex. 5.1.6 (308)).

Sequences of operators can converge strongly, or weakly, without converging in the uniform topology. Closed in the uniform topology, therefore, a concrete C^* algebra $\mathfrak{A}$ acting on a Hilbert space $\mathcal{H}$ might exclude elements of $\mathfrak{B}(\mathcal{H})$ that are the weak or strong limits of sequences of operators it does contain. We've just seen two such examples: the characteristic function χ_Δ is a bounded operator on the Hilbert space $L^2(X)$, but it's not in the uniform closure of the C^* algebra $C(X)$ acting on that Hilbert space. And the weak closure of a representation of the CARs for an infinite spin chain acting on $\mathcal{H}^+$ captures the polarization observable, but its uniform closure excludes that observable.

4.4 The representation of a C^* algebra

Austerity measures are a time-honored response to undesired non-uniqueness. Hole argument got you down? Then dispense with the physical properties distinguishing between solutions of the Einstein Field Equations related by hole diffeomorphisms. Fed up with skepticism about the external world? Then reduce physical properties (about which you might be wrong) to phenomenal ones (about which you can't be). Just so, an austere response to the surfeit of unitarily inequivalent Hilbert space realizations of the CCRs characterizing a theory of QM$_\infty$ might be to interpret that theory in terms of a structure *all Hilbert space realizations share*. That structure turns out to be the structure of an *abstract* C^* algebra. To articulate and compare responses to non-uniqueness, austere or otherwise, we need to understand the sorts of relations that obtain between such an algebra and the concrete Hilbert space quantizations whose structures it distills. We need the notion of the *representation* of a C^* algebra.

Characterized abstractly, a C^* algebra is a type of Banach space. Characterized concretely, it is a uniformly closed subalgebra of $\mathfrak{B}(\mathcal{H})$ for some $\mathcal{H}$. A C^* algebra that is conceived in abstraction admits a concrete Hilbert space representation.

4.4 THE REPRESENTATION OF A C^* ALGEBRA

Definition 4.17 (representation of a C^* algebra). A *representation* of a C^* algebra $\mathfrak{A}$ is a morphism (see Def. 4.8) $\pi : \mathfrak{A} \to \mathfrak{B}(\mathcal{H})$ from the C^* algebra into (although not necessarily onto) the algebra $\mathfrak{B}(\mathcal{H})$ of bounded operators on a Hilbert space $\mathcal{H}$.

Example 4.18 (representation of the C^* algebra $C(X)$). A concrete representation $(\pi, \mathcal{H})$ of $C(X)$ is obtained by letting $C(X)$ act multiplicatively on $L^2(X)$, the Hilbert space of square-integrable functions $\psi : X \to \mathbb{C}$. That is, $\pi(f), f \in C(X)$, is the operator on $L^2(X)$ that acts as follows on an arbitrary $|\psi\rangle \in L^2(X)$:

$$f|\psi(x)\rangle = |f(x)\psi(x)\rangle := |f(\psi)\rangle \tag{4.15} \spadesuit$$

$\pi(C(X))$ is a C^* algebra. It's a * algebra because the *-morphism π preserves *-algebraic structure. $L^2(X)$'s Hilbert space operator norm ($\|f\| = \sup_{\langle \psi|\psi\rangle = 1} \big||f(\psi)\rangle\big|$) supplies a C^* norm, and $\pi(C(X))$ is closed with respect to that norm because * morphisms are continuous. This illustrates a general fact.

Fact 4.19. A representation $\pi(\mathfrak{A})$ of a a C^* algebra $\mathfrak{A}$ is itself a C^* algebra (Bratteli and Robinson 1987, prop. 2.3.1 (42)).

Example 4.20 (some representations of the CAR algebra). (See §3.3.1.) Assign each site $k \in \mathbb{Z}$ in a doubly infinite chain a trio of spins $\sigma(x)_k, \sigma(y)_k, \sigma(z)_k$ obeying the Pauli relations (3.17). These spins generate a C^* algebra $\mathfrak{A}_{CAR}$ known as the *CAR algebra*. Conceived as an abstract C^* algebra, $\mathfrak{A}_{CAR}$ is isomorphic to the uniform closure of a Hilbert space representation of the extended Pauli relations for $n = \infty$.

π_+ *representation*. The representation π_+ maps each element of $\mathfrak{A}_{CAR}$ to a bounded operator on the Hilbert space $\mathcal{H}^+$, which is spanned by a basis consisting of a countable subset of maps from $\mathbb{Z}$ to ± 1. The basis vector $|s_0^+\rangle$ corresponds to the sequence $s_0^+ = \ldots, +1, +1, +1, \ldots$; other basis vectors correspond to sequences differing from s_0^+ in a finite number of sites. Where $[s]_j$ denotes the jth element of the sequence s, represent $\sigma(z)_n \in \mathfrak{A}_{CAR}$ by the bounded operator on $\mathcal{H}^+$ that acts as follows on an arbitrary basis vector:

$$\pi_+(\sigma(z)_n)|s\rangle = [s]_n|s\rangle \tag{4.16}$$

$\pi_+(\sigma(x)_n)$ and $\pi^+(\sigma(y)_n)$ cooperating with these to represent $\mathfrak{A}_{CAR}$ can be introduced. To simplify notation a bit, I'll write "$\sigma(z)_n^+$" for $\pi_+(\sigma(z)_n)$. Closed in $\mathcal{H}^+$'s uniform topology, $\pi_+(\mathfrak{A}_{CAR})$ is a concrete C^* algebra.

π_- *representation*. The representation π_- acts just like the representation π_+, but with the roles of $+1$ and -1 swapped. For example, $\mathcal{H}^-$ is spanned by a basis consisting of the doubly infinite sequence $s_0^- = \ldots, -1, -1, -1, \ldots$ and all sequences differing therefrom in a finite number of places; etc., etc. $\pi_-(\mathfrak{A}_{CAR})$ is another concrete C^* algebra. $\spadesuit$

4.4.1 Faith and reducibility

Definition 4.21 (faithful representation). A representation $(\pi, \mathcal{H})$ of $\mathfrak{A}$ is *faithful* iff $\pi(A) = 0$ implies $A = 0$ for all $A \in \mathfrak{A}$.

Where the C^* algebra $\mathfrak{A}$ is a subalgebra of $\mathfrak{B}(\mathcal{H})$, the natural inclusion mapping is a faithful representation of $\mathfrak{A}$ on $\mathcal{H}$. A representation fails to be faithful if it maps a non-zero element of the algebra to the zero operator on the representing Hilbert space. For instance, the representation $\pi_0 : \mathfrak{A} \to \mathfrak{B}(\mathcal{H})$ which maps each element of $\mathfrak{A}$ to the zero operator of $\mathcal{H}$ is monumentally unfaithful. Faithful representations are desirable because they don't sweep any of the original algebra's structure under the rug of triviality.

There is a sort of faithful representation that might seem to protest too much. It represents elements of the algebra over and over again as operators on the representing Hilbert space. Consider the "doubling" representation $\pi_D : \mathfrak{B}(\mathcal{H}) \to \mathfrak{B}(\mathcal{H} \oplus \mathcal{H})$ of the C^* algebra $\mathfrak{B}(\mathcal{H})$, which acts as follows on each $A \in \mathfrak{B}(\mathcal{H})$: $\pi_D(A) = A \oplus A$ (the Scholium on p. 39 covers the direct sum $\oplus$). The action of π_D restricted to the first summand of $\mathcal{H} \oplus \mathcal{H}$ is a representation of $\mathfrak{B}(\mathcal{H})$, as is π_D restricted to the second summand. Each $A \in \mathfrak{B}(\mathcal{H})$ is multiply represented in π_D: as A acting on the second summand of $\mathcal{H} \oplus \mathcal{H}$; as A acting on the first summand of $\mathcal{H} \oplus \mathcal{H}$; as $A \oplus A$ acting on the direct sum space. This excess makes the "doubling" representation *reducible*, in the sense that there is a non-trivial closed subspace of $\mathcal{H}$ that is invariant under $\pi(\mathfrak{A})$:

Definition 4.22 ((ir)reducible representation). A representation $(\pi, \mathcal{H})$ of $\mathfrak{A}$ is *irreducible* just in case the only closed subspaces of $\mathcal{H}$ that are invariant under $\pi(\mathfrak{A})$ are $\{0\}$ and $\mathcal{H}$. A representation is *reducible* if $\mathcal{H}$ has a non-trivial (that is, different from $\{0\}$ and $\mathcal{H}$) subspace invariant under $\pi(\mathfrak{A})$.[9] (Compare Def. 2.13.)

These are essentially the same notions of (ir)reducibility encountered in §2.3.4's explication of the Stone–von Neumann theorem. Here's a quick review of the notion of *subrepresentation*, introduced there: If a representation π of a C^* algebra $\mathfrak{A}$ in terms of bounded operators acting on a Hilbert space $\mathcal{H}_\pi$ is a reducible representation, there is a non-trivial closed subspace $\mathcal{K}$ of $\mathcal{H}_\pi$ invariant under the action of $\pi(\mathfrak{A})$. (In the "doubling" representation, $\mathcal{H} \oplus \mathcal{H}$ plays the role of $\mathcal{H}_\pi$; each of the summands of $\mathcal{H} \oplus \mathcal{H}$ is invariant under π_D.) By hypothesis of this invariance, $\pi(A) \in \mathfrak{B}(\mathcal{H}_\pi)$, the *representative* of an element A of $\mathfrak{A}$, maps vectors in $\mathcal{K}$ to vectors in $\mathcal{K}$. $\pi(A)$ thereby defines an operator on $\mathcal{K}$, known as "the restriction of $\pi(A)$ to $\mathcal{K}$" and written $\pi(A)|_\mathcal{K}$. The map $A \in \mathfrak{A} \to \pi(A)|_\mathcal{K}$ will furnish a representation of $\mathfrak{A}$ on $\mathcal{K}$ if π does on $\mathcal{H}$. The representation defined by restriction is a *subrepresentation* of π, call it $\sigma_\pi^\mathcal{K}$. Each of the summands of $\mathcal{H} \oplus \mathcal{H}$ carries a subrepresentation of the doubling representation. Notice that each subrepresentation is irreducible.

---

[9] Recall that a subspace $\mathcal{K}$ of $\mathcal{H}$ is invariant under $\pi(\mathfrak{A})$ if and only if for all $|\psi\rangle \in \mathcal{K}$ and all $A \in \mathfrak{A}$, $\pi(A)|\psi\rangle \in \mathcal{K}$.

4.4 THE REPRESENTATION OF A C^* ALGEBRA 85

For a subspace $\mathcal{K}$ of $\mathcal{H}$, let $\mathcal{K}^\perp$ denote its orthogonal complement. $\mathcal{K}^\perp$ contains all vectors orthogonal to every vector in $\mathcal{K}$, and $\mathcal{K} \oplus \mathcal{K}^\perp = \mathcal{H}$. Clearly, $\mathcal{K}^\perp$ is invariant under the action of π if $\mathcal{K}$ is. $\sigma_\pi^{\mathcal{K}^\perp}$, defined by restricting π's action to $\mathcal{K}^\perp$, is also a subrepresentation of π. Where $\{\mathcal{K}_a\}$ is a pairwise orthogonal family of closed subspaces of $\mathcal{H}_\pi$ such that (i) each $\mathcal{K}_a$ is invariant under π, and (ii) $\{\mathcal{K}_a\}$ spans $\mathcal{H}_\pi$, then π is a *direct sum* of its subrepresentations $\sigma_\pi^{\mathcal{K}_a}$. (In symbols $\pi = \oplus_a \sigma_\pi^{\mathcal{K}_a}$.) Any reducible representation π is a direct sum of subrepresentations (e.g. $\sigma_\pi^{\mathcal{K}^\perp}$ and $\sigma_\pi^{\mathcal{K}}$). (Verify that the doubling representation conforms to all of these claims.)

An irreducible representation has itself as its only (faithful) subrepresentation. Frugally, irreducible representations map no element of $\mathfrak{A}$ to more than one element of $\mathfrak{B}(\mathcal{H})$. Reducible representations, on the other hand, flout parsimony: in a reducible representation π of $\mathfrak{A}$, each element $A \in \mathfrak{A}$ has even more images than π has subrepresentations.

4.4.2 Unitary equivalence of representations

The notion of unitary equivalence generalizes straightforwardly to representations:

Definition 4.23. Two representations $(\pi_1, \mathcal{H}_1)$ and $(\pi_2, \mathcal{H}_2)$ of a C^* algebra $\mathfrak{A}$ are *unitarily equivalent* just in case there is an one-to-one invertible norm-preserving linear map $U : \mathcal{H}_1 \to \mathcal{H}_2$ such that $U\pi_1(A)U^{-1} = \pi_2(A)$ for all $A \in \mathfrak{A}$.

As Chapter 2 documents, we care about unitary equivalence because we're accustomed to think about quantum theories in terms that equate unitary equivalence with physical equivalence.

Although contrived, the following examples of unitarily inequivalent representations may be instructive.

Example 4.24. Let $\mathfrak{A}$ be a subalgebra of $\mathfrak{B}(\mathcal{H})$, where $\mathcal{H}$ is n-dimensional. The *identity representation* π_I of $\mathfrak{A}$ maps each $A \in \mathfrak{A}$ to itself. π_I and the direct sum representation $\pi_I \oplus \pi_I$ are both faithful representations of $\mathfrak{A}$. But they're not unitarily equivalent, because there is no unitary map between n-dimensional $\mathcal{H}$ and $2n$-dimensional $\mathcal{H} \oplus \mathcal{H}$. ♠

This example might give us pause. Chapter 2 trumpeted the adequacy of unitary equivalence as a criterion of physical equivalence for quantum theories. π_I and $\pi := \pi_I \oplus \pi_I$ are unitarily inequivalent representations of $\mathfrak{A}$. Proceedingly naively, because we haven't officially fixed terms for the discussion of such things, let's think of each as a quantum theory, whose observable set coincides with the self-adjoint elements of $\pi_I(\mathfrak{A})$ and $\pi(\mathfrak{A})$ respectively. Notwithstanding their unitary inequivalence, it may be rushing these theories to judgment to brand them baldly physically inequivalent. For their observable algebras $\pi_I(\mathfrak{A})$ and $\pi(\mathfrak{A})$ are isomorphic, which suggests that (once we equip the theories with a notion of state) the isomorphisms establishing their kinematic equivalence, kinematic* equivalence, and dynamical equivalence (Defs. 2.5, 2.9, and 2.10) won't be hard to come by. This isn't to say there's nothing to choose between the two representations. Parsimony, for instance, could favor irreducible π_I over reducible $\pi_I \oplus \pi_I$. Still, it might prompt worries that Chapter 2's use of unitary equivalence to explicate physical equivalence was somehow too blunt.

To assuage such worries, recall that unitary equivalence explicated physical equivalence for theories of *ordinary quantum mechanics*, which theories traffic in observable algebras isomorphic to $\mathfrak{B}(\mathcal{H})$. *The observable algebras of ordinary quantum mechanics are irreducible.* The worry-prompting example involves *reducible* observable algebras. A more general notion of physical equivalence for such algebras, one that includes unitary equivalence as a special case, will make its appearance in the next section.

The representation $\pi(\mathfrak{A})$ of a C^* algebra $\mathfrak{A}$ is closed in the uniform topology (Bratteli and Robinson 1987, prop. 2.3.1 (42)). But $\mathcal{H}$ makes available weaker topologies, i.e. topologies with respect to which sequences *that aren't uniformly convergent* might converge. Thus starting with $\pi(\mathfrak{A})$ and closing in a topology weaker than the uniform topology could eventuate in sets of observables richer than those contained in $\pi(\mathfrak{A})$ proper. Next we consider algebras obtained by following a recipe of this sort.

4.5 Von Neumann algebras

Concretely characterized,

Definition 4.25 (von Neumann algebra (topological)). A *von Neumann algebra* $\mathfrak{M}$ is a * algebra of bounded operators acting on some Hilbert space, which algebra is closed in the strong operator topology.[10]

Where π is a Hilbert space representation of a C^* algebra $\mathfrak{A}$, $\pi(\mathfrak{A})$ is a * algebra of bounded Hilbert space operators. One way to obtain a von Neumann algebra, then, is to close an algebra like $\pi(\mathfrak{A})$ in the strong operator topologies of the representation's Hilbert space.

Definition 4.26 (the commutant). Given an algebra $\mathfrak{D}$ of bounded operators on a Hilbert space $\mathcal{H}$, its *commutant* $\mathfrak{D}'$ is the set of all bounded operators on $\mathcal{H}$ that commute with every element of $\mathfrak{D}$. $\mathfrak{D}'$ is an algebra if $\mathfrak{D}$ is.

So, for example, the commutant of $\mathfrak{B}(\mathcal{H})$ consists of scalar multiples of the identity operator. Naturally enough, $\mathfrak{D}$'s *double commutant* $\mathfrak{D}''$ is $\mathfrak{D}'$'s commutant. Every element of $\mathfrak{B}(\mathcal{H})$ commutes with the identity operator, and so with every element of $\mathfrak{B}(\mathcal{H})'$. This makes $\mathfrak{B}(\mathcal{H})$ its own double commutant.

Here's why that's not a coincidence: von Neumann's celebrated double commutant theorem links the "topological" characterization of a von Neumann algebra in terms of closure with an "algebraic" characterization in terms of commutants. Von Neumann

---

[10] For more about von Neumann algebras, consult Kadison and Ringrose (1997a, chs. 5–6).

4.5 VON NEUMANN ALGEBRAS

showed that the strong and weak closures of a self-adjoint algebra $\mathfrak{D}$ of bounded Hilbert space operators coincide—and coincide as well with $\mathfrak{D}$'s double commutant. Thus,

Definition 4.27 (von Neumann algebra (algebraic)). A *von Neumann algebra* $\mathfrak{M}$ is a $*$-algebra of bounded operators such that $\mathfrak{M} = \mathfrak{M}''$.

We've just seen a splendid example: $\mathfrak{B}(\mathcal{H})$. Here is another noteworthy one.

Example 4.28 (maximal abelian subalgebra of $\mathfrak{B}(\mathcal{H})$). Suppose $\mathfrak{D} \subset \mathfrak{B}(\mathcal{H})$ is *maximal abelian*. That is, for all $A, B \in \mathfrak{D}, [A, B] = 0$, and no abelian subalgebra of $\mathfrak{B}(\mathcal{H})$ properly contains $\mathfrak{D}$. Then $\mathfrak{D} = \mathfrak{D}'$, which implies that $\mathfrak{D} = \mathfrak{D}''$. ♠

Non-example 4.29 ($C(X)$). The C^* algebra $C(X)$ acting on $L^2(X)$ is an abelian subalgebra of $\mathfrak{B}(L^2(X))$ that's not maximal abelian. The characteristic function $\chi_A \in \mathfrak{B}(L^2(X))$, for instance, clearly a member of both the commutant and the double commutant of $C(X)$, is not a member of $C(X)$, because it's not continuous. Hence $C(X)$ is not its own double commutant—which we knew already, because we knew that $C(X)$ wasn't weakly closed and we knew the double commutant theorem! ♠

Any set of bounded Hilbert space operators closed in the weak topology is closed in the uniform topology. So every von Neumann algebra is also a C^* algebra.

Let π be a representation of a C^* algebra $\mathfrak{A}$. $\pi(\mathfrak{A})''$, which according to the double commutant theorem is equivalent to $\pi(\mathfrak{A})$'s weak closure, is a von Neumann algebra, which I'll call *the von Neumann algebra affiliated with the representation* π. This maneuver enables us to transfer a relation of isomorphism, which holds (or not) between von Neumann algebras, to a relation, called *quasi-equivalence*, between the representations with which those von Neumann algebras are affiliated. The idea is that representations are quasi-equivalent whenever their affiliated von Neumann algebras are isomorphic. Officially,

Definition 4.30 (quasi-equivalence). Representations π and ϕ are quasi-equivalent if and only if there's a $*$-isomorphism α from $\phi(\mathfrak{A})''$ to $\pi(\mathfrak{A})''$ such that $\alpha[\phi(A)] = \pi(A)$ for all $A \in \mathfrak{A}$.

Unitarily equivalent representations are quasi-equivalent: the $*$-isomorphism is implemented unitarily. But unitary equivalence isn't necessary for quasi-equivalence: consider the von Neumann algebras $\mathfrak{B}(\mathcal{H})$ and $\mathfrak{B}(\mathcal{H}) \oplus \mathfrak{B}(\mathcal{H})$. Suppose $\mathcal{H}$ is finite-dimensional. The isomorphism $i: \mathfrak{B}(\mathcal{H}) \to \mathfrak{B}(\mathcal{H}) \oplus \mathfrak{B}(\mathcal{H}) : i(A) = A \oplus A$ establishing their quasi-equivalence is not a unitary map between $\mathcal{H}$ and $\mathcal{H} \oplus \mathcal{H}$. (This follows from a simple dimensional argument: if $dim(\mathcal{H}) = n$, $dim(\mathcal{H} \oplus \mathcal{H}) = 2n$, which disappoints hopes of a 1:1 map between them.) However, $\mathfrak{B}(\mathcal{H})$ is unitarily equivalent to each summand of $\mathfrak{B}(\mathcal{H}) \oplus \mathfrak{B}(\mathcal{H})$. Applied to representations and put roughly, quasi-equivalence is unitary equivalence up to multiplicity (see Bratteli and Robinson 1987, thm. 2.4.26 (80)). We will see presently that quasi-equivalent representations agree about which states are countably additive.

88   REPRESENTATION WITHOUT TAXATION

Digression 4.31 (von Neumann algebras, abstractly). Von Neumann algebras can also be characterized abstractly.[11] The dual B^* of a Banach space B is the vector space of linear functionals on B, equipped with the norm $\|f\| = \sup_{A \in B^+} |f(A)|$ (where B^+ is the unit ball of B, i.e. elements of norm 1 in B's uniform topology). Abstractly conceived,

Definition 4.32 (von Neumann algebra (abstract)). A von Neumann algebra $\mathfrak{M}$ is a C^* algebra such that there exists a Banach space M_* of which $\mathfrak{M}$ is the dual space (written $(M_*)^* = \mathfrak{M}$).

Here's a simple illustration:

Example 4.33. $\mathfrak{T}(\mathcal{H})$, the set of trace-class operators on $\mathcal{H}$, equipped with the norm $\|T\| = Tr(|T|)$, is a Banach space. The dual of $\mathfrak{T}(\mathcal{H})$ is $\mathfrak{B}(\mathcal{H})$: each element $A \in \mathfrak{B}(\mathcal{H})$ generates a linear functional a over $\mathfrak{T}(\mathcal{H})$ via $a(T) = Tr(AT)$ for all $T \in \mathfrak{T}(\mathcal{H})$ (Bratteli and Robinson 1987, prop. 2.4.3 (68)). A C^* algebra that's the dual of a Banach space, $\mathfrak{B}(\mathcal{H})$ is therefore a von Neumann algebra. ♠

The C^* algebra $\pi(\mathfrak{A})$ is closed in the uniform topology. The von Neumann algebra $\pi(\mathfrak{A})''$ is closed in the weak topology. The weak topology is more permissive than the uniform topology: nets and sequences of operators that don't converge uniformly may converge weakly. Thus the weakly closed von Neumann algebra $\pi(\mathfrak{A})''$ can contain bounded Hilbert space operators that do not make it into the uniformly closed C^* algebra $\pi(\mathfrak{A})$. As Ex. 4.16 might lead us to expect, projection operators provide an important example of elements excluded from C^* algebras but admitted to the von Neumann algebras that are their weak closures. A C^* algebra of bounded Hilbert space operators need only contain the projections 0 and I (the additive and multiplicative identities, respectively, for the algebraic operations), whereas von Neumann algebras are rich with projections. This richness abounds with interpretive consequences and constraints. For instance, projection operators are the basis of the classification of von Neumann algebras into strikingly different types (to be discussed in §7.1), only one of which has received much attention in the philosophy of ordinary QM. And a von Neumann algebra's projection operators form a *lattice* (to be discussed in §8.2.2) that promises to be a Rosetta Stone for the logic of quantum propositions.

Example 4.34 ($\pi(C(X))$ and $\pi(C(X))'' = \mathfrak{M}_f$). We've encountered the abstract C^* algebra $C(X)$ of continuous functions on a compact space X (Ex. 4.6); its representation $\pi(C(X))$ on the Hilbert space $L^2(X)$ (Ex. 4.18); and an argument that the weak closure of that representation is larger than its uniform closure (Non-ex. 4.29). Now let's put them all together. Where f is a bounded measurable function on X,

---

[11] Abstract von Neumann algebras are also called W^* *algebras*. Since every W^* algebra is isomorphic to a von Neumann algebra and conversely, I'll stick with eponymy.

let M_f be the operator on $L^2(X)$ corresponding to multiplication by f. (Notice that if f and g differ on a set of measure 0, M_f and M_g are the same operator on $L_2(X)$. So what $\mathfrak{M}_f$ really contains is equivalence (up to measure 0) classes of measurable functions.) The collection of such operators (with addition and multiplication defined pointwise) is a von Neumann algebra, call it $\mathfrak{M}_f$. $\mathfrak{M}_f$ coincides with $\pi(C(X))''$, and is a maximal abelian subalgebra of $\mathfrak{B}(L^2(X))$ (Kadison and Ringrose 1997a, ex. 5.1.6 (308)). $\mathfrak{M}_f = \pi(C(X))''$ includes characteristic functions for measurable subsets of X; these are its projection operators. These very operators are absent from $\pi(C(X))$, due to their discontinuity. ♠

What we've just seen: where $(\pi, \mathcal{H})$ is a representation of a C^* algebra $\mathfrak{A}$, the C^* algebra $\pi(\mathfrak{A})$ can be a proper subalgebra of the von Neumann algebra $\pi(\mathfrak{A})''$, which can in turn be a proper subalgebra of the von Neumann algebra $\mathfrak{B}(\mathcal{H})$.

Next, we muster the troops assembled so far to present the rudiments of the algebraic approach to QM.

4.6 States on C^* algebra

4.6.1 The GNS representation

The algebraic approach to quantum theories associates observables pertaining to a system with the self-adjoint elements of a C^* algebra $\mathfrak{A}$ appropriate to the system. The algebraic approach accommodates ordinary QM's picture of observables (see §2.1) as a special case: $\mathfrak{A}$ is set equal to $\mathfrak{B}(\mathcal{H})$ for some Hilbert space $\mathcal{H}$. Ordinary QM conceives states on $\mathfrak{B}(\mathcal{H})$ as normed, positive, countably additive linear functionals over that algebra. Provided $dim(\mathcal{H}) > 2$, Gleason's theorem places states, so conceived, in one-to-one correspondence with density operators ρ on $\mathcal{H}$ via the trace prescription.

By contrast, the algebraic approach to QM defines states directly in terms of the observable algebra $\mathfrak{A}$, which it does not constrain to coincide with some $\mathfrak{B}(\mathcal{H})$.

Definition 4.35 (state on a C^* algebra). An algebraic *state* ω on $\mathfrak{A}$ is a linear functional $\omega : \mathfrak{A} \to \mathbb{C}$ that is normed ($\omega(I) = 1$)) and positive ($\omega(A^*A) \geq 0$ for all $A \in \mathfrak{A}$).

Because $\omega(A)$ is real when A is self-adjoint, $\omega(A)$ may be understood as the expectation value of the observable (self-adjoint element) $A \in \mathfrak{A}$.

Here are two facts about states (in Def. 4.35's sense of normed, positive linear functionals) on a C^* algebra $\mathfrak{A}$:

Fact 4.36. A positive linear functional ω on a C^* algebra $\mathfrak{A}$ is (norm-)continuous. That is, if A_i converges to A in C^* norm, then $\omega(A_i)$ converges to $\omega(A)$ (Bratteli and Robinson 1987, prop. 2.3.11 (49)).

Fact 4.37. The set of normed positive linear functionals on a C^* algebra $\mathfrak{A}$ is convex. (I.e. if ω_1 and ω_2 are states, so is $\omega = \lambda\omega_1 + (1-\lambda)\omega_2$ for all $\lambda \in [0, 1]$.) Its extremal elements—that is, states ω which cannot be expressed as non-trivial convex combinations of other states—are pure states; all other states are mixed (Bratteli and Robinson 1987, corr. 2.3.12 (51)).

Unlike the states of ordinary QM, algebraic states are not required to be countably additive. The comment after Ex. 4.7 anticipates why: the requirement of countable additivity presupposes the strong operator topology, which is used to define a countably infinite sum of pairwise orthogonal projections. A concrete C^* algebra $\mathfrak{A}$ is guaranteed only to be uniformly closed. Thus $\mathfrak{A}$ may not contain the strong limits of elements it does contain, and so may not contain all the elements in terms of which the countable additivity requirement is stated. In a C^* algebraic setting, a requirement of countable additivity won't always be well posed. What's more, in the most general case, the requirement would be otiose. The projection operators guaranteed to appear in $\mathfrak{A}$ are the additive and multiplicative identities 0 and I. With respect to that set of projection operators, countable additivity reduces to linearity and positivity.

Because von Neumann algebras are strongly closed, a von Neumann algebra containing a countably infinite set of pairwise orthogonal projection operators will also contain the operator defined in terms of a strong limit as their sum. Such a von Neumann algebra thereby contains the entities in terms of which a countable additivity requirement is stated. So given a state—in the present and future sense of a normed, positive linear functional—on a von Neumann algebra $\mathfrak{M}$, we can ask a *further* question of it: does it define a probability measure over projection operators in $\mathfrak{M}$ that's countably additive? States on $\mathfrak{M}$ for which the answer is affirmative are known as "normal."

Definition 4.38 (normal state on a von Neumann algebra). A state ω on a von Neumann algebra $\mathfrak{M}$ is a normal state *with respect to that algebra* iff for each countable set $\{E_i\} \in \mathfrak{M}$ of pairwise orthogonal projection operators, $\omega(\sum_i E_i) = \sum_i \omega(E_i)$.

Gleason's theorem, concerning as it does countably additive probability measures, is a theorem (not about states in general but) about normal states. Section 2.1's "normal conception of states" as density operator states was aptly named: states on $\mathfrak{B}(\mathcal{H})$ conforming to the normal conception just are normal states with respect to $\mathfrak{B}(\mathcal{H})$. States in the algebraic sense (hereafter, simply "states") are immune to the ravages of Gleason's theorem.

The von Neumann algebra $\mathfrak{B}(\mathcal{H})$ can admit states other than the *normal* states determined by density operators in $\mathfrak{B}(\mathcal{H})$.

Example 4.39 (states and countable additivity). Here are two ways to characterize algebraic states on the von Neumann algebra $\mathfrak{M}_f(X)$, which acts on $L^2(X)$ (Ex. 4.34):

1. "normally": each unit vector $\psi \in L^2(X)$ determines a state ω_ψ on $\mathfrak{M}_f(X)$ via

$$\omega_\psi = \int_X \psi^*(x) f(x) \psi(x) dx \quad \text{for all } f \in \mathfrak{M}_f(X) \tag{4.17}$$

2. "pointedly": where λ is a point in the interval X, let's attempt to devise a state that assigns "located at λ" probability 1. Call our sought-after state ω_λ. The element of $\mathfrak{M}_f(X)$ answering to the proposition "located at λ" is the characteristic function χ_λ for the point λ. But because the point λ is a set of measure 0, χ_λ coincides with the 0 operator. Thus it will not do to characterize ω_λ as a state that maps χ_λ to 1, because no state does. Following Halvorson (2001), our characterization of ω_λ will be circuitous, and mediated by the notion of a state "converging" to a point. Let Δ_i be a countable set of ever-shrinking but always measurable subintervals of X centered at λ. (That is, $\Delta_i \subset \Delta_j$ if $i > j$; $\lambda \in \Delta_i$ for all i. For convenience, we will assume that $\Delta_0 = X$.) Then we will say that ω_λ *converges to* λ if it acts as follows on $\mathfrak{M}_f(X)$:

$$\text{If } \Delta_i \subseteq \Gamma \text{ for some } i \text{ then } \omega_\lambda(\chi_\Gamma) = 1. \tag{4.18}$$

Notice that this only partially defines ω_λ's action on $\mathfrak{M}_f(X)$: it tells us how ω_λ acts on the von Neumann algebra $\mathfrak{B}$ generated by the projection operators for the intervals Δ_i. Because ω_λ's action is normed, linear, and positive, ω_λ counts as a state on $\mathfrak{B}$. And the following fact assures us that there is a state on $\mathfrak{M}_f(X)$ that agrees with ω_λ on $\mathfrak{B}$:

Fact 4.40. Let $\mathfrak{B}$ be a C^* subalgebra of a C^* algebra $\mathfrak{A}$, and let ω be a state on $\mathfrak{B}$. Then there exists a state $\tilde{\omega}$ on $\mathfrak{A}$ that extends ω (i.e. $\tilde{\omega}$ and ω agree on $\mathfrak{B}$). (Bratteli and Robinson 1987, prop. 2.3.24 (60))

Now ω_λ isn't a normal state on $\mathfrak{M}_f(X)$. We'll show this by producing a countably infinite set of pairwise orthogonal projection operators in $\mathfrak{M}_f(X)$ with respect to which ω_λ violates countably additivity. Let $X = [0, 1]$, $\lambda = \frac{1}{2}$, and $\Delta_n \subset [0, 1]$ be the open interval $(\frac{1}{2} - \frac{1}{2^n}, \frac{1}{2} + \frac{1}{2^n})$. And let E_n be the projection operator for the interval Δ_n/Δ_{n+1}. That is, $E_n := \chi_{\Delta_n} - \chi_{\Delta_{n+1}}$. The countable set of projection operators E_n is pairwise disjoint. Moreover, as n goes to infinity, the sequence $S_n = \sum_{i=0}^{n} E_i$ converges in $L^2[0, 1]$'s strong operator topology to the identity operator. So ω_λ is countably additive only if

$$\omega_\lambda(I) = \lim_{n \to \infty} \sum_{i}^{n} \omega_\lambda(S_n) \tag{4.19}$$

But this condition fails: the l.h.s. must be 1, because ω_λ is a state, but the r.h.s. is 0, because for *each* S_n, $I = S_n + \chi_{\Delta_{n+1}}$. We know from (4.18) that $\omega_\lambda(\chi_{\Delta_{n+1}}) = 1$. Because ω_λ is linear, $\omega_\lambda(S_n) = 0$. Failing to be countably additive, the pointed state ω_λ fails to be normal with respect to $\mathfrak{M}_f(X)$.[12]

To endow the foregoing with physical significance you can think of the collection of characteristic functions χ_Δ as providing the spectral resolution of an

---

[12] For elaboration of the example, see Halvorson (2001).

ordinary QM position observable: a system whose state is given by $\psi \in L^2(X)$ has probability $\int_X \psi^* \chi_\Delta \psi \, dx$ of being located in Δ. Then ω_λ on $\mathfrak{M}_f(X)$ is an *exact position eigenstate* in the sense that it assigns *located in* Δ probability 1 if λ lies in Δ. ♠

Connections straightforward and surprising can be established between the sorts of states familiar from ordinary QM and states on C^* algebras, even abstract ones. Straightforwardly, a density operator ρ acting on a Hilbert space $\mathcal{H}$ carrying a representation $\pi : \mathfrak{A} \to \mathfrak{B}(\mathcal{H})$ of an algebra $\mathfrak{A}$ defines a state ω on $\mathfrak{A}$. Simply set $\omega(A) = \text{Tr}(\rho \pi(A))$ for all $A \in \mathfrak{A}$. What is more surprising is that we can travel in the other direction, from a state on an abstract algebra to its realization on a concrete Hilbert space. A definition will start us on our way:

Definition 4.41 (cyclic vector). $|\xi\rangle \in \mathcal{H}$ is cyclic with respect to the representation $\pi : \mathfrak{A} \to \mathfrak{B}(\mathcal{H})$ if and only if $\{\pi(\mathfrak{A}) |\xi\rangle\}$ is *dense* in $\mathcal{H}$. That is, any vector $|\phi\rangle \in \mathcal{H}$ can be approximated to arbitrary precision by linear combinations of elements of $\{\pi(\mathfrak{A}) |\Psi\rangle\}$.

So, for a schematic example, a set equipped with a norm is dense in its Cauchy completion. A less schematic example is furnished by the rational numbers, which are dense in the real numbers.

We are ready to see how a state on even an abstract C^* algebra determines a concrete Hilbert space representation of that algebra.

Fact 4.42 (GNS Theorem). Let ω be a state on a C^* algebra $\mathfrak{A}$. Then there exists a Hilbert space $\mathcal{H}_\omega$, a representation $\pi_\omega : \mathfrak{A} \to \mathfrak{B}(\mathcal{H}_\omega)$ of the algebra, and a cyclic vector $|\xi_\omega\rangle \in \mathcal{H}_\omega$ such that, for all $A \in \mathfrak{A}$, the expectation value the state ω assigns the algebraic element A is duplicated by the expectation value the vector $|\xi_\omega\rangle$ assigns the Hilbert space operator $\pi(A)$. In symbols, $\omega(A) = \langle \xi_\omega | \pi_\omega(A) | \xi_\omega \rangle$ for all $A \in \mathfrak{A}$. The triple $(\mathcal{H}_\omega, \pi_\omega, |\xi_\omega\rangle)$ is a *cyclic representation* because it contains a cyclic vector. It is unique up to unitary equivalence: that is, if $(\mathcal{H}, \pi)$ is a representation of $\mathfrak{A}$ containing a cyclic vector $|\psi\rangle$ such that $\omega(A) = \langle \psi | A | \psi \rangle$, then $(\mathcal{H}, \pi)$ and $(\mathcal{H}_\omega, \pi_\omega)$ are unitarily equivalent.

$(\mathcal{H}_\omega, \pi_\omega, |\xi_\omega\rangle)$ is called *the GNS representation* of the state, for Gel'fand, Naimark, and Segal, who showed how to construct it. The basic idea is simple. Recall that an algebra is a linear vector space and a Hilbert space is a linear vector space outfitted with an inner product, with respect to which it is complete. The key to the GNS construction is to build the Hilbert space $\mathcal{H}_\omega$ from the algebra $\mathfrak{A}$, with the state ω supplying the inner product: for $A, B \in \mathfrak{A}$, considered as vectors in $\mathcal{H}_\omega$, $\langle A | B \rangle := \omega(A^*B)$. In the GNS representation, $\mathfrak{A}$ acts on itself by left multiplication (i.e. for $B \in \mathfrak{A}$ and $|A\rangle \in \mathcal{H}_\omega$, $\pi(B)|A\rangle = |BA\rangle$) and the identity element I of $\mathfrak{A}$ plays the role of the cyclic[13] vector $|\xi_\omega\rangle$ implementing ω: $\langle I|A|I\rangle = \omega(IAI) = \omega(A)$. I am suppressing many complications; for these, see Bratteli and Robinson (1987, §2.3.3).

In ordinary QM, all and only pure states are vector states. But whether a state ω on an algebra $\mathfrak{A}$ is pure or mixed, it is implemented by a vector in its GNS representation. How can this be? The brusque and precise answer is:

Fact 4.43. A state ω on a C^* algebra $\mathfrak{A}$ is pure iff its GNS representation is irreducible (Bratteli and Robinson 1987, thm. 2.3.19 (57)).

Sketching part of the proof of this fact may usefully consolidate a number of the notions introduced to date:

"Argument" 4.44 (if π_ω is reducible, then ω is mixed). Let $\mathfrak{A}$ be a C^* algebra and suppose ω is a state on $\mathfrak{A}$ whose GNS representation $(\mathcal{H}_\omega, \pi_\omega, |\xi_\omega\rangle)$ is reducible. This means that $\mathcal{H}$ has a closed subspace invariant under the action of $\pi_\omega(\mathfrak{A})$. Call this subspace $\mathcal{H}_1$. Its orthogonal complement, the closed subspace $\mathcal{H}_1^\perp$, is also invariant under $\pi_\omega(\mathfrak{A})$. Now define a subrepresentation π_ω^1 of π_ω as follows: for each $A \in \mathfrak{A}$, $\pi_\omega^1(A)$ is the restriction to $\mathcal{H}_1$ of $\pi_\omega(A)$—i.e. the operator that does to each vector in $\mathcal{H}_1$ just what $\pi_\omega(A)$ did to that vector. Define a subrepresentation $\pi_\omega^\perp$ likewise. Let P_1 and $P_\perp$ be projection operators whose ranges are $\mathcal{H}_1$ and $\mathcal{H}_1^\perp$ respectively. Because π_ω^1 and $\pi_\omega^\perp$ are themselves representations of $\mathfrak{A}$, each of the following maps on every $A \in \mathfrak{A}$ defines a state on $\mathfrak{A}$:

$$\omega^1(A) = \frac{\langle \xi_\omega | \pi_\omega^1(A) | \xi_\omega \rangle}{|\langle \xi_\omega | P_1 | \xi_\omega \rangle|} \qquad \omega^\perp(A) = \frac{\langle \xi_\omega | \pi_\omega^\perp(A) | \xi_\omega \rangle}{|\langle \xi_\omega | P_\perp | \xi_\omega \rangle|}$$

We can express ω as a convex combination of ω^1 and $\omega^\perp$:

$$\omega = |\langle \xi_\omega | P_1 | \xi_\omega \rangle| \omega^1 + |\langle \xi_\omega | P_\perp | \xi_\omega \rangle| \omega^\perp$$

ω is mixed if and only if $\omega^1 \neq \omega^\perp$. Suppose, for reductio, that $\omega^1 = \omega^\perp$. Then $\omega^1 = \omega$, and $(\mathcal{H}_1, \pi_\omega^1, P_1|\xi_\omega\rangle)$ will be a GNS representation of ω. Here's why: it follows from how we've identified π_ω^1 and P_1 that if $\omega = \omega_1$, then $\omega(A) = \langle P_1 \xi | \pi_\omega^1(A) | P_1 \xi \rangle$, and $P_1|\xi\rangle$ must be cyclic for $\pi_\omega^1(\mathfrak{A})$ if $|\xi\rangle$ is cyclic for $\pi_\omega(\mathfrak{A})$. But then ω's GNS representation would not be unique up to unitary equivalence, for $(\mathcal{H}_1, \pi_\omega^1)$ and $(\mathcal{H}, \pi_\omega)$ can't be unitarily equivalent when $\mathcal{H} = \mathcal{H}_1 \oplus \mathcal{H}_1^\perp$ and neither summand is trivial. Thus $\omega^1 \neq \omega^\perp$ and ω is a mixed state on $\mathfrak{A}$. (For short argument for the converse, see Landsmann 1998, thm. 2.2.3.) ♠

The informative but metaphorical explanation of how the GNS representation of a mixed state can implement that state as vector state invokes the facts that vector states behave like mixed states in the presence of superselection rules (cf. §2.3.4) and that reducible representations have superselection sectors, a.k.a. subrepresentations. In light

---

[13] $|I\rangle$ is cyclic because $\{\pi(\mathfrak{A})\}|I\rangle$ is just $\mathfrak{A}$, and so dense in $\mathcal{H}_\omega$, which is $\mathfrak{A}$'s completion in the norm induced by ω.

of these facts, the reducibility of the GNS representation of a mixed algebraic state ω has the following significance: the vector state $|\xi_\omega\rangle$ behaves like a mixed state because it has components in each of π_ω's subrepresentations.

One might be tempted by the foregoing to suppose that, just as every mixed state is a convex combination of pure states, every reducible representation is a direct sum of irreducible representations. As we will see in Chapter 7, the assumption is not only wrong but falsified by many representations pertinent to QM_∞. The truth is:

Fact 4.45. Every non-degenerate[14]g representation of a C^* algebra $\mathfrak{A}$ is a direct sum of cyclic representation of $\mathfrak{A}$ (Bratteli and Robinson 1987, prop. 2.3.6 (46)).

Given that for any non-zero element A of a C^* algebra $\mathfrak{A}$, there exists a pure state ω on $\mathfrak{A}$ such that $\pi_\omega(A) \neq 0$ (Kadison and Ringrose 1997a, prop. 4.5.5 (280)), it follows from this and the mechanics of the GNS construction that:

Fact 4.46. Each C^* algebra has a faithful representation.

In particular, the representation of $\mathfrak{A}$ that is the direct sum, over pure states on $\mathfrak{A}$, of their GNS representations is faithful (Kadison and Ringrose 1997a, thm. 4.5.5 (281)).

4.6.2 Foliums

There is a sort of kinship question we might want to raise about GNS representations. Obviously, a state ω on a C^* algebra $\mathfrak{A}$ can, by the agency of the cyclic vector $|\xi_\omega\rangle$, be expressed in terms of ω's GNS representation. But what other states on $\mathfrak{A}$ can be expressed in terms of $(\mathcal{H}_\omega, \pi_\omega)$? In a clumsy and uninteresting sense, every state on $\mathfrak{A}$ can be so expressed, because every state on $\mathfrak{A}$ corresponds (via $\omega(\pi_\omega(A)) := \omega(A)$; π_ω needs to be faithful to ensure that the l.h.s. is well-defined) to a normed, positive linear functional over $\pi_\omega(\mathfrak{A})$. So we'd better refine our question to make it interesting. Which algebraic states on $\mathfrak{A}$ (and thus on $\pi_\omega(\mathfrak{A})$) admit natural and well-behaved extensions to the von Neumann algebra $\pi_\omega(\mathfrak{A})''$ affiliated with ω's GNS representation? It is intuitively plausible that states on $\mathfrak{A}$ (and thus on $\pi_\omega(\mathfrak{A})$) whose extensions to $\pi_\omega(\mathfrak{A})''$ are unique will be states with (roughly speaking) nice continuity properties with respect to the topologies used to construct $\pi_\omega(\mathfrak{A})''$ from $\pi_\omega(\mathfrak{A})$. Ultraweak continuity is one nice continuity property.

Definition 4.47. A state ω on a von Neumann algebra $\mathfrak{M}$ acting on a Hilbert space $\mathcal{H}$ is *ultraweakly continuous* iff for any sequence $A_i \in \mathfrak{M}$ that converges in $\mathcal{H}$'s ultraweak operator topology (Def. 4.13) to an operator $A \in \mathfrak{M}$, $\omega(A) = \lim_{n\to\infty} \omega(A_n)$.

Ultraweakly continuous states on $\mathfrak{M}$ have another familiar virtue: the demand for countably additivity is well posed for states on von Neumann algebras; ultraweakly continuous states on $\mathfrak{M}$ meet that demand.

---

[14] A representation is non-degenerate iff it has no subrepresentation that is the 0 representation.

Fact 4.48. A state ω on a von Neumann algebra is normal with respect to that algebra iff it's ultraweakly continuous (cf. Def. 4.13 and Bratteli and Robinson 1987, §2.4.3).

The following example features a state on a von Neumann algebra whose continuity properties are wicked.

Example 4.49 (normal states and continuity). Recall the CAR algebra and its representations π_+ and π_- (Ex. 4.20). Fix attention on the von Neumann algebra $\pi_+(\mathfrak{A}_{CAR})''$ affiliated with the former. Each normed vector in $\mathcal{H}^+$ generates a state in the algebraic sense on $\pi_+(\mathfrak{A}_{CAR})''$. BUT so does s_0^-, the basis vector in the π_- representation corresponding to the sequence all of whose entries are -1. Corresponding to s_0^- is a state $\omega_- : \pi_+(\mathfrak{A}_{CAR})'' \to \mathbb{C}$ that acts as follows on $\pi_+(\mathfrak{A}_{CAR})$: For all $A \in \mathfrak{A}_{CAR}$:

$$\omega_-(\pi_+(A)) = \langle s_0^- | \pi_-(A) | s_0^- \rangle \tag{4.20}$$

In particular, ω_- assigns each site in the infinite spin chain a value of -1 for the z-component of spin. Of course, $\pi_+(\mathfrak{A}_{CAR})''$ is bigger than $\pi_+(\mathfrak{A}_{CAR})$, and (4.20) doesn't say how ω_- acts on elements of latter that aren't elements of the former, e.g. $m(z)^+$. Fact 4.40 alerts us that ω_- admits an extension $\tilde{\omega}_-$ to $\pi_+(\mathfrak{A}_{CAR})''$. Such an extension $\tilde{\omega}_-$ can do whatever it wants to those added elements, provided it acts linearly, maps positive elements of $\pi_+(\mathfrak{A}_{CAR})''$ to non-negative numbers, and maps the identity to 1. $\tilde{\omega}_-$'s action on the complement of $\pi_+(\mathfrak{A}_{CAR})$ in $\pi_+(\mathfrak{A}_{CAR})''$ is not constrained by its action on $\pi_+(\mathfrak{A}_{CAR})$. In particular, $\tilde{\omega}_-$ on $\pi_+(\mathfrak{A}_{CAR})''$ is not *ultraweakly continuous*.

$\tilde{\omega}_-$'s defiance of ultraweak continuity is lurking in (4.20), and facts disclosed in §3.3.1 about the sequence of polarization observables m_z^N for subchains N sites long—a sequence which (the reader should assure herself) converges ultraweakly to the global polarization observable $m(z)^+$. The expectation value of the polarization observable m_z^N is just the average, over spins in the chain, of their spin-z expectation values. So (4.20) implies that for each m_z^N, $\tilde{\omega}(m_z^N) = -1$. In $\mathcal{H}^+$'s ultraweak topology, $\lim_{n \to \infty} m_z^N = I$. And because $\tilde{\omega}_-$ is a state, $\tilde{\omega}_-(I) = 1$. Thus m_z^N converges ultra weakly to I, but $\tilde{\omega}_-(m_z^N)$ does not converge to $\tilde{\omega}_-(I)$. $\tilde{\omega}$ is a state on $\pi_+(\mathfrak{A}_{CAR})''$, but not an ultraweakly continuous (that is, normal) one. ♠

A generalization of Gleason's theorem characterizes normal states of a von Neumann algebra in more familiar terms:

Fact 4.50 (normal state of a von Neumann algebra). A state ω on a von Neumann algebra $\mathfrak{M}$ acting on a Hilbert space $\mathcal{H}$ is normal iff there is a density operator ρ in $\mathfrak{B}(\mathcal{H})$ such that $\omega(A) = Tr(\rho A)$ for all $A \in \mathfrak{M}$ (Bratteli and Robinson 1987, thm. 2.4.21 (76–7)).

Remark, for future reference, that ρ needn't belong to $\mathfrak{M}$. For a Hilbert space $\mathcal{H}$ of dimension greater than 2, the normal states on the von Neumann algebra $\mathfrak{B}(\mathcal{H})$ are just the countably additive states on that algebra.

Given a representation π of a C^*-algebra $\mathfrak{A}$, we can now identify states on $\mathfrak{A}$ that admit well-behaved extensions to the von Neumann algebra $\pi(\mathfrak{A})''$ affiliated with that representation:

Definition 4.51 (π-normal states). Let $(\pi, \mathcal{H})$ be a representation of a C^* algebra $\mathfrak{A}$. A state ω on $\mathfrak{A}$ is π-*normal state* iff there is a density operator ρ in $\mathfrak{B}(\mathcal{H})$ such that $\omega(A) = Tr(\rho\pi(A))$ for all $A \in \mathfrak{A}$.

Where π_ω is the GNS representation determined by a state ω, π_ω-normal states are exactly those expressible as density matrices on ω's GNS representation. This is the very kinship relationship we sought. Grouping states on $\mathfrak{A}$ into clans whose members cohabit as density operator states on the same Hilbert space representation of $\mathfrak{A}$ carves the space of states on $\mathfrak{A}$ into *folium*s.

Definition 4.52 (folium). Where $\mathcal{S}$ is the set of positive linear functionals on a C^* algebra $\mathfrak{A}$, a *folium* $\mathcal{F}$ is a subset of $\mathcal{S}$ that is closed under convex combinations; is complete in $\mathcal{S}$'s uniform topology, wherein $\|s\| := \sup_{A \in \mathfrak{A}^+} s(A)$;[15]g and such that if φ lies in $\mathcal{F}$ and $A \in \mathfrak{A}$, then $\varphi_A := \varphi(A^*BA)$ for all $B \in \mathfrak{A}$ also lies in $\mathcal{F}$. ♠

Informally, a folium is a closed linear subspace of $\mathcal{S}$ invariant under something like Lüders' conditionalization.

We can use states to foliate $\mathcal{S}$ as follows:

Definition 4.53 (folium of a state). Let ω be a state on a C^* algebra $\mathfrak{A}$. ω's *folium* $\mathcal{F}_\omega$ is the set of all states expressible as density matrices on ω's GNS representation $(\pi_\omega, \mathcal{H}_\omega)$. Equivalently, $\mathcal{F}_\omega$ is the set of all π_ω-normal states on $\mathfrak{A}$.

Conversely, for any folium $\mathcal{F}$ there is some representation π of $\mathfrak{A}$ such that $\mathcal{F}$ contains all and only π-normal states. In a pinch, letting π be the direct sum over elements of $\mathcal{F}$ of their GNS representations will do.

Call states ω and φ on $\mathfrak{A}$ unitarily equivalent just in case their GNS representations are. Now if states ω and φ are unitarily equivalent, there's a one-to-one correspondence between states on $\pi_\omega(\mathfrak{A})$ expressible as density matrices on $\mathcal{H}_\omega$ and states on $\pi_\varphi(\mathfrak{A})$ expressible as density matrices on $\mathcal{H}_\varphi$. That is to say,

Fact 4.54. If ω and φ are unitarily equivalent, their folia coincide: $\mathcal{F}_\omega = \mathcal{F}_\varphi$.

Consider, by contrast, states ω and φ on $\mathfrak{A}$ whose folia have null intersection. No state on $\mathfrak{A}$ expressible as a density matrix on ω's GNS representation is so expressible on φ's GNS representation, and vice versa.

Definition 4.55 (disjoint states). States ω and φ on a C^* algebra $\mathfrak{A}$ such that $\mathcal{F}_\omega \cap \mathcal{F}_\varphi = \emptyset$ are *disjoint*.

---

[15] $\mathfrak{A}^+$ consists of elements of $\mathfrak{A}$ of unit C^* norm.

"Disjointness," a relation that can obtain between algebraic states, radicalizes the relation of orthogonality that can obtain between vectors implementing Hilbert space states. For the sake of simplicity, let's consider only pure states for this paragraph and the next. Like orthogonal states, disjoint states assign one another a "transition probability" of 0.[16] I'll say that quantum states (of any stripe) are *impossible relative to* one another just in case the transition probability between them is 0. Like orthogonal states, disjoint states are impossible relative to one another.

Disjointness radicalizes orthogonality in the following sense. If states $|\phi\rangle$ and $|\psi\rangle$ on $\mathfrak{B}(\mathcal{H})$ are orthogonal, there exists a host of pure states possible relative to both. Normed superpositions of $|\phi\rangle$ and $|\psi\rangle$ are examples. By contrast, if ω and ρ on $\mathfrak{A}$ are disjoint, no state possible relative to one is possible relative to the other. Here's an argument sketch: states possible relative to ρ lie in ρ's folium; states possible relative to ω lie in ω's folium; those folia have null intersection. Orthogonal states declare one another impossible. Disjoint states hold one another at an even greater distance: every state one of a pair of disjoint states declares possible, the other declares impossible.

We now have enough material on hand to discharge the obligation, incurred in Chapter 2, to motivate supplementing the criterion (ke) for the physical equivalence of theories (Def. 2.4). Modeling a theory as a pair consisting of a family of states and a collection of magnitudes in the domain of those states, (ke) requires of physically equivalent theories only that there exist bijections ι_s between those theories' sets of states and ι_q between those theories' collections of magnitudes, which bijections "preserve expectation values" (i.e. $\iota_s(\omega)(\iota_q(A)) = \omega(A)$ for all magnitudes A and states ω). Now, let $\mathfrak{A}$ be the algebra generated by the Weyl relations for an infinite system, and let ω and ϕ be *disjoint* pure states on it whose GNS representations of the Weyl relations are continuous. Consider two following quantum theories. The first (a.k.a. *the ω theorist*) identifies $\mathcal{F}_\omega$ as the one true folium of states, and $\pi_\omega(\mathfrak{A})''$ (which just is $\mathfrak{B}(\mathcal{H}_\omega)$) as the observable algebra; the second (a.k.a. *the ϕ theorist*) assigns $\mathcal{F}_\phi$ and $\pi_\phi(\mathfrak{A})'' = \mathfrak{B}(\mathcal{H}_\phi)$ these roles. Due to the continuity of the GNS representations π_ω and π_ϕ, $\mathcal{H}_\omega$ and $\mathcal{H}_\phi$ will be separable infinite-dimensional Hilbert spaces. Every separable infinite dimensional Hilbert space is isomorphic to every other, so (where $\mathcal{H}$ is that Hilbert space) each theorist takes states to be density operators on $\mathcal{H}$ and observables to be self-adjoint bounded operators on $\mathcal{H}$. The representations π_ϕ and π_ω *organize* those operators differently, but that's not an issue to which (ke) is sensitive, since it makes no mention of algebraic structuring. Now there's a very simple pair of bijections ι_s and ι_q between the theorists' observable sets and their state families satisfying (ke): two identity mappings. But it is premature to declare

---

[16] In algebraic QM, the transition probability between states ρ and ω on an observable algebra $\mathfrak{A}$ is given by $1 - \frac{1}{4}\|\omega - \rho\|^2$, where $\| \circ \|$ is the norm on the space of linear functionals on $\mathfrak{A}$. (See Roberts and Roepstorff 1969 for a suspiciously operationalist argument that transition probabilities, so defined, between mixed states are "unphysical." Although I'm not persuaded by the argument, its existence is part of why it's simpler to confine attention to pure states.) This quantity is 0 when ω and ρ are disjoint.

the theories physically equivalent on this account. For one thing, the cyclic vector $|\xi\rangle$ implementing ω in the ω theory gets mapped to a π_ϕ-normal state $\iota_s(|\xi\rangle)$ that corresponds to a state on $\mathfrak{A}$ *different from* ω (otherwise the folia $\mathcal{F}_\omega$ and $\mathcal{F}_\phi$ would intersect, contrary to the assumption that they're disjoint). Something gets lost in the translation manual (ke) sets up between the two theories. This something can be restored by supplementing (ke) with the demand that ι_q preserve the algebraic structure of each theory's collection of magnitudes. The criterion (KE) (Def. 2.5) results.

Provided we confine our attention to pure states/irreducible representations, pairs of those states/representations are either unitarily equivalent or disjoint. But when we extend our scope to include mixed states/reducible representations, the situation is more complex. Suppose that pure ω and mixed ρ are states on $\mathfrak{A}$, and that ρ's GNS representation π_ρ is an n-fold direct sum of ω's irreducible GNS representation π_ω: $\pi_\rho = \oplus_n \pi_\omega$. $\pi_\omega(\mathfrak{A})$ is unitarily equivalent to each of π_ρ's subrepresentation, but not to $\pi_\rho(\mathfrak{A})$ itself. Yet π_ρ and π_ω aren't disjoint: any state expressible as a density matrix on one is expressible as a density matrix on the other and vice versa. The folia of ω and ρ coincide. We need a notion of equivalence, finer-grained than unitary equivalence and suited to deal with the subtleties of subrepresentations, to capture this significant equivalence.

The discerning reader will have realized that we already have one, in the form of *quasi-equivalence*. States are quasi-equivalent when their GNS representations are. Quasi-equivalence turns out to coincide with the kinship relation that grouped states into folia according to the representations with respect to which they were normal:

Fact 4.56. States ω and φ on a C^* algebra $\mathfrak{A}$ are quasi-equivalent iff $\mathcal{F}_\omega = \mathcal{F}_\varphi$.

The coincidence of the folia of quasi-equivalent states has a simple explanation: von Neumann algebras affiliated with quasi-equivalent states are $*$-isomorphic; thus a state is normal on one such von Neumann algebra iff it's normal on the other.

Two handy rules governing the quasi-equivalence/disjointness of representations are (see Bratteli and Robinson 1987, 370):

Fact 4.57. τ and π are quasi-equivalent iff τ has no subrepresentation disjoint from π, and vice versa.

Fact 4.58. τ and π are disjoint iff they have no quasi-equivalent subrepresentations.

Because τ and π can be the GNS representations of states, these rules extend straightforwardly to the quasi-equivalence and disjointness of states on a C^* algebra.

4.7 Von Neumann factors

In general, the failure of two representations to be quasi-equivalent does *not* imply that they are disjoint. For their folia can intersect non-trivially without coinciding.

Consider disjoint irreducible representations π and φ, and the representation τ obtained by taking their direct sum. π and τ are not disjoint, because π is unitarily (and so quasi-) equivalent to a subrepresentation (π!) of τ. But neither are π and τ quasi-equivalent, for τ has a subrepresentation (φ) disjoint from π.

In the case, however, of *factor* representations—which crop up all over QM_∞—pairs of representations are either quasi-equivalent or disjoint. We'll work up to this result through a series of definitions.

Definition 4.59 (center of a von Neumann algebra). The *center* $\mathfrak{Z}_\mathfrak{M}$ of a von Neumann algebra $\mathfrak{M}$ consists of its intersection with its own commutant: $\mathfrak{Z}_\mathfrak{M} := \mathfrak{M} \cap \mathfrak{M}'$.

Definition 4.60 (factor von Neumann algebra). A *factor* von Neumann algebra $\mathfrak{R}$ is one whose center is trivial, in the sense that $\mathfrak{R} \cap \mathfrak{R}'$ contains only multiples of the identity. Example of a factor: the von Neumann algebra $\mathfrak{B}(\mathcal{H})$.

By extension, a representation π of a C^* algebra $\mathfrak{A}$ is a *factor representation* when its affiliated von Neumann algebra $\pi(\mathfrak{A})''$ is a factor, and a state ω on $\mathfrak{A}$ is a *factor state* when its GNS representation is a factor representation. Irreducible representations are factor representations—so all pure states are factor states. A reducible representation is a factor representation only if all of its subrepresentations are quasi-equivalent. Because quasi-equivalence is an equivalence relation, if any subrepresentation of one factor representation is quasi-equivalent to any subrepresentation of another factor representation, *every* subrepresentation of the first is quasi-equivalent to *every* subrepresentation of the second. The representations are quasi-equivalent. If two factor representations fail to be quasi-equivalent, no subrepresentation of one is quasi-equivalent to any subrepresentation of the other. That is, they're disjoint. Thus:

Fact 4.61. Factor representations are either quasi-equivalent or disjoint.

4.8 Conclusion and précis

The distinctions just laboriously drawn simply don't matter to ordinary QM. To turn the CARs for n degrees of freedom into a C^* algebra, find operators acting on a Hilbert space $\mathcal{H}$ to satisfy those CARs, then close under the formation of polynomials. Call the C^* algebra thereby obtained, CAR_n. CAR_n coincides with the von Neumann algebra $\mathfrak{B}(\mathcal{H})$ that is the n-fold tensor product of $\mathfrak{M}(2)$—the algebra of 2×2 complex matrices—with itself. In this instance of ordinary QM, "C^* algebra" and "von Neumann algebra" are just highfalutin' words for the collection of observables already in play—good old $\mathfrak{B}(\mathcal{H})$. What's more, it follows from the Jordan–Wigner theorem that every algebraic state on CAR_n is quasi-equivalent to every other. There is only one folium of algebraic states, those expressible as density operators on $\mathcal{H}$. States in the algebraic and ordinary senses coincide, and the radical specter of disjointness fades away.

Modulo some complications, similar conclusions hold for systems described by CCRs for n degrees of freedom.[17] Algebraic states on the relevant C^* algebra cohabit in a single folium, the folium of states implemented by density operators on the Schrödinger representation. Algebraic concepts add little (beyond, perhaps, a potential for pleonasm) to ordinary QM.

Algebraic concepts add a lot to the investigation of QM_∞, and will be repeatedly applied in the balance of this book. This chapter closes with a capsule review of ground covered so far.

A C^* **algebra** $\mathfrak{A}$ is an algebra with a involution and a norm. Although its elements needn't be bounded operators on a concrete Hilbert space, for any C^* algebra $\mathfrak{A}$, there's a Hilbert space $\mathcal{H}$ such that $\mathfrak{A}$ is isomorphic to a self-adjoint subalgebra of $\mathfrak{B}(\mathcal{H})$ closed in the uniform operator topology. One way to obtain a C^* algebra is to start with a collection of Hilbert space operators satisfying some interesting commutation relations, form polynomials of those operators, then take the uniform closure.

If we close in the possibly more permissive weak operator topology, we obtain a **von Neumann algebra** $\mathfrak{M}$, a weakly closed self-adjoint subalgebra of some $\mathfrak{B}(\mathcal{H})$. A von Neumann algebra $\mathfrak{M}$ can also be characterized algebraically: it is its own double commutant. That is, $\mathfrak{M} = \mathfrak{M}''$.

A **state** on a C^* algebra $\mathfrak{A}$ is a normed, positive, linear functional $\omega : \mathfrak{A} \to \mathbb{C}$. It is not required to be countably additive. A state ω determines, via the **GNS representation**, a triple $(\pi_\omega, \mathcal{H}_\omega, |\xi_\omega\rangle)$. π_ω is a **representation**, i.e. an algebraic-structure-preserving map $\pi_\omega : \mathfrak{A} \to \mathfrak{B}(\mathcal{H}_\omega)$, $\mathcal{H}_\omega$ is a Hilbert space, and $|\xi_\omega\rangle$ is a cyclic vector in $\mathcal{H}_\omega$ that reproduces ω's action on $\mathfrak{A}$ in the sense that $\langle \xi_\omega | \pi_\omega(A) | \xi_\omega \rangle = \omega(A)$ for all $A \in \mathfrak{A}$. This representation is unique up to unitary equivalence, and irreducible iff ω is pure.

$\pi_\omega(\mathfrak{A})''$ is a von Neumann algebra, the **von Neumann algebra affiliated with** ω's **GNS representation** π_ω. In general, $\pi_\omega(\mathfrak{A}) \subseteq \pi_\omega(\mathfrak{A})'' \subseteq \mathfrak{B}(\mathcal{H}_\omega)$. Because representations preserve algebraic structure, given any pair of faithful states ω and ϕ on $\mathfrak{A}$, $\pi_\omega(\mathfrak{A})$ and $\pi_\varphi(\mathfrak{A})$ will always be isomorphic C^* algebras. This isomorphism needn't extend to their affiliated von Neumann algebras.

Representations [states] π_ω and π_φ [ω and φ] are **unitarily equivalent** iff there's a unitary map $U : \mathcal{H}_\omega \to \mathcal{H}_\varphi$ such that $U\pi_\omega(A)U^{-1} = \pi_\varphi(A)$ for all $A \in \mathfrak{A}$. Representations [states] π_ω and π_φ [ω and φ] are **quasi-equivalent** iff there's an algebraic-structure preserving map $\alpha : \pi_\omega(\mathfrak{A})'' \to \pi_\varphi(\mathfrak{A})''$ such that $\alpha(\pi_\omega(A)) = \pi_\varphi(A)$ for all $A \in \mathfrak{A}$. Unitary equivalence is a special case of quasi-equivalence.

Quasi-equivalence can also be characterized by way of the notion of the **folium** of an algebraic state. The folium $\mathcal{F}_\omega$ of a state ω consists of all states on $\mathfrak{A}$ that can be implemented by density operators on ω's GNS representation. (These are also called

---

[17] The main complications are that the relevant C^* algebra, the *Weyl* algebra, is generated by Hilbert space operators satisfying the Weyl relations; that the closure step includes adding limit points; and that admissible states on this algebra are those whose GNS representations are continuous (see §2.3.4).

π_ω-**normal states**. They're countably additive with respect to the von Neumann algebra $\pi(\mathfrak{A})''$.) Representations [states] π_ω and π_φ [ω and φ] are **quasi-equivalent** iff $\mathcal{F}_\omega = \mathcal{F}_\varphi$, in which case the von Neumann algebras affiliated with the states are quasi-equivalent. If $\mathcal{F}_\omega \cap \mathcal{F}_\varphi = \emptyset$, representations [states] π_ω and π_φ [ω and φ] are **disjoint**.

5
Axioms for QM$_\infty$

> Just what is a quantum field theory... is a difficult question, since at present what we have, after thirty years of intensive effort, is a collection of partially heuristic technical developments in search of a theory; but it is a natural one to examine axiomatically.
>
> (Segal 1959, 341)

Axiomatic approaches to quantum theories press the apparatus introduced in the last chapter into service. This chapter opens with a declaration of sentiment concerning axiomatic approaches to theories of QM$_\infty$, then proceeds to sketch the axioms. Some of the most profound work done in the foundations of quantum field theory (QFT) unfolds in the axiomatic setting and continues discussions of locality, causality, and entanglement begun in the context of ordinary QM. The chapter closes with a semi-tendentious review of some of that work.

5.1 A declaration of sentiment

Philosophers of science are rarely guilty of accurate natural history of science. Few actual scientific theories are happily regarded as interpreted systems of axioms formulated in a first-order language; few functioning scientific laws resemble "All ravens are black;" few laboratories house infinitely extended Stern–Gerlach magnets. A partial explanation for the mismatch between sciences "in the field" and the sciences portrayed in philosophers' texts is that standard philosophical treads don't always find traction in the field. Given a theory T, Chapter 1 suggested, we confront the exemplary interpretive question of how exactly to establish a correspondence between T's models and worlds possible according to T. That is, we confront that question *if* T is the sort of thing that has models. "A collection of partially heuristic technical developments" isn't readily attributed a set of models about whose underlying ontology or principles of individuation philosophical questions immediately arise. This isn't to say that "a

5.1 A DECLARATION OF SENTIMENT

collection of partially heuristic technical developments" is unworthy of philosophical attention. It is in itself a philosophically provocative circumstance that such a collection can enjoy stunning empirical success.

The stunningly successful QFTs developed and employed by working physicists are calculationally fecund but (at present) mathematically rather suspect. Although typically lacking an explicit and well-defined Hilbert space setting, they set up calculations appropriate to Hilbert space operators and Hilbert space states. In the course of these calculations, characteristic infinities emerge, infinities which (what seems to some) the black art of renormalization spirits away. It hardly follows that physicists' QFTs are conceptually hamstrung or philosophically sterile.[1] Nor does it follow that physicists' QFTs will *never* admit expression mathematically explicit enough to engage the gears of the sort of interpretive inquiry sketched in Chapter 1. It is, after all, hardly unusual for a physical theory to debut before the mathematical tools adequate to its precise expression are fully developed.[2] But it does follow that a wide class of questions philosophers are wont to ask are difficult to pose with respect to physicists' QFTs.

Fortunately for philosophers, there is a tradition, dating to the early 1950s, of characterizing QFTs *axiomatically*. On the axiomatic approach, what a QFT *is* is something that satisfies a set of axioms governing associations between observable algebras and regions of spacetime. QFTs so understood are *eo ipso* the sort of model-theoretic entities philosophers can sink their teeth into. As Halvorson and Müger put it, "[Axiomatic QFT] is our best story about where QFT lives in the mathematical universe, and so is a natural starting point for foundational inquiries" (2007, 731–2). Fraser (2009) makes an even stronger recommendation: the interpretation of QFT should be based on axiomatic QFT, *rather than* physicists' QFTs, whose idealizations and lack of rigor muddy the interpretive waters.

QFTs understood axiomatically are less palatable to working physicists. With a few exceptions (such as the ϕ_2^4 theory, a carefully tuned (even "toy") interacting QFT set in two dimensions (see §11.1.3))[3], only *non-interacting* QFTs have been shown to satisfy standard axioms. The interacting QFTs that are the bread and butter of particle physics aren't (or aren't at present known to be) even examples of QFTs, according to axiomatic approaches.

Because the interpretive questions in which I'm interested are most readily addressed to theories admitting mathematically explicit formulations, most of the theories discussed in this book will be of mathematically explicable sort. (Non-linear, interacting theories lacking explicit and rigorous Hilbert space realizations will make appearances when, as in Chapters 11 and 14, their features are relevant to the arguments pursued.)

---

[1] For defenses of the conceptual cohesion of physicists' QFTs, see Wallace (2006) or t'Hooft (2007); for philosophical engagement with physicists' QFTs, see Teller (1995, chs. 6 and 7).

[2] Camp (2004) gives a real philosopher's take on this.

[3] Another exception is Gross–Neveu theory in two spacetime dimensions (Rivasseau 2003), and hopes exist for constructive non–abelian gauge theories in four spacetime dimensions (ibid.). I am grateful to Doreen Fraser for alerting me to these developments.

My attitude toward the axioms of axiomatic approaches is circumspect: I will use them to delimit a class of theories of QM_∞, a class from which I'll repeatedly draw illustrations meant to dramatize interpretive issues. We will see that there are mathematically explicit candidate theories of QM_∞ that violate pertinent axiom sets. I take there to be genuine interpretive questions about the physical status of these mavericks. It is an open question whether the class of axiom-satisfying theories of QM_∞ includes even *all* the physically significant, mathematically explicit theories of QM_∞ there are. So for now, the axioms laid on in this chapter have only the status of potential *constraints* on the interpretation of QM_∞.

5.2 Axioms for QFT

Axiomatic algebraic approaches to QFT set up an association between open bounded regions of a spacetime $(\mathcal{M}, g_{ab})$ (assumed to be time-orientable)[4] and C^* algebras of observables, subject to axioms expressing natural desiderata. Haag and Kastler's (1964) axioms were formulated with Minkowski spacetime in mind; Dimock's (1980) generalize these.[5] My exposition will start with Dimock's axioms, and then proceed to interesting special cases.

Axiomatic QFT associates with each open, bounded region $\mathcal{O} \subset \mathcal{M}$ a *local algebra* $\mathfrak{A}(\mathcal{O})$, a C^* algebra whose self-adjoint elements are the observables pertaining (in some sense) to the region $\mathcal{O}$. The adamantly operationalist original axiomatizers establish the association between regions and their local algebras by the interpretive maneuver of identifying elements of $\mathfrak{A}(\mathcal{O})$ with observables *measurable* by means of actions confined to the region $\mathcal{O}$.[6] But the association between local algebras $\mathfrak{A}(\mathcal{O})$ and regions $\mathcal{O}$ needn't be mediated by the notion of measurement or ideologies totemizing that notion. For instance, $\mathfrak{A}(\mathcal{O})$ can be understood as the algebra generated by a representation of the Weyl relations for the subspace $S_\mathcal{O}$ of the solution space S of the underlying classical theory spanned by solutions with compact support in $\mathcal{O}$.[7] Understanding the association this way leaves open the questions of how or whether

---

[4] The spacetime scholium starts on p. 65. $(\mathcal{M}, g_{ab})$ is time-orientable just in case there's a consistent way to choose which lobes of the lightcones associated with each point of $\mathcal{M}$ are future-directed. See Wald (1984, 189).

[5] For more on axiomatic approaches, see Araki (1999, ch. 4); Haag (1992). Different axiomatizations foreground different desiderata. For an illuminating axiomatization in which translation symmetries play a central role, see Halvorson and Müger (2007, §2). For a recent approach to axioms for QFT in curved spacetime, see Hollands and Wald (2010).

[6] "We have interpreted the elements of $\mathfrak{A}(\mathcal{O})$ as representing physical operations performable within $\mathcal{O}$" (Haag 1992, 105). Haag and Kastler (1964, 848) and Haag (1992, 2–4) offer additional statements of the operationalist creed.

[7] Briefly, the means of generation is the symplectic structure $\Omega_\mathcal{O}$ with which the classical theory's Lagrangian equips $S_\mathcal{O}$. The pair $(S_\mathcal{O}, \Omega_\mathcal{O})$ determines a set of Weyl relations; the algebra $\mathfrak{A}(\mathcal{O})$ is the norm closure of a representation of these Weyl relations. For details, see Kay and Wald (1991, 72–3).

to further interpret the association; the operationalism of the original axiomatizers is one interpretive option.

The axioms pertain to a QFT defined as map from open bounded regions $\mathcal{O} \subset \mathcal{M}$ to C^* algebras $\mathfrak{A}(\mathcal{O})$. (The definition is indifferent to whether these algebras $\mathfrak{A}(\mathcal{O})$ are abstract C^* algebras or concrete von Neumann algebras. Interpreters of QFT won't share this indifference.) The first axiom requires algebras associated with spacetime regions to reduplicate the inclusion relations between those regions:

1. **Isotony axiom.** Where $\mathcal{O}_1$ and $\mathcal{O}_2$ are open bounded regions of $\mathcal{M}$, if $\mathcal{O}_1 \subset \mathcal{O}_2$ then $\mathfrak{A}(\mathcal{O}_1)$ is a sub-algebra of $\mathfrak{A}(\mathcal{O}_2)$.

Isotony enables us to introduce the *quasilocal algebra* $\mathfrak{A}(\mathcal{M})$, which is the closure (in the C^* norm) of the union over all open bounded regions of $\mathcal{M}$ of those regions' local algebras. In symbols, $\mathfrak{A}(\mathcal{M}) = \overline{\cup_{\mathcal{O} \subset \mathcal{M}} \mathfrak{A}(\mathcal{O})}$. $\mathfrak{A}(\mathcal{M})$ is the algebra for all of spacetime. (We still might be able to enlarge it, by closing in topologies more permissive than that afforded by the C^* norm.) And so I will call a state on the algebra $\mathfrak{A}(\mathcal{M})$ a *global state*.

An automorphism of a C^* algebra $\mathfrak{A}$ is an isomorphism from $\mathfrak{A}$ to itself. An isometry of the spacetime $(\mathcal{M}, g_{ab})$ is a symmetry of the metric g_{ab}; that is, a map from $\mathcal{M}$ to itself that preserves the metric. Not all spacetimes have non-trivial isometries. When a spacetime does, its isometries can be composed to obtain transformations that are also isometries. That is to say, the isometries of a spacetime form a group. An important example is the Poincaré group for Minkowski spacetime, whose isometries include translations, rotations, reflections, and Lorentz boosts. The next axiom demands that the isometries of a spacetime be reflected in automorphisms of that spacetime's quasi-local algebra.[8]

2. **Covariance axiom** (covariance). Where Γ is the isometry group of $(\mathcal{M}, g_{ab})$, there is a group $G = \{\alpha_\gamma, \gamma \in \Gamma\}$ of automorphisms of $\mathfrak{A}(\mathcal{M})$ such that $\alpha_\gamma(\mathfrak{A}(\mathcal{O})) = \mathfrak{A}(\gamma \mathcal{O})$ for all $\mathcal{O} \subset \mathcal{M}$ and all $\gamma \in \Gamma$.

Covariance demands that the algebra $\mathfrak{A}(\gamma \mathcal{O})$ associated with the image of a region $\mathcal{O}$ under the action of an element γ of the spacetime isometry group coincides with the algebra obtained from $\mathfrak{A}(\mathcal{O})$ by acting on it with the corresponding element α_γ of the automorphism group. Thus if two regions are connected by a spacetime symmetry, that symmetry induces an algebraic structure preserving correspondence between elements of the local algebras associated with those regions. The thought, crudely put, is that if the spacetime metric doesn't care about the difference between regions $\mathcal{O}$ and $\gamma \mathcal{O}$, then neither should the algebras the quantum theory associates with those regions.

Of course, the quantum theory can incorporate *states* that aren't invariant under its environing spacetime's isometries. Where the automorphism $\alpha_\gamma : \mathfrak{A} \to \mathfrak{A}$ implements

---

[8] Dimock's (1980) covariance axiom, stated in terms of isometry classes of spacetimes, is somewhat subtler than this. I'm suppressing subtlety for the sake of simplicity.

an isometry γ, the state ω on the algebra $\mathfrak{A}$ is invariant under γ if and only if it is invariant under α_γ, where this invariance is understood as follows:

Definition 5.1 (invariant states). A state ω on a C^* algebra $\mathfrak{A}$ is *invariant under an automorphism* $\alpha : \mathfrak{A} \to \mathfrak{A}$ of that algebra iff for all $A \in \mathfrak{A}$

$$\omega(A) = \omega(\alpha(A)) := \omega \circ \alpha(A)$$

An individual state's failure to be invariant under an automorphism α_γ implementing a spacetime isometry γ doesn't imply that the quantum theory itself makes distinctions (e.g. between regions $\mathcal{O}$ and $\gamma\mathcal{O}$, considered in abstraction from particular states on the quasilocal algebra $\mathfrak{A}(\mathcal{M})$) its spacetime setting is unequipped to recognize.

Failing to be invariant under a one parameter automorphism group α_x, $x \in \mathbb{R}$, a state ω may nevertheless lend α_x the following aid and comfort:

Definition 5.2 ((strong) unitary implementability w.r.t. ω). Where ω is a state on a C^* algebra $\mathfrak{A}$, a one parameter group α_x, $x \in \mathbb{R}$, of automorphisms of $\mathfrak{A}$ is *(strongly) unitarily implementable* on ω's Gelf'and–Neimark–Segal (GNS) representation $(\mathcal{H}_\omega, \pi_\omega, |\xi_\omega\rangle)$ iff there exists a (strongly continuous) unitary group $U(x)$ acting on $\mathcal{H}_\omega$ such that for all $A \in \mathfrak{A}$:

$$\alpha_x(A) = U(x)\pi_\omega(A)U^{-1}(x)$$

If ω is α_x-invariant, it *must* lend that automorphism group the aid and comfort of strong unitary implementability:

Fact 5.3. If ω is α_x-invariant, then α_x is strongly unitarily implementable on ω's GNS representation (Araki 1999, thm. 2.33 (54) and corr. 2.34 (57)).

Regions $\mathcal{O}_1$ and $\mathcal{O}_2$ are *spacelike separated* just in case no point of $\mathcal{O}_1$ is connectable to any point of $\mathcal{O}_2$ by a *causal curve*, a curve whose tangent at any point is either timelike or lightlike. Intuitions about "locality" prohibit influences between operations in spacelike separated regions. These intuitions have found many expressions ("Einstein Locality," "separability," and "parameter independence," to name a few) in the literature about the Einstein–Podolsky–Rosen thought experiment and the Bell Inequalities (see Cushing and McMullin 1989 for a sample). Such intuitions are sometimes invoked to motivate the axiom

3. Microcausality axiom. If $\mathcal{O}_1$ and $\mathcal{O}_2$ are spacelike separated, every element of $\mathfrak{A}(\mathcal{O}_1)$ commutes with every element of $\mathfrak{A}(\mathcal{O}_2)$.

Section 5.3 will have more to say about the Microcausality axiom, as well as the fates of locality and entanglement in axiomatic QFT.

Another axiom intertwines the causal structure of the spacetime $(\mathcal{M}, g_{ab})$ with the inclusion structure of the local algebras. Its introduction will be preceded by a definition:

Definition 5.4 (domain of dependence). The *domain of dependence* $D(\mathcal{O})$ of a spacetime region $\mathcal{O} \subset (\mathcal{M}, g_{ab})$ is the set of points $p \in (\mathcal{M}, g_{ab})$ such that every inextendible causal curve through p intersects $\mathcal{O}$.

If the physics playing out in the spacetime is deterministic, data on $\mathcal{O}$ fixes data on $D(\mathcal{O})$. Even absent determinism, $D(\mathcal{O})$ is the region in the scope of only those causal influences that pass through $\mathcal{O}$. Haag and Kastler evocatively term $D(\mathcal{O})$ "the causal shadow" (1964, 848) of $\mathcal{O}$. The axiom:

4. Primitive Causality axiom. If $\mathcal{O}_1 \subset D(\mathcal{O}_2)$, then $\mathfrak{A}(\mathcal{O}_1) \subset \mathfrak{A}(\mathcal{O}_2)$.

ensures that a state on the local algebra associated with a region induces a state on the local algebra associated with that region's causal shadow. As Dimock explains, "The axiom embodies the basic dynamical principle that the past determines the future (in a certain local sense). If we know a state on $\mathfrak{A}(\mathcal{O}_2)$ then we know it on $\mathfrak{A}(\mathcal{O}_1)$" (1980, 230).

Scholium: Cauchy surfaces A *Cauchy surface* Σ in a spacetime $\mathcal{M}$ is a surface that every inextendible causal curve in $\mathcal{M}$ intersects exactly once. A spacetime $\mathcal{M}$ is *globally hyperbolic* iff it admits a Cauchy surface, in which case $\mathcal{M}$ can be foliated by a one (real) parameter family $\Sigma_t, t \in \mathbb{R}$, of Cauchy surfaces, each of which has the same topology: $\mathcal{M} = \Sigma_t \times \mathbb{R}$. Minkowski spacetime is globally hyperbolic; one nice foliation identifies Cauchy surfaces Σ_t with hyperplanes of simultaneity according to some inertial observer, with the time lapsed, according to the observer, between different hyperplanes of simultaneity determining the values of label t. ♠

The causal shadow of a Cauchy surface Σ in a spacetime $\mathcal{M}$ coincides with $\mathcal{M}$ itself. Where $\mathfrak{A}(\Sigma)$ is the observable algebra associated with a Cauchy surface,[9] Primitive Causality requires the global algebra $\mathfrak{A}(\mathcal{M})$ to be a subalgebra of $\mathfrak{A}(\Sigma)$. Because Isotony requires $\mathfrak{A}(\Sigma)$ to be a subalgebra of $\mathfrak{A}(\mathcal{M})$, it follows that $\mathfrak{A}(\Sigma) = \mathfrak{A}(\mathcal{M})$.

The final axiom amounts to a requirement that the others have an interesting instantiation:

5. Primitivity axiom. There exists an irreducible, faithful representation of $\mathfrak{A}_\mathcal{M}$.

In the less general but more studied case of Minkowski spacetime, we can appeal to the symmetries of that spacetime to place further demands on physically interesting realizations of the axioms. For instance, we can require there be an irreducible, faithful representation corresponding to a *Lorentz-invariant state* satisfying the spectrum condition, where this is understood as follows. A Lorentz-invariant state is a state invariant (in the sense of Def. (5.1)) under automorphisms complying with the Covariance axiom

---

[9] Because field operators have to be smeared by functions with support in a four-dimensional region of spacetime to construct the relevant algebra, we should be talking instead about a four-region $\tilde{\Sigma}$ that's a thin slab incorporating Σ.

by implementing the Poincaré group. Sketching the spectrum condition requires a few more preliminaries. Let ω be a Lorentz-invariant state on $\mathfrak{A}(\mathcal{M})$. In particular, ω will be invariant under the automorphism groups $\alpha_x, \alpha_y, \alpha_z, \alpha_t, x, y, z, t \in \mathbb{R}$ implementing the subgroup of the Poincaré group corresponding to spatial (x, y, z) and temporal (t) translations. From Fact 5.3 it follows that each group of translation automorphisms $\alpha_x, \alpha_y, \alpha_z, \alpha_t$ is strongly unitarily implementable on ω's GNS representation, by strongly continuous unitary groups we'll call $U(x), U(y), U(z), U(t)$. Stone's theorem implies that each of these unitary groups has an infinitesimal generator in the form of a positive, self-adjoint operator acting on $\mathcal{H}_\omega$. The infinitesimal generators of the unitaries implementing spacetime translations correspond to the four components of the relativistic energy-momentum observable: the operator P_x corresponding to the x-component of momentum is the infinitesimal generator for the group $U(x)$ of translations in the x direction, and so on.[10] According to the quantization algorithm, possible values of energy and momentum are confined to the spectra of these operators, which are pairwise commuting, so that they may be assigned eigenvalues simultaneously. Thus, quadruples of points in the spectra of the quartet of infinitesimal generators of spacetime translations specify energy-momentum vectors. A representation underwriting such a definition of an energy-momentum vector satisfies the *spectrum condition* just in case all of its normal states are such that these energy-momentum vectors lie in the closed forward lightcone. Representations satisfying the spectrum condition are representations unembarrassed by states of negative energy or spacelike four-momentum. Naturally enough, they are called *positive energy representations*. Notice that without Minkowski spacetime's substantial translational symmetries, the spectrum condition wouldn't even make sense.

The foregoing can be distilled into a single axiom:

6. Vacuum axiom. There exists a Lorentz-invariant state ω_0 over the quasilocal algebra $\mathfrak{A}(\mathcal{M})$ whose GNS representation is faithful, irreducible, and satisfies the spectrum condition.

Another axiom hinging critically on the translational symmetry of Minkowski spacetime is the

7. Weak Additivity axiom. For every closed, bounded region $\mathcal{O}$ of Minkowski spacetime $\mathcal{M}$, the closure in the C^* norm of the algebras $\mathfrak{A}(\mathcal{O} + a)$ for $a \in \mathbb{R}^4$ coincides with the quasilocal algebra $\mathfrak{A}(\mathcal{M})$.

Because $\mathbb{R}^4$ is just the aforementioned translational subgroup of Minkowski spacetime's isometry group, the Covariance axiom tells us how to understand $\mathfrak{A}(\mathcal{O} + a)$. Weak additivity ensures that the algebra for all of spacetime can be generated from the algebra for a local region by acting on that algebra with spacetime translations.

---

[10] P_x will generally be unbounded, but its spectral resolution will be given by bounded operators in $\mathfrak{B}(\mathcal{H}_\omega)$.

The general axioms (1–5) have non-trivial instantiations; including e.g. linear scalar fields in globally hyperbolic spacetimes. See Dimock (1980) for details. The Haag–Kastler axioms (1–3, 6–7) also have non-trivial instantiations, including models of *interacting* quantum fields in two or three spacetime dimensions (see e.g. Glimm and Jaffe 1972).

5.3 Entanglement and locality in QFT

Section 5.4 will present axioms for quantum statistical mechanics (QSM). But first, the present section pauses to address the important issues of nonlocality and entanglement in the setting of axiomatic QFT.

5.3.1 Motivating the Microcausality axiom

Section 5.2's Microcausality axiom is conventionally motivated by appeal to the conviction that modern physics is a local affair which leaves no room for action at a distance. Such "locality" is often asserted to be a consequence of the special theory of relativity (STR) and its supposed prohibition on superluminal signal transmission—so often that the "nonlocality" at issue in the violation of the Bell Inequalities is taken to raise questions about the very possibility of "peaceful coexistence" between QM and STR (Redhead 1986). Considered in the context of a suspected enmity between quantum phenomena and relativity theory, axiomatic QFT's Microcausality axiom is noteworthy for its *independence* from the Covariance axiom as it applies to Minkowski spacetime. Designed for Minkowski spacetime, the Haag–Kastler (1964) axioms include both a Lorentz covariance (i.e. covariance under the Poincaré group[11]) axiom and a Microcausality axiom. The Haag–Kastler axioms are logically independent in the sense that for each axiom there exist models that violate that axiom but satisfy the rest.

Soberly viewed, all the special theory of relativity (STR) can be taken to demand of a theory set in Minkowski spacetime is that it exhibit Lorentz covariance. And this demand is met by QFTs in its scope, provided they satisfy the Covariance axiom—*even if they violate the Microcausality axiom*! The existence of Lorentz-covariant QFTs establishes peaceful coexistence, whether those QFTs obey Microcausality or not. To demand our QFTs satisfy the Microcausality axiom, then, is to demand something more of them than mere consistency with STR. But what is this something more, and why should we demand it?

Here is how Haag explicates the Microcausality axiom: "The main principle expressed by it is the causal structure of events. Two observables associated with spacelike separated regions are compatible. The measurement of one does not disturb the measurement of the other. The operators representing these observables must commute" (1992, 107). The explication links Microcausality violation to

---

[11] Many presentations instead demand covariance under the identity-connected component of that group, which excludes the reflection symmetries of Minkowski spacetime. The distinction could matter e.g. discussions of CP violation. I don't think it matters to me, and I'll ignore it in the text.

disturbance-at-a-distance. But it is difficult to reconstruct the details of the linkage. Here is a reconstruction that I take to fail:

[*Microcausality Motivation?*] Suppose that $|\Psi\rangle$ is a global state; that any machination performable in the region $\mathcal{O}_1$ corresponds to acting on $|\Psi\rangle$ with an element $A \in \mathfrak{A}(\mathcal{O}_1)$; and that any machination performable in the region $\mathcal{O}_2$ corresponds to acting on $|\Psi\rangle$ with an element $B \in \mathfrak{A}(\mathcal{O}_2)$. If $AB|\Psi\rangle \neq BA|\Psi\rangle$, then the global state after machinations performed in both $\mathcal{O}_1$ and $\mathcal{O}_2$ would depend on the order in which those machinations were performed. But relativity theory teaches us that there is no fact of the matter about the temporal order of spacelike separated events. So it had better be that $AB|\Psi\rangle = BA|\Psi\rangle$—no matter what the pre-machination global state $|\Psi\rangle$ is. So A and B had better commute.[12]

I take this motivation to fail because I take one of its presuppositions to be incompatible with *other* axioms of axiomatic QFT. The presupposition is that the way to model an operation performed in the spacetime region $\mathcal{O}$ is by the change of state wrought by acting on the global state $|\Psi\rangle$ with some element A of $\mathfrak{A}(\mathcal{O})$. My tendentious contention is that, whatever one thinks of measurement in non-relativistic QM, there's no obvious way to square this model with the idea—an idea informing the Vacuum axiom (p. 108)—that the quasilocal algebra $\mathfrak{A}(\mathcal{M})$ admits a state. If $|\Psi\rangle$ is that global state, it's a state for all space and all time, and the notion of changing it by an operation executed within spacetime is nonsense. If $|\Psi\rangle$ is not that global state, we need an account of what state $|\Psi\rangle$ is, an account that lends coherence to the idea that acting on $|\Psi\rangle$ with an element of a local algebra $\mathfrak{A}(\mathcal{O})$ corresponds to carrying out a local operation performable in $\mathcal{O}$.

One strategy for understanding the transition $|\Psi\rangle \to A|\Psi\rangle$ as a model of a local operation is to suppose $|\Psi\rangle$ to be a state *at a time*, i.e. on a spacelike hyperplane Σ without edges. Minkowski spacetime is foliated by a one parameter family $\{\Sigma_t\}$ of such hyperplanes, translations between which are isometries of the Minkowski metric. (In fact, it's foliated by as many such families as it admits inertial observers.) Members of this family are thus natural candidates for the times at which states-at-a-time are states. In terms of these candidates, the effect of the local operation represented by $A \in \mathfrak{A}(\mathcal{O})$ and occurring between t and t' might be to map the pre-operation state $|\Psi\rangle$ on $\mathfrak{A}(\Sigma_t)$ to a post-operation state $A|\Psi\rangle$ on $\mathfrak{A}(\Sigma_{t'})$.

The Primitive Causality axiom (p. 107) undermines this strategy. Members of the family $\{\Sigma_t\}$ are also Cauchy surfaces. Supposing $|\Psi\rangle$ to be a state at a time on some hyperplane $\Sigma \in \{\Sigma_t\}$, talk of local operations changing $|\Psi\rangle$ is no more lucid than before. Primitive Causality implies that the algebra for all of spacetime $\mathfrak{A}(\mathcal{M})$ coincides with the algebra $\mathfrak{A}(\Sigma)$. Any state on one is automatically a state on the other, and so not subject to alteration by activity within the spacetime for which it's a state. *This* notion of

---

[12] I've proceeded as though the global state were a Hilbert space vector, but the reasoning can be expressed in a purely algebraic format. Simply interpret "acting on the algebraic state ω on $\mathfrak{A}$ with $A \in \mathfrak{A}$" as inducing a transition to the state ω_A, where $\omega_A(X) := \omega(A^*XA)/\omega(A^*A)$ for all $X \in \mathfrak{A}$.

a state-at-a-time can't underwrite the $|\Psi\rangle \to A|\Psi\rangle$ model of local operations invoked by the present reconstruction of Haag's motivation of the Microcausality axiom.

This doesn't eliminate the possibility that more nuanced and subtle models of local operations might sustain the claim that the Microcausality axiom is necessary to rule out action-at-a-distance. But it does suggest that it might be worth considering other strategies for motivating Microcausality. And an alternative strategy can be found in the accomplished and growing literature on the hierarchy of independence conditions in axiomatic QFT (for reviews, see Summers 1990; Rédei 1998, ch. 11; Rédei and Summers 2009). Results in that literature can be interpreted to establish that Microcausality is *sufficient* to rule out action-at-a-distance.

One result (a result known as the "no-signaling theorem" (see e.g. Rédei 2010)) concerns the following setup. Let $\mathcal{O}, \mathcal{O}_1, \mathcal{O}_2$ be open, bounded spacetime regions such that $\mathcal{O}_1$ and $\mathcal{O}_2$ are spacelike separated and each are subregions of $\mathcal{O}$. Let ω be a state on $\mathfrak{A}(\mathcal{O})$. Consider a self-adjoint element $A \in \mathfrak{A}(\mathcal{O}_1)$ with spectral projections $\{P_i\}$. We can use these projections to define a map $M^A : \mathfrak{A}(\mathcal{O}_1) \to \mathfrak{A}(\mathcal{O}_1)$ which acts as follows on an arbitrary element B of that algebra:

$$M^A(B) = \sum_i P_i B P_i \qquad (5.1)$$

This expresses the projection postulate, applied to a *non-selective* measurement of A. For it takes a "pre-measurement" state ϕ on $\mathfrak{A}(\mathcal{O}_1)$ to a "post-measurement" state $\phi \circ M^A : \phi \circ M^A(B) := \phi(M^A(B))$ for all $B \in \mathfrak{A}(\mathcal{O}_1)$. And $\phi \circ M^A$ is just a statistical mixture of possible A measurement outcomes weighted by their Born Rule probabilities—the mixture into which ϕ would collapse upon the occasion of an A measurement, according to the projection postulate.[13]

What the no-signaling theorem establishes is that whenever the projection postulate map M^A on $\mathfrak{A}(\mathcal{O}_1)$ extends to a map[14] $M^A_{ext} : \mathfrak{A}(\mathcal{O}) \to \mathfrak{A}(\mathcal{O})$, the extended map coincides with the identity map on $\mathfrak{A}(\mathcal{O}_2)$. It follows that states ψ on $\mathfrak{A}(\mathcal{O}_2)$ are invariant under the action of the extended map: for any such ψ and all $B \in \mathfrak{A}(\mathcal{O}_2)$, $\psi(B) = \psi(M^A_{ext}(B))$. Glossing this as "a non-selective measurement in region $\mathcal{O}_1$ has no influence on the spacelike separated region $\mathcal{O}_2$," we justify labeling the result a no-signaling theorem. Because the result is a consequence of the Microcausality axiom, we have thereby shown that axiom to be sufficient to rule out superluminal signals—or at least superluminal signals of the sort the result addresses.

There, of course, is the rub. There are a variety of potential superluminal signals about which the no-signaling theorem is silent. M^A implements a *non-selective*

---

[13] Notice that this approach makes the move against which I've just railed: it assimilates acting physically in a spacetime region $\mathcal{O}$ to acting on a state of the quasilocal algebra with an element of $\mathfrak{A}(\mathcal{O})$. This time the $|\Psi\rangle$ involved in the transition $|\Psi\rangle \to A|\Psi\rangle$ purporting to model an operation local to a region $\mathcal{O}$ is a state on the algebra $\mathfrak{A}(\mathcal{O})$ associated with that region. Still, the assimilation seems to me to be in tension with the very idea of a state of the quasilocal algebra.

[14] Technically, a unit-preserving, completely positive map, also known as a selective operation.

measurement. So the signaling ruled out by the no-signaling theorem is (something like) the sort that would be mediated by what the Bell inequality literature labels "parameter dependence": the dependence of statistics in one spacetime region on the occurrence of non-selective measurements in a spacelike-separated region. By contrast, "outcome dependence" makes statistics in one spacetime region vary with the outcome of a (*eo ipso* selective) measurement in a spacelike separated region. It is by now standard to insist that because we can't control the outcome of quantum measurements, parameter dependence, and not outcome dependence, is the variety of distant influence we could exploit to manipulate and control goings-on at a spacelike remove from ourselves. But I join Maudlin (1994, ch. 5) in regarding an influence as no less causal for being beyond our control. Those who deny this should at least admit that local *selective* measurements performed on states entangled across spatially separated regions (such states, we'll see in the next subsection, are absolutely generic to QFT) raise questions about distant causation that the no-signaling theorem fails to settle.

Even supposing parameter dependence were the only threat to causal good behavior looming in the quantum realm, the no-signaling theorem discussed here fails to disarm it. For the local operations whose inertness-at-a-distance it establishes are non-selective measurements of *discrete* observables. But the local algebras of axiomatic QFT abound with operations that aren't of this sort. Indeed, for every faithful state on the algebra for Minkowski spacetime (subject to the usual axioms), and every region $\mathcal{O}$ with non-empty spacelike complement, there's a "non-selective" operation on $\mathfrak{A}(\mathcal{O})$ with extensions that aren't causally inert (in the sense of the no-signaling theorem) at a distance (Accardi and Cecchini 1982). This is hardly the end of the story: there may be tiers in the independence hierarchy circumscribing notions of causal good behavior to which axiomatic QFT conforms (see Rédei and Valente 2009). But there is room to worry that Microcausality would be better motivated if it were demonstrably sufficient to rule out a wider variety of signaling than it at present is. That is, there is such room supposing we're inclined to worry about signaling, which I'm not sure we should be, once we've assured ourselves of Lorentz-covariance.

My own view is that the motivation for the Microcausality axiom has less to do with "locality" than it has to do with mereology: assuming (as the causal structure of relativistic spacetimes enjoins us to assume) that $\mathcal{O}$ and $\mathcal{O}'$ are different systems, each with its own C^* algebra, we represent their union by a tensor product of those algebras; elements of different components of a tensor product algebra commute.

Still, it might be fair to ask *why* we represent different elements of a composite system by components of a tensor product, and to regard "no superluminal causation" as the key to an informative answer in the case that the elements are spacelike-separated spacetime regions.[15] But my own inclination is to agree with Haag that Microcausality reflects the causal structure of the spacetime in which a QFT is set—how could it not,

---

[15] I'm grateful to Hans Halvorson for emphasizing this to me.

when that causal structure appears in the antecedent of the axiom?—but to resist the suggestion that this invocation is mediated by the spectre of some "disturbance at a distance" prohibited by STR. Fortunately, we can agree that Microcausality is an axiom without agreeing about why it should be!

5.3.2 Entanglement in QFT

Phenomena accurately described by ordinary QM violate the Bell Inequalities, signaling that at least one of the assumptions generating those inequalities must be false. Some of these assumptions can be cast in a form strongly reminiscent of STR's folkloric ban on superluminal causal influence. This has lent urgency to the question of whether QM and STR can "peacefully coexist."

Urgent though the question is, as stated it lacks precision. STR is not readily understood as a theory about causes and their admissable configurations. STR rather requires of spacetime theories formulated in Minkowski spacetime that they be Lorentz-covariant.[16] Non-relativistic QM, which is not a spacetime theory, is not subject to STR's requirements. So the question of whether STR and QM are capable of peaceful coexistence can be posed only after extensive precarious and heroic interpretive work—work addressing exceedingly non-trivial questions such as how to understand relativistic constraints on causal action in a stochastic setting—has been completed.

By contrast, axiomatic QFT is quite explicitly set in a spacetime; when that spacetime is Minkowski, the requirement of Lorentz-covariance takes the form of an unambiguous axiom. Thus no interpretive heroics are required to settle the question of what it takes for an axiomatic QFT to be Lorentz-covariant. It's sufficient for a QFT to be set in Minkowski spacetime and to satisfy the standard axioms. In the exquisitely controlled context of such a QFT, it is still possible to ask: to what extent, if any, do quantum weirdnesses persist, despite the manifest Lorentz-covariance of the theory?

The answers are surprising. One striking example is the Reeh–Schlieder theorem, which follows from the standard axioms.[17] Consider an algebra $\mathfrak{A}(\mathcal{O})$ of observables associated with an open bounded region $\mathcal{O}$ of spacetime, for instance a physics laboratory over the course of a fall afternoon, and the Minkowski vacuum state $|0\rangle$ on the quasilocal algebra $\mathfrak{A}(\mathcal{M})$. The theorem states that the set of states obtained by acting on $|0\rangle$ with elements of $\mathfrak{A}(\mathcal{O})$ is *dense* in the set of states on $\mathfrak{A}(\mathcal{M})$. In other words, any state of an axiom-satisfying QFT on Minkowski spacetime can be approximated arbitrarily closely by acting on its vacuum state by polynomial combinations of observables in $\mathfrak{A}(\mathcal{O})$. If it were appropriate to model events in the region $\mathcal{O}$ as applications

---

[16] Maudlin (1994) discusses STR's real and imagined implications, and the constraints they place on the interpretation of QM.

[17] Redhead (1995b) illuminates the Reeh–Schlieder theorem by explicating analogies between how Minkowski vacuum and spin-singlet states stand to algebras of observables pertaining to the component systems.

of elements of $\mathfrak{A}(\mathcal{O})$ to the global vacuum state, this would mean that operations in local regions, such as physics laboratories on fall afternoons, could produce arbitrary approximations of arbitrary global states!

The Reeh–Schleider theorem also implies that no observable associated with a finite spacetime region has $|0\rangle$ as an eigenstate. Vacuum correlations are thus omnipresent; their signature is that there is no local region with which the vacuum state associates a pure state. What's more, the correlations instituted by typical states have the capacity to violate Bell inequalities, even maximally (for results, see Summers and Werner 1987; Halvorson and Clifton 2000; for an account aimed at philosophers, see Butterfield 1995). And QFT entertains states of a sort Clifton et al. (1998) call "superentangled." A superentangled state is one that, given any pair of spacetime regions, correlates any observable from one region with some observable from the other. Clifton et al. show that these ubiquitously correlated states are dense in the set of states on $\mathfrak{A}(\mathcal{M})$. Results like this establish beyond a shadow of a doubt that Lorentz-covariance is no proof against quantum spookiness.

Research continues apace on issues of entanglement and the independence of local algebras in axiomatic QFT. In this book, I have no aspiration (or frankly hope) of contributing results to this growing literature, which I take to be in very good hands. The literature establishes that QFT is a vibrant and pertinent setting for the continuation of foundational investigations launched in the context of ordinary QM. My aim is the complementary one of suggesting that QFT serves as well to instigate novel questions of interpretation distinct from those familiar from foundational investigations of ordinary QM.

5.4 Axioms for QSM

In the non-relativistic case—paradigmatically, in the setting of the thermodynamic limit of statistical mechanics—axiomatic versions of the algebraic approach (for which see Sewell 2002, ch. 2; Primas 1983) associate with each open bounded region V of *space*—usually taken to be Euclidean three-space $\mathbb{R}^3$—a C^* algebra $\mathfrak{A}(V)$. A straightforward analog of the Isotony axiom ensures that these local algebras form a net whose C^* inductive limit is the quasilocal algebra $\mathfrak{A}$; a Covariance axiom demands that the symmetry group of the space be implemented by an automorphism group of this algebra. Insofar as Euclidean space lacks the rich causal structure of relativisitic spacetimes, the remaining axioms are accordingly attenuated. Locality becomes:

Independence axiom (independence). For all V_1 and V_2 such that $V_1 \cap V_2 = \emptyset$, every element of $\mathfrak{A}(V_1)$ commutes with every element of $\mathfrak{A}(V_2)$

and Primitive Causality disappears entirely.

Whereas in the relativistic context the quasilocal algebra $\mathfrak{A}(\mathcal{M})$, encompassing as it does the whole of the spacetime, already has a temporal dimension built into it, the quasilocal algebra $\mathfrak{A}$ for Euclidean three-space does not. One way[18] to add a temporal dimension is to introduce a one real parameter group α_t of C^* automorphisms of the quasilocal algebra $\mathfrak{A}$. On the model of the Heisenberg picture of ordinary QM, understand $\alpha_t(A)$ as the "evolute" of an element $A \in \mathfrak{A}$ through a time t. A C^* algebra $\mathfrak{A}$ equipped with a dynamical automorphism group α_t becomes a C^* *dynamical system*.

QFT supports analogs of these Heisenberg picture "dynamical" automorphisms, but I must beg to understand these obliquely. Assume our QFT is obtained by quantizing a classical theory whose solutions form a symplectic vector space (S, Ω). A representation of the Weyl relations will be a map W from S to unitary Hilbert space operators satisfying the Weyl relations over (S, Ω). A representation generates, via uniform closure, the Weyl algebra $\mathfrak{W}$ characteristic of the QFT. A symplectic symmetry of the classical theory—a map $s : S \to S$ preserving the symplectic product Ω—"lifts" to an automorphism α of $\mathfrak{W}$ by acting as follows on the Weyl unitaries generating that algebra: for each $\phi \in S$, set $\alpha(W(\phi))$ equal to $W(s\phi)$. Arageorgis, Earman, and Ruetsche (2002) discuss symmetries obtained in this way for quantized Klein–Gordon theory set in a globally hyperbolic spacetime. Where Σ_1 and Σ_2 are Cauchy surfaces and ι an identity map between them, $\alpha^{\iota}_{\Sigma_1, \Sigma_2}$ is the automorphism of $\mathfrak{W}$ corresponding to the symmetry of the classical solution space that maps a solution with initial data ϕ on Σ_1 to the solution whose initial data on Σ_2 is the first solution's Σ_1 initial data, transferred to Σ_2 by the identity map ι. $\alpha^{\iota}_{\Sigma_1, \Sigma_2}$ is the Heisenberg picture "dynamical" automorphism we've been after. My plea, in keeping with my resolute reading of the Primitive Causality axiom, is to decline to understand $\alpha^{\iota}_{\Sigma_1, \Sigma_2}$ as inducing a change of state from Σ_1 to Σ_2. If ω is the state on $\mathfrak{A}(\Sigma_1)$, then Primitive Causality requires ω—and not $\omega \circ \alpha^{\iota}_{\Sigma_1, \Sigma_2}$[19]—to be the state on $\mathfrak{A}(\Sigma_2)$. It is, however, perfectly consistent with Primitive Causality to take $\alpha^{\iota}_{\Sigma_1, \Sigma_2}$ to provide a criterion of identity over time for observables. According to this criterion, the element $\alpha^{\iota}_{\Sigma_1, \Sigma_2}(A)$ of $\mathfrak{A}(\Sigma_2)$ is the evolute of the element A of $\mathfrak{A}(\Sigma_1)$.

Example 5.5 (instantiating the axioms). (See Emch 2007, 1101–2.) This example will try to make plausible the (true) claim that the algebra for the infinite spin chain presented in §3.3.1 satisfies the Isotony, Independence, and Covariance axioms. The positive and negative integers $\mathbb{Z}$ label sites on an infinite one-dimensional lattice; each site k is equipped with a trio of Pauli spins $\sigma(x)_k, \sigma(y)_k, \sigma(z)_k$; spins associated with different sites commute. The Pauli spins generate the C^* algebra $\mathfrak{M}(2)$ of complex 2×2 matrices; let $\mathfrak{M}(2)_k$ be the copy of this algebra generated by $\sigma(x)_k, \sigma(y)_k, \sigma(z)_k$ and associated with the site k. Each finite subset $V \in \mathbb{Z}$ identifies a "region" of the spin chain: that constituted by sites whose labels lie in V. Define the "local" C^* algebra for the region V as follows:

$$\mathfrak{A}(V) := \otimes_{k \in V} \mathfrak{M}(2)_k$$

---

[18] As Chapter 12 will explain, *not* the only way.
[19] defined by $\omega \circ \alpha^{\iota}_{\Sigma_1, \Sigma_2}(A) = \omega(\alpha^{\iota}_{\Sigma_1, \Sigma_2}(A))$ for all $A \in \mathfrak{A}$

So defined, $\mathfrak{A}(V)$ is a copy of the C^* algebra $\mathfrak{M}(2^{n_V})$, where n_V is the number of sites in V. Given this principle of construction, $\mathfrak{A}(V_1)$ will be a subalgebra of $\mathfrak{A}(V_2)$ whenever V_1 is a subset of V_2. The Isotony axiom is satisfied, and the quasilocal algebra $\mathfrak{A}_{CAR}$ for the infinite spin chain, a C^* algebra, can be obtained by closing $\cup_{V \in \mathbb{Z}} \mathfrak{A}(V)$ in C^* norm. ($\mathfrak{A}_{CAR}$ is, of course, the *CAR algebra* familiar from §3.3.1.) Because Pauli spins associated with different sites commute, $[\mathfrak{M}(2)_k, \mathfrak{M}(2)_j] = 0$ if $i \neq j$. This guarantees that $[\mathfrak{A}(V_1), \mathfrak{A}((V_2)] = 0$ whenever $V_1 \cap V_2 = 0$, in compliance with the Independence axiom. As for covariance, the symmetries of the lattice $\mathbb{Z}$ are translations, and they form an additive group. Acting on a subset V of $\mathbb{Z}$, a arbitrary member g of this group yields a subset gV of the same cardinality. (More simply put, shifting a set of points doesn't change how many of them there are.) Hence $\mathfrak{A}(V)$ and $\mathfrak{A}(gV)$ are each copies of $\mathfrak{M}(2^{n_V})$. It should be plausible that $\mathfrak{A}_{CAR}$ admits an automorphism α_g taking each element of $\mathfrak{A}(V)$ to a counterpart in $\mathfrak{A}(gV)$, thereby satisfying the Covariance axiom. ♠

5.5 Conclusion

In this chapter, operator algebraic notions have served to frame axioms for QFT and QSM. Whether conformity with such axioms is an essential feature of theories of QFT and/or QSM is an interpretive question left open for now. The next chapter uses the algebraic apparatus to present a set of interpretive responses to the alarming circumstance that theories of QM_∞ fall outside the scope of the Stone–von Neumann and Jordan–Wigner theorems.

6

Interpreting QM$_\infty$
Some Options

> At the overall foundational level, the logical basis of the subject remains unsettled, more than six decades after its heuristic origin. There is not even a general agreement about what constitutes a quantum field theory, in precise terms.
>
> (Baez, Segal, and Zhou 1992, xiii)

Until the 1950s, Bratteli and Robinson report, "most people tacitly believed" (1997, 218) that representations of the CCRs for a given field theory were unique up to unitary equivalence. Explicit and physically reasonable counterexamples shook this faith. An early and striking one is due to van Hove (see Emch 1972, 17–28). One model of the interaction between mesons and "recoil-less" nucleons couples a neutral scalar field to an external classical field. Two physically interesting states of this model are the so-called "dressed" and "bare" vacuum states—respectively, the lowest energy state of the total Hamiltonian and the no-particle state of the free Hamiltonian. Van Hove demonstrated that the representation containing the dressed vacuum is unitarily inequivalent to the representation containing the bare vacuum. The bare vacuum representation, moreover, supports no operator corresponding to the total Hamiltonian. Writing in 1955, Wightman and Schweber give voice to an apparent moral of the von Hove model: "[unitarily inequivalent representations] are not pathological phenomena whose construction requires mathematical trickery...but occur in the most elementary examples of field theory" (1955, 824).[1]

Investigations of the formal structure of standard realizations of quantum field theoretic CCRs conducted in the 1950s and 1960s cast the proliferation of unitarily inequivalent representations in an intriguing light. Consider a classical field theory corresponding to the symplectic vector space (S, Ω). To quantize this theory is to

---

[1] Haag (1955) announced a general theorem establishing the unitary inequivalence of dressed and bare vacuum representations of non-trivially interacting QFTs; Hall and Wightman (1957) furnished a precise statement as well as a proof. Haag's theorem returns in Ch. 11.

find a representation of the Weyl relations over (S, Ω). Each concrete Hilbert space representation of the Weyl relations over (S, Ω) gives rise to an abstract C^* algebra, the *Weyl algebra*, which is representation-independent. This Weyl algebra is constructed by forming polynomials of operators satisfying the Weyl relations, then closing in the uniform topology. No matter what Hilbert space representation it starts with, this procedures yields the same (up to $*$-isomorphism) Weyl algebra. Thus independent of its Hilbert space antecedents, the Weyl algebra $\mathfrak{W}$ over (S, Ω) can be regarded as an *abstract* C^* algebra, one whose algebraic structure is shared by all its Hilbert space representations. For a theory of QM_∞ whose canonical observables obey CARs, one can likewise abstract from those CARs a single C^*-algebraic structure common to every Hilbert space representation of them. This C^* algebra is *the CAR algebra*, isomorphic to the uniform closure of a concrete representation of the CARs.

Hilbert space representations of Weyl algebras and CAR algebras in QM_∞ can have structure in excess of their C^*-algebraic structure. Concrete representations make available operator topologies—for instance, the weak topology, which defines convergence for sequences of operators in terms of the Hilbert space inner product—absent from the abstract algebra, whose natural topology is furnished by the C^* norm. Where $\mathfrak{A}$ is an abstract algebra of observables, the representation $\pi : \mathfrak{A} \to \mathfrak{B}(\mathcal{H})$ may have as its image only a proper subset of its affiliated von Neumann algebra $\pi(\mathfrak{A})''$, which in turn may be only a proper subset of $\mathfrak{B}(\mathcal{H})$. There may be bounded, self-adjoint operators on concrete Hilbert space representations of QM_∞, algebras without direct correlate in those algebras.

Haag and Kastler voice a suspicion prompted by this circumstance: "The relevant object is the abstract algebra and not the representation. The selection of a particular... representation is a matter of convenience without physical implications" (1964, 852). One aim of this chapter is to suggest how to understand the suspicion as a strategy for interpreting theories of QM_∞. So that we might keep tabs on what such a strategy is a strategy for, §6.1 develops a framework for *individuating* theories of QM_∞. Then §§6.2–6.7 present a *non-exhaustive* range of approaches to interpreting such theories. One extreme in this range is the "conservative" approach, of extending the ordinary QM-adapted interpretive principles (OQS), (OQM), and (OQD) (cf. § 2.1) to theories of QM_∞, thereby interpreting them in terms set by a concrete Hilbert space representation. Another extreme, radical in the sense that it breaks from the dominant tradition of interpreting ordinary QM, is a purely algebraic approach, prompted by Haag and Kastler's suspicion that concrete Hilbert spaces set the wrong terms for understanding theories of QM_∞. Sections 6.2–6.7 will sketch preliminary evaluations of these interpretive approaches, which evaluations provide a partial guide to the rest of the book.

For the most part, the interpretive strategies this chapter will describe count as what §1.5 called *pristine* interpretations. They adhere to the methodological ideal of pristine interpretation, which requires the interpreter of a physical theory to appeal only to its laws and his own insight into matters logically antecedent to particular applications of theory—including matters of mathematics, metaphysics, and the nature of science—when saying what worlds are possible according to the theory. Pristine interpretation forswears appeal to extranomic or *a posteriori* considerations, considerations such as initial or boundary conditions, dynamical details, or exigencies pressing upon physicists attempting to frame new theories, when equipping a theory with content. The worlds possible according to a pristinely interpreted theory can be identified in advance of, and without recourse to, those adulterating details. These details are of course relevant to the mundane business of *applying* the theory to particular problems. But on the pristine picture, this business lacks foundational interest.

When I call an interpretation "pristine," I mean to draw attention to a feature, not a virtue, of it. Indeed, I think a less charitable synonym for "pristine" is "extremist": the pristine interpreter maintains allegiance to "metaphysical" principles of interpretation, come what may. I regard the pristine interpreters introduced in this chapter—the Hilbert Space Conservative, the Algebraic Imperialist, and the Universalist—as *caricatures* whose exaggerated stances interfere with the proper aim of interpretation, which is to make sense of how theories succeed as well as they do. Thus this chapter aims to sow doubts, which later chapters cultivate, about the adequacy of pristine interpretations.

Although it is a thesis of this book that no single pristine interpretive strategy can meet the demands of interpreting theories of QM_∞, I take the most interesting questions to be not those settled but those prompted by the thesis. Why don't pristine interpretive strategies succeed, and what should their failure tell us about the interpretation of physical theories and its place in the galaxy of philosophical engagement with science? These questions are addressed explicitly in Chapter 15.

6.1 What are we interpreting?

How are we to individuate the theories of QM_∞ we're setting out to interpret? What makes such a theory the theory it is? There is a consensus among the community of people who work with such theories. Fulling reports that "Most theoretical physicists, following Schwinger, regard the *action principle* as fundamental. Theories are defined by Lagrangians" (1989, 126). Indeed, by fixing the symplectic structure of the classical theory, the Lagrangian fixes the commutation relations its quantizer seeks to represent. But not all interesting quantum theories do or need descend from real or imagined classical Lagrangians. Haag observes,

The idea that one must first invent a classical model and then apply to it a recipe called "quantization" has been of great heuristic value. In the past two decades, the method of passing from a classical Lagrangian to a corresponding quantum theory has shifted more and more away from the canonical formalism to Feynman's path integral. This provides an alternative (equivalent?) recipe. There is, however, no fundamental reason why a quantum theory should not stand on its own legs, why the theory could not be completely formulated without regard to an underlying deterministic principle.... (1992, 6)

Fulling proposes a liberalization of Schwinger's conventional wisdom, a liberalization which identifies a quantum theory, not with an underlying Lagrangian, but with the constitutive features that would be fixed by a Lagrangian, were one available:

> There is, however, another point of view, more consistent with the spirit of axiomatic field theory. *Any* commutation or anticommutation relations consistent with the dynamics can define a possible model... In this approach a formal theory consists of equations of motion plus commutation rules (or some more general algebraic relations). They need not determine eachother, but it is a nontrivial requirement that they be mutually consistent. (1989, 126)

While agreeing that quantum theories are to be defined by their characteristic commutation relations and equations of motion, Fulling does not insist that these emanate from the same source, or that that source be a Lagrangian.

I propose here to adopt the common core of the conventional wisdom and its liberalization. That is, I propose in the first instance to identify a theory of QM_∞, specified up to kinematics, by appeal to its fundamental (anti)commutation relations. (The matter of dynamics I will postpone until §§6.2.2 and 6.4.4, which involve a theory's dynamics in constraints on admissible kinematics.) The interpretive positions I consider here all, in one way or another, recognize the physical significance of the C^* algebra generated by elements satisfying these relations.[2] And so I will identify a theory of QM_∞ as a *theory of canonical type* $\mathfrak{A}$, where $\mathfrak{A}$ is the C^* algebra generated by the (anti)commutation relations defining the theory.

This principle of identification admits illustrations and demands codicils. Illustrations first: the Weyl algebra over $(\mathbb{R}^{2n}, \Omega)$ and the Weyl algebra over $(\mathbb{R}^{2m}, \Omega)$ won't be isomorphic if $m \neq n$; nor will the C^* algebras generated by canonical anticommutation relations (CARs) for m and $n \neq m$ be isomorphic; nor will C^* algebras generated by canonical commutation relations (CCRs) and CARs be isomorphic. And this is as it should be: mechanics for n degrees of freedom is a distinct (albeit related) theory from mechanics for $m \neq n$ degrees of freedom; theories of n and $m \neq n$ spin systems are likewise distinct; so too are theories of mechanical and spin systems. The non-isomorphism of the algebras expressing aforementioned theories' canonical types reflects these distinctions.

Now the codicils. Although the principle of individuation proposed succeeds in distinguishing between some theories we would like to regard as distinct, it is still all together too coarse-grained. Spins arranged in an infinite two-dimensional lattice $\mathbb{Z}^2$ obey the same CARs, and thus induce a C^* algebra of the same canonical type, as spins in an infinite one-dimensional chain. But surely the physical contents of theories set in one and in two dimensions are distinct. This distinction can be accommodated by recognizing that part of the presentation of a theory of canonical type $\mathfrak{A}$ is an association between subalgebras of $\mathfrak{A}$ and subregions of the region whose

---

[2] As Fulling's mention of "more general algebraic relations" indicates, not all positions do. Bob Wald has proposed making do with * algebras.

observable algebra is $\mathfrak{A}$. This association mediates a finer-grained criterion of identity for theories conceived as triples $(\mathfrak{A}, \mathcal{M}, \{\mathfrak{A}(\mathcal{O})\})$, where $\mathfrak{A}$ is a C^* algebra generated by CCRs/CARs for a space(time) $\mathcal{M}$ (assumed to be equipped with a metric), and $\{\mathfrak{A}(\mathcal{O})\}$ gives the association between subregions $\mathcal{O} \subset \mathcal{M}$ and subalgebras $\mathfrak{A}(\mathcal{O}) \subset \mathfrak{A}$. Something like the following mouthful might do the trick as a criterion of identity for theories so specified:

> $(\mathfrak{A}, \mathcal{M}, \{\mathfrak{A}(\mathcal{O})\})$ and $(\mathfrak{A}', \mathcal{M}', \{\mathfrak{A}'(\mathcal{O}')\})$ are theories of the same canonical type only if there's an isometry $i : \mathcal{M} \to \mathcal{M}'$ and an isomorphism $\phi : \mathfrak{A} \to \mathfrak{A}'$ such that whenever $(\mathcal{O}') = i((\mathcal{O}))$, $\mathfrak{A}'((\mathcal{O}')) = \phi(\mathfrak{A}((\mathcal{O}))$.

Because the coarser-grained criterion will suffice for most of the questions pursued in this book, I won't try to specify or defend any particular finer-grained criterion. But I will note that such a criterion ought to imply the non-identity of theories set in non-isometric space(time)s.

Chapter 1's account of interpretation split the interpreter's task into what it called the structure-specifying and the semantic phases. In the structure-specifying phase, the interpreter identifies the structures by which the theory would represent physical goings-on; in the semantic phase, she undertakes to articulate success conditions for representations so configured. These success conditions are, in van Fraassen's phrase, accounts of what the world would have to be like in order for the theory to be true. Often the task of structure specification can be resolved into subtasks: identifying the theoretical kinematics, i.e. the physical magnitudes and possible states the theory recognizes; identifying the theoretical dynamics, i.e. the theory's account of how states and magnitudes change over time.

Let us model the kinematic subtask of the structure-specifying phase of the interpretation as follows. Associate with a theory a pair $(\mathcal{Q}, \mathcal{S})$, whose first entry denotes the set of physical magnitudes the theory recognizes, and whose second entry denotes its set of states. I'll call $(\mathcal{Q}, \mathcal{S})$ a *kinematic template*, and call its completion on behalf of a theory a *kinematic pair* for that theory. It is part (but by no means all) of an interpretation of a theory of canonical type $\mathfrak{A}$ to specify a kinematic pair for that theory.

Given that a state is a map from magnitudes to their expectation values, an ideology about how such maps should behave could be sufficient to determine $\mathcal{S}$ given $\mathcal{Q}$. For instance, a theory of ordinary QM equates $\mathcal{Q}$ with the self-adjoint part of $\mathfrak{B}(\mathcal{H})$ for some separable $\mathcal{H}$, and adopts the ideology that states should be normalized, linear, positive, and countably additive. It follows from Gleason's theorem that ordinary QM must equate $\mathcal{S}$ with $\mathfrak{T}^+(\mathcal{H})$, the set of density operators on $\mathcal{H}$ (provided $\dim(\mathcal{H}) > 2$). I persist in taking the kinematic template for a theory to be an ordered pair $(\mathcal{Q}, \mathcal{S})$ because ideologies of state are discretionary, even optional. Imposing a *universal* rule for generating $\mathcal{S}$ given $\mathcal{Q}$ would obliterate a certain sort of interpretational freedom.

Sections 6.2–6.6 catalog some pristine approaches to the task of specifying a kinematic pair for a theory of canonical type $\mathfrak{A}$. Presented with a system whose CAR

(CCR) algebra is $\mathfrak{A}$, a pristine interpretation aspires to deploy only $\mathfrak{A}$, technical subtlety, and philosophical perspicacity in specifying a kinematic pair. Because a theory's kinematic laws determine its canonical algebra $\mathfrak{A}$, that algebra is a resource allowed the pristine interpreter. Non-nomic contingencies are not. On the supposition that the pristine interpreter's guiding metaphysical principles fail to draw significant distinctions between different integers or between commutation and anti-commutation, one would expect the pristine interpreter to adopt a *uniform interpretive strategy* for theories of different but related types: her interpretations of theories with non-ismorphic Weyl algebras $\mathfrak{A}_n$ and $\mathfrak{A}_m$ will be distinct, but they will arise from a stable set of interpretive principles.

I've already admitted that I consider the extremism of pristine approaches ill-judged. Section 6.7 introduces more pragmatic approaches to interpreting theories of QM_∞; subsequent chapters will try to argue that it's saner to be pragmatic than extremist.

6.2 Hilbert Space Conservatism

The interpreter I'll label the *Hilbert Space Conservative* completes the kinematic template $(\mathcal{Q}, \mathcal{S})$ on behalf of a theory of QM_∞ the same way it's standardly completed on behalf of a theory of ordinary QM. Our characterization of that way will be somewhat roundabout. The Hilbert Space Conservative maintains that the observables proper to a theory of canonical type $\mathfrak{A}$ are the self-adjoint elements of $\pi_\omega(\mathfrak{A})''$, the von Neumann algebra affiliated with the GNS representation of a pure state ω on $\mathfrak{A}$. It follows that $\pi_\omega(\mathfrak{A})''$ coincides with $\mathfrak{B}(\mathcal{H}_\omega)$, where $\mathcal{H}_\omega$ is the concrete separable Hilbert space carrying ω's GNS representation.[3] The Hilbert Space Conservative identifies possible states of the system with normal states on $\pi_\omega(\mathfrak{A})''$, i.e. with ω's folium $\mathcal{F}_\omega$. Because these states coincide with $\mathfrak{T}^+(\mathcal{H}_\omega)$, the recapitulation of ordinary QM is complete. When the Hilbert Space Conservative offers $(\pi_\omega(\mathfrak{A})'', \mathcal{F}_\omega)$ as the kinematic pair for a theory of QM_∞, she's just dressing the kinematic pair $(\mathfrak{B}(\mathcal{H}), \mathfrak{T}^+(\mathcal{H}))$ typical of ordinary QM up in algebraic garb.

The Hilbert Space Conservative also transfers ordinary QM's content coincidence criterion of physical equivalence (see §1.2) to the setting of QM_∞. Theories specified up to kinematic pairs $(\pi_\omega(\mathfrak{A})'', \mathcal{F}_\omega)$ and $(\pi_{\omega'}(\mathfrak{A})'', \mathcal{F}_{\omega'})$ are physically equivalent, the Hilbert Space Conservative maintains, just in case $(\pi_\omega(\mathfrak{A})'', \mathcal{H}_\omega)$ and $(\pi_{\omega'}(\mathfrak{A})'', \mathcal{H}_{\omega'})$ are unitarily equivalent, from which it follows that $\mathcal{F}_\omega = \mathcal{F}_{\omega'}$.

Conservatism, thus stated, adheres to the ideal of pristine interpretation insofar as the *type* $(\mathfrak{B}(\mathcal{H}), \mathfrak{T}^+(\mathcal{H}))$ of kinematic pair it associates with a theory of QM_∞ is a type it identifies without reference to extra-nomic features of the theory or its application. The most extreme form of Conservatism is a Conservatism of *principle* which maintains of each theory of QM_∞ that it ought be attributed a kinematic pair

---

[3] Because the GNS representation of a pure state ω is irreducible (Bratteli and Robinson 1987, thm. 2.3.19 (57)) and $\pi_\omega(\mathfrak{A})''$ is isormorphic to $\mathfrak{B}(\mathcal{H}_\omega)$ if π_ω is irreducible (Sakai 1971, thm. 1.12.9).

6.2 HILBERT SPACE CONSERVATISM

of the Conservative type. Thus committed to an overarching and uniform strategy for interpreting quantum theories, principled Conservatism is adamantly pristine. By contrast, a Conservatism of *opportunity* picks its spots, interpreting some, but not other, theories of QM_∞ in terms of a Conservative kinematic pair. Whether such an approach is pristine or not depends on how it identifies its opportunities. Insofar as the opportunistic Conservatism keys the content of a theory of QM_∞ to its "extra-nomic" features, it departs from the ideal of pristine interpretation.

Confronted with a particular theory of QM_∞, the Conservative (of any stripe) faces a daunting task. In ordinary QM, the uniqueness theorems assure us, once $\mathfrak{A}$ is fixed, all representations are unitarily equivalent. This is not so in QM_∞. The canonical algebra $\mathfrak{A}$ individuating the theory to be interpreted admits continuously many unitarily inequivalent Hilbert space representations. Thus there are, for the Hilbert Space Conservative, as many candidate *tokens* of the admissible type—kinematic pairs $(\pi_\omega(\mathfrak{A})'', \mathcal{F}_\omega)$—as there are disjoint pure states ω on $\mathfrak{A}$. Committed to unitary equivalence as a criterion of physical equivalence, the Hilbert Space Conservative must regard each of these as rival quantum theories. At most one of them can give the kinematic structure of *the* theory she's set out to interpret. Thus the burden of the Hilbert Space Conservative is to articulate and motivate a principle of privilege that enables her to enthrone a single unitary equivalence class of representations as physical, and to consign the rest to the dustbin of mere mathematical artifacts.

Invoking a principle of privilege is tantamount to selecting a pure state ω on $\mathfrak{A}$. Depending on how this selection is motivated and carried out—is ω pristinely identified by appeal to principles of metaphysics or desiderata determined by general accounts of the methodology of science? Is ω a brutally factive initial condition?—Hilbert Space Conservatism remains pristine or sullies itself. Next we consider some candidate principles of privilege, and the adulterations they may represent.

6.2.1 Some privileging strategies

The generic strategy is to announce a principle distinguishing physical states from unphysical ones, then hope that the privileged states fall into the folium of some pure state ω. If all and only π_ω-normal states are physical, then it is reasonable—in a sense §6.4's Harmony principle will articulate—to expand one's observable set to include not only self-adjoint elements of the C^* algebra $\pi_\omega(\mathfrak{A})$ but also self-adjoint elements of the von Neumann algebra $\pi_\omega(\mathfrak{A})''$ affiliated with ω. A kinematic pair of the sort desired by the Hilbert Space Conservative follows.

The "Symmetry Mongering" privileging strategy invokes space(time) symmetries Λ implemented by automorphisms α_λ of the quasilocal algebra $\mathfrak{A}$ to declare a state ω on $\mathfrak{A}$ *physical*$_{SM}$ only if each symmetry α_λ can be implemented (strongly) unitarily (Def. 5.2) in π_ω. For an early instance, Robinson (1966, 483) proposes a principle of this sort, which in the case of QFT set in Minkowski spacetime is encoded (in conjunction with the requirement that the vacuum be Lorentz-invariant) in the Vacuum axiom (see §5.2) of axiomatic QFTs.

In a spacetime equipped with global *timelike* isometries, representations of the CCRs for the Klein–Gordon field can exhibit energetics respecting those isometries. The respect is expressed by representations of states that are physical$_{SM}$, and takes roughly the following form: a privileged Hamiltonian operator can be identified as the infinitesimal generator of the (strongly continuous) family of unitaries acting on those representations to implement time translations that are symmetries of the metric. Each respectful representation has a vacuum state which is not only itself invariant under these timelike isometries, but also the lowest-energy eigenstate of the privileged Hamiltonian. What's more, each n-particle state of a respectful representation is an eigenstate of the privileged Hamiltonian, with an eigenvalue matching the energy of the corresponding n-particle classical state. Thus each respectful representation intimates a particle interpretation. (The precise nature of this intimation, and the threat posed to it by the apparent availability of unitarily inequivalent particle interpretations, is the topic of Chapters 9 and 10, which will also characterize "respect" more precisely.) Kay and Wald (1991) show that in suitably symmetric spacetimes, respectful representations of the Klein–Gordon CCRs are unique up to unitary equivalence. In the case that the respectful representation π_ω is irreducible, to demand respect is to attribute QFT a kinematic pair $(\pi_\omega(\mathfrak{A})'', \mathcal{F}_\omega)$ of the form $\big(\mathfrak{B}(\mathcal{H}), \mathfrak{T}^+(\mathcal{H})\big)$ favored by the Hilbert Space Conservative.

Another promising principle of privilege for the Hilbert Space Conservative declares a state ω on $\mathfrak{A}$ *physical$_H$* only if it satisfies the Hadamard condition, which means that it supports a procedure for assigning an expectation value to a stress-energy observable (see §1.4). Wald (1994, §4.6) announces a principle of this sort. This principle of privilege lends aid and comfort to the Hilbert Space Conservative in spacetimes with compact Cauchy surfaces (a.k.a. closed universes), where all Hadamard vacuum states are unitarily equivalent (Wald 1994, 97). Where ω is a pure Hadamard state for a closed universe, the demand that states be physical$_H$ issues in a kinematic pair $(\pi_\omega(\mathfrak{A})'', \mathcal{F}_\omega)$ of the sort the Hilbert Space Conservative seeks. But in open universes, this is not so.

6.2.2 Underprivileged?

Notice that each of the foregoing principles of privilege is of limited application. Symmetry-mongering works only in duly symmetric spacetimes; Hadamardism appeals only insofar as we're obligated to recognize a stress-energy observable, and eventuates in the sort of kinematic pair the Hilbert Space Conservative seeks only in spacetimes with compact Cauchy surfaces. Neither principle of privilege offers the Conservative solace in general space times, or in the thermodynamic limit of QSM, which is generally set in Euclidean three-space and fails to aspire to contact with the general theory of relativity. This suggests that the viability of Hilbert Space Conservatism depends on *what sort of theory* of QM$_\infty$—QFT in Minkowski spacetime? Or in a curved spacetime?? With or without compact Cauchy surfaces??? QSM????—one means to interpret. Later chapters will develop the suggestion that principled Hilbert Space Conservativism has no feasible uniform privileging strategy: no strategy that can be announced in advance of examining particular theories and their applications for singling out the favored folium of physical representations.

This suggestion prompts the worry that when Hilbert Space Conservatism is tenable, it may not be pristine. If the Conservative appeals to accidental features of the interpretive setting to single out a kinematic pair $(\mathfrak{B}(\mathcal{H}), \mathfrak{T}^+(\mathcal{H}))$ for a theory of QM_∞, she might violate the ideal of pristine interpretation. Insisting on the availability of particle talk or on the sustainability of a certain version of the project of semi-classical quantum gravity can underwrite privileging strategies supportive of Hilbert Space Conservatism. Depending on how such insistence is justified—options include appeals to analytic metaphysics as well as to socially negotiated patterns of practical physics—it too could be pristine or adulterating. For now, the possibilities to keep in mind are, first, that the Hilbert Space Conservative must appeal to metaphysically unprincipled extranomic features to privilege a kinematic pair $(\mathfrak{B}(\mathcal{H}), \mathfrak{T}^+(\mathcal{H}))$ for a theory of canonical type $\mathfrak{A}$, and, second, that the Hilbert Space Conservative has no uniform pristine strategy for equipping *each* theory of QM_∞ with a kinematic pair embodying Conservative values. Should this second possibility obtain, a pristine Conservatism of principle would not be a tenable strategy for interpreting theories of QM_∞, although there may be instances of theories of QM_∞ which an opportunistic Conservatism interprets felicitously.

6.2.3 Overprivileged?

Hilbert Space Conservatism has an obvious vulnerability. Interpreting a theory of canonical type $\mathfrak{A}$, the Conservative withholds physical significance from all states on $\mathfrak{A}$ except those lying in some privileged folium $\mathcal{F}_\omega$. The Conservative thus runs the risk of depriving QM_∞ of states it needs to discharge its explanatory tasks. Chapter 12 shows how real this risk is. As that chapter relates, the account of equilibrium appropriate to the thermodynamic limit of QSM generalizes the Gibbs notion of equilibrium by identifying a condition that every Gibbs equilibrium state satisfies, but which continues to make sense in circumstances, endemic to QM_∞, where the standard Gibbs state is undefined. In QM_∞, this *KMS condition* explicates the concept of equilibrium. States on $\mathfrak{A}$ satisfying the condition with respect to a given inverse temperature β and a given dynamics $\alpha_t : \mathfrak{A} \to \mathfrak{A}$ are known as (α_t, β)-KMS states. They are standardly taken to be equilibrium states at inverse temperature β with respect to the dynamics α_t.

Now in general, an (α_t, β)-KMS state and a (α_t, β')-KMS state on $\mathfrak{A}$ are disjoint if $\beta \neq \beta'$. If ω on $\mathfrak{A}$ is pure, its folium $\mathcal{F}_\omega$ can't contain both of a pair of disjoint states. (In Chapter 12 we'll see something even weirder: if ω is pure, $\mathcal{F}_\omega$ can't contain *any* equilibrium state at a non-zero temperature!) Confining physical possibility to states in a single folium $\mathcal{F}_\omega$, the Conservative allows that at most *one* equilibrium state is possible. Declaring equilibrium at all but at most one temperature to be impossible is not a promising route to interpreting QSM.

Another worry for Hilbert Space Conservatism is that the privileging strategy of symmetry mongering can be incompatible with the phenomenon of *broken symmetry*. Here is one definition:

Definition 6.1 (broken symmetry). Where an automorphism $\alpha : \mathfrak{A} \to \mathfrak{A}$ implements a symmetry of a theory of canonical type $\mathfrak{A}$, that symmetry is broken in a state ω on $\mathfrak{A}$ iff α fails to be unitarily implementable (in the sense of Def. 5.2) in ω's GNS representation π_ω.

The infinite spin chain of §3.1 provides a heuristic example of broken symmetry. Rotation is a symmetry of Euclidean three-space, and ought to be a symmetry of any dynamics unfolding therein. Intuitively, the ground state $[s_k]_j = +1$ of the representation $(\pi_+, \mathcal{H}^+)$ of the CARs can be obtained from the ground state $[s_k]_j = -1$ of the representation $(\pi_-, \mathcal{H}^-)$ by rotation. For the former is a state in which all spins point in the positive z direction; the latter a state in which they all point in the opposite direction. Because the representations $(\pi_-, \mathcal{H}^-)$ and $(\pi_+, \mathcal{H}^+)$ are unitarily inequivalent, the rotation that carries one ground state to another is not unitarily implementable.

Broken symmetry in the sense of Def. 6.1 is impossible in any state declared physical by the symmetry-mongering privileging strategy (supposing that the symmetry broken and the symmetry mongered coincide). No matter what principle of privilege is in play, the Hilbert Space Conservative withholds physical significance from each of a pair of different states of broken symmetry, simply because if one such state resides in the folium of physical states, the other doesn't. Chapters 12–14 will treat these, and a variety of other putative physical phenomena whose representation and explanation is impeded by Hilbert Space Conservatism, in greater detail.

To articulate a kinematic pair for a theory of canonical type $\mathfrak{A}$ is not thereby to complete even the structure-specifying stage of interpreting that theory. The question of dynamics remains. Whatever the dynamical details of a system assigned kinematic pair $(\pi_\omega(\mathfrak{A})'', \mathcal{F}_\omega)$, it is crucial for the Hilbert Space Conservative that those dynamics arise from observables in $\pi_\omega(\mathfrak{A})''$, and that $\mathcal{F}_\omega$ be closed under those dynamics. Otherwise, physically significant magnitudes—those generating dynamics—and physically significant states—those dynamically accessible from states granted antecedent physical significance—escape the clutches of the Conservative's preferred kinematic pair. Chapter 12 considers cases where this might not be so: instances of QM_∞ dynamics that outrun the resources of Hilbert Space Conservatism.

Along with the problems of underprivilege, these putative instances of overprivilege—of Conservatism withholding significance from states that merit it—reinforces the suspicion that Conservatism is not a promising *universal* strategy for interpreting QM_∞. The suspicion is that kinematic pairs appropriate to theories of QM_∞ can differ in kind from those typical of ordinary QM. After a detour chronicling the power of privilege as represented by an approach to the foundations of QFT known as *DHR Selection Theory*, §6.4 presents a historically significant development of this suspicion.

6.3 DHR selection theory

The problem of privilege for the Hilbert Space Conservative provides an occasion to sketch an approach to the foundations of QFT that deserves more attention than it will receive here: DHR selection theory. DHR selection theory shows just how far a little privilege can take you. D, H, and R are Doplicher, Haag, and Roberts, who originally proposed the selection criterion for physically reasonable states from which the approach takes its name (Doplicher, Haag, and Roberts 1971; 1974). Introduced to account for superselection phenomena, the approach has proven so resourceful that, some contend, DHR selection theory holds the key to understanding QFT (Halvorson and Müger 2007, 844). An accurate expression and more extensive development of DHR selection theory requires mathematical tools beyond the scope of this work (for these, see for starters Halvorson and Müger 2007, §7.1 ff., or Ojima 2003, §3.1). We'll settle here for a brief overview, which will set the stage for a necessarily cursory account of where the DHR approach to QFT fits into the landscape of interpretations of QM_∞ surveyed in this book. Lest the hopes of any would-be Conservatives be unduly stoked, I will emphasize at the outset that DHR selection theory finds its most powerful expression when states *disjoint* from its privileged vacuum are physically significant. That is, it finds its most powerful expression when physical significance extends beyond a single folium—a circumstance ruled out by Hilbert Space Conservatism. Thus although both are committed to some form of privilege, the Conservative and the advocate of DHR selection theory make poor allies.

6.3.1 Motivation and set up

To work our way into DHR selection theory, let us think of superselection rules on the following terms. Suppose a representation π of the algebra $\mathfrak{A}$ of observables decomposes into a direct sum $\oplus \pi_i$ of irreducible representations of $\mathfrak{A}$ that acts on the direct sum Hilbert space $\oplus \mathcal{H}_i$. Such a representation sustains a superselection rule denying the status of pure states to superpositions between vectors from distinct "superselection sectors" $\mathcal{H}_i$.

We will think of this superselection structure as arising from other ingredients. The first is a *field algebra* $\mathfrak{F}$ of concrete observables acting on a Hilbert space $\mathcal{H}$. The term of art "field algebra" signals that $\mathfrak{F}$ may contain self-adjoint elements that don't correspond to "physical" observables. Our second ingredient is a *gauge group* G acting on $\mathfrak{F}$. We take elements of G to be transformations between physically equivalent situations. We blend these ingredients in a *stipulation* that the physical observables are the *gauge-invariant* elements of $\mathfrak{F}$: the algebra of physical observables, which we'll call $\mathfrak{A}$, is the subalgebra of $\mathfrak{F}$ each of whose members is invariant under the action of each element of G. We gently warm this blend of $\mathfrak{F}$, G, and $\mathfrak{A}$ with theorems about representations of field systems with gauge symmetries (for an account, see Halvorson and Müger 2007, §9). The result: the concrete Hilbert space $\mathcal{H}$, the Hilbert space on which we supposed the field algebra $\mathfrak{F}$ to act, decomposes into a direct sum $\oplus \mathcal{H}_i$ of

Hilbert spaces $\mathcal{H}_i$, each of which transforms irreducibly under the action of the gauge group G and each of which carries an irreducible representation π_i of the algebra $\mathfrak{A}$ of physical observables. Thus given a field algebra $\mathfrak{F}$ with gauge group G acting on a Hilbert space $\mathcal{H}$, we can *reconstruct* the superselection structure of the physical algebra $\mathfrak{A}$ of $\mathfrak{F}$'s gauge-invariant elements. The reconstruction underwrites as well a criterion of physical significance for representations of $\mathfrak{A}$: representations π_i figuring in the direct sum $\oplus \mathcal{H}_i$ are the physically significant ones.

Now, the idea of the DHR approach is to shuffle the ingredients in this explication of superselection. Instead of starting with the field algebra $\mathfrak{F}$ and its gauge group G, and constructing the observable algebra $\mathfrak{A}$ and its superselection structure (in the form of a direct sum of physically significant representations), DHR selection theory starts with the observable algebra $\mathfrak{A}$, *endowed with a privileged representation*, and constructs the field algebra $\mathfrak{F}$ and gauge group G from them. It moves, its advocates would urge, from the phenomena of superselection (in the form of physical observables and physical states obeying superselection rules with respect to those observables) to the structures (e.g. $\mathfrak{F}$, G) that lie beneath those phenomena.

In slightly more detail: DHR selection theory begins with an algebra $\mathfrak{A}$ of physical observables for a Minkowski spacetime $\mathcal{M}$ and a privileged ("vacuum") state ω_0 on (and so representation π_0 of) that algebra. It subjects $\mathfrak{A}$ to a set of axioms impoverished (but also strengthened) relative to those announced in §5.2. Associations between subalgebras of $\mathfrak{A}$ and subregions of M are required to obey (among other requirements) Isotony, Microcausality, and *Haag Duality*. Haag Duality requires that the commutant of the algebra pertaining to a region $\mathcal{O}$ *coincide with* the algebra for $\mathcal{O}$'s spacelike complement. This is stronger than Microcausality, which requires *only* that the algebra for $\mathcal{O}$'s spacelike complement be a subalgebra of the commutant of the algebra for $\mathcal{O}$. In the DHR approach, other standard assumptions invoking the symmetries of Minkowski spacetime are suspended. In particular, "the [Lorentz] covariance and spectral properties of the vacuum sector, when present, are merely spectators" (Doplicher 1992, 392).

We will see that the privileged vacuum state ω_0 is the linchpin of the DHR approach. But DHR selection theory "effectively ignores the question of how to choose a vacuum representation" (Halvorson and Müger 2007, 842). Indeed it jettisons some of the apparatus—the axioms of Lorentz covariance and the spectrum condition—that in Minkowski spacetime underwrite the privilege of the Minkowski vacuum state. The DHR approach is not *incompatible* with these "spectator" axioms. It just doesn't put them to work. So any ground the spectator axioms might provide for privileging the standard Minkowski vacuum is ground DHR selection theory has no further reason to stand on.

Based on no articulate prior principles, DHR selection theory's *ab initio* privilege of a vacuum representation won't help the Hilbert Space Conservative solve her problem of privilege. But the *ab initio* privilege is just a starting point of DHR selection theory. It could transpire that DHR selection theory forms a powerful explanatory apparatus, to which the specification of a privileged vacuum representation stands as a postulate that is vindicated by the success of the theory based on it. In this case, the invocation of a privileged vacuum might be defended by *a posteriori* arguments, including the argument that the invocation figures in the simplest, strongest, most perspicuous systematization of phenomena we have. Such an argument strikes me as a reasonable response to the problem of privilege. The question, to which we will return, is whether it is available to the DHR theorist or (through DHR theory) to the Hilbert Space Conservative.

Given an algebra of observables and a privileged vacuum state, DHR's next step is to announce the "selection criterion" from which the approach derives its name. The intuitive idea, which won't be elaborated here, is that physically reasonable states are those that differ from the privileged vacuum only locally. I'll call these "DHR states" or (since states determine representations via the GNS construction) "DHR representations."

Here is where things get a little fancy. It can be shown that reasonable states meeting the selection criterion form a category, which coincides with the category of "localized transportable endomorphisms" of $\mathfrak{A}$. An endomorphism ρ of $\mathfrak{A}$ is a map from $\mathfrak{A}$ to itself preserving algebraic structure. An endomorphism ρ can differ from an automorphism in that $\rho(\mathfrak{A})$ can turn out to be a proper subset of $\mathfrak{A}$. Very roughly, an endomorphism ρ is *localized* in a region $\mathcal{O}$ if ρ leaves observables pertaining to the spacelike complement of $\mathcal{O}$ fixed; an endomorphism ρ localized in $\mathcal{O}$ is *transportable* if for any other region $\mathcal{O}_1$, there's a unitary $U_1 \in \mathfrak{A}$ that "translates" ρ into an endomorphism ρ_1 localized in $\mathcal{O}_1$. (The "translation manual" is $\rho_1(A) = U^{-1}\rho(A)U$ for all $A \in \mathfrak{A}$.) It's not ludicrous to suppose that a state ω satisfying the DHR selection criterion of differing from the privileged vacuum ω_0 only locally is a state that can be obtained by intertwining the privileged vacuum with a localized, transportable endomorphism ρ: that is, $\omega(A) = \omega_0(\rho(A))$ for all $A \in \mathfrak{A}$. The unludicrous supposition can be made precise in such a way that it's true, and explains the coincidence of the categories of DHR representations and localized transportable endomorphisms.

The payoff of considering localized transportable endomorphisms is that the category of endomorphisms exhibits structures that collections of representations lack. For instance, representations can't have other representations as inverses, but endormophisms can—the crux of an illuminating DHR-theoretic account of the nature of antimatter (Baker and Halvorson 2010). The equivalence of the categories of DHR representations and localized transportable endomorphisms makes accounts such as these illuminating of QFT *as we know it*, i.e. via states and representations.

The central result of DHR selection theory is the Doplicher–Roberts reconstruction theorem, which states that given the category of DHR representations of a physical observable algebra $\mathfrak{A}$, one can *uniquely* reconstruct the field algebra $\mathfrak{F}$ and its gauge group G, which will always turn out to be a compact Lie group. This recovery may contain clues to the nature of gauge symmetry as well as the origin of superselection. When the category of DHR representations is non-trivial in the sense that it contains elements unitarily inequivalent to one another, operations definable on the category identify representations exhibiting Bose–Einstein statistics, representations exhibiting Fermi–Dirac statistics, as well as conjugate pairs of representations, corresponding to matter/antimatter pairs.

In light of these impressive successes, let us confront a question raised earlier. Granting that the DHR approach enjoys theoretical success constituting a defense of its initial choice of a privileged vacuum representation, is this defense available to the Hilbert Space Conservative as a solution to her principle of privilege? The envisioned solution justifies confining physical possibility to the the folium of ω_0 on the grounds that privileging ω_0 enables the valuable theoretical successes of DHR selection theory.

I agree with Halvorson and Müger that invoking DHR selection theory, the Hilbert Space Conservative exposes her own position as severely limited in explanatory reach. The structure DHR selection theory puts to most significant explanatory use is a structure available only when the collection of DHR representations is non-trivial. "If we committed ourselves to one representation and ignored others, we would have no field operators, no gauge group, no definition of Bose and Fermi fields, no definition of antiparticles, etc." (Halvorson and Müger 2007, 844). When all states meeting the selection criterion are unitarily equivalent to the privileged vacuum state ω_0, there is no non-trivial superselection structure, and DHR selection theory has little explanatory success to boast of. Limiting physically reasonable states to the folium of a privileged vacuum representation ω_0, whether other states satisfy the DHR selection criterion or not, the Hilbert Space Conservative deprives DHR selection theory of the explanatory success we imagine her invoking to justify her choice of ω_0. On the other hand, recognizing the physical significance of unitarily inequivalent states satisfying the selection criterion and enabling DHR's significant explanatory successes, the Hilbert Space Conservative abandons a central tenet of her Conservatism: that physical possibility is confined to the folium of a single irreducible representation.

In the face of Hilbert Space Conservatism's inability to secure the explanatory success of DHR selection theory, Halvorson and Müger describe an interpretive stance they call *Representational Realism*. Whereas the Hilbert Space Conservative proceeds as though the central interpretive question were: "Which representations are physically reasonable?", Halvorson and Müger draw from DHR selection theory the moral that a selection criterion, a way of picking out the set of physically reasonable representations, is a meager first step toward understanding QFT. "The fundamental insight of DHR theory is that the set of representations itself has structure, and it this structure that explains phenomena" (2007, 844). For a representational realist, the content of a QFT of canonical type $\mathfrak{A}$ set in a spacetime $\mathcal{M}$ is given by (i) the axiom-governed association between subregions $\mathcal{O} \subset \mathcal{M}$ and sualgebras $\mathfrak{A}(\mathcal{O})$ of $\mathfrak{A}$, (ii) a "dynamics" in the form of a representation of $\mathcal{M}$'s translation group as an automorphism group of $\mathfrak{A}$, and (iii) "the symmetric tensor $*$-category $DHR(\mathfrak{A})$ of DHR representations" (Halvorson and Müger 2007, 844).

Both the insights of DHR theory and the interpretive stance they motivate deserve further attention than philosophers of physics have given them, and than they'll get here. The attention they'll get here will take the wet-blanket form of indicating some *prima facie* reservations about the ultimate adequacy of the approach.

6.3.2 Reservations

Granted that the structure of DHR representations *itself* is explanatory, there is still room to worry that there will be explananda not admitting explanans of the envisioned form. Candidates for such explananda include phenomena whose presentation requires representations that fail the DHR selection criterion. Here are some:

The structure of DHR representations of the algebra $\mathfrak{A}$ of physical observables for the free Klein–Gordon field in Minkowski spacetime is trivial. That is, by the lights of the DHR selection criterion, the only physically reasonable representations are those unitarily equivalent to the Minkowski vacuum state. In Chapters 9 and 10 we'll encounter representations, disjoint from the Minkowski vacuum representation, that promise to do explanatory work. Chapter 9 introduces the Rindler representation of the algebra $\mathfrak{A}(\mathcal{R})$ associated with a "wedge" of Minkowski spacetime; this representation promises to accommodate the putative phenomena of *Fulling quanta*: the quanta of radiation detected, according to the Unruh effect, by observers accelerating through the Minkowski vacuum state. Demoting the Rindler representation from physical relevance, the DHR selection criterion would stymie such explanations. "Coherent" representations make an appearance in Chapter 10. Central to quantum optics, coherent representations of $\mathfrak{A}$ can be unitarily inequivalent to the Minkowski vacuum representations. In particular, so-called infrared coherent representations, disjoint from the Minkowski vacuum, are invoked in explanations of certain scattering phenomena. Again, the DHR selection criterion would stymie these explanations.

So there is a sense in which the DHR selection criterion is stingy. Whether it's too stingy is a matter of whether the explanatory maneuvers it blocks are ones we ought to be undertaking to begin with.

My presentation of DHR selection theory sets it squarely in Minkowski spacetime. Reservations attend the particularity of that setting. Don't we want to interpret QFT in curved spacetimes, and other theories of QM_∞ in good old-fashioned Euclidean three-space? There is in principle no obvious impediment to adapting the DHR framework to more general settings. However, there may be in-practice impediments, hinging on answers to questions like the following: does the coincidence of states satisfying the selection criterion and the category of localized transportable endomorphisms hold only for Minkowski spacetime? If it holds for more general spacetimes, how general? For the spacetimes for which it holds, is a reconstruction theorem available? What plays the role of the privileged vacuum state in more general spacetimes? All these questions may have clear and reassuring answers (see e.g. Brunetti and Ruzzi 2007), but I don't know what they are.

Even supposing that the DHR approach can be adapted to perfectly general spacetimes, another reservation remains. Ojima expresses it as follows: "the symmetry arising from this beautiful theory [viz. DHR selection theory] is, however, destined to be unbroken, excluding the situation of SSB [Spontaneous Symmetry Breaking]" (2003, 237). The problem is that the DHR approach stipulates $\mathfrak{A}$ be *invariant* under the action of the gauge group G. It follows that in any representation complying with this stipulation, G is unitarily implementable. But broken gauge symmetry—vacuum states whose folia foil the unitary implementability of some gauge symmetries—is supposed by many to be crucial to the astonishing empirical success of modern QFT (see Chapter 14). This reservation, too, may be surmountable: Ojima (2003) charts a reconciliation between DHR selection theory and broken gauge symmetry.

6.4 Algebraic Imperialism

Irving Segal's 1947 "Postulates for general quantum mechanics" pioneered the C^* algebraic approach to quantum theories. Addressing the apparent foundational problem of unitarily inequivalent representations of the CCRs for QFT, Segal presses the approach into service of an interpretive strategy Arageorgis (1995, 132) has dubbed "Algebraic Imperialism":

The proper sophistication, based on a mixture of operational and mathematical considerations, gives however a unique and transparent formulation ... the canonical variables are fundamentally elements in an abstract algebra of observables, and it is only relative to a particular state of this algebra that they become operators in Hilbert space. (Segal 1959, 343)

Notice how Segal adheres to the ideal of pristine interpretation. "The unique and transparent formulation" the Imperialist will recommend is *"based on a mixture of operational and mathematical considerations"*—considerations of the very sort the ideal endorses.

Algebraic Imperialism holds that the extra structure one obtains along with a concrete representation of (e.g.) the Weyl algebra is extraneous—the Weyl algebra, and not any of its concrete but disparate Hilbert space realizations, encapsulates the physical content of QFT. As Segal puts it, "[T]he important thing here is that the observables form some algebra, and not the representation Hilbert space on which they act" (1967, 128). Kastler is even more explicit. "The choice of [an abstract algebra 'without reference to some particular realization as a [norm]-closed operator algebra on some Hilbert space'] as a frame for quantum mechanics implies that the specification of a special representation is physically irrelevant, all the physical information being contained in the algebraic structure of the abstract algebra $\mathfrak{A}$ alone" (1967, 180).

Inspired by the representation-independence of the Weyl algebra, Algebraic Imperialism construes a theory of QM_∞ admitting unitarily inequivalent Hilbert space representations in terms of the abstract algebraic structure every such representation shares. Physical magnitudes pertaining to a theory of canonical type $\mathfrak{A}$ are given by the self-adjoint part of the abstract C^* algebra $\mathfrak{A}$; possible states are states on $\mathfrak{A}$. Where $S_\mathfrak{A}$ is the set of these states, the Imperialist equips a theory of canonical type $\mathfrak{A}$ with the kinematic pair $(\mathfrak{A}, S_\mathfrak{A})$.

6.4 ALGEBRAIC IMPERIALISM

Algebraic Imperialism and Hilbert Space Conservatism are clearly rival accounts of the content of a theory of QM_∞. For there can arise states ω and ω' on a canonical algebra $\mathfrak{A}$ which are disjoint. Both are, according to the Algebraic Imperialist, possible states for a theory of canonical type $\mathfrak{A}$. But they can't both lie in the folium of some pure state ω on $\mathfrak{A}$. So for the Conservative, they can't both be possible according to a theory of canonical type $\mathfrak{A}$.

6.4.1 Apologetic Imperialism

Unitary equivalence is a criterion of physical equivalence ill-adapted to Algebraic Imperialism. Consider a concrete representation π of the abstract C^* algebra $\mathfrak{A}$ and an arbitrary state ω on that algebra. ω directly determines a state on the concrete C^* algebra $\pi(\mathfrak{A})$—the state that assigns each concrete representative $\pi(A)$ what ω assigns its abstract counterpart $A \in \mathfrak{A}$. In virtue of this one-to-one correspondence, the collection $\mathcal{S}_\mathfrak{A}$ of states on $\mathfrak{A}$ can be considered a collection of states on $\pi(\mathfrak{A})$. Now consider pure states $\omega, \omega' \in \mathcal{S}_\mathfrak{A}$. Because the Imperialist regards how an abstract algebraic structure is concretely represented to be a matter without physical significance, she'll take there to be no physical difference between the kinematic pairs $(\pi_\omega(\mathfrak{A}), \mathcal{S}_\mathfrak{A})$ and $(\pi'_\omega(\mathfrak{A}), \mathcal{S}_\mathfrak{A})$. By her lights, each has exactly the physical content captured by the kinematic pair $(\mathfrak{A}, \mathcal{S}_\mathfrak{A})$. But if ω and ω' are unitarily inequivalent, the kinematic pairs $(\pi_\omega(\mathfrak{A}), \mathcal{S}_\mathfrak{A})$ and $(\pi'_\omega(\mathfrak{A}), \mathcal{S}_\mathfrak{A})$ are unitarily inequivalent as well, for by hypothesis there is no unitary map between $\pi_\omega(\mathfrak{A})$ and $\pi_{\omega'}(\mathfrak{A})$. If kinematic pairs must be unitarily equivalent to be physically equivalent, pairs with observable sets $\pi_\omega(\mathfrak{A})$ and $\pi_{\omega'}(\mathfrak{A})$, pairs whose content the Imperialist would identify, fail to be physically equivalent.

Recognizing a mismatch between their account of physical content and the standard use of unitary equivalence as a criterion of physical equivalence, early Imperialists (e.g. Haag and Kastler 1964, 851; see also Robinson 1966, 488) urged, instead of unitary equivalence, *weak equivalence* as a criterion of physical equivalence. It seems to me that these advocates regarded their attribution of simultaneous physical significance to disjoint states as requiring some sort of apology, an apology they couched in terms of weak equivalence. So I will call them *apologetic Imperialists*.

Weak equivalence is a relation that holds (or not) between Hilbert space representations of a C^* algebra $\mathfrak{A}$. Two representations are weakly equivalent exactly when no finite set of expectation values specified to finite accuracy can locate a state in one, rather than the other, representation. More precisely,

Definition 6.2 (weak equivalence). Representations $(\pi_1, \mathcal{H}_1)$ and $(\pi_2, \mathcal{H}_2)$ of a C^* algebra $\mathfrak{A}$ are weakly equivalent iff, given any π_1-normal state ω on $\mathfrak{A}$ and given any finite set of elements $A_i \in \mathfrak{A}$ and any finite set of non-zero margins of experimental error ϵ_i, there exists a π_2-normal state ω' that reproduces within those margins of errors ω's assignment of expectation values to those observables.

Now the apologizing Imperialist "define[s] two representations to be *physically equivalent* if and only if they are weakly equivalent" (Robinson 1966, 488).

A rough and ready operationalism suggests a justification of the equation of physical and weak equivalence. Kastler contends: "We want the results of any finite set of measurements on a physical state to be equally well describable in terms of a density matrix on $\mathcal{H}_1$ or a density matrix on $\mathcal{H}_2$. As measurements are never totally accurate, 'equally well' is to be understood as 'to any desired degree of accuracy' "(1964, 180–81). Any actual measurement performed on a system in the state ω_1 will fix the expectation values of some finite collection of observables A_i, $i = 1, 2, \ldots, n$, within experimental error margins $\varepsilon_i > 0$. Accordingly, such a measurement cannot distinguish ω_1 from any other state ω_2 satisfying $|\omega_1(A_i) - \omega_2(A_i)| < \varepsilon_i$ for all i. Such states are, the apologetic Imperialist observes, for all practical purposes, physically equivalent. Weakly equivalent representations are such that any state normal with respect to one is, for all practical purposes, physically equivalent to a state normal with respect to the other. The apologetic Imperialist takes this to sanction the conclusion that weakly equivalent representations are for all practical purposes physically equivalent as well. Haag and Kastler conclude: "The fact that no actual measurement can be performed with absolute precision implies that the realistic notion of 'physical equivalence' is far less stringent than that of unitary equivalence" (1964, 849).

Fell's theorem consolidates weak equivalence as a criterion of physical equivalence hospitable to algebraic approaches. Fell's theorem implies that all faithful representations of the Weyl algebra are weakly equivalent (see Wald 1994, 81). Defining physical equivalence as weak equivalence, the apologetic Imperialist can conclude, "All faithful representations of [the Weyl algebra] are physically equivalent" (Robinson 1966, 488)—and chastise the Conservative for elevating surplus structure to physical significance: "It is in this new notion of equivalence in field theory that the algebraic approach has its greatest justification. All the physical content of the theory is contained in the algebra itself; nothing of fundamental significance is added to a theory by its expression in a particular representation" (ibid.).

Notice that the observables with respect to which weakly equivalent representations are practically indistinguishable are elements of the abstract algebra $\mathfrak{A}$ or their representatives. Consider, for some concrete representation π_1 of $\mathfrak{A}$, an observable in the affiliated von Neumann algebra $\pi_1(\mathfrak{A})''$ but without correlate in $\mathfrak{A}$. I will call such an observable *parochial* to the representation π_1. Examples of parochial observables include the total number operators (or more properly the projection operators in their spectral resolution) encountered in suitable representations of the Weyl algebra for the free Klein–Gordon field. There is a π_1-normal state, the vacuum state, which assigns the total number operator N_{π_1} affiliated with the π_1 representation expectation value 0. But there is no normal state in any representation π_2 disjoint from π_1 that assigns N_{π_1} an expectation value within any finite ϵ of 0 (Clifton and Halvorson 2001b, prop. 11 (450)). Because every faithful representation of the Weyl algebra is weakly equivalent to every other, apologetic Imperialists would hail π_1 and π_2 as physically equivalent... even though all π_2-normal states assign an observable parochial to π_1—its total number operator N_{π_1}—"expectation values"[4] arbitrarily *far* from that assigned by the π_1 vacuum state. Of course, this is a compelling criticism of weak equivalence as an explication of physical equivalence only for those inclined—as Imperialists are *not*—to regard parochial observables as physical.

Haag and Kastler develop an argument for withholding physical significance from "global" observables such as the total number operator or the net polarization of an infinite spin chain. These observables share the feature that they're defined in each representation in terms of weak limits of functions of canonical observables. Haag and Kastler focus on the example of global charge. Every state within an irreducible representation of the algebra $\mathfrak{A}$ of "conventional field theory" (1964, 854) is an eigenstate *with the same eigenvalue* of an observable representing the total charge present in spacetime; states in disjoint representations determine distinct eigenvalues of this observable. (The polarization of the infinite spin chain (reviewed in §3.3.1) is an example of this sort of thing.) Construing global charge as a superselection observable, Haag and Kastler refer to disjoint irreducible representations as "sectors."[5] Indeed, we can label disjoint irreducible representations of $\mathfrak{A}$ by their associated charges: every state in π_k is an eigenstate of global charge associated with eigenvalue k.

Now, π_k and $\pi_{k'}$ for $k \neq k'$ aren't unitarily equivalent. Every π_k-normal state assigns the global charge observable a different determinate value from every $\pi_{k'}$-normal state. Yet the representations π_k and $\pi_{k'}$ are weakly, and therefore physically, equivalent, according to the apologetic Imperialist. How can this be, given the states' adamant disagreement about the value of the global charge observable?

It can be, Haag and Kastler assure us, "because the distinction between different sectors cannot be made by means of experiments in finite regions" (1964, 854). Consider the inequivalent irreducible representations π_{-1} and π_3. Every state in the former corresponds to a determinate global charge of -1; every state in the latter corresponds to a determinate global charge of 3. Haag and Kastler's "particle behind the moon" argument for earthbound physicists is that the apparent physical difference between these representations can be undone "by adding 4 elementary particles of negative charge in a remote region of space—behind the moon, say" (ibid.; see also Kastler 1964, 180–81). But such an addition makes no difference to earth physics: "the effect of this added charge on the expectation values of the quasilocal quantities [the elements of earth's local algebra] tends to 0 as the regions is moved to ∞" (Haag and Kastler 1964, 854). Hence the difference between the π_{-1} and π_3

---

[4] The scare quotes because, strictly speaking, the state is outside the domain of the observable.

[5] The superselection angle needn't concern us here, but you can think of it this way: The "standard representation" π of the field algebra $\mathfrak{A}$ is reducible; its disjoint irreducible subrepresentations (the π_i in $\pi = \oplus_i \pi_i$) each correspond to a different value c_i of global charge. Thus, where I_i is the identity element of the subrepresentation π_i, the charge observable takes the form $C = \oplus_i c_i I_i$. Commuting with every other element of $\pi(\mathfrak{A})''$, C is a superselection observable; the Hilbert space on which each subrepresention acts is a superselection sector.

representations fails the test of significance: that a difference has to have a laboratory manifestation to be physical. Differences in global charge (and other global quantities) notwithstanding, Haag and Kastler maintain weakly equivalent representations to be physically equivalent.

The particle-behind-the-moon argument, along with other justifications apologetic Imperialists offer for weak equivalence as a criterion of physical equivalence, is avowedly operationalist. But these justifications ultimately fail, *even by operationalist lights*. For, as Summers (1999) has observed, even operationalists ought to take states to be predictive instrumentalities. Operationalists should therefore expect the equivalence of states to extend to their predictions concerning future measurements. But Fell's theorem offers us no assurance that a π_1-normal state for all practical purposes indistinguishable from a π_2-normal state with respect to some set $\{A_i\}$ of algebraic observables will continue to mimic the first state's predictions with respect to an expanded or altered set of observables. The practical indistinguishability of states that secures the weak equivalence of the representations bearing them is only a backward-looking or static sort of indistinguishability. Before recognizing a normal state on one representation as a suitable counterpart to a normal state on another, an operationalist should demand not only backward-looking but also forward-looking indistinguishability: the indistinguishability of putative counterpart states conceived as predictive instrumentalities. Thus the appeal to Fell's theorem should not convince even the operationalist that weak equivalence is an acceptable notion of physical equivalence.

The Hilbert Space Conservative identifies the kinematic pair for a theory of canonical type $\mathfrak{A}$ in terms of a Hilbert space representation, then adopts a criterion of physical equivalence for kinematic pairs applicable to Hilbert space representations. The apologetic Algebraic Imperialist identifies the kinematic pair for a theory of canonical type $\mathfrak{A}$ in terms of an abstract algebra, then adopts a criterion of physical equivalence applicable to... Hilbert space representations (albeit that portion of them bearing a representation of an abstract algebra). This is odd. A second, and related, oddity of apologetic Algebraic Imperialism is that weak equivalence isn't straightforwardly a content coincidence criterion of physical equivalence. To claim that $(\pi_1, \mathcal{H}_1)$ and $(\pi_2, \mathcal{H}_2)$ are weakly equivalent is not to offer isomorphisms between their state spaces and observable sets. It's rather to make an existence claim, relativized to finite sets of algebraic observables and error margins, to the effect that, for any state in one folium, there exists some state on the second indistinguishable from it.

6.4.2 Bold Imperialism

Apologies for algebraic liberality invoking weak equivalence fall flat. Perhaps the best way for an Imperialist to articulate his criterion of physical equivalence is the bold way. Just as his notion of algebraic state deposes the Hilbert space notion of state, so too a criterion of physical equivalence suited to the algebraic approach should depose the Hilbert space criterion of unitary equivalence. Defining physical equivalence directly

6.4 ALGEBRAIC IMPERIALISM   137

in terms of the Imperialist kinematic pairs ($\mathfrak{A}, \mathcal{S}_\mathfrak{A}$) and ($\mathfrak{A}', \mathcal{S}_{\mathfrak{A}'}$), the bold Imperialist maintains such pairs to be physically equivalent only if $\mathfrak{A}$ and $\mathfrak{A}'$ are related by an isomorphism $\alpha : \mathfrak{A} \to \mathfrak{A}'$ identifying the canonical elements generating $\mathfrak{A}$ with those generating $\mathfrak{A}'$. This is a content coincidence criterion of physical equivalence: α establishes the bijection between observables ($A \in \mathfrak{A} \to \alpha(A) \in \mathfrak{A}'$) and states ($\omega \in \mathcal{S}_\mathfrak{A} \to \omega \circ \alpha^{-1} \in \mathcal{S}_{\mathfrak{A}'}$) securing content coincidence. The bold Imperialist would not flinch in the face of the observation that he thereby elevates a continuum of states giving rise to unitarily inequivalent GNS representations to physical significance. For, he would maintain, he has no reason to take unitary equivalence seriously as a notion of physical equivalence to begin with.

Once ($\mathfrak{A}, \mathcal{S}_\mathfrak{A}$) is identified as the appropriate kinematic pair for a theory of QM_∞, the bold Imperialist's criterion of physical equivalence follows straightforwardly. The challenge for the bold Imperialist is to justify his choice of kinematic pair.

Segal, possibly the boldest Imperialist of them all, rises to the challenge by an operationalist ploy of his own. All hands, Segal suggests, should grant that the canonical observables pertaining to a system are operationally meaningful, as are bounded functions of finite sets of canonical observables. All hands should go along with the suggestion. Conceding significance to canonical observables is a *sine qua non* of doing physics at all. What's more, as Haag observes, there is a straightforward sense in which bounded functions of operationally meaningful observables are themselves operationally meaningful. "The change from the operator A to the operator $F(A)$ does not mean that we change the apparatus; it only labels the measuring results differently, assigning the value $F(a)$ to the result formerly labeled by the value a" (1992, 5).

With this common ground established, Segal's crucial move is to grant physical significance to the uniform closure but not (where they differ) the weak closure of this set of antecedently meaningful observables. He accepts that "uniform approximation is operationally meaningful" because where f_n uniformly approximates f, f's "expectation value in any state is simply the limit of the expectation values of the approximating bounded functions" (1959, 348–9). If you don't believe this, imagine, for reductio, that there's some state ω on $\mathfrak{A}$ such that $\omega(f_i)$ fails to converge to $\omega(f)$, even though $f_i \in \mathfrak{A}$ converges to $f \in \mathfrak{A}$ in $\mathfrak{A}$'s C^* norm. If f_i converges in $\mathfrak{A}$'s norm to f, then for any representation π of $\mathfrak{A}$, $\pi(f_i)$ converges to $\pi(f)$ in that representation's uniform topology (this is because $*$-morphisms are continuous: see §4.4). But our hypothesis implies that the sequence fails to converge uniformly in ω's GNS representation: if the sequence converged uniformly, it would converge strongly, but $\pi_\omega(f_i)$ doesn't converge to $\pi_\omega(f)$ in $\mathcal{H}_\omega$'s *strong* topology. The cyclic vector $|\Omega\rangle$ implementing ω spoils strong convergence: by hypothesis, $|\omega(f_n - f)| = |\langle\Omega|(\pi(f_n) - \pi(f))|\Omega\rangle|$ won't go to zero as n goes to ∞, but that quantity must go to zero if $\pi_\omega(f_i)$ is to converge strongly to $\pi_\omega(f)$. We have our contradiction, and can conclude that when f_i converges in to f in C^* norm [equivalently, when $\pi(f_i)$ converges uniformly to $\pi(f)$ in a representation π of $\mathfrak{A}$], $\omega(f_n)$ converges to $\omega(f)$. Segal takes this to secure the operational significance of uniform closure, because it licenses us to relabel a sequence of f_n measurements as an f measurement.

To see why Segal would deny operational significance to weak closure, imagine a sequence of $A_i \in \mathfrak{A}$ that does not converge in $\mathfrak{A}$'s C^* norm. Suppose further that in some representation π of $\mathfrak{A}$, the sequence $\pi(A_i)$ *does* converge *in $\mathcal{H}_\pi$'s weak topology*, to an operator A which, if self-adjoint, will be an observable parochial to π. We know that if ω is π-normal, then it's ultra-weakly continuous on the von Neumann algebra $\pi(\mathfrak{A})''$ (see §4.6.2; Bratteli and Robinson 1987, thm. 2.4.21 (76)). For our purposes, what matters is that if a state is *weakly continuous* on a von Neumann algebra, then it's ultra-weakly continuous. Conversely, a non-ultra-weakly continuous state is not weakly continuous. Thus, if a state ϕ on $\mathfrak{A}$ is not π-normal—and so not weakly continuous on $\pi(\mathfrak{A})''$—there are sequences $\phi(A_i)$, $(A_i \in \mathfrak{A})$, which fail to converge, even though the sequence $\pi(A_i)$ converges in $\mathcal{H}_\pi$'s weak topology.

Segal's point is that if $\mathfrak{A}$ admits states ϕ that are not π-normal, there are elements of $\pi(\mathfrak{A})$'s weak closure (e.g. A) whose expectation values in ϕ aren't fixed by ϕ's expectation value assignments to sequences of observables conceded significance by all hands (e.g. A_i). The operationalization of the antecedently significant observables doesn't extend automatically to $\pi(\mathfrak{A})$'s weak closure. Uniform limits, by contrast, *are* operationally significant because, if $\pi(f_i)$ converges uniformly to some limit $\pi(f)$, $\omega(f_i)$ perforce converges as well.

The key difference, then, between weak and uniform limits is that the world can get itself into a condition—the condition of a state not normal with respect to the representation whose weak topology is used to take a weak limit—where the expectation value of weak limit floats free of any set, even an infinite one, of laboratory machinations, even perfectly accurate ones. With uniform limits, this is not so. Thus Segal would vest uniform, but not weak, limits with operational, and so physical, significance. It follows that he regards as significant the C^*, but not (where they differ) the von Neumann algebras generated by elements satisfying the canonical relations defining a theory. It follows as well that he regards as equivalent kinematic pairs espousing isomorphic C^* algebras. Therein lies his bold Imperialism.

Segal's derivation of Imperialism from operationalism impresses me as deeper and less opportunistic than the apologetic Imperialists' plea for weak equivalence. Segal rests his case on a principle like:[6]

> (SEGOP)   Let $\mathcal{P}$ be the set of bounded functions of finite sets of canonical observables for some system. X is an observable pertaining to the system only if there is a sequence $X_i \in \mathcal{P}$ such that, for any normed positive linear functional $\omega : \mathcal{P} \to \mathbb{C}$, $\omega(X_i)$ converges. When this condition is met, $\lim_{i \to \infty} \omega(X_i)$ is the expectation value of X in the state ω.

Whereas the normal conception of states is an ideology that fixes the state space for a theory given its observable set, (SEGOP) is an ideology that given the state space of a theory of canonical type $\mathfrak{A}$ (*in the form of a space of normed, positive linear functionals*

---

[6] A complication I'm suppressing is the need to add codicils to (SEGOP) to reflect the requirement that observables have real expectation values.

over $\mathfrak{A}$) fixes its observable set. Reflecting the conviction that physical magnitudes are those for which there exist measurement protocols, (SEGOP) has operationalist roots. But the principle makes minimal appeal to the depressing actualities of laboratory life—its finitude, imperfect material conditions, or inaccuracy. Indeed, a confirmed metaphysician who forswears operationalism might nevertheless accept (SEGOP) as an explication of the parsimony desideratum discussed in §1.2. This is the desideratum that an interpretation attribute to worlds possible according to a theory no more complexity or structure than the theory demands to discharge its explanatory duties. The insight underlying the explication is that in theories satisfying (SEGOP), the worlds possible and the magnitudes characterizing those worlds march in lockstep.

Of course, the catch is that built into (SEGOP) is a substantive assumption, italicized in the foregoing paragraph, about which states are possible for a theory of canonical type $\mathfrak{A}$. The possible states are those that correspond to positive linear functionals on the set of bounded functions of the system's canonical observables. This space of possible states corresponds to $\mathcal{S}_\mathfrak{A}$, states in the algebraic sense on $\mathfrak{A}$. (SEGOP)'s rulings won't sway those who, like the Conservative, reject this assumption.

6.4.3 Instantiation and necessity

Still, I think there's a less tendentious principle that preserves the spirit of (SEGOP) without baldly presupposing (as (SEGOP) does) the Imperialist conception of states. The motivation for the less tendentious principle invokes the idea that some relations a theory imposes between physical magnitudes aren't merely contingent but lawlike, in the sense that they hold in every world possible according to theory. We have, after all, been calling the CARs or CCRs generating the canonical algebra $\mathfrak{A}$ "kinematical laws." Here I'll catalog a variety of ways relationships posited by physical theories can be lawlike, offer an account of what it is for a state to *instantiate* such lawlike relationships, and extract from the discussion (SEGOP)'s less tendentious cousin, the Harmony principle, a criterion of adequacy for the kinematic pair one would associate with a theory imposing such lawlike relationships.

Early in his *Physics*, Aristotle tells us what physics is about: natural things, understood as things with *natures*. "The obvious difference between [natural] things and things which are not natural," he says, "is that each of the natural ones contains within itself *a source of change and of stability*" (II.i, italics mine). Attributing the Philosopher uncanny quantum mechanical prescience, let us suppose that by "a source of change and of stability," he means a Hamiltonian. Thus a natural quantum mechanical system is one that can be assigned a Hamiltonian.

In QM$_\infty$, dynamics for the quasilocal algebra pertaining to an infinitely extended system in its entirety (e.g. a QSM system in $\mathbb{R}^3$) can be defined as the limit of dynamics for subalgebras pertaining to subsystems. For instance, where subalgebras $\mathfrak{A}(V)$ of $\mathfrak{A}$ describe subsystems occupying finite volumes $V \in \mathbb{R}^3$, dynamics for a theory of canonical type $\mathfrak{A}$ can be obtained by describing dynamics for $\mathfrak{A}(V)$, then taking the limit as $V \to \infty$. Schematically

$$H = \lim_{V \to \infty} H_V \qquad (6.1)$$

An infinite system for which that limit is ill-defined, lacking a principle of change and stability, fails to be a natural system. It violates Aristotle's demand that the relationships between magnitudes include dynamical relationships, that is, relationships between magnitudes at different times. Insofar as a system must meet this demand to even be natural, the relationship demanded is lawlike. Every system possible according our theory of nature is a system meeting the demand.

Another sort of lawlike interweaving of magnitudes *defines* one magnitude in terms of limits of sequences of other magnitudes. The z-component m_z of the global polarization of the infinite spin chain is an example (see §3.3.1). m_z is defined as the $N \to \infty$ limit of the polarization of N-site lengths of the chain:

$$m_z = \lim_{N \to \infty} m_z^N \qquad (6.2)$$

An infinite system for which that limit is ill-defined lacks a polarization observable, in spite of supporting the observables in whose terms a polarization observable is defined. Such a system violates the demand that the relationships between magnitudes include definitional relationships, i.e. relationships identifying one magnitude in terms of its functional and/or limiting dependence on other magnitudes. Insofar as definitions apply to every system possible according to the theory offering those definitions, the relationships demanded are lawlike.

The axioms of QFT effect yet another sort of lawlike interweaving of magnitudes. Recall the spectrum condition of §5.2, according to which the unitary groups $U_j(b)$ implementing spacetime translations in the vacuum representation are strongly continuous; their infinitesimal generators P_j such that $U_j(b) = e^{-iP_j b}$ define a four-momentum whose spectrum must lie in the forward light cone. This axiom presupposes the strong continuity of the groups $U_j(b)$. It presupposes, that is, that

$$P_j = \lim_{b \to 0} \frac{U_j(b) - I}{b} \qquad (6.3)$$

A representation in which the family of unitary operators implementing spacetime translations fails to be strongly continuous cannot satisfy the Vacuum axiom. Such a representation violates the demand that the relationships between magnitudes include relationships presupposed or required by axioms of the theory governing the behavior of those magnitudes. Insofar as every QFT system possible according to axiomatic QFT satisfies the axioms, the relationships demanded are lawlike.

Finally, the laws, dynamic and otherwise, of a theory can impose particular relationships on physical magnitudes. The CARs and CCRs are non-dynamical examples of relationships required by law. A dynamic example involving a limiting relationship is the Schrödinger equation, whose integral form requires the time evolution of a system with Hamiltonian H to be described by a *strongly continuous* family of time evolution operators $U(t)$ whose infinitesimal generator is H. This law of quantum mechanical time development requires that some physical magnitudes are obtained as limits of others. In particular, the Schrödinger equation requires that:

$$H = \lim_{t \to 0} \frac{U(t) - I}{t} \tag{6.4}$$

A representation in which the unitaries implementing time translations fail to be strongly continuous thereby fails to satisfy the Schrödinger equation. Such a representation violates the demand that the relationships between magnitudes include relationships required by physical law.

The foregoing catalog of sorts of lawlike relationships encountered in physics, cursory and incomplete as it is, prompts a question. Suppose that physical possibilities are those which satisfy physical laws. When those laws impose relationships between magnitudes, physical possibilities obey those laws by instantiating those relationships. Suppose as well that each state recognized by a quantum theory codes a situation that's physically possible, according to that theory. The question is: What does it take for the possibility coded by a quantum state to instantiate lawlike relationships of the foregoing sorts? For relationships which require or presuppose that some magnitudes are limits of others, here is a natural answer:

Instance. Let ω be a state in the algebraic sense on $\mathfrak{A}$, and suppose $A_i \in \mathfrak{A}$. If $A = \lim_{i \to \infty} A_i$ is or is presupposed by a lawlike relation, then ω instantiates that law only if $\omega(A) = \lim_{i \to \infty} \omega(A_i)$.

Instance requires expectation values which quantum states assign to magnitudes to stand to one another in the same limiting relationships the magnitudes themselves do.

It should not occasion alarm that this requirement resembles the "silly" (Mermin 1993) one von Neumann imposes on hidden variable theories in the course of obtaining his notorious No-Go result against them. Von Neumann requires the *actual* values hidden states attribute to (even incompatible) observables to mirror the functional relationships between those observables. John Bell explains why the requirement is silly: "There is no reason to demand [this] individually of the hypothetical dispersion free states, whose function is to reproduce the measurable properties of quantum mechanic when averaged over" (1987, 4). All **Instance** demands, by contrast, is that expectation values reflect functional and limiting relationships. This exonerates it from the charge of silliness traditionally leveled at von Neumann's demand. Such an exoneration is not an argument that **Instance** gives an ultimately satisfying analysis of the instantiation of physical laws by quantum systems. I take that question to require much more discussion than it receives here. Because **Instance** is at least a natural starting point for such an analysis, it's worth investigating its consequences.

With this prolegomenon accomplished, we can present the less tendentious principle in the spirit of (SEGOP). A principle of harmony, it requires an interpretation to admit as physical all and only states that instantiate the lawlike relations the interpretation deems significant. Cast in terms of kinematic pairs, the principle is:

Harmony (Q, S) is an admissible kinematic pair for a theory only if (i) to each lawlike relationship posited by the theory there correspond elements of Q standing in that relationship, and (ii) S includes all and only states instantiating the relationships in (i).

To give the Harmony principle teeth, of course, we need to get a handle on what lawlike relationships the theory imposes. That's liable to be a slippery matter, greased up as it is with questions about the theory's legitimate explanatory projects and the proper role therein of laws. But this, it strikes me, is as it should be: the questions "which states are physical?" and "which observables are physical?" *ought to be* entangled, and entangled as well with questions about the nature and identity of physical law and the point of interpreting physical theory. And it's not like we've gotten *nowhere*: the Harmony principle sets a tenet of interpretive good taste (maybe even a test of internal coherence) for an interpretation. We'll see it in action in §6.5.

Harmony can't adjudicate between competing interpretations that meet its scruples. We'll have to fall back on our vaguer criteria to call those contests. Understanding Segal's argument for the operational significance of uniform closure as a case that Imperialism satisfies the Harmony principle, we'll turn next to a preliminary evaluation of Imperialism as an interpretation of QM_∞.

6.4.4 Imperialism: excess and deficiency

The Hilbert Space Conservative was liable to recognize too few states to support QM_∞'s explanatory agendas. The Imperialist is prone to the complementary shortfall of recognizing too few observables. As the discussion of weak equivalence suggests, the abstract algebra $\mathfrak{A}$ of observables an Imperialist attributes a quantum field theoretic system typically excludes observables constituting a particle notion. The abstract algebra $\mathfrak{A}$ for a system at the thermodynamic limit of QSM excludes such significant observables as global charge, temperature, and chemical potential. With these exclusions in view, later chapters will assess Imperialism's prospects of making sense of QFT as particle physics and QSM as thermodynamics.

Offering $(\pi_\omega(\mathfrak{A})'', \mathcal{F}_\omega)$ as a kinematic pair for a theory of canonical type $\mathfrak{A}$, Hilbert Space Conservatism faced possible dynamical challenges in the form of dynamically significant magnitudes absent from $\pi_\omega(\mathfrak{A})''$, or dynamical evolutions that start, but do not stay, in $\mathcal{F}_\omega$. Provided that the dynamics for a theory of canonical type $\mathfrak{A}$ can be implemented by a one real parameter family of automorphisms α_t of $\mathfrak{A}$, Imperialism needn't be fazed by such exotica. Still, the Imperialist faces a complementary dynamical worry. As the last section noted, in QM_∞, dynamics for an infinitely extended system assigned quasilocal algebra $\mathfrak{A}$ can be defined as the limit of dynamics for subsystems assigned local algebras $\mathfrak{A}(V)$. The catch for the Imperialist is that this limit may fail to converge in $\mathfrak{A}$'s C^* norm (provided by the uniform topology of a Hilbert space carrying a representation of $\mathfrak{A}$)—converge, that is, to a family of dynamical automorphisms α_t on $\mathfrak{A}$. Such a breakdown of uniform convergence needn't leave the system without dynamics, provided there are representations π_ω of $\mathfrak{A}$ in which the local dynamics have a *weak* limit as $V \to \infty$ (see Chapter 12 and Sewell 2002, §2.4.5). System dynamics are well-defined *in those representations*, but not on $\mathfrak{A}$.

This can pose two challenges to the Imperialist. First, insofar as such dynamics are defined at all, they're defined in terms of parochial observables he is unprepared to recognize as physical. Second, he recognizes as physical states—particularly, states outside the folia where local dynamics have a well-defined weak limit—for which dynamics are ill-defined. But, Aristotle would urge, surely to be physical is to admit of dynamical development. Chapters 12–14 elaborate these challenges to Imperialism.

6.5 Mixed strategies

By a "mixed" strategy for interpreting QM_∞, I mean one that mixes and matches elements of Imperialist and Conservative kinematic pairs. Two mixed strategies for interpreting a theory of canonical type $\mathfrak{A}$ are discussed in Clifton and Halvorson (2001b, 430 ff.). *Mixed strategy A* merges the Hilbert Space Conservative's conception of states with the Algebraic Imperialist's conception of observables. That is, the strategy confines physically relevant states to the folium of some (not necessarily) pure state ω on $\mathfrak{A}$, and identifies observables with self-adjoint elements of the C^* algebra $\mathfrak{A}$ common to all Hilbert space representations. Thus mixed strategy A offers $(\mathfrak{A}, \mathcal{F}_\omega)$ as the kinematic pair for a theory of canonical type $\mathfrak{A}$. *Mixed strategy B* merges the Hilbert Space Conservative's conception of observables with the Algebraic Imperialist's conception of states. For some (not necessarily pure) state ψ on $\mathfrak{A}$, mixed strategy B identifies observables with self-adjoint elements of the von Neumann algebra affiliated with ψ's GNS representation, and identifies states with states in the algebraic sense on $\mathfrak{A}$. Thus mixed strategy B offers $(\pi_\psi(\mathfrak{A})'', \mathcal{S}_\mathfrak{A})$ as the kinematic pair for a theory of canonical type $\mathfrak{A}$.[7]

I claim that both mixed strategies violate the Harmony principle. No *contentious* claims about which relationships are physical laws mediate the claim. Bracketing for now questions about which relationships hold of messy and interesting *physical* necessity, I will concentrate on the straightforward variety of lawlike relationship between magnitudes one gets *for free* along with an algebra: the relationships between elements of an algebra constitutive of its algebraic structure. Thus I will argue that the variety of lawlike relationships with respect to which the interpretations violate Harmony are those lawlike relationships involved in the defining of one magnitude in terms of others. The crux of the argument is that each strategy's observable algebra contains interdefinable magnitudes whose defining relationships aren't instantiated by all and only the states that strategy entertains.

---

[7] Clifton and Halvorson label strategy A "conservative about states, liberal about observables" and strategy B "liberal about states and conservative about observables" (2001b). Their "conservatives," when faced with an interpretive choice between more and less, always opt for less. My Conservative, faced with an interpretive choice between carrying on in the old way and innovating, always opts for the old way.

Let's start with mixed strategy B, the strategy which offers on behalf of a theory of canonical type $\mathfrak{A}$ the kinematic pair $(\pi_\psi(\mathfrak{A})'', \mathcal{S}_\mathfrak{A})$, where ψ is some state on $\mathfrak{A}$. On this strategy, π_ψ is the representation whose affiliated von Neumann algebra $\pi_\psi(\mathfrak{A})''$ contains the physical magnitudes. So if a state ω on $\mathfrak{A}$ is disjoint from ψ, some self-adjoint elements of the von Neumann algebra affiliated with ω are disqualified from the status of physically significant observables. The challenge for mixed strategy B is to justify this disqualification. In a way, the challenge is deeper than the Hilbert Space Conservative's problem of privilege. The Conservative can at least observe Harmony-satisfying consonance between the states she recognizes and the observables she recognizes. Segal has emphasized that mixed strategy B enjoys no such consonance. In $\mathcal{S}_\mathfrak{A}$ are states whose values on $\pi_\psi(\mathfrak{A})$ fix their values on $\pi_\psi(\mathfrak{A})''$, and states for which this is not so. In particular, there can be in $\pi_\psi(\mathfrak{A})''$ observables which are ultraweak limits of sequences of other observables in $\pi_\psi(\mathfrak{A})''$ even though the elements of $\mathfrak{A}$ which are the pre-images under π_ψ of those sequences don't converge in C^* norm. (The global polarization of an infinite chain of spin is an example of such an observable.) If ω is disjoint from ψ, no state in ω's folium is an ultraweakly continuous state on $\pi_\psi(\mathfrak{A})''$—otherwise it would be a state in ψ's folium, contrary to the assumption that ω and ψ are disjoint. So according to mixed strategy B, there is some sequence $\pi_\psi(A_i)$ of physically significant observables that converges ultraweakly to a physically significant observable $A \in \pi_\psi(\mathfrak{A})''$, and some physically significant state ω such that $\omega(\pi_\psi(A_i)) := \omega(A_i)$ fails to converge. Mixed strategy B entertains states which fail to instantiate lawlike relationships, specifically, those between A_i and A. This violates Harmony. In a sense Segal has made precise, mixed strategy B recognizes more observables than its state set can articulate.

Mixed strategy A also violates Harmony. Again, take the lawlike relationships at issue in the Harmony principle to be the definitional relationships interweaving magnitudes in strategy A's observable algebra $\mathfrak{A}$. Because $\mathfrak{A}$ is a C^* algebra, its magnitudes are interwoven by the taking of uniform limits. The states on $\mathfrak{A}$ instantiating such definitional relationships are the uniformly continuous ones, i.e. the states ϕ such that if A_i converges uniformly to A, $\phi(A_i)$ converges to $\phi(A)$. The states mixed strategy A deems physical are states in the folium of some favored algebraic state ω. These states will sure enough be uniformly continuous and thus instances of lawlike definitional relationships strategy A recognizes. *But so will states disjoint from ω*, because *every* state on $\mathfrak{A}$ is uniformly continuous. Strategy A violates Harmony by withholding physical significance from states that meet the basic qualification of instantiating the interdefinitional relationships it recognizes.

I take this to be a *prima facie* case against the mixed strategies. But they are not beyond rehabilitation. In particular, an account of prevailing lawlike relationships beyond the interdefinitional ones given along with an observable algebra might restore harmony between these strategies' observable sets and their state spaces. I can't, for my part, imagine a plausible version of such an account.

6.6 Universalism

An interpretive stance I call "Universalism" construes the content of a quantum theory of canonical type $\mathfrak{A}$ by means of the extravagance of the *universal representation* π_U of $\mathfrak{A}$.[8] The universal representation is the direct sum, over states on $\mathfrak{A}$, of their GNS representations:

$$\pi_U = \oplus_{\omega \in S_\mathfrak{A}} \pi_\omega(\mathfrak{A}) \tag{6.5}$$

π_U acts on the direct sum Hilbert space $\mathcal{H}_u = \bigoplus_{\omega \in S_\mathfrak{A}} \mathcal{H}_\omega$.

$\pi_U(\mathfrak{A})''$, the von Neumann algebra affiliated with this universal representation—a.k.a. the *universal enveloping von Neumann algebra*—is the weak closure of π_U. One version of Universalism offers $\pi_U(\mathfrak{A})''$ as the algebra of observables for a theory of canonical type $\mathfrak{A}$, and normal states on $\pi_U(\mathfrak{A})''$ as its possible states.[9] $\pi_U(\mathfrak{A})''$ has the nice feature that its normal states coincide with $S_\mathfrak{A}$ (Emch 1972, 120–21). Thus the Universalist kinematic pair is $(\pi_U(\mathfrak{A})'', S_\mathfrak{A})$.

The Conservative commits invidious privilege. The Universalist (like her religious counterpart) is adamantly noncommittal. The Imperialist haughtily ignores representation-specific observables that could be wage-earning members of physical theories. The Universalist welcomes any observable that can make it into the weak closure of the universal representation.[10] Mixed strategies run afoul of the Harmony principle, a principle of interpretive good taste Universalism promises to obey. Notwithstanding this litany of accomplishments, Kastler is not converted. $\pi_U(\mathfrak{A})''$, he remarks, "contains all the operators of the standard theory—but it has no interesting algebraic structure" (1964, 184; see also Takesaki 2002, 157).

Michael Redhead reports that Clark Glymour once branded contextualist responses to No Go results like the Kochen–Specker argument—responses asserting that as many physical magnitudes correspond to a self-adjoint degenerate Hilbert space operator as that operator has distinct eigenbases—the "deOckhamization of QM" (Redhead 1989, 135). One might complain that Universalism threatens the deOckhamization of QM_∞. π_U represents each element A of $\mathfrak{A}$ *continuously* many times: once as $\pi_U(A)$ and once for each proper subrepresentation of π_U. Of course, by the lights of the Universalist, there are exactly as many observables in $\pi_U(\mathfrak{A})''$ as there should be. But each state in Universalism's state set $S_\mathfrak{A}$ is also implementable in continuously many ways. For instance, the pure state ϕ can be expressed as its GNS vector in the summand $\mathcal{H}_\phi$ of the universal Hilbert space, but also as a vector in each summand $\mathcal{H}_\psi$ where ψ is a pure state in ϕ's folium. And that doesn't begin to exhaust the possibilities.

---

[8] See Kronz and Lupher (2005) for one version of this proposal. For more on the universal representation, see Kadison and Ringrose (1997a, §10.1).

[9] Ch. 10 will discuss other versions.

[10] This may be fewer observables than you imagine. It follows from theorem 10.3.5 of Kadison and Ringrose (1997a, 738) that $\pi_U(\mathfrak{A})'' \neq \oplus_{\phi \in S} \pi_\phi(\mathfrak{A})''$. So "portmanteau" operators like $\pi_\phi(X) \oplus_{\psi \neq \phi} 0$—the sort of operators we'd expect to implement π_ϕ-parochial observables on $\mathcal{H}_u$—won't necessarily make it into $\pi_U(\mathfrak{A})''$. Ch. 10's examination of strategies for "universalizing" the particle notion will discuss this in more detail.

To dismiss Universalism on these grounds would be bigotry. After all, in ordinary QM, continuously many different vectors implement any given pure state. Less ephemeral complaints about Universalism derive from its capacity to sustain dynamics. Recall that a possible dynamical challenge for the Imperialist took the form of dynamics on local subalgebras $\mathfrak{A}(V)$ of the quasilocal algebra $\mathfrak{A}$ whose convergence to a global dynamics on $\mathfrak{A}$ is representation-dependent. For systems subject to such representation-dependent dynamics, not every state in the folium of the universal representation will sustain a global dynamics. Only those states in the folium of a representation for which the global dynamics is well-defined will do. Thus the dynamical details of a theory of QM_∞ can draw distinctions—between dynamics-sustaining states on $\mathfrak{A}$ and dynamics-stymying ones—that seem both physically significant and unavailable to the Universalist. Cast in terms of the Harmony principle, there can be lawlike relationships—including the ones Aristotle would build into the very idea of being a natural system—that the Universalist's states fail to instantiate. Discussions of phase structure and broken symmetry in Chapters 12–14 will develop this sort of challenge to Universalism.

6.7 Unpristine approaches

The interpretive options discussed in this chapter are largely pristine. They equip a theory of canonical type $\mathfrak{A}$ with a kinematic pair whose type is identified in terms of $\mathfrak{A}$ and broadly philosophical commitments to considerations such as parsimony, plenitude, and meaning. The Universalist and the Imperialist take $\mathfrak{A}$ on its own to fix not only the type of kinematic pair assigned the theory, but also the token. All these interpretations operate in the ambit of the modal toggle picture of a physical theory, a fellow traveller of what §1.2 called the standard account of theoretical content. On that picture, a physical theory specified up to a kinematic pair sorts logically possible worlds into those that are (according to its kinematics) physically possible and those that are not. These physically possible worlds are, as it were, instantaneous; the theory's dynamical laws tell us which physically possible worlds are the time developments of which others. If the theory is lucky enough to be true, uninteresting contingency tells us which trajectory the actual world lies on.

Different pristine interpretations offer different toggling mechanisms: a Hilbert space structure of observables, an abstract algebraic structure, a universal representation structure, and so on. But they deploy these toggling mechanisms, as it were, *a priori*, before the messy business of applying the theory in question to individual problems begins. A pristine interpretation models this business as a selection from the worlds possible according to the theory (i.e. the theory's content, as preconfigured by the interpretation), the world most relevant to present circumstances.

Mathematical physicists discussing algebras and their representations might be taken to suggest a rival construal of the relation between the content of a theory and its application. Primas puts the point this way:

6.7 UNPRISTINE APPROACHES   147

To select a particular representation means to concentrate the attention to a particular phenomenon and to forget effects of secondary importance. With such a choice (which is possible if we adopt an appropriate topology in our mathematical formalism) we lose irrelevant information so that the system becomes simpler. To decide which representation to use is the same as to decide what is relevant for the problem at hand. This cannot be done a priori but only when the relevant patterns have been decided upon. (1983, 174–5)

The idea that different applications call for different representation-dependent topologies echoes a 1953 remark of Friedrichs—"Different limit processes may be appropriate for different experimental situations" (1953, 142). Kadison puts the thought evocatively:

Mathematically, a representation [of an abstract algebra] distinguishes a certain "coherent" family of states from among [the full set of algebraic states], and at the same time, in effect, "coalesces" some of the algebraic structure. (1965, 186)

The representation π of a canonical algebra $\mathfrak{A}$ for a theory of QM_∞ "coalesces" the algebraic structure of the von Neumann algebra $\pi(\mathfrak{A})''$ obtained by closing $\pi(\mathfrak{A})$ in the weak topology of the representation's Hilbert space. That algebraic structure wasn't in (merely uniformly closed) $\mathfrak{A}$ to begin with. Acting on that algebra with the catalyst of a particular representation issued in $\pi(\mathfrak{A})''$ as a sort of precipitate. If different circumstances call for different representations, the algebraic structure coalesced will differ as well.

These comments suggest an alternative to the ideal of pristine interpretation. The doctrine of *unpristine* interpretation allows that the contingent application of theories does not *merely select* among some preconfigured set of their contents, but *genuinely alters* their contents. It follows that there can be an *a posteriori*, even a pragmatic, dimension to content specification, and that physical possibility is not monolithic but kaleidoscopic. Instead of one possibility space pristinely associated with a theory from the outset, many different possibility spaces, keyed to and configured by the many settings in which the theory operates, pertain to it. Following Kadison, I call this the *coalescence approach* to interpreting physical theories.

The balance of this book aims to elaborate and defend the coalesence approach. The argument for it unfolds as follows. Considering a variety of circumstances wherein unitarily inequivalent representations arise in QM_∞, I assess the pristine interpretations cataloged here and find them wanting. Pristine approaches, I will suggest, fail *because they are pristine*—that is, because they forswear appeal to the very contingencies that shed light on how particular theories of QM_∞ represent and explain. This general account of the failure of pristine approaches suggests the general idea of a coalescence approach. A more refined account of coalescence will emerge in the course of searches for particular remedies to particular failures of pristine accounts. The concluding chapter revisits the scientific realism debate with coalescence in mind.

7

Extraordinary QM

One aim of this book is to challenge the imposition of the ideal of pristine interpretation on the rough and tumble of how physical theories do what they ought to do, which is (among other things) unify and explain phenomena. And one thesis of this book is that QM_∞ violates ground rules that have typically governed philosophers' typically pristine engagement with ordinary QM. Thus the plot of this chapter and the next, which chronicle, respectively, formal difference between theories of ordinary QM and QM_∞, and their interpretive consequences.

In ordinary QM, the observables pertaining to a quantum system are the self-adjoint elements of $\mathfrak{B}(\mathcal{H})$, the collection of bounded operators acting on a separable Hilbert space $\mathcal{H}$, a.k.a. (we'll see in §7.1) a *Type I von Neumann algebra*. An *atom E* in a von Neumann algebra $\mathfrak{M}$ is a non-zero projection operator *minimal* in $\mathfrak{M}$ in the sense that no other non-zero projection in $\mathfrak{M}$ has a range that is a proper subspace of E's.[1] Thus an atom of $\mathfrak{B}(\mathcal{H})$ is a one-dimensional projection operator. Atoms correspond to the endpoints of measurement collapse, according to textbook interpretations of ordinary QM. Atoms specify the "value states" a modal interpreter allots a system described by a non-degenerate density operator, and the actual state of a system described by a mixture subject to an ignorance interpretation. Atoms are what most familiar interpretations of ordinary QM use to characterize the occurrent conditions of quantum systems, as well as to explicate quantum probabilities. Atoms are key players in the interpretation of ordinary QM.

A theoretical phenomenon endemic to QM_∞ but without precedent among the sorts of quantum theories philosophers are accustomed to discussing is the *atomless* von Neumann algebra, an observable algebra that *has no atoms*. Sections 7.1 and 7.2 will explain why such algebras are mathematically possible, divulging along the way further useful facts about C^* and von Neumann algebras, such as the rudiments of a classification of von Neumann algebras into different types. Sections 7.3–7.5

---

[1] I abuse terminology here. As §8.2.2 will relate, "atom" is, properly speaking, a lattice-theoretic notion that happens to be instantiated by minimal projection operators.

will describe how the possibility of atomless von Neumann algebras comes to be instantiated in QFT and in the thermodynamic limit of QSM, divulging along the way further useful facts about algebraic approaches to those settings, such as how to extend the notion of equilibrium to the thermodynamic limit.

Relying as they do on atoms, familiar strategies for interpreting ordinary QM flounder when they encounter atomless von Neumann algebras. Chapter 8 will document both the reliance and the floundering.

7.1 Typing von Neumann factors

Every von Neumann algebra $\mathfrak{M}$ is a subalgebra of $\mathfrak{B}(\mathcal{H})$ for some $\mathcal{H}$, but it doesn't follow that every $\mathfrak{M}$ has projection operators corresponding to every subspace of the Hilbert space on which it acts. Let $\mathcal{P}(\mathfrak{M})$ be the set of projection operators in the von Neumann algebra $\mathfrak{M}$. It's a consequence of von Neumann's double commutant theorem that $\mathfrak{M} = \mathcal{P}(\mathfrak{M})''$. A standard typology of von Neumann algebras is based on the sorts of projections $\mathcal{P}(\mathfrak{M})$ contains.[2]

The *range* of a projection E in von Neumann algebra $\mathfrak{M}$ acting on a Hilbert space $\mathcal{H}$ is the linear span of $\{|\psi\rangle \in \mathcal{H} : E|\psi\rangle = |\psi\rangle\}$. Thus the range of E is a subspace of $\mathcal{H}$.

Definition 7.1. Two projections E and F in $\mathfrak{M}$ are *equivalent* (written $E \sim F$) just in case their ranges are isometrically embeddable into one another, by a partial isometry[3][g] that's an element of $\mathfrak{M}$.

Equivalence so construed is manifestly relative to $\mathfrak{M}$. Happily enough, it is also an equivalence relation.

Definition 7.2. When E's range is a subspace of F's (written $E \leq F$), E is said to be a *subprojection* of F. Equivalent criteria are that $FE = EF = E$ and that $|E|\psi\rangle| \leq |F|\psi\rangle|$ for all $|\psi\rangle \in \mathcal{H}$.

When E's range is a *proper* subspace of F's, we write $E < F$. Along with the notion of equivalence introduced in the last paragraph, the subprojection relation defines the relation *weaker than* (written $\preceq$), which imposes a partial order on projections in a von Neumann algebra.

Definition 7.3. E is *weaker than* F iff E is equivalent to a subprojection of F. Because $\preceq$ is a partial order, $E \preceq F$ and $F \preceq E$ together imply that $E \sim F$.

Now, by obvious analogy with Cantor's definition of infinity, we can say a projection $E \in \mathfrak{M}$ is *infinite* iff there's some projection $E_0 \in \mathfrak{M}$ such that $E_0 < E$ and $E \sim E_0$.

---

[2] The classification originates with Murray and von Neumann (1936); Sunder (1987, ch. 2) gives a compact overview.

[3] i.e. a 1:1 map that preserves the vector norm.

In this case, E_0's range is both a proper subset of, and isometrically embeddable into, E's range. A projection E which is not infinite is *finite*.

Two final terms of art:

Definition 7.4 (abelian projection). A non-zero projection $E \in \mathfrak{M}$ is *abelian* iff the von Neumann algebra $E\mathfrak{M}E$ (in which E serves as the identity), acting on the Hilbert space $E\mathcal{H}$, is abelian.

In ordinary QM, the abelian projections are the one-dimensional ones. Such projections are trivially abelian: every operator on a one dimensional subspace is a function of the identity operator for that subspace.

Definition 7.5 (minimal projection (atom)). A non-zero projection $E \in \mathfrak{M}$ is *minimal* iff E's only subprojections in $\mathfrak{M}$ are 0 and E itself.

It follows that if a projection is minimal, it is abelian.

The foregoing ideas frame a typology of von Neumann algebras. The typology applies in the first instance to von Neumann algebras $\mathfrak{R}$ which are factors;[4] on such algebras, the weaker than relation $\preceq$ imposes a total order (Kadison and Ringrose 1997b, prop. 6.26 (408)).

Definition 7.6 (Type I). A Type I factor contains an abelian projection.

In an arbitrary von Neumann algebra, if a projection is minimal, then it is abelian. For von Neumann factors, the converse holds: abelian projections are minimal, i.e. atoms. Type I factors are the stuff of ordinary QM. Every factor of Type I_n is isomorphic to $\mathfrak{B}(\mathcal{H}_n)$, the algebra of $n \times n$ complex-valued matrices. A Type I_n factor with a physical application is the CAR algebra for $\log_2 n$ spin systems. Every Type I_∞ factor is isomorphic to the algebra of infinite dimensional square complex-valued matrices. A Type I_∞ factor with a physical application is the weak closure of the Schrödinger representation of the Weyl algebra for a particle on the real line. Another Type I_∞ factor arises as follows: start with the CAR algebra $\mathfrak{A}_{CAR}$ for the infinite spin chain (Ex. 3.3.1). Let $\mathfrak{A}_k$ be the subalgebra of $\mathfrak{A}$ pertaining to site k. Thus $\mathfrak{A}_k$ is just $\mathfrak{M}(2)$, the algebra of 2×2 complex-valued matrices generated by the Pauli relations. Introduce a state ϕ on $\mathfrak{A}_{CAR}$ whose restriction to each $\mathfrak{A}_k$ is the pure normal state ψ on $\mathfrak{M}(2)$. (Naively, ϕ is the infinite tensor product state $\psi \otimes \psi \otimes \psi \ldots$.) Take ϕ's GNS representation and close in that representation's weak topology. The resulting von Neumann algebra $\pi_\phi(\mathfrak{A}_{CAR})''$ is a Type I factor (for details about this and subsequent factors constructed from the spin chain, see Emch 1984, 401–2).

---

[4] Recall, $\mathfrak{R}$ is a factor iff $\mathfrak{R} \cap \mathfrak{R}'$ contains only multiples of the identity. Note the convention that "$\mathfrak{R}$" (and variants) refers to factors, and "$\mathfrak{M}$" (and variations) to von Neumann algebras not assumed to be factors. For more on how to extend this classification to von Neumann algebras that aren't factors, see Kadison and Ringrose (1997a, ch. 5; 1997b, ch. 6).

Not all von Neumann algebras have atoms, as the following simple example illustrates.

Example 7.7 (an atomless von Neumann algebra). Let $\mathcal{H}$ be the separable Hilbert space $L_2([0,1])$ of square integrable functions on the unit interval $[0,1]$ equipped with the Lebesgue measure. Where f is a bounded measurable function on $[0,1]$, let M_f be the operator on L_2 corresponding to multiplication by f. The collection $\{M_f\} = \mathfrak{D}_Q$ (with addition and multiplication defined pointwise) is a maximal abelian von Neumann algebra acting on $\mathcal{H}$ (Kadison and Ringrose 1997b, ex. 5.1.6 (308)). $\mathfrak{D}_Q$ is atomless.

To see why, begin with a characterization of the projection operators in $\mathfrak{D}_Q$. For each Borel subset X of $[0,1]$, the characteristic function χ_X is a projection in $\mathfrak{D}_Q$, and every projection in $\mathfrak{D}_Q$ is the characteristic function for some Borel subset of $[0,1]$.[5] $\mathfrak{D}_Q$ is atomic only if it contains a minimal projection operator, i.e. a non-zero χ_S whose only proper subprojection is the zero operator. But consider what it takes for χ_S to be different from the zero operator. If S is a set of measure 0, the vector norm of $\chi_S|\psi\rangle$ for an arbitrary $\psi \in L^2([0,1])$ is given by

$$|\chi_S|\psi\rangle|^2 = \int_0^1 \psi^*(x)\chi_S\psi(x)dx = \int_S \psi^2(x)dx = 0 \qquad (7.1)$$

If S is a set of measure 0, χ_S maps every element of $L^2([0,1])$ to the zero vector.[6] Thus χ_Y is a non-zero projection operator only if the set Y is measurable. But if Y is measurable, χ_Y can't be an atom in $\mathfrak{D}_Q$. Every measurable set Y has a measurable proper subset. Let X be a measurable proper subset of Y. $\chi_X \neq 0$, because X is measurable. And χ_X is a subprojection of χ_Y because acting on an arbitrary element of $L^2([0,1])$ with $\chi_Y\chi_X$ is the same as acting on it with χ_X. We can rule out $\chi_X = \chi_Y$ by comparing their actions on elements of $L^2([0,1])$ with support on X's complement in Y. Thus χ_X is a non-zero projection operator in $\mathfrak{D}_Q$ that's a proper subprojection of χ_Y, which upsets χ_Y's claim to atomicity. Because χ_Y was an arbitrary non-zero projection in $\mathfrak{D}_Q$, it follows that *no* projection in $\mathfrak{D}_Q$ is an atom. ♠

It turns out that every atomless maximal abelian von Neumann algebra is isomorphic to $\mathfrak{D}_Q$ (Kadison and Ringrose 1997b, 665 ff.). Note that $\mathfrak{D}_Q$ is not a factor algebra:

---

[5] To elaborate: the definition of a projection operator ($A^2 = A = A^t$) implies that a function $f(x)$ acting multiplicatively on $L^2(0,1)$ is a projection operator only if $[f(x)]^2 = f(x) = \bar{f}(x)$ for almost all $x \in [0,1]$. This implies in turn that $f(x) = 0$ or 1 almost everywhere—thus it differs (up to measure 0) from a characteristic function for a measurable interval $S \subset [0,1]$. The converse follows from running the argument in reverse (Kadison and Ringrose 1997a, ex. 2.5.12 (117)).

[6] The collection $\mathcal{P}(\mathfrak{D}_Q)$ of projection operators in $\mathfrak{D}_Q$ induces a map from Borel subsets of $[0,1]$ to $\{0,1\}$. Sophisticates would say that the kernel of this map (i.e. the pre-image of 0 under χ_S) is just $\mathcal{N}$, the collection of measure 0 subsets of $[0,1]$. From this it follows (by something called the first isomorphism theorem for rings) that $\mathcal{P}(\mathfrak{D}_Q)$ is isomorphic to $\mathcal{B}([0,1])/\mathcal{N}$, the Boolean algebra of equivalence, up to measure 0, classes of Borel subsets of $[0,1]$. See Halvorson (2001) for details and an illuminating discussion of their significance.

because it's abelian, $\mathfrak{D}_Q$'s intersection with its own commutant is $\mathfrak{D}_Q$, and thus far from trivial!

There are, however, atomless factor algebras. They include Type II factors.

Definition 7.8 (Type II). A Type II factor contains a finite projection but no abelian projections.

Because the abelian projections in a factor algebra are exactly the minimal ones, a Type II factor is atomless. In a sense that can be made precise (Sunder 1987, §1.3), Type II factors have projections whose ranges are subspaces of *fractional* dimension. In a factor of Type II_1, the identity operator is finite; in a factor of Type II_∞, the identity operator is infinite. That old standby, the infinite spin chain, provides an example of a Type II_1 factor. Once again, let ϕ be a "product state" on $\mathfrak{A}_{CAR}$, but this time let ϕ's restriction to each $\mathfrak{A}_k$ be, not a pure state, but the homogeneous mixture of pure states proportional to the identity operator. Take ϕ's GNS representation and close in that representation's weak topology. The resulting von Neumann algebra $\pi_\phi(\mathfrak{A}_{CAR})''$ is a Type II_1 factor. Indeed, it's *the* Type II_1 factor, for any other Type II_1 factor is isomorphic to $\pi_\phi(\mathfrak{A}_{CAR})''$. (For further discussion of this example, see Rédei and Summers 2007, §6.1.)

A factor is Type III if it is neither Type I nor Type II:

Definition 7.9 (Type III). A Type III factor has neither finite nor abelian projections.

Type III factors are atomless. All their projections are infinite and therefore (Kadison and Ringrose 1997b, corr. 6.3.5 (414)) equivalent. The only (non-zero) projections in a Type III factor have infinite-dimensional ranges. It should come as no surprise that the infinite spin chain furnishes (infinitely many) examples of Type III factors. As before, let ϕ be a "product state" on $\mathfrak{A}_{CAR}$, this time one whose restriction to each $\mathfrak{A}_k$ takes the form $\rho = \frac{exp(-\beta\sigma)}{Tr(exp(-\beta\sigma))}$, where the inverse temperature β is a non-negative real number and σ is a non-degenerate self-adjoint element of $\mathfrak{A}_k$. Chapter 12 will describe dynamics with respect to which ρ is an equilibrium state on $\mathfrak{A}_k$. Take ϕ's GNS representation and close in that representation's weak topology. The resulting von Neumann algebra $\pi_\phi(\mathfrak{A}_{CAR})''$ is a Type III factor. Powers showed that Type III factors thus constructed from distinct βs are non-isomorphic: thus there are continuously many Type III factors. It took Alain Connes and the resources of modular theory (sketched roughly in §7.4) to offer an illuminating characterization of how distinct Type III factors differ.

Digression: the classification of Type III factors. From Takesaki's 1994 memoir:

[Connes] gave a seminar talk at Oslo on the structure theorem for factors of type III_λ, $0 < \lambda < 1$. A surprised Størmer asked him, "Have you proven this already?" Answered Connes, "It will be proven shortly." Indeed he did upon his return to Paris. He had, however, published too many results in Comptes Rendus, so he was unable to submit his work to Comptes Rendus. He then visited Bandol, a small town on the Mediterranean Coast east of Marseille, where D. Kastler and D. Robinson lived.... Also visiting at the time was R.T. Powers a world leader in the classification theory of ITPFI factors. Connes was trying to construct an AFD factor of type III which is not ITPFI. One day in the summer of '72, he finally proved the existence of such a factor, and wanted to talk to Powers. He knew that Powers was on the beach of Bandol at that time of day, and rushed down to the beach. He was then told that Powers was swimming in the sea, which did not stop him. He further rushed to the sea looking for Powers to report his exciting discovery. Well, I don't know how Powers and Connes managed to do mathematics in the middle of the sea, but it was clearly an early triumph of A. Connes. (Takesaki 1994, 238) ♠

Due to their intricacy and my limited exegetical skills, I'll construct no explicit examples of Type II or III factor algebras (for these, see Sunder 1987, §4.3). But §7.5 will describe physical settings to which such algebras apply.

Because atoms are defined in terms of the subprojection relation, which in turn is defined algebraically ($E \leq F$ iff $EF = FE = E$), we wouldn't expect a von Neumann algebra without atomic [or finite] projections to be *-isomorphic to a von Neumann algebra with such projections. And indeed,

Fact 7.10. Von Neumann factor algebras $\mathfrak{R}_1$ and $\mathfrak{R}_2$ are quasi-equivalent only if they are of the same type (Kadison and Ringrose 1997b, corr. 6.5.3 (424)).

There is a classification of von Neumann algebras that cross-categorizes the one just rehearsed. A von Neumann algebra $\mathfrak{M}$ is *semifinite* if, roughly speaking, it has enough finite projections (see Kadison and Ringrose 1997b, 422–5 for a precise definition). Factors of Types I and II$_1$ are semifinite. $\mathfrak{M}$ is *finite* if its identity element $I \in \mathfrak{M}$ is finite; otherwise, it's *infinite*. Factors of Types I$_n$ and II$_1$ are finite; factors of Types I$_\infty$, II$_\infty$, and III are infinite. $\mathfrak{M}$ is *purely (or properly) infinite* if all its non-zero projections are infinite. Factors of Type III are purely infinite.

If $\mathfrak{M}$ is not factorial, its center $\mathfrak{M} \cap \mathfrak{M}'$ needn't be trivial. Projection operators in $\mathfrak{M}$'s center are known as *central projections*. They figure in the extension of the foregoing classification to von Neumann algebras not assumed to be factors.

Definition 7.11 (classification of general von Neumann algebras (cf. Kadison and Ringrose 1997b, §6.5)). A von Neumann algebra $\mathfrak{M}$ is:

1. Type I if it contains an abelian projection that is a sub projection of no non-trivial (i.e. different from 0 and I) central projection.
2. Type II if it is not Type I, but does contain a finite projection that is a subprojection of no non-trivial central projection.
3. Type III if it is neither Type I nor Type II.

We can invoke this classification to determine that the non-factor algebra $\mathfrak{D}_Q$ is Type I. Because $\mathfrak{D}_Q$ is abelian, so is $I\mathfrak{D}_Q I$, revealing I to be an abelian projection. $\mathfrak{D}_Q$ is its own center, so *every* projection in $\mathfrak{D}_Q$ is a central projection. But the only central projection of which I is a sub projection is I itself, a trivial projection.

We will have occasion to refer later to the Type Decomposition theorem, which (to first approximation) states that an arbitrary von Neumann algebra $\mathfrak{M}$ can be decomposed into a direct sum of algebras of Type I, Type II$_1$, Type II$_\infty$, and Type III, which algebras are are known as $\mathfrak{M}$'s *central summands* (see Kadison and Ringrose 1997b, §6.5).

7.1.1 Dimension functions and subtypes

Disclaimer: The material in this informal and digressive subsection isn't central to later developments. It's offered in the hope that it might firm the reader's grip on how different types of factor von Neumann algebras are different.

Intuitively, a *dimension function* on a von Neumann algebra $\mathfrak{M}$ is a function that takes projections in $\mathfrak{M}$ as input and returns (upto rescaling) the dimension of their associated subspaces as output. Let $E_N \in \mathfrak{M} \subseteq \mathfrak{B}(\mathcal{H})$ be the projection operator whose range is the subspace $N \subseteq \mathcal{H}$. Let $\mathcal{P}(\mathfrak{M})$ denote the set of projection operators in the von Neumann algebra $\mathfrak{M}$, known (for reasons outlined in §8.2.2) as $\mathfrak{M}$'s *projection lattice*. Officially,

Definition 7.12. A *dimension function* for $\mathfrak{M}$ is a map D from $\mathcal{P}(\mathfrak{M})$ to the non-negative reals such that:

(i) $M \sim N$ iff $D(E_M) = D(E_N)$;
(ii) if $M \perp N$, then $D(E_M \oplus E_N) = D(E_M) + D(E_N)$; and
(iii) E_N is finite iff $D(E_N) < \infty$

for all $E_M, E_N \in \mathcal{P}(\mathfrak{M})$.

(i) expresses the desideratum that a dimension function assign equivalent subspaces the same dimension; (ii) the desideratum that the dimension assigned the direct sum of orthogonal subspaces be the sum of the dimensions assigned those subspaces individually; and (iii) the desideratum that a dimension function declare a subspace finite if and only if it is.

For a straightforward example of a dimension function, set $\mathfrak{M} = \mathfrak{B}(\mathcal{H})$ for an n (finite!) dimensional $\mathcal{H}$. Then the function $D(E) = Tr(EI)$, where I is the identity operator on $\mathcal{H}$, defines a dimension function on $\mathcal{P}(\mathfrak{M})$. Pleasingly, $D(E) = k$ when E's range is k-dimensional.

We care about dimension functions because, if a von Neumann algebra $\mathfrak{R}$ is a factor, $\mathcal{P}(\mathfrak{R})$ admits a dimension function which is unique up to rescaling (Sunder 1987, thm. 1.3.1 (27)). Thus the range of a dimension function D on a factor $\mathfrak{R}$ is a guide to $\mathfrak{R}$'s type:

If $\mathfrak{R}$ is ...	possible values of $D(E)$ are ...
Type I$_n$	$0, x, 2x, \cdots, nx; x > 0$
Type I$_\infty$	$0, x, 2x, \cdots \infty; x > 0$
Type II$_1$	$[0, x]$

Type II$_\infty$ $[0, \infty]$

Type III$_\lambda, \lambda \in [0, \infty]$ $\{0, \infty\}$

The scaling factor x, when it appears, is conventionally normalized to 1.

7.2 Atomlessness and normality

Here I characterize the formal extent of atomlessness. I also identify a feature of atomless von Neumann algebras without precedent among the Type I von Neumann algebras addressed by interpretations of ordinary QM. Alarmingly, *if $\mathfrak{M}$ lacks atoms, it lacks pure normal states as well.*

By definition, if $\mathfrak{M}$ is a Type II or III factor, it lacks atoms. Atoms are minimal projection operators, and factors with minimal projection operators are Type I. In general, a maximal abelian subalgebra of a countably decomposable[7,8] von Neumann algebras $\mathfrak{M}$ is completely atomless, unless $\mathfrak{M}$ contains a Type I central summand. This follows from from the fact (Kadison and Ringrose 1997b, ex. 6.9.28 (449)) that if $\mathfrak{M}$ is a countably decomposable von Neumann algebra with no central summand of Type I, and if $\mathfrak{D}$ is a maximal abelian subalgebra of $\mathfrak{M}$—i.e. an algebra of pairwise commuting element of $\mathfrak{M}$ properly contained in no larger such algebra—then for any positive integer n, each non-zero projection E in $\mathfrak{D}$ contains n non-zero orthogonal projections which are equivalent relative to $\mathfrak{M}$. Let $n = 2$ and let F, G be projections required by the theorem to exist, i.e. $E = F + G$. Clearly, $F < E$. Thus no non-zero projection E in $\mathfrak{D}$ is minimal in $\mathfrak{M}$. But we know from a Zorn's Lemma argument that any projection in $\mathfrak{M}$ is part of some maximal abelian subalgebra of $\mathfrak{M}$.[8] So we can conclude that no projection in $\mathfrak{M}$ is minimal.

Ordinary QM deals in Type I factors, whose normal states (which are implemented by density operators) include pure states (which correspond to projection operators). This familiar situation is not perfectly general. If $\mathfrak{M}$ is atomless, *it admits no pure normal states*! To see why this is so, consider, for starters, a Type II or a Type III factor algebra $\mathfrak{R}$. Suppose, for contradiction, that there exists a pure normal state ω on $\mathfrak{R}$. Because ω is pure, its GNS representation $\pi_\omega : \mathfrak{R} \to \mathfrak{B}(\mathcal{H}_\omega)$ is irreducible. By a series of short steps,[9] it follows that $\mathfrak{R}$ is unitarily equivalent to $\mathfrak{B}(\mathcal{H}_\omega)$. In other words, $\mathfrak{R}$ is unitarily

---

[7] $\mathfrak{M}$ is countably decomposable if each orthogonal family of subprojections of $\mathfrak{M}$'s identity element is countable. The von Neumann algebras typical of QM$_\infty$ are countably decomposable.

[8] Not to be mysterious: Zorn's Lemma concerns non-empty partially ordered sets. It states that if S is such a set, and if every totally ordered subset of S has a least upper bound, then S has a maximal element. The ingredients of the argument are: each projection in $\mathfrak{M}$ belongs to an abelian subalgebra of $\mathfrak{M}$, and the set of abelian subalgebras of $\mathfrak{M}$ to which a projection belongs can be partially ordered by set-theoretic inclusion.

[9] The steps are: because π_ω is irreducible, $\pi_\omega(\mathfrak{R})''$ is unitarily equivalent to $\mathfrak{B}(\mathcal{H}_\omega)$ (Sakai 1971, thm. 1.12.9). But a von Neumann algebra is unitarily equivalent to the GNS representation of its normal states (Bratteli and Robinson 1987, thm. 2.4.24 (79)). So $\mathfrak{R}$ is unitarily equivalent to $\pi_\omega(\mathfrak{R})$. This exposes $\pi_\omega(\mathfrak{R})$ as a von Neumann algebra, and so identical to its double commutant $\pi_\omega(\mathfrak{R})''$. So $\mathfrak{R}$ is unitarily equivalent to $\pi_\omega(\mathfrak{R})$, which is identical to $\pi_\omega(\mathfrak{R})''$, which is unitarily equivalent to $\mathfrak{B}(\mathcal{H}_\omega)$. Eliminating the middlemen, $\mathfrak{R}$ is unitarily equivalent to $\mathfrak{B}(\mathcal{H}_\omega)$.

equivalent to a Type I factor. But this is impossible, because $\mathfrak{R}$ was stipulated to be a Type II or III factor, and Fact 7.10 implies that only a Type I factor can be unitarily equivalent to a Type I factor. The surprising conclusion: there is no such thing as a pure normal state on a Type II or III factor von Neumann algebra.

We've seen that factors of Type II and III host no pure normal states. This lack of hospitality derives from the absence from those factors of atoms. Here is a sketch of a simple argument that if a von Neumann algebra $\mathfrak{M}$ is non-atomic, it lacks pure normal states as well.[10]

"Argument": atomless $\mathfrak{M}$ lacks pure normal states. Although not every normal state on a von Neumann algebra $\mathfrak{M}$ is implemented by a density operator *in that algebra*, we can set up a correspondence between projection operators in $\mathfrak{M}$ and normal states on $\mathfrak{M}$.

Definition 7.13. The *support projection* S_ω of a state ω on $\mathfrak{M}$ is the "smallest" projection in $\mathfrak{M}$ assigned probability 1 by ω. Thus S_ω is the meet of all projections $E \in \mathfrak{M}$ such that $\omega(E) = 1$.[11]

The support projection of a normal state is unique (Takesaki 2002, lemma III.3.6 (134)). We need one more definition:

Definition 7.14. The *left kernel* K_ω of of a state ω on a C^* algebra $\mathfrak{A}$ contains all and only those elements A of $\mathfrak{A}$ such that $\omega(A^*A) = 0$.

Thus K_ω reflects the set of positive operators ω maps to 0. The argument that an atomless $\mathfrak{M}$ lacks pure states makes use of a general fact about pure states on C^* algebras and their left kernels:

Fact 7.15. A state ω on a C^* algebra $\mathfrak{A}$ is pure iff ω is the only state on $\mathfrak{A}$ that vanishes on K_ω (Kadison and Ringrose 1997b, thm. 10.2.8 (731)).

Given a state ω on an atomless von Neumann algebra $\mathfrak{M}$, it's easy to construct a different state ϕ on $\mathfrak{M}$ that vanishes on K_ω. ω's support projection S_ω can't be a minimal projection in $\mathfrak{M}$ because $\mathfrak{M}$ has no minimal projections. So there is some non-zero projection $E \in \mathfrak{M}$ such that $E < S_\omega$. Consider a state ϕ whose support projection is E. (Any vector state in E's range will do.) ϕ and ω are different states because there are elements of $\mathfrak{M}$ (notably, $S_\omega - E$, on which ϕ but not ω vanishes) they map to different expectation values. Because $S_\phi = E < S_\omega$, ϕ maps every element of K_ω to 0. (Intuitively, the vector implementing ϕ is orthogonal to everything the vector implementing ω is orthogonal to.) Thus ϕ also vanishes on ω's left kernel. Given Fact (7.15), this refutes ω's claim to be pure. ♠

---

[10] See Clifton and Halvorson (2001b) for details.
[11] For non-normal states, there is no smallest such projection. (This is a consequence of Takesaki 2002, thm. III.3.8 (134).)

7.2 ATOMLESSNESS AND NORMALITY 157

The following example illustrates the truth that no pure state on an atomless von Neumann algebra is normal.

Example 7.16 (a pure state on $\mathfrak{D}_Q$ that isn't normal). Consider the projection lattice $\mathcal{P}(\mathfrak{D}_Q)$ of the abelian von Neumann algebra $\mathfrak{D}_Q$ acting multiplicatively on $L^2[0,1]$ (Ex. 7.7). Recall that $\mathcal{P}(\mathfrak{D}_Q)$ consists of projection operators in $\mathfrak{D}_Q$, which are characteristic functions χ_S for measurable subsets S of $[0,1]$. Now, ω on $\mathcal{P}(\mathfrak{D}_Q)$ is pure just in case $\omega(\chi_S) \in \{0,1\}$ for all S.[12] We'll describe such a pure state by describing a state that "converges to (a point) λ", and invoking some facts (the contexts of which §8.2.2 will elaborate). Recall from Ex. 4.7 that a state ω_λ converges to λ if it acts as follows on $\mathcal{P}(\mathfrak{D}_Q)$: Let $\{\Delta_i\} \subset [0,1]$ be a nested set of ever-shrinking but always measurable open neighborhoods around $\lambda \in [0,1]$.

$$\text{If } \Delta_i \subseteq \Gamma \text{ for some } i \text{ then } \omega_\lambda(\chi_\Gamma) = 1. \tag{7.2}$$

Notice that this only partially defines ω_λ's action on $\mathcal{P}(\mathfrak{D}_Q)$. Now the facts. The first concerns ultrafilters, and tells us everything we'll need, for our purposes, to know about them.

Fact 7.17. Each ultrafilter of a Boolean lattice generates a two-valued homomorphism on that lattice.

We care about this fact because $\mathcal{P}(\mathfrak{D}_Q)$ happens to be a Boolean lattice, and a two-valued homomorphism on $\mathcal{P}(\mathfrak{D}_Q)$ happens to be a map from it into $\{0,1\}$ that corresponds to a pure state. So if we have a two-valued homomorphism on $\mathcal{P}(\mathfrak{D}_Q)$, we have a pure state on it. Now we invoke a theorem whose proof invokes Zorn's Lemma.

Fact 7.18 (Ultrafilter Extension Theorem). Any subset S of a Boolean lattice possessing the *finite meet property*—that if $x_1, ..., x_n \in S$ then $x_1 \cap ... \cap x_n \neq 0$—is contained in some ultrafilter (Bell and Machover 1977, corr. 3.8).

This establishes that the map (7.2) extends to a pure state on $\mathcal{P}(\mathfrak{D}_Q)$. For consider a countable set $F_\lambda = \{\chi_{\Delta_i}\} \subset \mathfrak{D}_Q$ of our characteristic functions for (ever-shrinking but always measurable) open neighborhoods around $\lambda \in [0,1]$. If $\chi_{\Delta_j} \subset F_\lambda$ for an index set of natural numbers J, then $\cap_{j \in J} \chi_{\Delta_j} = \chi_{\Delta_n}$, where n is the largest integer in J. Because $\chi_{\Delta_n} \neq \emptyset$, F_λ possesses the *finite meet property*. It follows from Facts 7.17, 7.18, and the suppressed definition of an ultrafilter, that there is a pure state ω on $\mathfrak{D}_Q$ such that $\omega(\chi) = 1$ for all $\chi \in F_p$. This pure state coincides with ω_λ's action on $\mathfrak{D}_Q$ insofar as (7.2) defines that action. So ω_λ extends to a pure state on $\mathfrak{D}_Q$.

---

[12] Because a pure state ω on an abelian algebra is multiplicative (i.e. $\omega(XY) = \omega(X)\omega(Y)$ (Kadison and Ringrose 1997b, prop. 4.4.1 (269)) and because for any projection operator E, $E^2 = E$.

This establishes that ω_λ is a pure state on an atomless von Neumann algebra. It was the work of Ex. 4.7 to demonstrate that ω_λ fails to be normal. Thus ω_λ is an example of a pure *non-normal* state on an atomless von Neumann algebra.[13] ♠

The failure of an atomless von Neumann algebra to admit any pure normal states is strange. And it secures the even stranger possibility of what Clifton and Halvorson (2001b) call "intrinsically mixed states." An intrinsically mixed state on a C^* algebra $\mathfrak{A}$ is a mixed state on $\mathfrak{A}$ that's *disjoint from every pure state on* $\mathfrak{A}$. For an example, consider ω on $\mathfrak{A}$ such that the von Neumann algebra $\pi_\omega(\mathfrak{A})''$ affiliated with ω's GNS representation is a Type III factor. I will call such a state a *Type III factor state*. Now, if ω were pure, its GNS representation would be irreducible and its affiliated von Neumann algebra would be Type I, contrary to our assumption. Thus ω must be a mixed state on $\mathfrak{A}$. So far so good. But here's the weird thing. ω *must be disjoint from every pure state on* $\mathfrak{A}$. For any pure state ψ on $\mathfrak{A}$ has an irreducible GNS representation whose affiliated von Neumann algebra $\pi_\psi(\mathfrak{A})''$ is unitarily equivalent to $\mathfrak{B}(\mathcal{H}_\psi)$, a Type I factor. So a pure state ψ is a Type I factor state. By assumption, our mixed state ω is a Type III factor state. Belonging to different types, ω and ψ are not quasi-equivalent (Fact 7.10). Since ω and ψ are factors, the only other option is that they're disjoint (Fact 4.60). So if ω is a Type III factor state on a $\mathfrak{A}$, it's disjoint from every pure state on $\mathfrak{A}$. That is to say, ω is intrinsically mixed.

Suppose that ω on $\mathfrak{A}$ is intrinsically mixed, and let π_ω be its GNS representation. Because ω is mixed, π_ω is reducible (Fact 4.42). What's more, *each* of π_ω's subrepresentations is reducible too, as is each of π_ω's subrepresentations' subrepresentations, and so on, *ad infinitum*. Suppose, for reductio, that π_ω had an irreducible subrepresentation σ_i acting on a Hilbert space $\mathcal{H}_i$. Then every normed vector $|\psi\rangle$ in $\mathcal{H}_i$ would define a state ψ on $\mathfrak{A}$ (via $\psi(A) = \langle\psi|\sigma_i(A)|\psi\rangle$ for all $A \in \mathfrak{A}$). We know $|\psi\rangle$ is cyclic because every vector in an irreducible representation is cyclic (Bratteli and Robinson 1987, prop. 2.3.8 (47)). This makes the triple $(\sigma_i, \mathcal{H}_i, |\psi\rangle)$, with σ_i irreducible, unitarily equivalent to ψ's GNS representation. Because its GNS representation is irreducible, ψ is a pure state (Fact 4.42). A π_ω-normal state, ψ belongs to ω's folium. But this is contrary to the assumption that ω, as intrinsically mixed, is *disjoint* from every pure state on $\mathfrak{A}$. Thus, on an atomless $\mathfrak{M}$, a mixed state is never a mixture of pure states—it is mixed "all the way down."

Of course, none of these disclosures need matter to the interpreter of quantum theories if atomless von Neumann algebras were idle mathematical curiosities, without physical application. The following sections establish that they most assuredly are not. As a prelude, §7.3 explicates the notion of equilibrium appropriate to the thermodynamic limit of QSM, and connects that notion to modular theory. Modular theory in turn affords additional insight into how different types of von Neumann factors differ from one another. The axioms governing QM_∞ imply the existence of physical

---

[13] For a proof that $\mathfrak{D}_Q$ has no pure normal states, see Halvorson (2001).

7.3 KMS states

The notion of a *KMS state*, which generalizes the (perhaps familiar) notion of a Gibbs equilibrium state, will figure prominently in the sequel. In ordinary QM, the canonical (Gibbs) equilibrium state $\rho_\beta : \mathfrak{B}(\mathcal{H}) \to \mathbb{C}$ is determined by the system dynamics (in the guise of Hamiltonian H generating those dynamics), and the equilibrium temperature. For inverse temperature β $(:= \frac{1}{kT}$, where k is Boltzmann's constant and T temperature), ρ_β is:

$$\rho_\beta = \frac{\exp(-\beta H)}{Tr[\exp(-\beta H)]} \quad (7.3)$$

For a simple example, consider a spin $\frac{1}{2}$ system in an external magnetic field of strength $B \in \mathbb{R}$ oriented in the z-direction. The Hamiltonian for this system couples B to the z-component of spin:

$$H = -B\sigma(z) \quad (7.4)$$

Plugging (7.4) into the recipe (7.3) for a Gibbs equilibrium state at inverse temperature β, we obtain:

$$\rho = \frac{e^{\beta B\sigma(z)}}{Tr(e^{\beta B\sigma(z)})} \quad (7.5)$$

We encountered a state of (7.5)'s form before (§7.1): tensor producted with itself *ad infinitum* to describe an infinite spin chain, it defines a state ϕ on the CAR algebra $\mathfrak{A}_{CAR}$ whose affiliated von Neumann algebra $\pi_\phi(\mathfrak{A}_{CAR})''$ is a Type III factor. (For more on ρ considered as an equilibrium state of a single spin system, see Emch 1972, 89–91.)

A result due to von Neumann exhibits the Gibbs state's credentials as an equilibrium state. Provided that H has a discrete spectrum that is bounded from below, ρ_β is not only an entropy-maximizing state but also a Helmholtz free energy-minimizing state (for a precise statement and elaboration, see Emch 2007, 1090–94).

For realistic, finite quantum systems, the Gibbs state is well-defined and unique (Ruelle 1969). But in order for the quantity $Tr[\exp(-\beta H)]$ to be finite for all β—in order, that is, for the trace prescription applied to a state defined by (7.3) to yield sensible expectation values—H must satisfy the antecedent of the von Neumann result, i.e. it must have a spectrum that's discrete and bounded from below. This presupposes in turn a physical model in which dynamics take the ordinary QM form of a continuous unitary group acting on a concrete Hilbert space. A system governed by an indiscrete Hamiltonian has no Gibbs equilibrium state. A C^* dynamical system $(\mathfrak{A}, \alpha_t)$ also lacks a Gibbs equilibrium state if α_t can't be strongly unitarily implemented on a relevant

Hilbert space representation of $\mathfrak{A}$. Systems thereby bereft of Gibbs equilibrium states include infinite systems encountered at the thermodynamic limit of QSM, where it is thought we must go, if we are to accommodate phase structure and symmetry breaking (see Chapter 12). To extend the notion of equilibrium to systems for which the Gibbs condition (7.3) is undefined, we must generalize the explication of equilibrium afforded by the Gibbs state. We must also motivate the claim that the generalization is a generalization of the notion of *equilibrium*, for the justification that appeals to von Neumann's result is unavailable in the absence of a pure, discrete Hamiltonian.

When all the molecules in the vessel occupy its upper left-hand corner, their positions are highly correlated. The positions of molecules in an equilibrium distribution are not. Naively, the sorts of correlations they enforce can distinguish equilibrium from non-equilibrium states. So one place to look for a generalizable feature of the Gibbs state is in the correlations it defines. We can think of the KMS condition as such a feature.[14]

Let $B(t)$ denote the "evolute" of an observable $B \in \mathfrak{B}(\mathcal{H})$ through a time t. In the Heisenberg picture of ordinary quantum mechanics, $B(t) := \exp(iHt) B \exp(-iHt)$,[15] where H is the system Hamiltonian. Suppose we want to keep track, not only of instantaneous correlations but also of how time evolution affects the correlations between arbitrary observables $A, B \in \mathfrak{B}(\mathcal{H})$ as the latter evolves in time. The expectation values $\langle AB(t) \rangle$ and $\langle B(t)A \rangle$ each give us a way to do so. As $B(t)$ and A will in general fail to commute, these ways will in general differ. But in the case that the expectation values are calculated for a canonical equilibrium state, the difference between $\langle AB(t) \rangle$ and $\langle B(t)A \rangle$ will be tractable:

$$\langle B(t)A \rangle_{\rho_\beta} = Tr(\rho_\beta B(t)A) = \frac{Tr\left(\exp(-\beta H) \exp(iHt) B \exp(-iHt) A\right)}{Tr\left(\exp(-\beta H)\right)}$$

$$\langle AB(t) \rangle_{\rho_\beta} = Tr(\rho_\beta AB(t)) = \frac{Tr\left(\exp(-\beta H) A \exp(iHt) B \exp(-iHt)\right)}{Tr\left(\exp(-\beta H)\right)}$$

Comparing the expressions above, we see that replacing t in the formula for $\langle AB(t) \rangle_{\rho_\beta}$ with $t + i\beta$ yields the formula for $\langle B(t)A \rangle_{\rho_\beta}$—or at least it would if the naive substitution of a complex for a real variable were well-taken. Suppressing technical niceties which validate the substitution,[16] we can say: the time correlations the Gibbs state defines for pairs of observables satisfies the *KMS condition*, named for Kubo (1957) and Martin and Schwinger (1959), who first remarked its connection to equilibrium. Particularized to ordinary QM, this condition reads:

---

[14] Here I follow Emch and Liu (2002, 350–51).

[15] Don't sweat the sign of the exponents. This section adopts the sign convention common to most expositions of modular theory, the topic of §7.4.

[16] These are that there's some function $f_{AB}(z)$ of a complex variable that is analytic for $0 \leq Im(z) \leq \beta$ and continuous on the boundaries of the strip $\{z \in \mathbb{C} : 0 \leq Im(z) \leq \beta\}$. It's $f_{AB}(t)$ that keeps track of $\langle B(t)A \rangle_{\rho_\beta}$.

Definition 7.19 (KMS (ordinary QM)). A state ρ on $\mathfrak{B}(\mathcal{H})$ is a *KMS state* (at inverse temperature β) with respect to the evolution generated by $H \in \mathfrak{B}(\mathcal{H})$ iff for all $A, B \in \mathfrak{B}(\mathcal{H})$, $\langle B(t)A \rangle_\rho = \langle AB(t+i\beta) \rangle_\rho$.

In ordinary QM, the KMS condition is not just a feature of Gibbs equilibrium states but a defining feature, in the sense that a state ρ on $\mathfrak{B}(\mathcal{H})$ satisfies the KMS condition with respect to H and β iff ρ is the Gibbs state ρ_β (see Emch and Liu 2002, thm. 10.2.2 (351)).

The Hilbert space apparatus which frames the KMS condition so stated is incidental. We can generalize the observable algebra $\mathfrak{B}(\mathcal{H})$ appearing in Definition 7.19 to a C^* algebra $\mathfrak{A}$, which can, but needn't, be a Type I von Neumann algebra. Taking ρ to be a state on that algebra, we can generalize the expectation values $\langle \circ \rangle_\rho$ to the values $\rho(\circ)$ assigned by that state. Finally, we can generalize the dynamics considered, from Heisenberg picture dynamics implemented unitarily on a fixed Hilbert space, to C^* dynamics, in the form of a one-parameter family of automorphisms α_t of $\mathfrak{A}$. We obtain a condition which applies to states on C^* dynamical systems, without prejudice to the capacity of those systems to satisfy expectations borne of ordinary QM:

Definition 7.20 (KMS (general)). Where $\mathfrak{A}$ is C^* algebra and α_t a one-parameter family of automorphisms on that algebra, a state ρ on $\mathfrak{A}$ is a *KMS state* (at inverse temperature β) with respect to α_t iff for all $A, B \in \mathfrak{A}$, $\rho(\alpha_t(B)A) = \rho(A\alpha_{t+i\beta}(B))$.

When $\beta = \infty$ (i.e. for 0 temperature), the KMS condition reduces to the requirement that ρ be invariant under the action of α_t, along with analytic continuity requirements. The latter ensure that α_t is strongly unitarily implementable on ρ's GNS representation, with the cyclic vector $|\xi_\rho\rangle$ implementing ρ in that GNS representation the ground state of the Hamiltonian generating the unitary evolution (see Sewell 2002, 118).

There is a host of reasons to herald the KMS condition as an explication of equilibrium (see Emch 2007, 1122–44.) If the C^* dynamical system $(\mathfrak{A}, \alpha_t)$ admits a standard Gibbs state at inverse temperature β, the (α_t, β)-KMS state is unique and coincides with that Gibbs state (Bratteli and Robinson 1997, ex. 5.3.31 (119)). There are models of infinite systems (e.g. the Bose gas and the Heisenberg ferromagnet) for which states we have independent reason to regard as thermal equilibrium states satisfy the KMS condition. Moreover, KMS states exhibit stability features, including invariance under the action of the dynamical group α_t, typical of equilibrium states. And, it has been argued, a state that satisfies the KMS condition with respect to β acts like a thermal reservoir: any finite system coupled to it also reaches thermal equilibrium at β (Sewell 1986). As a criterion for thermal equilibrium, the KMS condition's credentials are impressive.

The next section catalogs an uncanny connection between the KMS condition for equilibrium and the apparatus of modular theory, a powerful tool for exploring the depths of the theory of von Neumann algebras.

7.4 A modicum of modular theory

Disclaimer. This is another section that, in the interest of preparing the reader for further adventures in the literature, develops technicalia in more detail than subsequent discussions presuppose. For readers at present content with their level of preparation, I herewith summarize this section's relevant results:

The Short Version. A faithful normal state ω on a von Neumann algebra $\mathfrak{M}$ defines a one (real) parameter family of automorphisms σ_t^ω on $\mathfrak{M}$. σ_t^ω is known as a *modular group*. ω satisfies the KMS condition (Def. 7.20) with respect to the flow defined by σ_t^ω at inverse temperature $\beta = 1$. Facts 7.29 and 7.30 will be invoked in Chapters 8 and 12.

The KMS condition appears in a different guise in the drama of Tomita–Takesaki modular theory, some of which—the parts which promise to help us come to terms with the exotica of QM_∞—I sketch here.[17]

A vector $\xi \in \mathcal{H}$ is *cyclic* for a von Neumann algebra $\mathfrak{M}$ (no longer assumed to be a factor) acting on $\mathcal{H}$ iff $\{\mathfrak{M}\}\xi$ is dense $\mathcal{H}$. ξ is *separating* just in case $A\xi = 0 \Rightarrow A = 0$ for all $A \in \mathfrak{M}$. If ξ is cyclic for $\mathfrak{M}$, then it's separating for $\mathfrak{M}$'s commutant $\mathfrak{M}'$, which is also a von Neumann algebra; and vice versa.

Given a cyclic and separating vector ξ for a von Neumann algebra $\mathfrak{M}$ acting on a Hilbert space $\mathcal{H}$, we can use that vector to transfer the algebraic adjoint operation on $\mathfrak{M}$ to an operator on $\mathcal{H}$.[18]

Introduce a map $S_0 : \mathcal{H} \to \mathcal{H}$ that acts as follows

$$S_0(A\xi) = A^*\xi \tag{7.6}$$

Thanks to ξ's cyclicity, S_0 will be defined on a dense subset of $\mathcal{H}$. Extend S_0 to an operator S defined on all of $\mathcal{H}$ by closure (see Sunder 1987, ch. 2.3). S is known as a *conjugation map*.

The conjugation map S will have a polar decomposition[19] $S = J\Delta^{\frac{1}{2}}$ where J is an anti-unitary operator[20g] called the *modular conjugation*; Δ is an invertible, positive self-adjoint but generally unbounded operator called the *modular operator*; and the pair (J, Δ) is unique. The story thus far: *A faithful normal state on a von Neumann algebra $\mathfrak{M}$ associates with that algebra a unique pair of operators (J, Δ) that define a conjugation map on the Hilbert space on which $\mathfrak{M}$ acts.*

The pair (J, Δ) defined by a faithful, normal state ω on $\mathfrak{M}$ can divulge information about ω, about $\mathfrak{M}$, or about both. For instance, if $\mathfrak{M}$ is abelian, the conjugation

---

[17] See Sunder (1987, ch. 2) for a brief exposition. For more about applications of modular theory to QFT, see Borchers (2000).

[18] In general we will have such a vector. Unless otherwise announced, the von Neumann algebras here encountered are σ-finite, i.e. they contain an at most countable set of orthogonal projections. Now, $\mathfrak{M}$ is σ-finite iff it admits a faithful normal state, in which case it is isomorphic with an algebra $\pi(\mathfrak{M})$ which admits a cyclic and separating vector ξ_ρ (Bratteli and Robinson 1987, prop. 2.5.6 (86)).

[19] For more about polar decompositions, see Zhu (1993, ch. 12).

[20] i.e. $J = J^*, J^2 = I$, and $\langle J\psi | J\phi \rangle = \langle \psi | \phi \rangle^* = \langle \phi | \psi \rangle$.

7.4 A MODICUM OF MODULAR THEORY

operator S is identical to the modular conjugation J, and the modular operator Δ is the identity (see Sunder 1987, ch. 2, for an argument). To see what (J, Δ) might disclose about ω requires a few preliminaries.

Definition 7.21. A *weight* w on a von Neumann algebra $\mathfrak{M}$ is a linear map from positive elements of $\mathfrak{M}$ to the positive reals.

As positive operators, projection operators, as well as unitary operators, are in the domain of a weight. The algebraic notion of a state on a von Neumann algebra is a weight that is normalized, in the sense that $w(I) = 1$.

Definition 7.22. A *trace* on $\mathfrak{M}$ is a weight w such that $w(A^*A) = w(AA^*)$ for all $A \in \mathfrak{M}$ (which implies that $w(AB) = w(BA)$ for all $A, B \in \mathfrak{M}$ (Sunder 1987, 38)).

A trace is a *tracial state* if and only if it's normalized. Notice that an $(\alpha_t, 0)$-KMS state—i.e. a KMS state at inverse temperature 0—is a tracial state (see Def. 7.20).

Example 7.23 (weights and traces on $\mathfrak{B}(\mathcal{H})$). Let $\dim(\mathcal{H}) = n$. For the von Neumann algebra $\mathfrak{B}(\mathcal{H})$, a density operator W defines a weight via the trace prescription: $w_W(A) := Tr(WA)$. w_X is a tracial weight if and only if $w_X(A^*A) = Tr(XA^*A) = Tr(XAA^*) = w_X(AA^*)$ for all $A \in \mathfrak{B}(\mathcal{H})$. This happens when X is a multiple of the identity operator, in which case (ignoring scaling factors) $Tr(IAA^*) = Tr(AIA^*) = Tr(IA^*A)$, where the second equality uses the invariance of the trace under cyclic permutations. Notice that w_I, restricted to projection operators in $\mathfrak{B}(\mathcal{H})$, gives a dimension function on that von Neumann algebra, which up to rescaling is unique. (Because a dimension function is a tracial weight, and because a dimension function on a factor is unique up to rescaling, $\frac{w_I}{n}$ is the unique tracial state on $\mathfrak{B}(\mathcal{H})$.) When $\mathfrak{B}(\mathcal{H})$ is Type I_∞, the weight w_I still yields a dimension function when restricted to projections in $\mathfrak{B}(\mathcal{H})$, but can no longer be normalized to define a tracial state. ♠

The foregoing example illustrates the general fact that only finite von Neumann algebras have tracial states (Takesaki 2002, thm. V.2.6 (312)).

This brings us at last to an example of what the pair (J, Δ) defined by a faithful, normal state ω on $\mathfrak{M}$ might tell us about ω: *the modular operator Δ induced by a tracial state is the identity* (Kadison and Ringrose 1997b, corr. 9.2.15 (611)). This might induce us to hope that in general, the modular operator Δ a faithful normal state ω picks out for a von Neumann algebra $\mathfrak{M}$ tells us something about ω or $\mathfrak{M}$ or both—to hope, for instance, that Δ's departure from the identity provides some measure of how far ω is from tracial or (because only finite von Neumann factors admit tracial states) how far $\mathfrak{M}$ is from finite.

This hope is borne out by Tomita–Takesaki modular theory, which also bears an intimate connection to the KMS condition discussed in the last section. The key result is the Tomita–Takesaki Theorem, which exhibits remarkable properties of the elements of the polar decomposition of the conjugation map:

Fact 7.24 (Tomita–Takesaki Theorem). Where J and Δ are the modular conjugation and modular operator associated with a von Neumann algebra $\mathfrak{M}$ by a faithful normal state ω,

(i) $J\mathfrak{M}J = \mathfrak{M}'$
(ii) $\Delta^{it}\mathfrak{M}\Delta^{-it} = \mathfrak{M}$ for all $t \in \mathbb{R}$, where Δ^{it} is unitary for each t and strongly continuous. ♠

We already knew that J was an anti-unitary map from the Hilbert space on which $\mathfrak{M}$ acts to itself. (i) reveals that J is also an anti-isomorphism from $\mathfrak{M}$ onto its commutant.

Δ^{it} is a family of operators built by the agency of spectral theory from the positive operator Δ. (ii) establishes that each member of this family is unitary and therefore bounded, even when Δ is unbounded. It thereby underwrites the introduction of the all-important *modular automorphism group* σ_t^ω, which acts as follows on each $A \in \mathfrak{M}$:

$$\sigma_t^\omega(A) := \Delta^{it}A\Delta^{-it} \tag{7.7}$$

Thus *modular theory extracts from a faithful normal state ω on a von Neumann algebra $\mathfrak{M}$ a privileged and strongly continuously one-parameter family of automorphisms—i.e. a "flow"*[21]—on $\mathfrak{M}$.

We can work the other way. Given a one (real) parameter family σ_t of automorphisms for $\mathfrak{M}$, we can ask: is there a faithful normal state on $\mathfrak{M}$ that has σ_t as its modular automorphism group? The answer is affirmative for a state ω satisfying the (hauntingly familiar) *modular condition*[22] imposed by σ_t:

Definition 7.25 (modular automorphism group). σ_t is the modular automorphism group ω defines for $\mathfrak{M}$ iff for each $A, B \in \mathfrak{M}$ and for all t

$$\omega(\sigma_t(B)A) = \omega(A\sigma_{t+i}(B)) \tag{7.8}$$

This is just the KMS condition (Def. 7.20) for inverse temperature $\beta = 1$.[23]

Suppose ω satisfies (7.8) with respect to σ_t. Then ω is invariant under the action of σ_t (Sunder 1987, lemma 2.5.4 (64)), and that group is the *only* flow with respect to which ω satisfies the modular condition (Sunder 1987, thm. 2.5.11 (68)). So every faithful normal state on a von Neumann algebra satisfies the modular/KMS condition with respect to exactly one flow, although not necessarily one naturally read as a group of time translations.[24] For von Neumann factors, the converse holds as

---

[21] Perchance (thinking of the parameter t as time) a time evolution (see Connes and Rovelli 1994)?

[22] As before, the statement of this condition suppresses continuity conditions on correlation functions $f_{AB}(t)$.

[23] What about other inverse temperatures? If ω satisfies condition (7.8) for σ_t such that $\sigma_t(A) = \Delta^{it}A\Delta^{-it}$, then ω satisfies condition (7.8) for $\sigma_{\beta t}$ such that $\sigma_{\beta t}(A) = \Delta^{i\frac{t}{\beta}}A\Delta^{-i\frac{t}{\beta}}$, provided $\beta > 0$. Thus a $(\sigma_t, 1)$ KMS state is also a $(\sigma_{\beta t}, \beta)$ KMS state, for any positive β.

[24] See Earman and Ruetsche (2005) for examples of modular groups with good claims to be time translations.

7.4 A MODICUM OF MODULAR THEORY

well: if σ_t is the modular automorphism group a faithful normal state ω defines for a von Neumann factor $\mathfrak{R}$, then ω is the *only* state on $\mathfrak{R}$ satisfying the modular/KMS condition for σ_t. But mark well that not every flow on a von Neumann algebra is a modular automorphism group for some faithful normal state. For instance, no faithful state on an abelian von Neumann algebra $\mathfrak{D}$ is a KMS state with respect to any automorphism group different from the identity (Bratteli and Robinson 1997, 113). This is due to the fact, noted above, that the modular operator for a faithful, normal state on an abelian algebra is the identity.

The notion of a modular automorphism group also assists in the classification of von Neumann factors. The assistance, which I'll sketch here, takes the form of defining for each factor $\mathfrak{R}$ a commodity, its *modular spectrum*, that reveals $\mathfrak{R}$'s type. We begin with a definition.

Definition 7.26 (modular invariants). Where σ_t^ω is the modular automorphism group a faithful normal state ω defines for a factor $\mathfrak{R}$, $\mathfrak{R}$'s *modular invariants* $\mathfrak{R}^{\sigma^\omega}$ are those elements of $\mathfrak{R}$ invariant under the action of σ_t^ω: $\mathfrak{R}^{\sigma^\omega} := \{A \in \mathfrak{R} : \sigma_t^\omega(A) = A \text{ for all } A \in \mathfrak{R}\}$.

A fact to which we'll have occasion to refer is:

Fact 7.27. A is invariant under σ_t^ω iff $\omega(AB) = \omega(BA)$ for all $A, B \in \mathfrak{R}$ (Kadison and Ringrose 1997b, 617).

It follows from Def. 7.22 that $\mathfrak{R}^{\sigma^\omega} = \mathfrak{R}$ if and only if ω is tracial. Section 8.3.3 will describe how the coincidence of $\mathfrak{R}^{\sigma^\omega}$ and $\mathfrak{R}$ for tracial states constitutes a well-known problem for modal interpretations.

The notion of modular invariants helps us keep tabs on another significant class of states: ergodic states. In classical statistical mechanics, ergodic states are ones for which space averages equal time averages. I will put off until Chapter 12 any attempt to link the following definition to more familiar conceptions of ergodicity.

Definition 7.28 (ergodic state). ω is *ergodic* iff it's an extremal element of the convex set of σ_t^ω-invariant states.

Fact 7.29. If ω is ergodic, then $\mathfrak{R}^{\sigma^\omega} = \{\mathbb{C}I\}$.

So for both tracial and ergodic states, there's something uninteresting about their associated sets of modular invariants. Section 8.3.3 will describe how the coincidence of $\mathfrak{R}^{\sigma^\omega}$ and $\{\mathbb{C}I\}$ for ergodic states constitutes a less-appreciated problem for the extension of modal interpretations to the setting of QM_∞.

Back to our task of enlisting modular theory in the classification of von Neumann algebras. The *modular spectrum* $\mathcal{S}(\mathfrak{R})$ of a von Neumann factor $\mathfrak{R}$ is given by $\cap_\omega sp(\Delta_\omega)$, where the intersection is over faithful normal states on the algebra, and $sp(\Delta_\omega)$ is the spectrum of the modular operator for the state ω. When Δ_ω is the identity operator—which it will be when ω is tracial or $\mathfrak{R}$ abelian—its spectrum is 1. As ω tends toward ergodic, $sp(\Delta_\omega)$ tends toward the positive reals. Thus $sp(\Delta_\omega)$ and $\mathcal{S}(\mathfrak{R})$ are, respectively, the devices we'd hoped would gauge ω's departure from traciality, and $\mathfrak{R}$'s departure from finitude.

An important result gives an instance of such gauging.

Fact 7.30 (modular automorphisms of semifinite factors are inner). If factor algebra $\mathfrak{R}$ is semi-finite (i.e. Type I or Type II$_1$), each σ_t^ω it admits is an *inner automorphism*, i.e. an automorphism implemented by a unitary $U_t^\omega \in \mathfrak{R}$ in the sense that $\sigma_t^\omega(A) = U_t^\omega A U_t^{*\omega}$ for all $A \in \mathfrak{R}$. It follows that $\mathcal{S}(\mathfrak{R}) = 1$. Conversely, if $\mathfrak{R}$ admits an inner modular automorphism (i.e. one determined by a faithful, normal state) then it is semi-finite (Sunder 1987, prop. 3.4.6 (111)).

The classification of Type III factors afforded by the modular spectrum identifies Type III$_1$ factor algebras as those whose modular spectra are the positive reals (see Halvorson and Müger 2007, 737–40, for a discussion). Impressionistically, when $\mathfrak{R}$ is a Type III$_1$ factor, $\mathcal{S}(\mathfrak{R})$ is as far as it can be from $\{0, 1\}$, which suggests that Type III$_1$ factors are as far as von Neumann algebras can be from commutative. Insofar as non-commutivity is the mark of the quantum, Type III$_1$ factors are the apotheosis of quantum physics, if they're physical at all. The next section documents that they're physical.

7.5 Extraordinarily physical

In ordinary QM, observable algebras are Type I factors. In QM$_\infty$, this needn't be so. Physical applications demanding observable algebras of other sorts abound. This section catalogs a few such applications; the next chapter argues that these instances of extraordinary QM slip through the clutches of many standard approaches to interpreting ordinary QM.

7.5.1 Non-atomic von Neumann algebras in QFT

Let ω be the Minkowski vacuum state for the mass m Klein–Gordon field. Where $\mathfrak{A}(\mathcal{O})$ is the algebra associated with a region $\mathcal{O}$ of Minkowski spacetime, [Isotony] ensures that $\mathfrak{A}(\mathcal{O}) \subset \mathfrak{A}(\mathcal{M})$. Restricting ω to elements of $\mathfrak{A}(\mathcal{O})$ defines a state, call it $\omega_\mathcal{O}$, on that local algebra. The analogy with the mixed reduced states the entangled spin singlet state implies for each of its subsystems might induce us to expect $\omega_\mathcal{O}$ to be a mixed state. For many choices of $\mathcal{O}$ it is also intrinsically mixed. For instance, if $\mathcal{O}$ is a region with a non-empty spacelike complement, the von Neumann algebra $\pi_{\omega_\mathcal{O}}(\mathfrak{A}(\mathcal{O}))''$ affiliated with $\omega_\mathcal{O}$'s GNS representation is a Type III factor (Araki 1964), and therefore atomless. Results by Buchholz, D'Antoni, and Fredenhagen (1987) indicate that local algebras are Type III not just for free scalar fields, but generic quantum fields of physical interest. The *global* von Neumann algebra will be Type I if the representation satisfies the vacuum axiom, or even if it lacks a vacuum state but harbors a "mass gap." Still, there are massless theories of both bosons and fermions with no mass gap and no vacuum state in which the global von Neumann algebra satisfies a positive energy condition but is Type II or Type III (see Doplicher, Figliolini, and Guido 1984; Buchholz and Doplicher 1984; Borek 1985).

So even global algebras of physical interest in QFT can be Type III. But the availability of atomless global algebras is not needed to secure the foundational relevance of the Type III local algebra obtained by restriction. These algebras arise from states to which any adequate interpretation of QFT is beholden, notwithstanding their failure to be states of isolated systems. For the non-isolated systems of which they're states are the very systems—corresponding to the localized regions of spacetime in which we find ourselves and our laboratory apparatus—of which we need to make sense in order to interpret QFT. If the algebras appropriate to such systems are atomless, interpretive maneuvers mediated by atoms will not avail. Next we encounter atomless von Neumann algebras that crop up in QM_∞ by means other than restriction.

7.5.2 Non-atomic von Neumann algebras in QSM

KMS states at finite temperatures in the thermodynamic limit of QSM correspond to Type III factors for a wide variety of physically interesting systems: Bose and Fermi gases, the Einstein crystal, the BCS model (see Emch 1972, 139–40; Bratteli and Robinson 1997, corr. 5.3.36 (130)). The exceptions are KMS states at temperatures at which phase transitions occur (if there are any for the systems in question); then KMS states are direct sums/integrals of Type III factors, but still intrinsically mixed; their affiliated von Neumann algebras are atomless. Unlike most of the Type III factor states noted above for QFT, these atomless algebras aren't obtained by restriction. They're algebras for the entire system under consideration.

Systems in equilibrium at *infinite* temperature ($\beta = 0$) are also of interest in QSM. An $(\alpha_t, 0)$-KMS state is an α_t-invariant tracial state, i.e. a *chaotic* state, perched at the apex of a hierarchy of increasingly random behaviors (the lowest rung of which, ergodicity, we've touched upon and will return to). Chaotic states in the thermodynamic limit of QSM are Type II_1 factor states (see Takesaki 2003, §XIV.1). It follows that they're intrinsically mixed, and that their affiliated von Neumann algebras are atomless.

There is, however, some vestige of ordinary quantum mechanics left at the thermodynamic limit. If a state ω of a C^* dynamical system $(\mathfrak{A}, \alpha_t)$ is α_t-invariant, then α_t is strongly unitarily implementable in ω's GNS representation (Fact 5.2). Let H be the self-adjoint generator of the family of unitaries implementing α_t on the Hilbert space $\mathcal{H}_\omega$ of that GNS representation. Then ω is a *ground state* iff H's spectrum is strictly positive. A ground state of a C^* dynamical system is a KMS state at inverse temperature ∞ of that system. And extremal elements of the set of (α_t, ∞)-KMS states are Type I factor states (Bratteli and Robinson 1997, thm. 5.3.37 (131)).

7.6 Interlude

This residue of ordinary QM clinging to the thermodynamic limit of QSM is anomalous. Type II and III factor states are the norm in the thermodynamic limit; they're the typical states of local regions of spacetime in QFT. The next chapter argues that conventional accounts of quantum probability, interpretations developed with ordinary QM in mind, founder in the waters of QM_∞, whose murky depths conceal atomless von Neumann algebras.

8
Interpreting Extraordinary QM

The last chapter chronicled notable *formal* differences between the algebras of observables characteristic of ordinary QM and those endemic to QM_∞. About those differences, this chapter asks: so what? Do these formal differences matter to the *interpretation* of QM? I will suggest that they do. The atoms typically absent from the algebras of QM_∞ are the linchpins on which many familiar interpretations of ordinary QM turn. Exported to QM_∞, those interpretations lack traction. Spinning frictionlessly, they fail to discharge their fundamental interpretive duties.

Foremost among these is the duty to make sense of quantum probabilities. I distinguish two aspects of this duty. One is to explain how quantum systems come to occupy conditions—call them *quantum states*—that offer specifiable (and ergo testable) probability assignments. Another is to explicate the probabilities these quantum states assign: to explain, for instance, what these probabilities are probabilities for, and why they are obedient to the Born rule. These aspects overlap with the traditional problems of preparation and measurement. Section 8.1 treats the former. It sketches an account of state preparation that can be adapted to both collapse and no-collapse interpretations of QM, isolates the role in that account of atoms, and argues that the atomlessness of QM_∞ observable algebras foils accounts of state preparation based on the ordinary QM strategy.

This suggests that the process of state preparation in QM_∞ needs to be reconceived, if it is to be recognized as possible at all. Accepting that it's possible, §8.2 turns to the question of how to understand the probabilities assigned by quantum states. It introduces a family of pristine approaches to interpreting ordinary QM under the heading of the Maximal Beable Approach (MBA). Briefly, to take the MBA to interpreting a quantum theory is to identify a *maximal beable subalgebra* of that theory's observable algebra, i.e. a maximal set of that theory's observables that can (subject to constraints arising from metaphysical scruples) be attributed determinate values simultaneously. Section 8.2 identifies the central role discharged by atoms in MBA interpretations of ordinary QM, among which it numbers collapse, "Bohmian," modal, and relative state formulations. Confronting the MBA with QM_∞'s atomlessness, §8.3

considers its options. One is to continue to pursue the MBA, albeit non-atomically. I'll argue that while such pursuit is formally possible, in the sense that even atomless von Neumann algebras have maximal beable subalgebras, it shirks key interpretive tasks. The second response to atomlessness is Rob Clifton's, and takes the form of a modal interpretation adapted to atomless von Neumann algebras. Clifton's interpretation abandons the MBA's maximality requirement. A consequence of this abandonment is that Clifton's modal interpretation often has *nothing* to say about systems in its scope. The chapter closes with a review of how theories of QM_∞ upset the presuppositions of ordinary quantum semantics, including the ideal of pristine interpretation.

8.1 Preparation

Part of the task of interpreting quantum probability is making sense of our capacity to bring quantum systems into conditions—call them *quantum states*—we can understand to assign probabilities to the outcomes of future interactions. To address this question is to grapple with the quantum mechanical preparation problem, a poor relative of the quantum measurement problem.

The textbook account of state preparation casts it as "pre-measurement," and solves the preparation problem using the same legerdemain by which it would spirit the measurement problem away. To prepare a spin $\frac{1}{2}$ system in the "up" eigenstate $|+\rangle$ of the z-component of spin, simply measure $\sigma(z)$. Systems exiting your Stern–Gerlach magnet in the "up" trajectory have, courtesy of measurement collapse, been deposited in the corresponding eigenstate $|+\rangle$.

But a "pre-measurement" model of preparation needn't invoke collapse, so long as it can help itself to Lüders' Rule. In ordinary QM, Lüders' Rule is a rule for extracting, from a state W, probabilities conditional on an event E. Where W is a density operator and E a projection operator in $\mathfrak{B}(\mathcal{H})$ such that $Tr(WE) \neq 0$, these conditional probabilities are encoded by a density operator state W_E defined as follows:

$$W_E(A) := \frac{Tr(AEWE)}{Tr(EW)} \quad \text{for all } A \in \mathfrak{B}(\mathcal{H}) \tag{8.1}$$

Bub (1977) identifies robust support for W_E's claim to encapsulate the probabilities W assigns conditional on E. For notational convenience, let ω and ω_E be states on $\mathfrak{B}(\mathcal{H})$ determined by W and W_E respectively. ω_E is the unique normal state on $\mathfrak{B}(\mathcal{H})$ with the property that for any projection $F \in \mathcal{P}(\mathfrak{B}(\mathcal{H}))$ such that $F \leq E$, $\omega_E(F) = \frac{\omega(F)}{\omega(E)}$. The definition of *classical* conditional probabilities makes this property's significance clear. To conditionalize a classical probability function ρ on an event E, one follows the rule: $\rho_E(F) = \frac{\rho(F\&E)}{\rho(E)}$. But to follow the rule in the quantum context is to break others. Where E and F are incompatible events represented by non-commuting projections and ρ is a quantum state, the quantity $\rho(F\&E)$ appearing in the classical rule lacks an orthodox definition: no event is the joint event "E and F". But in the case that E and F commute—as they will (see Def. 7.2) when $F \leq E$—there is no impediment to following the classical rule. The event $E\&F$ is just the event F, and applying the classical rule to the state ω yields $\omega_E(F) = \frac{\omega(E\&F)}{\omega(E)} = \frac{\omega(F)}{\omega(E)}$. Exhibiting the property under discussion, W_E satisfies the *classical* definition of conditional probability *where that definition makes quantum sense.*

When E is a one-dimensional projection operator or atom, W_E reduces to E wherever it is defined. Supposing E to be a measurement outcome, we see that an arbitrary pre-measurement state Lüders conditionalized on that outcome coincides with the state produced by a measurement collapse yielding that outcome. Reposed in terms of Lüders' rule, the collapse-ridden account of state preparation is:

Recipe 8.1 (preparation by pre-measurement). To prepare the pure state E on $\mathfrak{B}(\mathcal{H})$:

Step 1. Subject a system in an arbitrary pre-measurement state W to a measurement of the atom E.

Step 2. Obtain the outcome *Yes*.

Step 3. Attribute the system initially in the state W the Lüders conditionalized state $W_E = E$. ♠

The collapse postulate offers an explaination of why the Lüders conditionalized state W_E so successfully encapsulates the subsequent behavior of the system. It is just the state E, in which the system is left by measurement collapse.

Interpretations denying measurement collapse cannot take W_E *literally* as the post-measurement state. But they can take W_E to summarize a set of probabilities *conditional on E* implicit in W. Because measurements subsequent to the E measurement occur in the scope of the condition, W_E is the appropriate predictive instrumentality to use when considering those measurements—appropriate because Lüders' Rule is the correct expression for quantum conditional probabilities, not because collapse has occurred. Collapse-free interpretations are thereby entitled to follow every step of the preparation recipe above. They demur to collapse-ridden interpretations not about whether the recipe succeeds, but about why.[1]

Successfully following the preparation recipe, we start with a system in an arbitrary, and generally unknown, quantum state W, and wind up with a system we are warranted in attributing a specific, and known, quantum state E. Lüders' rule is not the only ingredient essential to the preparation recipe. Another is the presence in $\mathfrak{B}(\mathcal{H})$ of what I'll call *witnesses*, elements of $\mathfrak{B}(\mathcal{H})$ with the feature that their measurements can have outcomes warranting the attribution of a particular quantum state to the system measured. More generally, a witness to a state on a von Neumann algebra is defined as follows:

Definition 8.2 (witness). A projection $E_\phi \in \mathfrak{M}$ is a *witness* for a state ϕ just in case for any normal state ω on $\mathfrak{M}$, if "$\omega(\phi) \neq 0$"[2] then $\omega_{E_\phi}(A) := \frac{\omega(E_\phi A E_\phi)}{\omega(E_\phi)} = \phi(A)$ for all $A \in \mathfrak{M}$.

---

[1] For more on Lüders' Rule without collapse, see van Fraassen (1991); Ruetsche (2003).
[2] To remove the scare quotes, call upon the *support projection* S_ϕ of a state ϕ (Def. 7.13) and understand "$\omega(\phi)$" as "$\omega(S_\phi)$".

$\omega_{E_\phi}(A)$ is the state ω Lüders conditionalized on a *Yes* outcome of an E_ϕ measurement. E_ϕ's status as a witness for ϕ assures us that, *no matter what normal state a system occupies before an E_ϕ measurement yielding a "Yes" outcome, it can be assigned the state ϕ after that measurement*. E_ϕ's status as a witness for ϕ ensures that following the preparation recipe eventuates in a system we may warrantedly attribute the state we take ourselves to have prepared.

Now, let us re-express the preparation recipe in terms of an arbitrary von Neumann algebra:

Recipe 8.3 (preparation by pre-measurement (general)). To prepare the state ϕ on $\mathfrak{M}$:

Step 1. Subject a system in an arbitrary pre-measurement state ω to a measurement of a witness $E_\phi \in \mathfrak{M}$ for ϕ.

Step 2. Obtain the outcome *Yes*.

Step 3. Attribute the system initially in the state ω the Lüders conditionalized state ω_{E_ϕ}.

ω_{E_ϕ} coincides with ϕ just because E_ϕ is a witness for ϕ. Only if ϕ has a witness E_ϕ can we follow this recipe. In the case of ordinary QM, normal pure states correspond to atoms in $\mathfrak{B}(\mathcal{H})$, which atoms also serve those pure states as witnesses. Thus in ordinary QM, atoms are crucial to accounts of preparation by pre-measurement.

Atomless von Neumann algebras of the sort endemic to QM$_\infty$ spoil the recipe. They spoil it because no normal state ϕ on an atomless algebra $\mathfrak{M}$ has a witness. Only pure normal states have witnesses, and an atomless $\mathfrak{M}$ has no pure normal states. The argument that only pure normal states have witnesses draws on some apparatus introduced in §7.2.

"Argument" 8.4 (only pure normal states have witnesses.). Folia are closed under Lüders conditionalization (see §4.6.2). Since a witness for ϕ is an element of $\mathfrak{M}$ such that Lüders conditionalization on the witness turns an arbitrary normal state on $\mathfrak{M}$ into the state ϕ, only normal states have witnesses. So what we need to show is that no mixed states have witnesses. Suppose ϕ is a mixed state on a von Neumann algebra $\mathfrak{M}$, and assume, for reductio, that it has a witness E_ϕ. From Def. 8.2, if ρ is a normal state on $\mathfrak{M}$ that assigns ϕ a non-zero probability, $\phi(E_\phi) = \frac{\rho(E_\phi E_\phi E_\phi)}{\rho(E_\phi)} = 1$ (because a projection F is idempotent: $FF = F$). Let S_ϕ be ϕ's support projection, the smallest projection made true by ϕ (Def. 7.13). Since $\phi(E_\phi) = 1$, $S_\phi \leq E_\phi$. Because ϕ is mixed, S_ϕ is not minimal in $\mathfrak{M}$ (Digression 7.2). So there is some projection $E \in \mathfrak{M}$ such that $E < S_\phi$, which implies that $E < E_\phi$. Now let ψ be a normal state on $\mathfrak{M}$ with support projection E. Because E_ϕ is a witness for ϕ,

$$\frac{\psi(E_\phi E E_\phi)}{\psi(E_\phi)} = \phi(E)$$

Since $E < E_\phi$ and $\psi(E) = 1$, the l.h.s. becomes $\frac{\psi(E)}{\psi(E_\phi)} = 1$. But the r.h.s. is less than 1, because S_ϕ is the smallest projection ϕ maps to 1 and $E < S_\phi$. We have our contradiction and can conclude that no mixed state has a witness. (See also Ruetsche and Earman 2010.) ♠

Atomless von Neumann algebras are in a witness protection program. They admit no pure normal states, but only pure normal states have witnesses. Because the preparation by pre-measurement strategy succeeds in preparing only states with witnesses, we follow the preparation by pre-measurement recipe to no avail when $\mathfrak{M}$ is an atomless von Neumann algebra.

How then are we to account for the preparation of states on that algebra? The recipe covers preparation procedures that take normal states to normal states, so we might aim instead to explicate the preparation of states that aren't normal. But this does not seem promising. The point of preparation is, after all, to bring a system into a state from which we can extract probabilistic predictions that enable us to make sense of the natural world. Insofar as the probabilities assigned by non-normal states lack the coherence of countable additivity and may also lack the coherence of instantiating natural laws (see §§4.6.1 and 6.4.2), accounting for the preparation of non-normal states threatens to miss the point. For another thing, the account of preparation by "pre-measurement" may not extend intact to non-normal states. The very notion of a witness supposes that the state of a system prior to a preparation interaction is normal. Once non-normal states are countenanced, this supposition is unmotivated.

Of course, none of the foregoing rules out the possibility that the preparation problem in QM_∞ will succumb to a strategy dramatically distinct from the preparation by pre-measurement strategy familiar from ordinary QM. Nor does it rule out compelling arguments that QM_∞ is of a cast so different from that of ordinary QM that no analog of the latter's preparation problem afflicts the former. Although I can't rule out these possibilities, neither can I imagine how to instantiate them. Thus the next section unfolds in the scope of an assumption I assign a low credence: the assumption that we have a handle on how to prepare typical QM_∞ systems in particular quantum states.[3] The question the section addresses is: does the absence of atoms from the observable algebras pertaining to such systems impede techniques, developed with ordinary QM in mind, for understanding the probabilities those states assign?

8.2 The MBA and ordinary QM

This section and the next describe a widespread approach to interpreting ordinary QM: the maximal beable approach, or MBA. Isolating the pivotal role atoms in $\mathfrak{B}(\mathcal{H})$ play in familiar instantiations of the MBA, they assess the MBA's prospects of success when it comes to the atomless von Neumann algebras of QM_∞, and suggest that those prospects are limited.

---

[3] The thermodynamic limit of QSM suggests an exception: arguably we can prepare equilibrium states simply by waiting long enough.

8.2.1 Characterizing the MBA

Pronouncements expressing the orthodox "Copenhagen" interpretation of ordinary QM abound in problematic and ambiguous notions, prominently "measurement" and its would-be explicators: "subjectivity," "classicality," and so on. John Bell imagines an alternative interpretive strategy:

> it is interesting to speculate on the possibility that a future theory will not be *intrinsically* ambiguous and approximate. Such a theory could not be fundamentally about "measurements," for that would again imply incompleteness of the system and unanalyzed interventions from outside. Rather it should again become possible to say of a system not that such and such may be *observed* to be so, but that such and such *be* so. The theory would be not be about "*observ*ables" but about "*be*ables". (Bell 1987, 41)

Drawing upon the pristine interpretive resource of the theoretical apparatus, suppose that the properties pertaining to a system correspond to self-adjoint elements of some $\mathfrak{B}(\mathcal{H})$, or more precisely to the assignment of eigenvalues to those elements. The lesson of the Bell–Kochen–Specker theorem is that the option of taking a density operator state W on $\mathfrak{B}(\mathcal{H})$ to describe an ensemble of systems, each of which exhibits a determinate eigenvalue for each observable pertaining to it, is exercised at (what most commentators' metaphysical scruples reckon to be[4]) considerable cost, if the dimension of $\mathcal{H}$ is greater than 2. Bell imagines another option:

> Could not one just promote *some* of the observables of the present quantum theory to the status of beables? The beables would then be represented by linear operators in the state space. The values which they are allowed to *be* would be eigenvalues of those operators. For the general state the probability of a beable *being* a particular value would be calculated just as was formerly calculated the probability of *observing* that value. (Bell 1987, 41)

The project Bell envisions is the *maximal beable approach* (MBA) to interpreting quantum theories: to first approximation, the approach of identifying the largest set of quantum observables pertaining to a system that can (subject to constraints arising from metaphysical scruples) enjoy determinate values simultaneously.

Notice that Bell has the MBA aspire to *explicate* quantum probabilities. Successfully executing the approach, we come to understand not only what quantum probabilities are probabilities for—eigenvalue assignments to determinate observables—but also how to calculate those probabilities: "just as was formerly calculated the probability of *observing* [those] values." The explication of quantum probabilities Bell imagines is integral to understanding QM as an *empirical* theory—a theory whose empirical commitments include *explicit* probability assignments to *specified* eventualities.

Halvorson and Clifton (1999) translate the interpretive approach Bell conjures into the formal apparatus of operator theory. Their translation has the virtue of generality: it expresses the MBA as an approach to interpreting quantum theories couched as C^*

---

[4] Famously, Bell's metaphysical scruples are exceptional; see e.g. Bell (1987, 8–9).

algebras, without prejudice to whether those algebras are abstract or concrete, weakly closed or merely closed in C^* norm. Their first requirement is that beables form an *algebra*. After all, if A and B are beables, so ought to be their products and linear combinations. Notice that there are many ways to ground this expectation "metaphysically"—i.e. antecendently to the consideration of eclectic specifics of theoretical application. For instance, an operationalist, insisting that to be is to be measurable, might observe that if A and B are measurable, a simple pen and paper operation renders $A + B$ measurable as well. Non-operationalists as well are tempted by the principle that beables are closed under the taking of products and linear combinations. Such a principle is implicit in the FUNC rule made famous by Redhead's exposition of Bell–Kochen–Specker arguments (Redhead 1989, 121).

Halvorson and Clifton further suppose that the collection of beables exhibits another sort of closure: closure under the taking of uniform limits. This supposition may also be grounded "metaphysically"—for instance, by appeal to SEGOP (cf. §6.4.2). It follows from the closure suppositions that the beables reside in a C^* algebra.

Now, the MBA's mission is to characterize those C^* algebras that can be algebras of beables. Halvorson and Clifton's characterization allows beable algebras to be state-dependent, and is mediated by the notion of a dispersion-free state[5]:

Definition 8.5. A state ω on a subalgebra $\mathfrak{D}$ of a C^* algebra $\mathfrak{A}$ is **dispersion-free** iff $\omega(A^2) = (\omega(A))^2$ for each self-adjoint $A \in \mathfrak{D}$.

(Note that because ω is a state on $\mathfrak{D}$ in the algebraic sense, it needn't be countably additive, even if $\mathfrak{D}$ contains enough projection operators for a requirement of countable additivity to make sense.) To appreciate why freedom from dispersion is significant, assume that A is a self-adjoint Hilbert space operator and that $\omega(A)^2 = \omega(A^2)$. It follows that ω assigns probability 1 to one of A's eigenvalues.[6] The venerable EPR reality criterion warrants the conclusion that, with respect to the state ω, A corresponds to an "element of reality." A has a determinate value; in Bell's parlance, A is a beable with respect to ω. So is every other self-adjoint element of an algebra $\mathfrak{D}$ satisfying Def. 8.5 with respect to ω.

A state that's dispersion-free on an algebra assigns to each observable in that algebra one of its eigenvalues. A dispersion-free state thereby specifies a "pattern of beable instantiation," i.e. an assignment of values to determinate observables. Extending nomenclature usually associated with the modal interpretation, I'll call such an assignment a *value state*.

If ω is dispersion-free on $\mathfrak{D}$, $\mathfrak{D}$ is an algebra of beables on which ω defines a value state. But this is a special case of beable instantiation, one that suppresses the

---

[5] I simplify their exposition somewhat.

[6] The missing steps: where $\{E_i\}$ are A's spectral projections and $\{a_i\}$ their associated eigenvalues, $A = \sum_i a_i E_i$. Thus $\omega(A)^2 = \omega(A^2)$ only if $a_i a_j \omega(E_i) \omega(E_j) = 0$ when $i \neq j$. But this implies that there's some i such that $\omega(E_i) = 1$.

phenomenon of quantum probability. The general predicament of a quantum system, according to the MBA, is to occupy a quantum state ψ that determines, not a unique value state on its beable algebra $\mathfrak{D}$, but a probability distribution over possible value states on $\mathfrak{D}$. Such a ψ is not dispersion-free on $\mathfrak{D}$. But provided ψ can be understood as a statistical *mixture of dispersion-free states*, beable instantiation is insured.

Definition 8.6 (mixture of dispersion-free states). A state ψ on a subalgebra $\mathfrak{D}$ of a C^* algebra $\mathfrak{A}$ is a *mixture of dispersion-free states* on $\mathfrak{D}$ iff there exists a probability measure μ on the space Λ of dispersion-free states ω_λ on $\mathfrak{D}$ such that for all $A \in \mathfrak{D}$:

$$\psi(A) = \int_\Lambda \omega_\lambda(A) d\mu(\lambda) \tag{8.2}$$

Because each dispersion-free state ω_λ on $\mathfrak{A}$ specifies a value state, one can understand a system described by the quantum state ψ to *really be* in one of those value states, with μ giving the probability measure over the options. Because μ is a perfectly straightforward classical probability measure, it is innocent of quantum weirdness, such as interference, and poses no interpretive challenges peculiar to QM.

We can now present Halvorson and Clifton's analysis of the notion of a *maximal beable algebra*:

Definition 8.7 (maximal beable algebra w.r.t. ψ). $\mathfrak{D}$ is a beable algebra for a state ψ iff ψ corresponds to a mixture of dispersion-free states on $\mathfrak{D}$. $\mathfrak{D}$ is a *maximal* beable algebra only if it is properly contained in no larger algebra that can be so understood.

Thus to pursue the MBA is to proceed as follows:

Given a state on an algebra of observables, characterize those subalgebras of "beables," that are maximal with respect to the property that the state's restriction to the subalgebra is a mixture of dispersion-free states. Such *maximal beable subalgebras* could then represent maximal sets of observables with simultaneously determinate values distributed in accordance with the state's expectation values. (Halvorson and Clifton 1999, 2442; see also Bub 1997, 117, 119)

Maximal beable subalgebras, so defined, exist for any state (Halvorson and Clifton 1999, 2447). In the case of *faithful states*, maximal beable subalgebras are readily characterized.[7,8]

Fact 8.8. If ω is a faithful state on $\mathfrak{A}$, then it can be represented as a mixture of dispersion-free states on a subalgebra $\mathfrak{D}$ of $\mathfrak{A}$ iff $\mathfrak{D}$ is abelian (Clifton 2000, 172).

Remark how pristine the MBA is. It is articulated by technical fortitude, and informed by metaphysical scruples. These include but are not limited to the following: to be instantiated is to take a value that is not disjunctive or interval-valued or otherwise fuzzy, but dispersion-free; properties are closed under algebraic operations; plenitude is so desirable that if an interpretation can add to the list of properties instantiated

---

[7] Recall that a state ω on $\mathfrak{A}$ is faithful iff $\omega(A) = 0$ implies $A = 0$ for all positive $A \in \mathfrak{A}$.

8.2 THE MBA AND ORDINARY QM

in a quantum world, it should. The next section discusses another way the MBA is metaphysically scrupulous: the worlds possible according to it are worlds with a classical semantic structure.

8.2.2 Lattice entertain you

This section articulates an attractive consequence of taking the MBA. Suppose $\mathfrak{D}$ is a maximal beable algebra. The attractive consequence is that a value state, in the MBA's sense of a map taking each observable in $\mathfrak{D}$ to one of its eigenvalues, corresponds to a maximal collection of determinate-value-attributing propositions that can receive truth valuations obedient to the classical truth tables. Thus the MBA is a scheme for casting not only quantum probabilities but also quantum semantics in a classical mold.

Consider a von Neumann algebra $\mathfrak{M}$ acting on a Hilbert space $\mathcal{H}$. Each projection operator $E \in \mathfrak{M}$ corresponds to a family of propositions through the agency of the spectral theorem. Pick some self-adjoint $A \in \mathfrak{M}$ that has E in its spectral resolution, and rewrite E to express this relation: $E = E_\Delta^A$. The projection operator E_Δ^A corresponds to the proposition, "Observable A has a value in the interval Δ." To be sure, E will be in the spectral resolution of many *other* observables, such as a bounded function $f(A)$ of A. What follows will imply that using one of these other observables to identify E with a proposition—"Observable $f(A)$ has a value in the interval $f(\Delta)$"—eventuates in a proposition truth functionally equivalent to the original. Expressions like "E", which I've been using indiscriminately to refer to a projection operator on $\mathcal{H}$, as well as to the subspace of $\mathcal{H}$ that's the range of E, will henceforth also sometimes refer to the proposition corresponding to E. I will generally trust to context to disambiguate.

Let $\mathcal{P}(\mathfrak{M})$ denote the collection of projection operators in the von Neumann algebra $\mathfrak{M}$. $\mathcal{P}(\mathfrak{M})$ is partially ordered by the relation $\leq$ of subspace inclusion. This partial order enables us to define for each pair of elements $E, F \in \mathcal{P}(\mathfrak{M})$, their *greatest lower bound* (a.k.a. *meet*) $E \cap F$ as the projection whose range is the largest closed subspace of $\mathcal{H}$ that is contained in both E and F; and their *least upper bound* (a.k.a. *join*) $E \cup F$ as the projection whose range is the smallest closed subspace of $\mathcal{H}$ that contains both E and F. A *lattice* is a partially ordered set every pair of elements of which has both a least upper bound and a greatest lower bound. Thus the foregoing definitions render $\mathcal{P}(\mathfrak{M})$ a lattice.

A lattice S has a zero element 0 such that $0 \leq a$ for all $a \in S$ and a unit element 1 such that $a \leq 1$ for all $a \in S$. The zero operator (the projection operator for the null subspace) is the zero element of $\mathcal{P}(\mathfrak{M})$ and the identity operator I is the unit element. The *complement* of an element a of a lattice S is an element $d' \in S$ such that $d' \cup a = 1$. A lattice is complemented if each of its elements has a complement. It is *orthocomplemented* if these complements are such that $d'' = a$ and $a \leq b$ if and only if $b' \leq d'$. The reader should note similarities between the behavior of the complement in an orthocomplemented lattice and the behavior of the negation operator in standard sentential logic.

$\mathcal{P}(\mathfrak{M})$ is an orthocomplemented lattice, with the complement E' of $E \in \mathcal{P}(\mathfrak{M})$ supplied by the projection $E^\perp := I - E$, whose range is the orthogonal complement of E's range.

All a, b, c in a *distributive* lattice S satisfy the distributive law: $a \cup (b \cap c) = (a \cup b) \cap (a \cup c)$; $a \cap (b \cup c) = (a \cap b) \cup (a \cap c)$. Notoriously, $\mathcal{P}(\mathfrak{M})$ need not be distributive. But in the special case that $\mathfrak{M}$ is abelian, $\mathcal{P}(\mathfrak{M})$ is distributive. Indeed, it's a *Boolean lattice* (a.k.a. a *Boolean algebra*), i.e. a distributive complemented lattice. The simplest Boolean lattice is the set $\{0, 1\}$, where each element is the other's complement and meet and join are defined as follows: $a \cap b = min\{a, b\}$; $a \cup b = max\{a, b\}$. Call this lattice B_2. Notice that B_2's elements can be put into one-to-one correspondence with the truth-values *false* (0) and *true* (1). A *Boolean homomorphism* between Boolean lattices B and B_2 is a map $h : B \to B_2$ preserving the Boolean operations:

$$h(a \cup b) = h(a) \cup h(b)$$
$$h(a \cap b) = h(a) \cap h(b)$$
$$h(a') = h(a)'$$
$$h(0) = 0$$
$$h(1) = 1$$

(NB the second and third (equivalently the second and fourth) are sufficient to define a Boolean homomorphism; the remaining properties are consequences.)

Construing B as a lattice of propositions, we can construe lattice operations—meet ($\cap$), join ($\cup$), and complement ($'$)—as logical operations—conjunction (&), disjunction (v), and negation ($\sim$), respectively. Given this construal, a two-valued homomorphism $h : B \to B_2$ on a Boolean lattice B amounts to a *truth valuation on B respecting the classical truth tables* for disjunction, conjunction, and negation.

If a Boolean lattice B has an atom—i.e. a minimal non-trivial element—a, that atom generates a two-valued homomorphism h_a on B as follows:

$$h_a(b) = 1 \text{ if } a \leq b \tag{8.3}$$
$$h_a(b) = 0 \text{ otherwise}$$

If B is finite, *all* its two-valued homomorphisms are determined in this way (Bell and Machover 1977, corr. 5.3). Although not all Boolean lattices have atoms, even atomless Boolean lattices can admit two-valued homomorphisms—just not ones generated, on the model of (8.3), by atoms. To the nature and significance of such two-valued homomorphisms, §8.3 will return.

Brought home to von Neumann algebras, the upshot of this brief excursion through lattice theory is this: when $\mathfrak{D}$ is abelian, a two-valued homomorphism on the lattice of propositions $\mathcal{P}(\mathfrak{D})$ is a classical truth valuation. Each atom in $\mathcal{P}(\mathfrak{D})$ determines such a truth valuation.

We can now articulate the connection promised between value states of a maximal beable subalgebra and classical truth valuations. Thanks to Fact 8.8, we know that a maximal beable algebra $\mathfrak{D}$ for a faithful state ω on a von Neumann algebra $\mathfrak{M}$ is *eo ipso* also a maximal abelian subalgebra of $\mathfrak{M}$. Because $\mathfrak{D}$ is abelian, $\mathcal{P}(\mathfrak{D})$ is a Boolean lattice. A value state on $\mathfrak{D}$ takes every element of the projection lattice $\mathcal{P}(\mathfrak{D})$ to a value in its spectrum. A two-valued homomorphism h on $\mathcal{P}(\mathfrak{D})$ corresponds to a classical truth valuation on that lattice. It also corresponds to a value state on $\mathfrak{D}$: each observable A in $\mathfrak{D}$ will have a spectral resolution $\{E_i^A\}$ in $\mathcal{P}(\mathfrak{D})$; h assigns A the eigenvalue corresponding to E_n^A such that $h(E_n^A) = 1$. (Properties of h catalogued above assure that this recipe eventuates in a unique value state.) By subjecting value states on maximal beable algebras to widely recognized desiderata like FUNC, we can move in the other direction: a value state on a maximal beable algebra $\mathfrak{D}$ (assumed to be abelian), when restricted to $\mathcal{P}(\mathfrak{D})$, determines a two-valued homomorphism—and hence a classical truth valuation—on that Boolean lattice. Thus the MBA honors the metaphysical scruple of configuring "a maximal set of co-obtaining properties" in such a way that the semantics of assertions concerning properties is classical.

8.2.3 A scheme and some instances

You can execute the MBA in three easy steps:

Recipe 8.9 (maximal beable recipe). Given a system in a faithful state ω on a von Neumann algebra $\mathfrak{M}$,

Step 1. Identify a maximal abelian subalgebra $\mathfrak{D}$ of $\mathfrak{M}$.
Step 2. Characterize two-valued homomorphisms of $\mathcal{P}(\mathfrak{D})$. These are possible value states of the system.
Step 3. Use ω to define a probability distribution over the homomorphisms identified in Step 2.

To see the MBA in action, we need look no farther than ordinary QM. There the von Neumann algebra $\mathfrak{M}$ figuring in the maximal beable recipe is just $\mathfrak{B}(\mathcal{H})$. I will focus on faithful states. (Bub 1997 gives a comprehensive account of the MBA in the presence of non-faithful states.) My excuse for this limitation is that we are ultimately interested in the feasibility of extending semantics already devised for ordinary QM to QM$_\infty$, where most physically interesting states *are* faithful. It follows from Fact 8.8 that a maximal beable subalgebra of $\mathfrak{B}(\mathcal{H})$ for a faithful state is a *maximal abelian subalgebra* of $\mathfrak{B}(\mathcal{H})$, i.e. an abelian subalgebra of $\mathfrak{B}(\mathcal{H})$ not properly contained in any other abelian subalgebra of $\mathfrak{B}(\mathcal{H})$. There is a simple procedure for generating a maximal abelian subalgebra of $\mathfrak{B}(\mathcal{H})$: start with a complete set $\{E_i\}$ of orthogonal one-dimensional projection operators (atoms!) in $\mathfrak{B}(\mathcal{H})$, take their linear combinations, then close in the weak topology. The result will be an abelian von Neumann algebra consisting of every element of $\mathfrak{B}(\mathcal{H})$ that has $\{E_i\}$ as a spectral resolution.[8] This is the maximal abelian subalgebra of $\mathfrak{B}(\mathcal{H})$ *generated by* $\{E_i\}$.

Familiar interpretations of ordinary QM can be seen as variations on the foregoing atomic theme. I don't claim that this is the only, or the most comprehensive, way of seeing them. But I would like to suggest that is a way of seeing them that isolates interpretive features that might be generalized to the setting of QM_∞. **Collapse interpretations** specify a maximal beable algebra $\mathfrak{D}$ for a system subject to the measurement of a non-degenerate observable A. $\mathfrak{D}$ is the maximal abelian subalgebra of $\mathfrak{B}(\mathcal{H})$ generated by A's eigenprojections E_i^A. Each E_i^A determines a two-valued homomorphism on $\mathcal{P}(\mathfrak{D})$ and thus a value state on $\mathfrak{D}$. The probability that a system in pre-measurement state W collapses to the value state coded by E_i^A is $Tr(WE_i^A)$—which explains the empirical adequacy of Born Rule probabilities.

A **"Bohmian" interpretation** dictates a prefered determinate observable. For simplicity, suppose that it's discrete and non-degenerate and call it R.[9] R's eigenprojections E_i^R are a complete set of orthogonal atoms; they generate a maximal abelian subalgebra $\mathfrak{D}_R$ of $\mathfrak{B}(\mathcal{H})$. $\mathfrak{D}_R$'s value states are encoded by the atoms E_i^R; the probability that a system attributed quantum state W occupies the value state coded by E_i^R is the Born Rule probability $Tr(WE_i^R)$.

Modal interpretations also fit the maximal beable mold. Suppose W is faithful and non-degenerate. Then $\mathfrak{D}_W$, the maximal abelian subalgebra of $\mathfrak{B}(\mathcal{H})$ generated by W's spectral projections E_i^W, is the beable subalgebra; "value states" of the system stand in one-to-one correspondence with W's spectral projections, with the eigenstate–eigenvalue link dictating what's true of a system whose value state is coded by E_i^W; the probability that the system is in the value state coded by E_i^W is the Born Rule probability $Tr(WE_i^W)$.

One way to exhibit a **relative state formulation** as a variation on the MBA is to suppose that the universe admits a preferred decomposition into subsystems, and that the Hilbert space for each subsystem has a preferred basis (dictating the character of the "worlds" corresponding to different branches of the universal wave function). Let W be the reduced state of some preferred subsystem induced by the universal wave function, and let $\{E_i\}$ be the complete set of eigenprojections onto the preferred basis for that subsystem. Then proceed as with the modal interpretation, only with this epicycle: the E_i's keep track not of value states (mutually exclusive possible conditions of the system) but of *worlds*, of which there are many, and to none of which the system is confined *simpliciter*. In the world kept track of by E_i, what's true of the system is given by applying the eigenstate–eigenvalue link to E_i. The profligate metaphysics

---

[8] See Beltrametti and Casinelli (1981, §3.2) for an argument.

[9] Bohm, of course, supposes R to be position, an operator with a continuous spectrum. Technical difficulties attendant upon this complication—difficulties not with articulating Bohmian mechanics but with shoe-horning it into a framework where properties correspond to self-adjoint operators on a separable Hilbert space (see §3.1.1)—motivate the shudder quotes around "Bohm" above. Value states for Bohm proper are non-normal states on $\mathfrak{B}(\mathcal{H})$, such as the state ω_λ on $L^2([0,1])$, that "converges" to λ (see Ex. 7.16).

precludes a straightforward epistemic interpretation of quantum probabilities; there are other options, for instance, that Born rule probabilities encapsulate the degrees of belief of a rational observer in a universe truly described by a relative state formulation (Wallace 2003; for another approach to relative state formulation probabilities, see Barrett 1999).

8.3 The MBA and QM$_\infty$

Let $\mathfrak{M}$ be von Neumann algebra, not assumed to be a Type I factor. To answer central interpretive questions, many familiar interpretations of ordinary QM exploit a feature of $\mathfrak{B}(\mathcal{H})$ that $\mathfrak{M}$ may not necessarily share. $\mathfrak{B}(\mathcal{H})$ has *atoms*. A tactic widespread in the interpretation of ordinary QM is to use a complete set of orthogonal atoms in $\mathfrak{B}(\mathcal{H})$ to specify a maximal beable subalgebra $\mathfrak{D}$ of $\mathfrak{B}(\mathcal{H})$ (Step 1 of the maximal beable recipe). Each atom in $\mathfrak{D}$ determines a two-valued homomorphism on its projection lattice $\mathcal{P}(\mathfrak{D})$, and thus an eigenvaluation on $\mathfrak{D}$ (Step 2). The atom defining a homomorphism determines (via the trace prescription) the probability a normal state W on $\mathfrak{B}(\mathcal{H})$ assigns the "maximal set of co-obtaining properties" corresponding to that homomorphism (Step 3). This atomic strategy resurrects a classical probability structure (the trace prescription restricted to $\mathcal{P}(\mathfrak{D})$) and a classical semantic structure ($\mathcal{P}(\mathfrak{D})$ understood as a lattice of propositions) from the ashes of QM's assault on our time-honored intuitions.

Alas, von Neumann algebras encountered in QM$_\infty$ are typically atomless. An atomless von Neumann algebra $\mathfrak{M}$ lacks candidate endpoints of collapse; it lacks analogs to precise values of hidden variables; it lacks "bookkeeping devices" for value states; it lacks correlates to "worlds." What is the interpreter of theories of QM$_\infty$ to do?

One option is to grasp at straws. Atoms are minimal projection operators, and factors with minimal projection operators are Type I. This on its own needn't plunge atomic variations on the MBA into despair. For what they require are atoms in the projection lattice, not of $\mathfrak{M}$ entire, but of their favored maximal beable (thus, abelian) subalgebra $\mathfrak{D}$ of $\mathfrak{M}$. So they can pin their hopes on the possibility that $\mathfrak{D}$ has atoms even if $\mathfrak{M}$ does not. Unfortunately, this possibility does not survive scrutiny. Suppose, for reductio, that E is an atom in $\mathfrak{D}$. Because $\mathfrak{M}$ contains no minimal projections, there exists a projection operator $F \in \mathfrak{M}$ such that $F < E$. Because E is an atom in $\mathfrak{D}$, $F \notin \mathfrak{D}$. But then $\mathfrak{D}$ is not a maximal abelian subalgebra of $\mathfrak{M}$, which is our contradiction. $\mathfrak{D}$ is non-maximal because $\mathfrak{D}^F := \mathfrak{D} \cup F$ is an abelian subalgebra of $\mathfrak{M}$ that properly contains $\mathfrak{D}$.

"Argument" 8.10. That $\mathfrak{D}^F$ is abelian follows from two simple facts:

(i) If E and G are projections such that $[E, G] = 0$, then either
  (a) $E = G$, (b) $EG = GE = 0$, (c) $E < G$, or (d) $G < E$.
(ii) If $G < E$, then $GE = EG = G$.

Because E is posited to be an atom in the abelian algebra $\mathfrak{D}$, (i) implies for any $G \in \mathcal{P}(\mathfrak{D})$ that either (a) $E = G$, or (b) $GE = EG = 0$, or (c) $E < G$. (d) is ruled out by the assumption that E is an atom in $\mathfrak{D}$. So let G be an arbitrary projection in $\mathfrak{D}$. If (a) $E = G$, then $[G, F] = [G, E] = 0$. Since G was arbitrary, it follows that $F \notin \mathfrak{D}$ commutes with every element of $\mathcal{P}(\mathfrak{D})$, and hence with every element of $\mathfrak{D}$, which therefore is not maximal abelian. If (b) $EG = GE = 0$, then (using E's atomicity) $FG = FEG = 0 = GEF = GF$. Once again, $F \notin \mathfrak{D}$ commutes with every element of $\mathfrak{D}$, which therefore is not maximal abelian. Finally, if (c) $E < G$, then $F < E$ and the transitivity of $<$ imply that $F < G$. By fact (ii), $[G, F] = 0$. Again, $\mathfrak{D}$ is not maximal abelian. ♠

8.3.1 Hope

We've encountered an in-principle impediment to extending familiar interpretations of ordinary QM, interpretations geared toward Type I factor von Neumann algebras $\mathfrak{B}(\mathcal{H})$, to *arbitrary* von Neumann algebras $\mathfrak{M}$. When $\mathfrak{M}$ is atomless, the devices familiar interpretations use—to identify maximal beable subalgebras of $\mathfrak{B}(\mathcal{H})$, to define two-value homomorphisms on the projection lattices of those algebras (more prosaically, to characterize the possibilities quantum probabilities are probabilities for), and to extract a probability distribution over these homomorphism from the quantum state—simply aren't available.

But this needn't panic the advocate of the MBA. The ingredients the maximal beable recipe calls for, I will argue here, *will* be available, even when $\mathfrak{M}$ is atomless. Those ingredients are maximal beable subalgebras, two-valued homomorphisms thereon, and probability distributions thereover.

First, $\mathfrak{M}$ harbors the maximal abelian subalgebras requisite for Step 1 of the maximal beable recipe. As Halvorson and Clifton observe (1999, 2447), this follows from a Zorn's Lemma argument. The set D of abelian subalgebras of $\mathfrak{M}$ can be partially ordered by set-theoretic inclusion. Every totally ordered subset of D has a least upper bound. So Zorn's Lemma implies that D has a maximal element: a maximal abelian subalgebra, call it $\mathfrak{D}$, of $\mathfrak{M}$. This gets us, without the mediation of atoms, the existence of maximal beable subalgebras of arbitrary von Neumann algebras.

And once we have this, we're home free. By Fact 8.8, a faithful state ω on an abelian $\mathfrak{D}$ can be expressed as a mixture of dispersion-free states on $\mathfrak{D}$. Now if ω_λ is dispersion-free on $\mathfrak{D}$, $\omega_\lambda(A) \in Sp(A)$ for all $A \in \mathfrak{D}$. In particular, ω_λ takes every element of the projection lattice $\mathcal{P}(\mathfrak{D})$ to the spectrum $\{0, 1\}$ characteristic of a projection operator. $\mathcal{P}(\mathfrak{D})$ is a Boolean lattice on which ω_λ thereby defines two-valued homomorphism. This gets us, without the mediation of atoms, the existence of two-valued homomorphisms on the projection lattices of $\mathfrak{D}$. Such homomorphisms code value states in the sense favored by the MBA, completing Step 2 of the recipe. Finally, Def. 8.6 gives us a way of taking a faithful state on $\mathfrak{M}$ to correspond to a probability distribution over these value states:

$$\omega(A) = \int_\Lambda \omega_\lambda(A) d\mu(\lambda) \quad \text{[MDFS]}$$

This completes Step 3 of the recipe. Regarding the MBA in these general terms suggests that, despite the proclivities of variations developed for ordinary QM, the MBA does not presuppose that $\mathfrak{M}$ contains atoms.

8.3.2 Tempered

This isn't to say that the program can get along without them. Each value state ω_λ is a dispersion-free state on a maximal abelian subalgebra $\mathfrak{D}$ of $\mathfrak{M}$. Each ω_λ is thus a pure state on $\mathcal{P}(\mathfrak{D})$ (see Ex. 7.16 for an argument). If ω_λ were a $\mathfrak{D}$-*normal* state, its action on $\mathcal{P}(\mathfrak{D})$ would uniquely determine a pure normal state on $\mathfrak{D}$. But we've seen that if $\mathfrak{M}$ is completely atomless, so are its maximal beable subalgebras. And we know (see pp. 155 ff.) that completely atomless von Neumann algebras lack pure normal states. Thus no dispersion-free state ω_λ in the expression [MDFS] above is a normal state on $\mathfrak{D}$. These ω_λ break the "natural" requirement of countable additivity.

Perhaps the MBA can navigate this bump by observing that normality is a virtue states exhibit in exercising their capacity to assign probabilities. It's the virtue of assigning countably additive probabilities. It's a virtue of a piece with preparability: a virtue of a state conceived as a predictive instrumentality. But this isn't how the MBA conceives states like ω_λ. (Given the pessimism of §8.1, this might be a good thing.) The MBA isn't using dispersion-free states ω_λ to assign probabilities. Instead those states are defining two-valued homomorphisms on $\mathcal{P}(\mathfrak{D})$. The states ω_λ are also *receiving* (on behalf of those homomorphisms and the maximal patterns of beable instantiation they determine) probabilities from ω. Their non-normality needn't hinder them in either of these roles.

Still, it prompts a question. What are the maximal patterns of beable instantiation coded by these dispersion-free states ω_λ like? This is a fair question, because to answer it is to characterize the value states by means of which the MBA lends content to a theory of QM$_\infty$. It is also a vexed question. To see why, we turn once again to our old friend, the maximal abelian algebra $\mathfrak{D}_Q$ and its pure state ω_λ—the state that "converges" to λ (see Examples 7.7 and 7.16). The naive gloss on ω_λ is that it assigns "located at λ" probability 1. This seems like it ought to be maximally specific information about the value state ω_λ codes. But it isn't. *Even by the lights of the MBA, there are truths to be had that we don't know enough about ω_λ to get.* For there are countably many distinct pure states on $\mathfrak{D}_Q$ that "converge" to λ (Halvorson 2001, prop. 2). Thus, for an arbitrary $\chi \in \mathfrak{D}_Q$—an arbitrary beable, according to the MBA—we don't know whether $\omega_\lambda(\chi) = 1$ or $\omega_\lambda(\chi) = 0$.

Halvorson aptly describes the aporia befalling the MBA in the presence of non-atomic von Neumann algebras. "Although we 'know' that there are ultrafilters (i.e. pure states) on [atomless $\mathfrak{D}$], we do not know this because someone has constructed an example of such an ultrafilter... We are told that there is some pure state

ω on $\mathfrak{D}$, but we are not given a recipe for determining the value $\omega(A)$ for an arbitrary element $A \in \mathfrak{D}$" (Halvorson 2001, 41). Contrast this "ineffability" of maximal beable subalgebras of atomless von Neumann algebreas with the fruits of the pursuit of the MBA in the context of ordinary QM. Where $\mathfrak{D}$ is a maximal abelian subalgebra of $\mathfrak{B}(\mathcal{H})$, and E is an atom in $\mathfrak{D}$, the state $\omega_E := \omega_E(A) = \mathit{Tr}(AE)$ for all $A \in \mathfrak{D}$, is a pure state defining a two-valued homomorphism on the projection lattice $\mathcal{P}(\mathfrak{D})$. For each $F \in \mathcal{P}(\mathfrak{D})$, we know whether $\omega_E(F)$ is 0 or 1. Knowing the truth value of each proposition we take to be determinately truth-valued, we know exactly what a value state coded by E_i^A is like. We also know what probability the normal quantum state W assigns *this* value state: it's $\mathit{Tr}(WE)$. Also contrast a pure state (q, p) of a classical system with phase space $\mathbb{R}^2$. Functions $f : \mathbb{R}^2 \to \mathbb{R}$ make up the beable algebra. For every element of this algebra, we can say what value the pure state (q, p) assigns it.[10] Once again, the MBA sets terms for characterizing a value state, which terms it is able to meet. In the presense of non-atomic von Neumann algebras, this is not so.

The ineffability of pure states on atomless $\mathcal{P}(\mathfrak{D})$ suggests that we have no handle, analogous to the one supplied in the Type I case by applying the eigenstate–eigenvalue link to the atom generating a two-valued homomorphism, on how to decode the facts these pure states encode. It also suggests that we have no handle, analogous to the one supplied in the Type I case by applying the trace prescription to the system state and the atom generating a two-valued homomorphism, on how to assign those facts probabilities, or even use those facts to explicate probabilities assigned other facts. Confronted with the atomless von Neumann algebras of QM_∞, then, the MBA shirks two key interpretive tasks. First, it fails to explicitly characterize the value states possible according to the theory. It tells us that there exist two-valued homomorphisms, defined by pure states, to which these value states correspond. But *judged by its own lights* it doesn't complete the task of describing those value states. Second, it fails to explicate the probabilities the theory assigns to these value states. Thus the MBA fails to equip QM_∞ with empirical content, in the form of specific probability assignments to explicitly characterized value states possible for systems described by atomless von Neumann algebras.[11]

The MBA might defend itself against this criticism by demanding: what's so great about effability? One might think that maximal specificity is fine when you can get it, but not an essential component of an adequate interpretation. After all, an interpretation might be able to say *enough* about the worlds in which a theory of QM enjoys empirical success to make compelling sense of that success, without saying all there is to say about those worlds. I myself am not inclined to think that utter effability

---

[10] Do we know what probability a statistical density ρ assigns this pure state? Well, yes: it's 0. But we also know how to amalgamate (q, p) with other utterly effable pure states into regions ρ assigns non-zero probability. We can't make analogous claims for intrinsically mixed ω on $\mathfrak{M}$ and the non-normal pure state ω_λ on $\mathfrak{M}$ from which ω is disjoint.

[11] But see Halvorson (2001) for a discussion of the non-constructability of pure states on atomless algebras, along with ways to avoid the difficulties that non-constructability creates.

8.3 THE MBA AND QM$_\infty$

is a *sine qua non* for anything. But I also think that the dispersion-free states ω_λ by which the MBA codes value states are prone to a more telling failure than effability. They can be lawless.

The lawlessness at issue is the sort outlined in §6.4.2: states entertained by the MBA can fail to instantiate lawlike relationships between observables constitutive of physics. In particular, pure states on non-atomic von Neumann algebras—the very states the MBA deploys to code value states for most theories of QM$_\infty$—can lack continuity features presupposed or implied by physical law. Consider, for illustration, a point system moving along an extremely large but compact subset $[-z, +z]$ of the real line, and associated with the algebra of observables $\mathfrak{B}(L^2([-z, +z]))$. Suppose that this system is far from either endpoint, and in the present epoch enjoys a particularly simple dynamics: it drifts at unit velocity in the positive z direction. Thus a system whose initial state is given by, say, a Gaussian centered at $p \in [-z, +z]$ will evolve after a time t into a state given by Gaussian centered at $p + t$. Now consider a state ω_λ of this system that converges to λ, in the sense of (4.18). Example 7.7 should make it plausible that $\mathfrak{B}(L^2([-z, +z]))$ will have a non-atomic abelian subalgebra, the von Neumann algebra generated by characteristic functions for measurable subsets of $[-z, +z]$, and that ω_λ corresponds to a pure state on this non-atomic algebra. ω_λ is lawless in the sense that it fails to instantiate the law of quantum mechanical time development given by Schrödinger's equation.

Schrödinger's equation requires the family $U(t)$ of unitary time evolution operators to be strongly continuous in t, so that $U(t)$ is generated by the self-adjoint operator corresponding to the Hamiltonian of the system. Section 6.4.2 suggested that to instantiate this law, a state ω must be such that $\omega(U(t))$ is continuous in t. ω_λ's failure to instantiate this law with respect to the drift evolution follows from two claims. First, where $U(t)$ implements the drift evolution, acting on a state converging to λ with $U(t)$ yields a state converging to $\lambda + t$: $U(t)\omega_\lambda = \omega_{\lambda+t}$. Second, for $t \neq 0$, ω_λ and $\omega_{\lambda+t}$ are "orthogonal". Thus:

$$\omega_\lambda(U(t)) " = " \langle \omega_\lambda | U(t) | \omega_\lambda \rangle = \langle \omega_\lambda | \omega_{\lambda+t} \rangle = \begin{cases} 0 \text{ if } t \neq 0 \\ 1 \text{ for } t = 0 \end{cases} \quad (8.4)$$

The "jump" in $\omega_\lambda(U(t))$ that occurs at $t = 0$ keeps that state from instantiating the continuity required by the Schrödinger's equation.[12]

Given that states such as $|\omega_\lambda\rangle$ look like good candidates for the determinate configurations by which the Bohm theory supplements orthodox ordinary QM, this example might seem to suggest that the Bohm theory is lawless. But I would resist the suggestion. By the lights of the Bohm theory, the state of a quantum particle on the real line is given by the pair $(|\psi\rangle, q)$, where $|\psi\rangle$ is its orthodox QM wave

---

[12] Of course, the calculation (8.4) is nonsense, because it treats ω_λ like a vector in $L^2([-z, +z])$, which it isn't. But the argument can be patched up, for instance, by noting that it follows from $U(t)$'s continuity that $\chi_\Delta(t) := U(t)^{-1} \chi_\Delta U(t)$ should also be continuous, and exhibiting discontinuities in $\omega_\lambda(\chi_\Delta(t))$.

function and q is its determinate position or configuration. This configuration is the quantity which the Bohm theory might represent by a state such as $|\omega_\lambda\rangle$, *if* the Bohm theory were beholden to represent it in terms of a state on a quantum observable algebra. In Bohmian mechanics, the wave function evolves in accordance with the Schrödinger equation, and the configuration evolves in accordance with the Bohm theory's very own guidance equation. I take two circumstances to protect the Bohm theory from charges of lawlessness based on the foregoing example. The first is that the Bohm theory needn't use $|\omega_\lambda\rangle$ to represent the configuration of the system. q will do. $|\omega_\lambda\rangle$ is a state on an algebra generated by the spectral projections of the orthodox quantum position observable. But there's no compulsion to understand the determinate configuration q in terms of the orthodox quantum position observable. Indeed, I take the most enthusiastic advocates of the Bohmian approach to understand q directly, and on its own terms, as a configuration. So I don't think the Bohm theory is committed to using states such as $|\omega_\lambda\rangle$ in the first place. But even if the Bohm theory used states like $|\omega_\lambda\rangle$ to code the determinate configurations of system, it will still attribute wave functions like $|\psi\rangle$ to those systems. As these wave functions will be normal states on the von Neumann algebra $\mathfrak{B}(L^2(-z,+z))$, they will splendidly instantiate Schrödinger evolution. This is what I take to be the second circumstance protecting the Bohm theory against charges of lawlessness.

Returning to the topic of QM_∞: the foregoing example of a pure state on a non-atomic von Neumann algebra which fails to instantiate lawlike relationships between elements of that algebra has QM_∞ analogs. For instance, in the thermodynamic limit of QSM, the good behavior of dynamics is representation-dependent. Chapter 12 will present models whose dynamical good behavior is confined to Type III representations π of the systems' C^* algebra $\mathfrak{A}$. Only with respect to these representations do finite volume dynamics have a (weak) limit as $V \to \infty$, and it is this limit that endows the entire system with dynamics. Because the von Neumann algebra $\pi(\mathfrak{A})''$ is Type III, its pure states aren't π-normal. Let ϕ be such a pure state. Restricted to $\pi(\mathfrak{A})$, ϕ defines a state (which we'll continue to call ϕ) on the C^* algebra $\mathfrak{A}$. But the local dynamics do not have a well-behaved infinite volume limit in the pure state ϕ's GNS representation. Thus insisting that value states are pure states, the MBA will favor value states for such models which fail to instantiate dynamical laws.

8.3.3 Interpretation unbound?

The MBA rests on metaphysical scruples, for instance, the scruple that a possible world is a *maximal* set of co-obtaining properties. Perhaps these scruples are to blame for the approach's putative inadequacy to QM_∞. Rob Clifton's (2000) strategy for interpreting the exotic von Neumann algebras encountered in QFT dispenses with the maximality scruple. To oversimplify, given a state ω on a von Neumann algebra $\mathfrak{M}$, Clifton identifies a beable algebra $\mathfrak{D}$ that isn't a maximal abelian subalgebra of $\mathfrak{M}$ but *is* the largest abelian subalgebra of $\mathfrak{M}$ characterizable, by means Clifton deems admissible, in terms of ω and $\mathfrak{M}$.[13] In other words, $\mathfrak{D}$ can be embedded in larger abelian subalgebras of $\mathfrak{M}$, but ω doesn't tell us which ones.

The merits of refraining from plenitude deserve further debate. A demerit of Clifton's particular strategy will be the focus of this section (see also Clifton 2000; Earman and Ruetsche 2005). Because the present discussion interacts with digressive material from earlier sections, herewith:

The Short Version. Clifton's prescription has distressingly little to say about many states of physical interest, for the beable algebra $\mathfrak{D}$ those states fix via Clifton's prescription contains only multiples of the identity observable. These states include ergodic states, which are dense in the set of normal states on a Type III factor algebra.

Where ω is a faithful normal state on a von Neumann algebra $\mathfrak{M}$, let $\mathfrak{M}^{\sigma\omega}$ be ω's modular invariants (Def. 7.26). Clifton's modal interpretation identifies $\mathfrak{D}$ with the center of $\mathfrak{M}^{\sigma\omega}$. (The specification of the algebra of determinate observables for unfaithful states is more intricate. I can ignore its details because physically interesting states in QM_∞ are generally faithful.) This is just a fancy way to express the modal interpretation of ordinary QM, with the center epicycle added to contend with non-factor algebras (see Earman and Ruetsche 2005).

Let us reflect upon what befalls Clifton's prescription in ordinary QM when W is degenerate. Consider the worst case, that W acting on $\mathfrak{B}(\mathcal{H})$ for an n-dimensional $\mathcal{H}$ is the tracial state $\frac{1}{n}I$. Then for all $B \in \mathfrak{B}(\mathcal{H})$, $Tr(WAB) - Tr(WBA) = \frac{1}{n}(Tr(AB) - Tr(BA)) = 0$. Given Fact (7.27), W's modular invariants coincide with $\mathfrak{B}(\mathcal{H})$. Because $\mathfrak{B}(\mathcal{H})$ is irreducible, its center is trivial. According to Clifton's prescription, then, *no (non-trivial) observables are determinate* on an ordinary QM system in a tracial state.

Tracial states also put old-fashioned modal interpretations of ordinary QM in difficulty. When W is highly degenerate, continuously many different sets of pairwise orthogonal projections conspire in its spectral resolution; for Hilbert spaces of dimension higher than 2, the modal interpretation assigns all those projections determinate values on pain of reinviting Bell–Kochen–Specker type contradictions. One way out of this jam is to assign only multiples of the identity determinate values when W is maximally degenerate, and take comfort in the fact that maximally degenerate states form sets of measure 0 in the state space. Clifton's QFT-adapted modal prescription duplicates the first part of this response. But it may not be capable of the second.

When it comes to modular invariants, the opposite of tracial states are the *ergodic states* characterized by the property that their only modular invariants are multiples of the identity (Fact 7.29). The center of the set of modular invariants of an ergodic state consists again of multiples of the identity. Thus Clifton's prescription declares no non-trivial observable determinate in an ergodic state. Clifton's (2000) challenge to his own modal interpretation as applied to algebraic QFT is that ergodic states, unlike maximally degenerate states in ordinary QM, amount to a set of states dense in the

---

[13] Kitajima (2004) establishes Clifton's result in full generality.

state space for a Type III factor. *Every* state of a Type III$_1$ factor is ergodic, and many equilibrium states are Type III$_1$ factor states. Even though Clifton's revamped modal interpretation is geared toward QM$_\infty$, these states reduce it to silence.

8.4 Conclusion

Clifton argues that his strategy for identifying $\mathfrak{D}$ is the only strategy satisfying certain constraints. Halvorson (personal communication) has questioned the propriety of one of these constraints. So perhaps other versions of a leaner, meaner interpretive game will fare better in the face of QM$_\infty$ exotica than Clifton's has.

However the leaner, meaner game turns out, there are lessons to learn from the impetus to play. One is that atoms, and the interpretive strategies they underwrite, are not a perfectly general feature of the sorts of von Neumann algebras that arise in physical applications. It follows that assumptions widespread in the semantics of ordinary QM—that atoms are state witnesses, fact encoders, and probability bearers—are upset by QM$_\infty$. The MBA can be freed of these assumptions and extended to QM$_\infty$. But due to the non-constructability of pure states on atomless von Neumann algebras, the extension shirks the main tasks of quantum semantics: the *characterization* of worlds possible according to a quantum theory, and the *explication* of the probabilities that theory assigns those worlds. It appears that we can't interpret QM$_\infty$ by *simply* extending or even adapting our favorite semantics for ordinary QM to infinite quantum systems. *Suppose* one associates with a theory of QM$_\infty$ a kinematic pair of the sort familiar from ordinary QM, i.e. a kinematic pair consisting of a von Neumann algebra and its normal states. Even then interpreting QM$_\infty$ is adamantly not the uninteresting mechanical exercise of applying one's favored semantics for ordinary QM to theories that fall outside the scope of the uniqueness theorems.

But where might the MBA have gone wrong? At its most general, the MBA makes a resource of the formal apparatus of operator algebras and their self-adjoint elements. In particular, the MBA assimilates a physical property to a dispersion-free state on an element of a "beable" algebra. Perhaps algebras are the wrong formal apparatus. Perhaps the salient algebra has been misidentified. Or perhaps the aspiration to explicate the metaphysical commodity *property* so directly in terms of formal apparatus is misguided.

The MBA also rests on a set of metaphysical scruples, for instance, that a possible world is a *maximal* set of co-obtaining properties. Perhaps these scruples are to blame for the approach's inadequacy to QM$_\infty$. Rob Clifton's strategy relaxes them, but comes to grief.

The suggestion I find interesting is that the MBA goes wrong by allowing itself *only* the resources of formal apparatus and metaphysical scruples. It goes wrong by adhering to the ideal of pristine interpretation. This chapter is by no stretch of the imagination an argument for that conclusion. The chapters which follow try to be. Whereas this chapter has focused on what Chapter 1 labeled the semantic phase of interpretation, they will turn to the structure-specifying phase: the phase of equipping theories with kinematic and dynamic structures, of heralding magnitudes and states as "genuinely physical," and and doing likewise for the time developments they admit. Those chapters exhibit circumstances that might demand *unpristine* approaches to the structure-specifying phase of interpreting QM_∞.

9

Is Particle Physics Particle Physics?

> Neutrinos they are very small
> They have no charge and have no mass
> And do not interact at all.
> The earth is just a silly ball
> To them, through which they simply pass,
> Like dustmaids down a drafty hall
> Or photons through a sheet of glass....
>
> —John Updike (1963, 3)

Section 3.3.2 sketched the standard quantization of the mass m Klein–Gordon field on Minkowski spacetime, which uses creation and annihilation operators, $a_k^\dagger$ and a_k, to represent the CCRs. A *particle interpretation* of this QFT supposes the former to "add" quanta of excitation, or *particles*, to the states on which they act; the latter, to "remove" them. The standard Minkowski vacuum $|0_M\rangle$ is the unique state destroyed (i.e. mapped to the zero vector) by every annihilation operator. This exposes the vacuum as the state from which there are no further particles to remove; i.e. it is the no-particle state. A dense set of states of the theory can be obtained from $|0_M\rangle$, by acting on it with polynomials of creation and annihilation operators. A dense set of states can thereby be understood in terms of their particle contents.

Or can they? Jack drifts inertially through Minkowski spacetime in its vacuum state $|0_M\rangle$. He observes no particles. This is hardly a surprise. The vacuum state $|0_M\rangle$ has no particles for Jack to observe. What is a surprise is that Jill, accelerating uniformly through the same spacetime in the same vacuum state, observes particles, such a variety of them that the distribution of their energies manifests a thermal spectrum at a temperature related to her acceleration—or so conventional presentations of *the Unruh effect* (after Unruh 1976) suggest. The Unruh effect raises questions about particles—if no particles are present in $|0_M\rangle$, how can Jill be seeing them? If Jill is seeing particles in $|0_M\rangle$, how can Jack persist in parsing the content of quantum field theory by means of a particle concept that declares them to be absent from $|0_M\rangle$? Is QFT, a.k.a. particle physics, really particle physics?—which will occupy the next three chapters.

This chapter will identify the challenge the Unruh effect poses to particle interpretations of QFT. Its first task is to explain how particle physics could fail to be particle physics. Section 9.1 circumscribes the class of QFTs (free bosonic fields) which will be the referent of the first occurrence of "particle physics" in the title of this chapter, and §9.2 explicates a sense of "particle interpretation" applicable to theories which are referents of the second occurrence. (When I need to distinguish a particle interpretation in §9.2's sense from other sorts of particle interpretations, I'll call it a *fundamental particle interpretation*.) Section 9.3 presents a preliminary case that a free bosonic field theory supports a fundamental particle interpretation.

The predicament of Jack and Jill is often invoked to undermine particle interpretations of QFT. Sections 9.4–9.9 suggest that the invocation, though confused, harbors a serious challenge to fundamental particle interpretations. The challenge is that there are more physically reasonable states than a fundamental particle interpretation can accommodate. The challenge rests on the availability of unitarily inequivalent representations of the Weyl algebra for a free bosonic field, representations which correspond to what §9.5 will call "incommensurable particle notions."

9.1 Particle physics

Quarks, leptons, bosons—the particles familiar from popular science articles, Nobel prize citations, and John Updike poems—are members of the guild of QFTs known as the Standard Model of High Energy Physics (HEP). These chapters address the status of particles from the much simpler theoretical perspective of a single QFT, the quantized Klein–Gordon field. Here I'll indicate some salient differences between the two theoretical settings, and attempt to legitimate the narrow focus.

To oversimplify enormously: HEP individuates elementary particle types by the values of fundamental parameters—for instance, mass, electric charge, and spin[1]—which characterize them. Thus a photon is a particle of type mass 0, spin 0, electric charge 0, etc.; an electron a particle of type mass .511 MeV, spin $\frac{1}{2}$, and charge -1, etc.; a charm quark is a particle of type mass 1.3 GeV, spin $\frac{1}{2}$, electric charge $\frac{2}{3}$, etc.; and so on. Different values of non-fundamental parameters can characterize different particles of the same type—mass m (spin 0 charge 0) bosons, say, can come in an infinite variety of energies and momenta. (In an attempt to forestall confusion, I will call particles of the same fundamental type but attributed different values for non-fundamental parameters, different *varieties* of that fundamental type.) HEP is set in Minkowski spacetime. Wigner showed that values of mass and (half-integer) spin label irreducible unitary Hilbert space representations of Minkowski spacetime's symmetry group, the

---

[1] Also color charge, flavor charge, isospin, . . .

Poincaré group.[2] Viewed from the stance of Hilbert Space Conservatism, this means the space of states of a mass m, spin s particle consists of density operators on the Hilbert space carrying the corresponding representation of the Poincaré group.

HEP associates each elementary particle with excitations of a corresponding quantum field. For instance, the particle type *mass m boson* corresponds to the quantization of a field ϕ that, in the absence of interaction with other fields or with itself, obeys the classical Klein–Gordon equation for mass m, which we first encountered in §3.3.2. This equation of motion is determined by a Lagrangian density of the form:

$$\frac{1}{2}[\dot{\varphi}(\vec{x})^2 - (\vec{\nabla}\varphi(\vec{x}))^2 - m^2\varphi(\vec{x})^2] \tag{9.1}$$

In a globally hyperbolic spacetime, a Lagrangian density imposes a symplectic structure of the space of solutions to the equations of motion it generates (Woodhouse 1992, thm. 7.2.2). The resolution to quantize classical Klein–Gordon theory by representing the *Weyl relations* over the symplectic vector space determined by (9.1) implies that the particle associated with the quantization is a boson, i.e. has integer spin. That the field ϕ is a scalar (as opposed to a tensor-valued or spinor-valued) field signals that the associated particle in fact has spin 0; the parameter m gives the mass of the associated particle. The absence from the Klein–Gordon Lagrangian of terms of the form $\psi^n \phi^m \ldots$, coupling ϕ to other fields, and of terms of quadratic or higher order in ϕ, coupling ϕ to itself, brands it as the Lagrangian for a free, non-self-interacting field. It also means that the space of solutions to the Klein–Gordon equation is linear: if ϕ and χ are solutions, so is $\phi + \chi$—a circumstance exploited, we shall see, in constructing a Hilbert space for the quantized theory.

Thus the quantized Klein Gordon (a.k.a. free bosonic) field is taken to describe one type of elementary particle, the mass m boson, in the absence of interactions. There are explicit and consistent Hilbert space representations of the Weyl relations determined by (9.1), and an articulate C^* algebraic structure all those Hilbert space representations share.

The Standard Model, by contrast, purports to describe not only every type of subatomic particle but also the interactions into which these types enter. Hence the Standard Model deploys Lagrangians involving numerous fields, variously coupled. The corresponding field theories are highly nonlinear. By the wizardry of powerful approximation techniques and symmetry-invoking heuristics, theoretical physicists extract phenomenally accurate predictions from the Standard Model. Still, there is *no* explicit Hilbert space realization of the Standard Model's underlying physics—that is, no explicit Hilbert space in whose terms are defined states and observables pertaining to systems described by the Standard Model. Nor, obviously, are there *many* Hilbert

---

[2] Other parameters in the Standard Model are linked not to spacetime symmetries but to internal symmetries.

space realizations sharing a common C^*-algebraic structure, much less anything that stands to the Standard Model as the Weyl algebra stands to the quantized Klein–Gordon field. Although the S-matrix that informs calculations of scattering cross-sections *looks* like it acts on a Hilbert space, the calculations are perturbative, and the S-matrix itself a black box, surrendering no hints about the details of a Hilbert space describing the field at times intermediate between the incoming ($t = -\infty$) and outgoing ($t = +\infty$) free field eras.

My apology for focusing on the exceedingly simple QFT constituted by free bosonic field theory, rather than the full-blown Standard Model, is multipart. First, the methodological directive to tackle simple cases before hard ones applies. Any grip we can manage on a theory of a single type of non-interacting particle could orient us when we confront theories of myriad types of interacting particles. Second, impediments (such as Haag's theorem, discussed in Chapter 11) to understanding QFT as particle physics arise in the context of interactions, but not for free field theories. This suggests (without demonstrating) that if the free bosonic field can't be understood as particle physics, then the full-blown Standard Model can't be either. And I intend to argue that the free bosonic field can't *always* be understood as particle physics.

Finally, it's worth emphasizing one sense in which the family of free bosonic field theories is more general than the Standard Model. The Standard Model is set in Minkowski spacetime. Free bosonic field theory can be pursued in a wide variety of curved spacetimes. Such pursuit serves the purpose of aspirational physics. Insofar as QFT on curved spacetime combines elements of quantum mechanics (in the form of quantum fields) and general relativity (in the form of curved spacetime), it could act as a signpost on the road to quantum gravity. Holding to the slogan that physically significant means significant for physics, I want to keep tabs on how well putative interpretations of QFT make sense of that theory as an arena for aspirational physics.

9.2 Particle interpretations

QFT is known far and wide as "particle physics." But even when they confine their attention to the putatively tractable case of the free bosonic field, interpreters of QFT don't pay the particle notion much respect. Particles, as we know them, are confined to finite, bounded regions of spacetime. A number of results (Malament 1996; Halvorson and Clifton 2002; De Bièvre 2007) suggest that such confinement cannot be had in QFT. Particles, as we know them, possess what the scholastics called "haecceity", Teller "primitive thisness" (1995, 17), and Redhead "transcendental identity" (1983, 9). An extensive literature (see Castellani 1998 for a sample) questions whether bosons and fermions can exhibit this and yet abide by the non-Maxwell–Boltzmann statistics characteristic of their aggregates. Particles, as we know them, are persistent in their presence and inert in their absence. Excitations of the quantum field are subject to creation and annihilation; even the vacuum fluctuates (Redhead 1983). Particles, as we know them, come in sets with determinate cardinality. States of the quantum field need not be eigenstates of the total particle number operator, and thus (on the usual way of thinking about quantum systems) need not be states describing a determinate number of particles (Teller 1995, ch. 2). These are all grounds—grounds already carefully mapped—for doubting that QFT is a theory of particles as we know them, particles as well-localized, enduring, individuable gobbets of mass-energy.

So suppose that we doubt that. We might continue to entertain a further question. Should interpreters of QFT construe the theory in terms of a *particle-like* concept? Redhead (1983), Huggett (1995), and Teller (1995), among others, begin to contour a concept that some of them argue is well-suited for the job. Let us momentarily call this the *quanta* concept. Quanta aren't strictly localized. Quanta lack haecceity but may possess individuality in some other sense. Quanta aren't fixities but what Redhead dubs "ephemerals," prone to creation, annihilation, and superposition. None of the grounds adduced above for doubting that QFT treats of particles, as we know them, are grounds for doubting that QFT treats of quanta, so understood. This section and the next sketch grounds for commitment to quanta—which, with the qualifications and concerns of the foregoing paragraphs understood, I will conform to common usage in phrasing as grounds for commitment to "particles"—and characterizes commitment so grounded as commitment to a *fundamental particle interpretation of QFT*.

When is a QFT particle physics? This is a question about how the theory ought to be interpreted. Since theories don't determine (although they may reasonably constrain) their interpretations, the question can't be answered by an interesting set of necessary and sufficient conditions. But it can be profitably rephrased: when does a QFT admit a particle interpretation? I will say that a QFT admits a *particle interpretation* when it can be attributed a kinematic and dynamic stucture susceptible to a *particle semantics*; that is, a semantics that characterizes worlds possible according to the QFT in terms of their particle contents. (Section 9.3 will make the nature of such characterization more precise.) Insofar as the characterization in question distinguishes between possible worlds, a particle semantics in the relevant sense also individuates worlds possible according to the theory in terms of their particle contents.

Now, the mass m free bosonic field is conventionally understood to describe a single type of particle, the mass m boson. Thus a particle semantics for the mass m free bosonic field would individuate and characterize its possible worlds in terms of varieties of mass m bosons. A theory resistant to a particle interpretation is one whose possibility space includes states which cannot be attributed particle contents, or states between which theoretically meaningful distinctions can be drawn, which distinctions are opaque to attributions of particle contents.

It is, of course, a further and substantial question whether a theory admitting a particle interpretation should be given one! Postponing that question, the next section builds a case that free bosonic field theory admits a particle interpretation.

9.3 Pro particles

Consider a specific QFT, say quantized mass m Klein–Gordon theory on Minkowski spacetime. On a particle interpretation, this theory is about particles of a specific type, say mass m bosons. Its (pure) states are (superpositions of) possible states of particles of this type, alone or in aggregate; these states assign expectation values to events involving particles of this type, events such as the creation of a particle of this variety in one state and the destruction of a particle of that variety in another. Those committed to the theory are committed to particles of the sort circumscribed by the theory, the particles with respect to which the theory's states are characterized and its empirical contents expressed. (Here please understand "commitment" so broadly that it is something even constructive empiricists can manage.)

9.3.1 Fock space, heuristically

One case for a particle interpretation of a QFT hinges on suggestive features of a *Fock space representation* of that theory's CCRs/Weyl Relations. The textbook heuristic for obtaining such a representation plays out in Minkowski spacetime, where the Klein–Gordon equation takes the form:

$$(\Box^2 - m^2)\varphi(x) = 0. \tag{9.2}$$

We are taking $\varphi(x)$, with x an element of the four-manifold of Minkowski spacetime, to be real-valued. In terms of a global inertial coordinate system $(\vec{x}, t)$ (where $\vec{x}$ designates the three spatial coordinates), (9.2) arises from the Lagrangian density

$$\mathcal{L} = \int \left(\frac{1}{2}[\dot{\varphi}(\vec{x})^2 - (\vec{\nabla}\varphi(\vec{x}))^2 - m^2\varphi(\vec{x})^2]\right) d^3x \tag{9.3}$$

where $\dot{\varphi}$ is the time derivative of φ.

Naively following the Hamiltonian quantization recipe (cf. §2.3.3), we note that the classical momentum density $\pi(x) \equiv \frac{\partial \mathcal{L}}{\partial \dot{\varphi}(x)} = \dot{\varphi}(x)$ is canonically conjugate to the configuration variable $\varphi(x)$. Casting $\pi(x)$ and $\varphi(x)$ as analogs of the canonical observables p and q for a classical Hamiltonian theory, we aim to quantize Klein–Gordon theory by promoting φ and π to Hilbert space operators $\hat{\varphi}$ and $\hat{\pi}$ obeying the *equal time canonical commutation relations* (ETCCRs)[3]:

$$[\hat{\varphi}(x), \hat{\varphi}(y)] = [\hat{\pi}(x), \hat{\pi}(y)] = 0, \quad [\hat{\varphi}(\vec{x}, t), \hat{\pi}(\vec{x}', t)] = -i\delta^3(\vec{x} - \vec{x}')\hat{I} \tag{9.4}$$

A little Fourier analysis reveals a path toward a representation of the ETCCRs. Re-expressing the Klein–Gordon field $\varphi(x)$ as a real-valued function on momentum space, we are struck by analogies between its Lagrangian and that of the simple harmonic oscillator. For the sake of simplifying exposition, let us substitute for the

---

[3] This chapter temporarily restores hats to Hilbert space operators to emphasize which variables are quantum and which classical.

spacelike Cauchy surface $\mathbb{R}^3$ of Minkowski spacetime a flat three-torus $\mathbb{T}^3$ whose sides are of length L. This imposes periodic boundary conditions on solutions to (9.2): because waveforms resident on $\mathbb{T}^3$ must wrap smoothly around the torus, they are constrained in each spatial dimension to have wavelengths $\frac{L}{n}$, where n is an integer. Then solutions to (9.2) can be Fourier-expanded thus:

$$\varphi(\vec{x}, t) = L^{\frac{3}{2}} \sum_{\vec{k}} \varphi_{\vec{k}}(t) e^{i\vec{k}\cdot\vec{x}} \tag{9.5}$$

Here $\vec{k}$ is a momentum three-vector of the "wrap-around" form $k = \frac{2\pi}{L}(n_1, n_2, n_3)$, with n_1, n_2, and n_3 integers; and the coefficients $\varphi_{\vec{k}}(t)$ are related to $\varphi(\vec{x}, t)$ by:

$$\varphi_{\vec{k}}(t) = L^{\frac{3}{2}} \int \varphi(\vec{x}, t) e^{-i\vec{k}\cdot\vec{x}} d^3x$$

Expressed in terms of $\varphi_{\vec{k}}(t)$, the Lagrangian (9.3) becomes (cf. Wald 1994, 31–2):

$$\mathcal{L} = \sum_{\vec{k}} \frac{1}{2} (|\dot{\varphi}_{\vec{k}}|^2 - \omega_{\vec{k}}^2 |\varphi_{\vec{k}}|^2) \tag{9.6}$$

where $\omega_{\vec{k}}^2 = \vec{k}^2 + m^2$. Comparing (9.6) with the Lagrangian (2.28) of the simple harmonic oscillator (reviewed in §2.3.5) suggests that we model the real Klein–Gordon field as an infinite collection of decoupled harmonic oscillators, and exploit this analogy in its quantization.

For details of the exploitation, see Peskin and Schroeder (1995, 19–24). Here are some of its highlights. Each Fourier mode $\varphi_{\vec{k}}$ is treated as its own oscillator, and accordingly assigned operators $\hat{a}_{\vec{k}}, \hat{a}_{\vec{k}}^\dagger$ interdefined with $\varphi_{\vec{k}}$ (and thence $\hat{\varphi}(\vec{x}, t)$ and $\hat{\pi}(\vec{x}, t)$) in such a way that $\varphi(\vec{x}, t)$ and $\hat{\pi}(\vec{x}, t)$'s obedience to the ETCCRs (9.4) implies and is implied by the commutation relations[4]

$$\left[\hat{a}_{\vec{k}}, \hat{a}_{\vec{k}'}\right] = 0 = \left[\hat{a}_{\vec{k}}^\dagger, \hat{a}_{\vec{k}'}^\dagger\right], \quad \left[\hat{a}_{\vec{k}}, \hat{a}_{\vec{k}'}^\dagger\right] = \delta(\vec{k} - \vec{k}')\hat{I}. \tag{9.9}$$

which we will call the *Heisenberg Relations*.

---

[4] Expanded in terms of $\hat{a}_{\vec{k}}, \hat{a}_{\vec{k}'}^\dagger$, the field operators $\hat{\varphi}(\vec{x}, t)$ assume the form

$$\hat{\varphi}(\vec{x}, t) = \sum_{\vec{k}} \frac{1}{\sqrt{2L^3 \omega_{\vec{k}}}} \left(\exp(i(\vec{k}\cdot\vec{x} - \omega_{\vec{k}} t)) \hat{a}_{\vec{k}} + \exp(-i(\vec{k}\cdot\vec{x} - \omega_{\vec{k}} t)) \hat{a}_{\vec{k}}^\dagger \right) \tag{9.7}$$

The foregoing impressionistic exposition obscures a mathematical nicety: the infinite sum on the r.h.s. of (9.7) fails to converge. Intuitively, the wildly oscillatory behavior of the high-frequency modes is to blame. Fortunately, mathematical rigor can be restored by "smearing" the field operator, formally defined at each point $x \in M$ by (9.7), by a smooth function f with compact support on M. The resulting *smeared field operator*

$$\hat{\varphi}(f) = \int f(\vec{x}, t) \hat{\varphi}(\vec{x}, t) d^4x \tag{9.8}$$

is an operator-valued distribution on M.

9.3 PRO PARTICLES

All physics owes its existence to the interweaving of physical magnitudes. Just so, a suggestive family of operators can be built from the collection of operators $\hat{a}_{\vec{k}}, \hat{a}^\dagger_{\vec{k}}$ satisfying the CCRs. Pursuing the analogy with the quantized harmonic oscillator, express the quantum Hamiltonian $\hat{H}$ and momenta $\hat{P}_i$ as follows:

$$\hat{H} = \sum_{\vec{k}} \omega_{\vec{k}} (\hat{a}^\dagger_{\vec{k}} \hat{a}_{\vec{k}} + \frac{1}{2}), \quad \hat{P}_i = \sum_{\vec{k}} \vec{k}_i \hat{a}^\dagger_{\vec{k}} \hat{a}_{\vec{k}} \qquad (9.10)$$

The lowest energy eigenstate of this Hamiltonian will be a state $|0\rangle$ such that:

$$\hat{a}_{\vec{k}} |0\rangle = 0 \quad \text{for all } \hat{a}_{\vec{k}} \qquad (9.11)$$

Alarmists will note that the energy eigenvalue associated with this ground state $|0\rangle$ is divergent, owing to the infinite sum over $\frac{1}{2}\omega_{\vec{k}}$. So will be every other energy eigenvalue, owing to the same infinite sum. To address this divergence, issue a disclaimer to the effect that only energy *differences* matter physically, and simply set this "zero point energy" $\sum_{\vec{k}} \frac{1}{2}\omega_{\vec{k}}$ to 0 when ascribing the "physical" energy (Peskin and Schroeder 1995, 22).[5]

Now consider the operators

$$\hat{N}_{\vec{k}} = \hat{a}^\dagger_{\vec{k}} \hat{a}_{\vec{k}} \qquad (9.12)$$

As with the analogous operators for the simple harmonic oscillator, $\hat{N}_{\vec{k}}$ eigenvalues are non-negative integers. Expressed in terms of these $\hat{N}_{\vec{k}}$, the Hamiltonian and momentum operators assume the forms $\hat{H} = \sum_{\vec{k}} \omega_{\vec{k}} (\hat{N}_{\vec{k}} + \frac{1}{2})$ and $\hat{P}_i = \sum_{\vec{k}} \vec{k}_i \hat{N}_{\vec{k}}$. Hence any eigenvector of any $\hat{N}_{\vec{k}}$ is also an eigenvector of energy and momentum. Suggestively, an $\hat{N}_{\vec{k}}$ eigenvector $|n_{\vec{k}}\rangle$ with eigenvalue $n_{\vec{k}}$ is also an eigenstate of energy and momentum, with physical values characteristic of $n_{\vec{k}}$ identical particles, each of energy $\omega_{\vec{k}}$ and momentum $\vec{k}$. This inspires the understanding of $\hat{N}_{\vec{k}}$ as a "number operator." $\hat{N}_{\vec{k}}$'s expectation value in a state gives a count of the quanta of excitation of variety $\vec{k}$ in that state (see Baez, Segal, and Zhou 1992, 59, for a list of other reasons favoring the identification of $\hat{N}_{\vec{k}}$ as a number operator). As Peskin and Schroeder observe, "It is quite natural to call these excitations *particles*, since they are discrete entities that have the proper relativistic energy-momentum relation" (1995, 22).

It follows from (9.11) and (9.12) that for all $\vec{k}$, the ground state $|0\rangle$ is an eigenstate of $\hat{N}_{\vec{k}}$ with eigenvalue 0. We can therefore regard $|0\rangle$ as the vacuum, or no-quantum, state. Introduce the operator

---

[5] This expedient corresponds to "normal ordering"—moving all creation operators to the left of all annihilation operators. Using the CCRs, the Hamiltonian can be rewritten $\hat{H} = \frac{1}{2}\sum_{\vec{k}} \omega_{\vec{k}} (\hat{a}^\dagger_{\vec{k}} \hat{a}_{\vec{k}} + \hat{a}_{\vec{k}} \hat{a}^\dagger_{\vec{k}})$, which becomes $\sum_{\vec{k}} \omega_{\vec{k}} \hat{a}^\dagger_{\vec{k}} \hat{a}_{\vec{k}}$ when normal ordered. The problematic $\frac{1}{2}\omega_{\vec{k}}$ vanishes. In curved spacetime, the normal ordering strategy fails, and other regularization procedures are called for (see Wald 1995).

$$\hat{N} = \sum_{\vec{k}} \hat{N}_{\vec{k}} \qquad (9.13)$$

Counting quanta of *all* varieties, $\hat{N}$ is the *total number operator*. Reassuringly, the vacuum state $|0\rangle$ is an eigenstate of $\hat{N}$ with eigenvalue 0.

Any state of the form $\hat{a}^{\dagger}_{\vec{k}}|0\rangle$ will also be an eigenstate of $\hat{N}_{\vec{k}}$ with eigenvalue 1, as well as an eigenstate of the Hamiltonian and momentum operators, with (physical) energy $\omega_{\vec{k}}$ and (physical) momentum $\vec{k}$. Other eigenstates of $\hat{N}_{\vec{k}}$ can be obtained by acting repeatedly on the vacuum with $\hat{a}^{\dagger}_{\vec{k}}$ and normalizing: $|n_{\vec{k}}\rangle = \frac{(\hat{a}^{\dagger}_{\vec{k}})^n |0\rangle}{\sqrt{n_{\vec{k}}!}}$. Thus $\hat{a}^{\dagger}_{\vec{k}}$ invites the title *"creation operator"*, an operator that acts on a state to add a quantum of excitation of variety $\vec{k}$ to that state.[6] By contrast, $\hat{a}_{\vec{k}}$ acts on a state $|n_{\vec{k}}\rangle$ to produce the eigenstate of $\hat{N}_{\vec{k}}$ associated with the next lowest eigenvalue: $\hat{a}_{\vec{k}}|n_{\vec{k}}\rangle = \sqrt{n_{\vec{k}}}|(n-1)_{\vec{k}}\rangle$. Removing a quantum of excitation of variety $\vec{k}$ from a state, $\hat{a}_{\vec{k}}$ is the *annihilation operator* for such quanta.

It is time to expose the Hilbert space on which this representation of the ETCCRS acts. It turns out to be the symmetric Fock space (Def. 3.3.2) over the single-particle Hilbert space $\mathcal{H}$ spanned by positive frequency solutions to the Klein–Gordon equation. Let's elaborate this, starting with $\mathcal{H}$. An arbitrary real solution to (9.2) can be expanded in terms of the positive and negative frequency modes (Eq. (9.7), and understand the overbar to denote complex conjugation):

$$u^{+}_{\vec{k}}(\vec{x}, t) = \frac{\exp(i\vec{k}\cdot\vec{x} - i\omega_{\vec{k}} t)}{\sqrt{2L^3 \omega_{\vec{k}}}}, \quad u^{-}_{\vec{k}}(\vec{x}, t) = \overline{u^{+}_{\vec{k}}(\vec{x}, t)} \qquad (9.14)$$

We can impressionistically justify labeling the modes $u^{+}_{\vec{k}}$ "positive frequency" by appeal to the quantization we seek. There the Hamiltonian (energy) operator is the infinitesimal generator of time evolution, which licenses us to regard $-i\partial/\partial t$ as an energy operator. The modes $u^{+}_{\vec{k}}$ are eigenfunctions of this operator, with eigenvalues $\omega_{\vec{k}} > 0$. They are also a complete orthonormal basis for $\mathcal{H}$ (see Wald 1994, 33–4). Considered as such, these modes are none other than eigenvectors $\hat{a}^{\dagger}_{\vec{k}}|0\rangle$ of the Hamiltonian. Our single-particle Hilbert space $\mathcal{H}$ is spanned by states of the form $\hat{a}^{\dagger}_{\vec{k}}|0\rangle$, i.e. positive frequency Fourier modes $u^{+}_{\vec{k}}(\vec{x}, t)$.

Let $[\mathcal{H} \otimes \mathcal{H}]$ denote the symmetrized tensor product of $\mathcal{H}$ with itself. $[\mathcal{H} \otimes \mathcal{H}]$ is spanned by symmetric linear combinations of states of the form $\hat{a}^{\dagger}_{\vec{k}}\hat{a}^{\dagger}_{\vec{k}'}|0\rangle$. The n-fold symmetrized tensor product space $[\otimes_n \mathcal{H}]$ is spanned by symmetric linear combinations of states of the form $\prod_{i=1}^{n} \hat{a}^{\dagger}_{\vec{k}_i}|0\rangle = \hat{a}^{\dagger}_{\vec{k}_1} \hat{a}^{\dagger}_{\vec{k}_2} \ldots \hat{a}^{\dagger}_{\vec{k}_n}|0\rangle$. The *symmetric Fock space* $\mathfrak{F}(\mathcal{H})$ over $\mathcal{H}$ is the completed direct sum $\oplus_{i=0}^{\infty}[\otimes_i \mathcal{H}]$ (where $\otimes_0 \mathcal{H}$, the vacuum or no particle state, is stipulated to be $\mathbb{C}$).

---

[6] Teller (1995, ch. 2) identifies the misleading overtones of the title.

9.3 PRO PARTICLES   199

It is on $\mathfrak{F}(\mathcal{H})$ that operators $\hat{a}^\dagger_{\vec{k}}, \hat{a}_{\vec{k'}}$ representing the ETCCRs act: the creation operator $\hat{a}^\dagger_{\vec{k}}$ can be defined in terms of a set of maps, one of the form $\otimes^n_1 \mathcal{H}$ to $\otimes^{n+1}_1 \mathcal{H}$ for each $n \geq 0$; likewise $\hat{a}_{\vec{k}}$ can be defined in terms of a set of maps, one of the form $\otimes^{n+1}_0 \mathcal{H}$ to $\otimes^n_0 \mathcal{H}$ for each $n \geq 0$ (see Wald 1994, A.2, for details; NB different authors use different normalization conventions in their definitions of these operators).

Let's tally up some features of $\mathfrak{F}(\mathcal{H})$ (see Strocchi 2008, 74–7). First, $\mathfrak{F}(\mathcal{H})$ carries an irreducible representation of the Heisenberg relations (9.9) in terms of the creation and annihilation operators $\hat{a}^\dagger_{\vec{k}}, \hat{a}_{\vec{k}}$. Second, a total number operator $\hat{N} := \sum_{\vec{k}} \hat{a}^\dagger_{\vec{k}} \hat{a}_{\vec{k}}$ exists and is well-defined on a dense domain of $\mathfrak{F}(\mathcal{H})$. This is sometimes called the *Fock condition*, and (given the first feature) is equivalent to a third feature of $\mathfrak{F}(\mathcal{H})$: it contains a vector $|0\rangle$ such that $a_{\vec{k}}|0\rangle = 0$ for all $\vec{k}$ (Strocchi 2008, prop. 2.1, 75). Annihilated by each annihilation operator, $|0\rangle$ is *the* Fock vacuum vector. What justifies the definite description is this: $|0\rangle$ is also an eigenvector of $\hat{N}$ with eigenvalue 0, and in an irreducible Fock space representation, that eigenvalue is associated with a one-dimensional eigenspace.

This package of features will be significant enough to what follows that we will collect them under a definition.

Definition 9.1 (irreducible Fock space representation). An irreducible Fock space representation π of the Weyl relations over (S, Ω) is an irreducible representation of those Weyl relations that induces a representation of the corresponding Heisenberg relations in terms of creation and annihilation operators that determine a total number operator defined on a dense domain of the Hilbert space on which π acts.

9.3.2 The particle notion as fundamental

Let π be an irreducible Fock space representation of the Weyl relations. Allied with π is a particle notion, the π-*particle notion*, circumscribed by the creation and annihilation operators, the vacuum vector, the variety number operators $\hat{N}^\pi_{\vec{k}}$, and the total number operator $\hat{N}^\pi$ conspiring in the irreducible Fock space representation. Like all good concepts, the π-particle notion has a logical anatomy. It affords inferences like the following: the total number of particles is the sum, over varieties $\vec{k}$, of the numbers of particles of variety $\vec{k}$; the number of particles of variety $\vec{k}$ in a state $|\psi\rangle$ is one more than the number of particles of variety $\vec{k}$ in a state $\hat{a}_{\vec{k}}|\psi\rangle$, and so on. The conceptual interrelations help make the π-particle notion the notion it is.

Fock space states admit of analysis in terms of their particle contents because the Fock space vacuum is *cyclic* with respect to creation and annihilation operators bearing a representation of the CCRs. That is, where $|0\rangle$ is the Fock space vacuum, and $\{P(\hat{a}, \hat{a}^\dagger)\}$ is the set of polynomial functions of its creation and annihilation operators, the set of states $\{P(\hat{a}, \hat{a}^\dagger)\}|0\rangle$ is dense in the Fock space. As Ashtakar and Magnon remark, "A 'particle interpretation' would be naturally built into [the Fock space] formalism: an element of [the single particle Hilbert space] $\mathcal{H}$ could be thought of as representing a quantum state of a single particle, and that of $\mathfrak{F}(\mathcal{H})$, as representing some superposition of *n* particle states for various values of *n*" (1975, 378). In a similar vein, Strocchi claims the creation and annihilation operators are "(algebraically) complete in the sense that their algebraic functions suffice for the description of the states of the system" (2005, 32).

The completeness at issue here deserves further explication. Begin by observing that the position and momentum observables of a theory of classical mechanics are taken to be *fundamental* because they are what Strocchi called "algebraically complete." Any pure state of the classical theory—i.e. any point in its phase space—can be specified by assigning values to those observables; different pure states correspond to different sets of values for fundamental observables (see §2.3.1). Please notice that "fundamental" is being used in the physicist's, not the metaphysician's, sense: classical magnitudes are fundamental in the physicist's sense when assigning them values is sufficient to fix the classical state. Although this may be grounds for, it is not the same as, claiming these magnitudes to be metaphysically fundamental or ontologically basic. What exactly the latter claim *means* is a matter I will skirt. But I will suggest, timorously, that whatever it means for magnitudes to be metaphysically fundamental, magnitudes must be fundamental in the physicist's sense to qualify. Underlying the suggestion is the idea that if there are physical states that can't be understood by means of a set $\mathfrak{Q}$ of magnitudes, that set can't be metaphysically fundamental. So $\mathfrak{Q}$ is metaphysically fundamental only if it is fundamental in the physicist's sense.

What, in the quantum context, does it take for a collection of magnitudes to be fundamental in the physicist's sense? I propose an answer that traces its lineage back to a notion Dirac introduced in his 1930 *Principles of Quantum Mechanics*: the notion of a complete set of *quantum* observables. This notion can be made precise along the following lines (see Prugovečki 1971, 311–12). Consider a separable Hilbert space $\mathcal{H}$.

Definition 9.2. A set of operators $\{A_1, ..., A_n\}$ on $\mathcal{H}$ is *complete* if:

(i) Each operator A_k in the set has a point spectrum $\{\lambda_i^k\}$;
(ii) For each *n*-tuple of eigenvalues $(\lambda_1, ... \lambda_n)$ of operators in the set, there's a vector $|\psi_{\lambda_1...\lambda_n}\rangle$ that's a simultaneous eigenvector of the operators associated with those eigenvalues;
(iii) These vectors $|\psi_{\lambda_1...\lambda_n}\rangle$ span $\mathcal{H}$.

A complete set of operators is a set of pairwise commuting operators with a common dense domain that spans $\mathcal{H}$. (Notice that if $\mathcal{H}$ is infinite dimensional, the index set for the collection of observables must be infinite as well.) It follows that $\mathcal{H}$ has a basis of unit vectors, *each element of which basis can be labeled by the n-tuple of eigenvalues it assigns the collection of operators* $\{A_i,\}$ *making up the complete set*. Vectors in the basis assigned the same *n*-tuple of eigenvalues are identical; vectors assigned different *n*-tuple of eigenvalues are distinct. Thus an eigenvalue assignment to the complete set *fixes* an element of this basis. Because the basis spans the space, each vector in the space can be characterized, and characterized uniquely, as a superposition of basis vectors whose identities are thus fixed.[7]

Now, suppose the pure states of a quantum theory correspond to unit vectors in $\mathcal{H}$. Then *every* pure state of the theory admits a characterization of the sort just described. And *every* mixed state of the theory can be characterized in terms of these pure states. The idea behind the account of fundamental quantum magnitudes developed below is that such a capacity for the characterization of all physical states qualifies a collection of observables for fundamental status.

In ordinary QM, where observable algebras take the form $\mathfrak{B}(\mathcal{H})$ and pure states correspond to vectors in $\mathcal{H}$, the idea reinforces and is reinforced by the maximal beable approach to interpreting quantum theories, discussed in the last chapter. The collection of simultaneous eigenprojections of a complete set of observables for $\mathcal{H}$ generates a maximal beable subalgebra of $\mathfrak{B}(\mathcal{H})$. Thus a complete set of operators for the Hilbert space in which an ordinary quantum theory is set determines a maximal beable interpretation of that theory. According to such an interpretation, the domain of quantum fact consists in the simultaneous possession by observables in the complete set of determinate values. This circumstance lends more plausibility to the suggestion I'm developing: that a collection of quantum magnitudes is fundamental only if it is complete.

Notice that in general a Hilbert space $\mathcal{H}$ will admit many distinct complete sets of observables, whereas (one might hope) the set of ontologically basic magnitudes is unique, if it exists at all. This is only part of why being fundamental in the physicist's sense is at most necessary, and not sufficient, for being metaphysically fundamental.

Alas, for present purposes, it won't do to gloss "fundamental" (in the physicist's sense) as "complete" (in the sense of Def. 9.2). Our present inquiry concerns QM_∞ and the possibility that quantum theories encountered there shouldn't be understood as assuming the concrete Hilbert space form familiar from ordinary QM. But Def. 9.2 adopts that form as its framework. To articulate the idea that a collection of quantum magnitudes is fundamental only if it is (in some sense) complete, we need a more abstract criterion of fundamentality than the Hilbert-space-adapted criterion provided by Dirac's notion of a complete set of observables. We also want this criterion of fundamentality to reproduce those features of Def. 9.2's account that recommend complete sets of observables as candidates for fundamental status in the context of ordinary QM.

---

[7] In the case that $\mathcal{H}$ is infinite dimensional, the simultaneous eigenvectors of the complete set will be dense in $\mathcal{H}$. There will be elements of $\mathcal{H}$ *not* expressible as *finite* superpositions of simultaneous eigenvectors of the complete set, elements defined as the limit points of infinite convergent sequences of such superpositions. Any practical characterization of such elements in terms of simultaneous eigenvectors of the complete set will involve "and so ons." I propose to regard this as good enough: for any ϵ and any "limit" vector ϕ, there will be a finitely characterizable vector ψ such that $\langle\psi|\phi\rangle \approx 0$. So the characterizations we can offer in practice will give us very good handles on the limit vectors.

Here are first steps toward such an account. Consider a quantum theory in the form of a kinematic pair $(\mathfrak{Q}, \mathcal{S})$, where $\mathfrak{Q}$ is a C^* algebra of observables and $\mathcal{S}$ is a collection of states on that algebra. We will say

Definition 9.3. A set of elements $\{A_1, ..., A_n\} \subset \mathfrak{Q}$ is *fundamental* with respect to the theory $\mathfrak{Q}, \mathcal{S}$ only if:

(i) Each magnitude A_i has a point spectrum $\{\lambda_i^k\}$;
(ii) For each n-tuple $(\lambda_1, \ldots \lambda_n)$ of elements of the spectra of magnitudes $\{A_i\}$, there's a unique state $\omega \in \mathcal{S}$ such that $\omega(A_i) = \lambda_i$. Call such a state a "D-state."
(iii) Every other state in $\mathcal{S}$ defines a probability distribution over D-states.

It will suffice for our purposes to apply this criterion in the scope of two further assumptions. The first is that D-states are pure states on $\mathfrak{Q}$. As such, each will be implemented by a vector state on an irreducible representation of $\mathfrak{Q}$. The second assumption is that there is some irreducible representation, call it π_ω, on which all D-states are implementable as vector states. Given these assumptions, there are two ways a state $\phi \in S$ can satisfy requirement (iii) by defining a probability distribution over D-states. One is for ϕ to be implemented by a vector in $\mathcal{H}_\omega$ which is a superposition of vectors implementing D-states. The other is for ϕ to be a convex combination of D-states, in which case ϕ would be implemented by a density operator on $\mathcal{H}_\omega$. Either way, ϕ is π_ω-normal. So in this special case, the collection of magnitudes $\{A_i\}$ is fundamental only if every state in S is π_ω-normal.

With these preliminaries out of the way, let us set to work on the particle question. When $\mathcal{H}_\pi$ is the Hilbert space housing an irreducible Fock space representation π, its variety number operators $\hat{N}_k^\pi = \hat{a}_k^\dagger \hat{a}_k$ form a complete set in the sense of Def. 9.2 with respect to $\mathcal{H}_\pi$.[8] Each has a point spectrum consisting of the natural numbers. For any n-tuple of eigenvalues of those operators, there's a vector in the Fock space that is a simultaneous eigenvector of the variety number operators associated with those eigenvalues. (To obtain this vector, think of the n-tuple as defining a map that takes each variety k to the $\hat{N}_k^\pi$ eigenvalue $n(k)$, then act on the vacuum $n(k)$ times with each creation operator $\hat{a}_k^\dagger$ and symmetrize.) And these vectors are dense in the Fock space—the circumstance that underwrites the "occupation number formalism" explicated in Teller (1995, ch. 3).

---

[8] Keep in mind that $\vec{k}$ labels wave number and so different values of $\vec{k}$ correspond to different eigenvalues of momentum. This ensures that $\hat{N}_k^\pi$ commutes with $\hat{N}_{k'}^\pi$ for all k, k', which commutativity helps qualify the collection of variety number operators as a complete set. Later k will give way to a generalized index that needn't be a wave number $\vec{k}$. For instance, in §9.5, this index will be an element of the space S of classical solutions. Then the variety numbers $\hat{N}_f^\pi, \hat{N}_g^\pi$ for solutions $f, g \in S$ will commute iff f and g are orthogonal—a rather special condition. Thus, in the more general setting, care must be taken to construct a collection of particle observables that constitutes a complete set, if such a construction is possible at all. (If it isn't, so much the worse for the prospects of extending the strategy of fundamental particle interpretation to the more general setting.) I gloss over this complication in the text, but will track it in future notes.

9.3 PRO PARTICLES 203

The completeness with respect to the Fock space $\mathcal{H}_\pi$ of the collection of variety number operators $\{\hat{N}^\pi_k\}$ qualifies the π particle notion as fundamental, according to Def. 9.3 only under an additional assumption. Concerning the QFT for which π provides an irreducible Fock space representation, the assumption is that the physically significant states of that QFT reside in π's folium. In this case, the particle observables $\{\hat{N}^\pi_k\}$ are fundamental to that QFT in much the same (physicist's!) sense as p and q are fundamental to classical mechanics: because $\{\hat{N}^\pi_k, \hat{N}^\pi\}$ is complete with respect to the Hilbert space $\mathcal{H}_\pi$ whose vectors correspond to pure states of the theory, each state in a dense set of pure states of the theory is a *unique* superposition of basis vectors, each of which is identified by the eigenvalues it assigns observables in that set.

It also follows that the total number operator $\hat{N}^\pi = \sum_k \hat{N}^\pi_k$ is defined on a dense set of physical states. Since inferential relations between propositions concerning observables constituting a particle notion are integral to that notion, we will regard $\hat{N}^\pi$ as an honorary element of the complete set of particle observables. The conclusion: when the space of states possible according to a QFT is confined to the folium of an irreducible Fock space representation π, the π-particle notion, the notion framed by the operators $\{\hat{N}^\pi_k, \hat{N}^\pi\}$, can stake a claim to fundamental status. We can understand states of the theory in terms of their particle contents. We have, that is, met a necessary condition for giving the theory a particle interpretation.

I call a particle interpretation of a QFT made available by the completeness of a particle notion with respect to that theory's family of physical states a *fundamental particle interpretation*, due to the analogy between particle observables in a QFT subject to such an interpretation and the fundamental position and momentum observables of a theory of classical mechanics.

The observables circumscribing a particle notion are complete with respect to the Fock space allied with that notion. But, it should be noted, they are not unique in this distinction. In a sense Arageorgis (1995, §5.4.5) makes precise, any Fock space representation of the CCRs for the free bosonic field is unitarily equivalent to a *field configuration representation*, i.e. a representation each of whose states can be understood in terms of wave functions over possible configurations of the classical field. This might suggest that whenever a set of particle observables enjoys fundamental status, so too does a set of field configuration observables. The choice between a field and a particle interpretation of a theory duly confined to a Fock space is a choice of semantics for fixed and agreed-upon theoretical structure. Section 9.9 will return to the issue of field-y semantics for QFT. What matters for present purposes is that the fixed structure in question is amenable to particle semantics—not that it's resistant to non-particle semantics. The availability of a (privileged) Fock space representation of the ETCCRs for the free bosonic field makes a particle interpretation of the theory available, but it doesn't compel such an interpretation.

For different values of mass, the standard quantizations of Klein–Gordon theory reside in unitarily inequivalent representations. So it is important to emphasize that even the most adamant particle interpreter does not maintain of this family (parameterized by values of m) of QFTs that a single particle notion suffices for its expression and articulation. The most adamant particle interpreter maintains rather that for each QFT in the family, there is a particle notion in whose terms to interpret that theory. For mass m and mass $m' \neq m$ Klein–Gordon theory, these particle notions needn't—applying to particles of different mass, they shouldn't!—coincide. An adamant particle interpreter guided by an antecedent metaphysical commitment to particles (albeit one alienable from common sense expectations about their localizability, etc.) adheres to the ideal of pristine interpretation.

9.4 Anti particles: an argument, and a loophole

The Unruh effect has evoked the suspicion that particle physics isn't particle physics after all. Many authors (e.g. Wald 1994, 166; Fulling 1972, 283; Unruh 1976, 886) give voice to this suspicion in the course of technical expositions whose primary purpose is not interpretive. So it is not surprising that the suspicion rarely takes the form of an explicit argument from the Unruh effect to the conclusion that one cannot, after all, understand states possible according to a specific QFT in terms of their particle contents. What follows is my best attempt to reconstruct such an argument. It is not a particularly auspicious attempt, but probing its inadequacies will reveal the genuine challenge the Unruh effect poses fundamental particle interpretations.

The Unruh effect supplies the first premise of the inauspicious reconstruction. Redhead observes of that effect, "the definition of the vacuum as the absence of particles depends on the state of motion of the observer. So

(UP1)   two mutually accelerating observers disagree on whether all the particles have been removed [from $|0_M\rangle$]!" (1995b, 77).

The next premise denies that there is any reason to favor one observer's assessment of $|0_M\rangle$'s particle contents over another's. Wald asks, "Which of these two observers is 'correct' in his assertion? The answer is, of course, that both observers are correct. It simply happens that the natural notion of 'particles' defined by accelerating observers...differs from the natural notion of particles defined by inertial observers..." (Wald 1994, 166). That is to say,

(UP2)   There is no basis for privileging one observer's assessment of $|0_M\rangle$'s particle contents over the other's.

Jack and Jill are competent physicists, so if anybody is onto the facts about particles, one of them is. This motivates the third premise:

(UP3)   If there were a fact about $|0_M\rangle$'s particle contents, there would be a basis for privileging one observer's assessment over the other's.

The argument concludes

(UC)   There is no fact about $|0_M\rangle$'s particle contents.

9.4 ANTI PARTICLES: AN ARGUMENT, AND A LOOPHOLE

The main grounds for a fundamental particle interpretation of QFT were that every state of that theory could be understood in terms of its particle contents. But if there is no fact about $|0_M\rangle$'s particle contents, $|0_M\rangle$ is a state that cannot be so understood. So the conclusion (UC) undermines the main grounds for commitment to a particle interpretation of quantized mass m Klein–Gordon theory in Minkowski spacetime.

Of course, the argument just rehearsed has an enormous loophole. It casts Jack's and Jill's assessments of $|0_M\rangle$'s particle content as contradictory. But need they be? Jack's and Jill's assessments could, after all, differ in something like the way Mary's and Martha's assessment of the temperature of a vat of liquid differ, when Mary uses a Celsius and Martha uses a Fahrenheit thermometer. Or Jack's and Jill's assessments of particle content could differ in the way Mary's and Martha's assessments of electric and magnetic field strengths at a point differ, when Mary and Martha are moving relative to one another. In each case, the translations between Mary's assessment and Martha's are so direct and straightforward that we needn't take their divergence to imply that there is no matter of fact assessed. We might instead attribute those matters of fact structure sufficient to account for the divergent assessments: temperature is measured on a scale whose gradation is matter of conventional choice; E and B field strengths are tractably relative to observers' states of motion, and so on. (Notice that such attributions can take the form of positive physics.) The loophole is big enough to hold the claim that the Unruh effect reveals, not the absence of matters of fact concerning particles, but the presence in those matters of fact of an unobvious, and physically interesting, complexity.[9]

The loophole is that Jack's and Jill's assessments could simply be different ways of expressing the same facts. One way to close this loophole would be to show that Jack's and Jill's disagreement reflected something like the *domain incommensurability* of their particle assessments. Expressed in the language of operator algebras, the idea is that if the states to which Jack's particle notion applies were disjoint from the states to which Jill's applies, there would be no prospect of a common realm of particulate fact which their assessments depict in different but compatible ways, and no prospect of a positive physics intertwining Jack's account and Jill's. Sections 9.5 and 9.6 argue that Jack's particle concept and Jill's particle concept are indeed incommensurable in this sense.

Notice that such incommensurability undermines the undermining argument just given! (UP1) claims that Jack takes $|0_M\rangle$ to describe an absence of particles and Jill takes $|0_M\rangle$ to describe the presence of a thermal distribution of them. It follows from the incommensurability of Jack's and Jill's particle notions that (UP1), so understood, is false: there is no state to which Jack and Jill attribute different particle contents because any state to which one's particle notion applies is a state to which the other's doesn't. But the incommensurability serves as a premise in a better argument against particle interpretations, which §9.9 makes explicit.

---

[9] Ashtekar and Magnon appreciate this possibility: "this 'observer-dependent' description has a certain aesthetic appeal: it sets up some sort of tie between what observers do and what they detect" (1975, 389).

9.5 Closing the loophole: incommensurable particle notions

This section marshalls algebraic resources to articulate and compare Jack's and Jill's particle concepts. The result of the comparison is the claim that those particle concepts are *incommensurable*.

9.5.1 Fock space without the heuristic

The frequency-splitting heuristic §9.3 sketched for quantizing the scalar field has its limitations. Although it lacks rigor, this correctable (see Wald 1994, §3.2) deficiency will not concern me here. I am concerned rather with another commodity the frequency-splitting heuristic lacks: generality. Adapted to the carefully controlled setting of Minkowski spacetime, the frequency-splitting procedure relies on special features of that setting, and so does not survive in the rough and tumble of arbitrarily curved (but still globally hyperbolic) spacetimes without those features. This section identifies those features and the spacetimes that have them. It also describes how and in what sense a particle notion may be associated with quantizations carried out in spacetime backgrounds lacking the special features presupposed by the frequency-splitting procedure.

In an arbitrary globally hyperbolic spacetime $(\mathcal{M}, g_{ab})$, we will continue to treat classical Klein–Gordon theory as a double (S, Ω). Here S consists of real, continuously differentiable solutions to the covariant Klein–Gordon equation (9.2) (with $\Box$ understood in terms of derivative operators ∇_a, $a \in \{1, 2, 3, 4\}$ compatible with the metric g_{ab}), which solutions have compact support on Cauchy surfaces of $(\mathcal{M}, g_{ab})$. We've dropped the simplifying assumption of the previous section that these Cauchy surfaces are toroidal, and with it, its consequence that the Fourier expansion of solutions assume the form of discrete sums. This space of solutions has the symplectic structure:

$$\Omega(\varphi, \chi) = \int_\Sigma (\varphi \nabla_a \chi - \chi \nabla_a \varphi) dV^a \qquad (9.15)$$

where Σ is a Cauchy surface (any Cauchy surface—(9.15) is independent of which), and V^a is the natural volume element for Σ: $n^a \sqrt{h} d^3 x$ (where n^a is the unit normal to Σ and h the determinate of the metric h_{ab} on Σ induced by g_{ab}). To quantize Klein–Gordon theory on curved spacetime is to find a representation of the Weyl relations over (S, Ω).

In order to follow the frequency-splitting procedure to obtain such a quantization, one must identify the positive frequency subspace of the classical solution space. This identification requires in turn the specification of a time parameter t with respect to which the positive frequency solutions are positive frequency. In a globally hyperbolic spacetime of topology $\Sigma \times \mathbb{R}$, consider a time parameter t that indexes a set of Cauchy surfaces foliating the spacetime: $\Sigma \times \mathbb{R} = \Sigma_t \times t$. In order for the frequency-splitting quantization procedure carried out with respect to this foliation to be unambiguous, solutions which are positive frequency on *any* Cauchy surface in this foliation must be positive frequency on *every* Cauchy surface in this foliation. Otherwise, a solution's status as positive frequency will vary from Cauchy surface to Cauchy surface. Such variance would mean that there isn't a single, unambiguous positive frequency subspace, but a family of them, indexed by the time parameter of the Cauchy surface on which their elements qualify as positive frequency. Applying the positive frequency heuristic to different members of the family could eventuate in different, possibly unitarily inequivalent quantum field theory constructions. (Section 11.2's discussion of "cosmological particle creation" in Friedman–Robertson–Walker (FRW) spacetimes will provide an example.)

A time parameter t under which the frequency-splitting decomposition exhibits the desired invariance is available whenever the globally hyperbolic spacetime $(\mathcal{M}, g_{ab})$ in which the quantization is set is *static*.

Scholium: Static and stationary spacetimes. A spacetime is *stationary* iff it hosts a one-parameter family of isometries whose orbits are timelike. Thus a globally hyperbolic spacetime that's stationary admits a foliation by Cauchy surfaces the timelike transformations between which are symmetry transformations of the metric. (For example, inertial time translations from one Cauchy slice to another in Minkowski spacetime are symmetries of the Minkowski metric.) The stationarity condition is equivalent to the presence of a global timelike *Killing field* V^a. (A vector field V^a is a Killing field for $(\mathcal{M}, g_{ab})$ iff $\nabla_a V_b + \nabla_b V_a = 0$. See Wald 1984, ch. 8 for more on this and other spacetime structures.) Integral curves of Killing vector field are orbits of isometries of the metric. Consider a *stationary frame* whose world lines are integral curves of this Killing field. With respect to a coordinate system adapted to this frame, and in particular, with respect to "Killing time" t such that $\partial/\partial t = V^a$, the metric components are time-independent. In a *static* spacetime, the global timelike Killing field is hypersurface orthogonal. With respect to a *static frame* (defined on the model of the stationary frame, above) the metric is not only time-independent but time-orthogonal. Intuitively, the spatial geometry isn't "rotating." ♠

In spacetimes that are stationary but not static, the positive frequency subspace of the classical solution space is not unambiguously defined. One can, however, use "Killing time" to identify a self-consistent and unambiguous subspace of the classical solution space on which the symplectic product defines an inner product which is everywhere positive-definite. From this, one can construct a single particle Hilbert space, and thence a Fock space circumscribing a particle notion (see Ashtekar and Magnon 1975). The construction assumes the Klein–Gordon field to have strictly positive mass ($m > 0$) and the Killing vector field V^a to be nowhere arbitrarily small or lightlike.[10]

---

[10] The formal requirement is that there exist a Cauchy surface Σ (with unit normal n^a) on which $V^a V_a > \epsilon V_a > \epsilon^2$ for some $\epsilon > 0$. See Wald (1994, §4.3).

Non-stationary spacetimes foil the Killing time quantization strategy. It does not follow that such spacetimes foil the constitution of a Fock space representation circumscribing (some sort of) particle notion. In a *generic* globally hyperbolic spacetime, classical Klein–Gordon theory can still be associated with a real symplectic vector space (S, Ω). And there are methods for constructing from such a (S, Ω) a single particle Hilbert space $\mathcal{H}$ in whose terms the Weyl relations constituting the quantization of the theory can be represented. Here I sketch one such method (see Florig and Summers 2000; Reents and Summers 1994).

Let us remind ourselves what we're after. A representation of the Weyl relations over (S, Ω) is a map W from S to the unitary operators on a Hilbert space $\mathcal{H}$ satisfying[11]:

$$W(f)W(g) = e^{\frac{\Omega(f,g)}{2}} W(f+g) \quad W(f)^* = W(-f) \tag{9.16}$$

for all $f, g \in S$. In the service of constructing such a representation, undertake the project of turning the real symplectic space (S, Ω) into a complex Hilbert space—a project made much easier by S's linearity. The project requires an account of what it is to multiply an element of S by a complex number. An Ω-*admissible complex structure* on S furnishes such an account. Such a complex structure is real linear map on $J : S \rightarrow S$ satisfying

1. $\Omega(Jf, Jg) = \Omega(f,g)$
2. $-\Omega(Jf, f) \geq 0$ for $f \neq 0$
3. $J^2 = -\mathbb{I}$

For all $f \in S$ and $\alpha, \beta \in \mathbb{R}$, J tells you what happens when you multiply f by the complex number $\alpha + i\beta$: you get $\alpha f + \beta J f$.

We can use the classical Klein–Gordon field on the torus to give an example of a complex structure. Recall that an arbitrary solution $f \in S$ can be Fourier decomposed in terms of positive and negative frequency modes $u_{\vec{k}}^+(\vec{x}, t)$ and $u_{\vec{k}}^-(\vec{x}, t)$. The symplectic product Ω on S (9.15) implies that $\Omega(u_{\vec{k}}^+, u_{\vec{\ell}}^+) = \Omega(u_{\vec{k}}^-, u_{\vec{\ell}}^-) = 0$ and that $\Omega(u_{\vec{k}}^+, u_{\vec{\ell}}^-) = \delta_{k\ell}$.[12] It follows that the map between positive and negative frequency subspaces defined by $J : u_{\vec{k}}^+ \rightarrow u_{\vec{k}}^-$ is an Ω-admissible complex structure (but by no means the only one).

Once we have an Ω-admissible complex structure J on (S, Ω), we can define a scalar product, guaranteed to be positive-definite, on S:

$$\langle f, g \rangle := -\Omega(Jf, g) + i\Omega(f, g) \tag{9.17}$$

Treating S as a linear vector space over the scalar field of complex numbers, and completing it with respect to the norm defined by $\langle f, g \rangle$, we obtain a complex Hilbert

---

[11] Notice that, for the foreseeable future, Hilbert space operators are going hatless to reduce clutter.

[12] This collection of circumstances renders $\{u_{\vec{k}}^+, u_{\vec{j}}^-\}$ a *symplectic orthonormal system* (see Florig and Summers 2000).

space on which $\langle f, g \rangle$ defines an inner product. Call this Hilbert space $\mathcal{H}_J$. Notice that whereas Ω on its own determines the imaginary part of the inner product on $\mathcal{H}_J$, Ω requires the collusion of a discretionary complex structure J to determine the real part of that inner product. This is a symptom of the disease epidemic in this book: the failure of the Weyl relations over (S, Ω) to determine, up to unitary equivalence, a unique Hilbert space representation.

Now that we have a complex Hilbert space, we can get down to the business of representing the Weyl relations in terms of unitary operators acting on that space. We'll do so by using J to define a state ω_J on the Weyl algebra $\mathfrak{A}$ generated by those unitaries, then taking ω_J's GNS representation. The result will be an irreducible Fock space representation of the Weyl relations: a natural setting, the last section argued, for a fundamental particle interpretation of the QFT whose canonical algebra is determined by those Weyl relations.

For each Weyl unitary $W(f)$, set:

$$\omega_J(W(f)) := e^{\frac{\Omega(Jf, f)}{4}} \qquad (9.18)$$

Because the unitaries $W(f)$ generate the Weyl algebra $\mathfrak{A}$, ω_J extends by linearity to a state on $\mathfrak{A}$. The GNS representation $(\pi_J, |\xi_J\rangle, \mathcal{H}_{\omega_J})$ of this state will have a number of suggestive features (see Roepstorff 1970, 303). First, it's irreducible. Second, in ω_J's GNS representation, the map $f \mapsto \pi_J(W(f))$ is strongly continuous. Stone's theorem ensures that for each $f \in S$ there exists a self-adjoint $\Phi(f)$ acting on the Hilbert space $\mathcal{H}_{\omega_J}$ that is the infinitesmal generator of the family of unitaries $W(tf)$:

$$W(tf) = e^{it\Phi(f)} \qquad (9.19)$$

It can be shown that there's a common dense domain of $\mathcal{H}_{\omega_J}$ on which these *field operators* $\Phi(f)$ satisfy the CCRs:

$$\Phi(f)\Phi(g) - \Phi(g)\Phi(f) = i\Omega(f, g)\mathbb{I} \qquad (9.20)$$

Moreover, we can use these field operators to define "annihilation" and "creation" operators:

$$a(f) = \frac{1}{\sqrt{2}}(\Phi(f) + i\Phi(Jf)) \qquad (9.21)$$

$$a^\dagger(f) = \frac{1}{\sqrt{2}}(\Phi(f) - i\Phi(Jf)) \qquad (9.22)$$

Coupled with the first feature, this means we have an irreducible representation of the Heisenberg relations. The cyclic GNS vector $|\xi\rangle$ stands to these as a vacuum state: $a(f)|\xi\rangle = 0$ for all $f \in S$.

A total number operator N^J can be defined on a dense domain of $\mathcal{H}_{\omega_J}$. (Where $W(f)$ is the unitary operator on $\mathcal{H}_{\omega_J}$ which is the Weyl representative of the solution $f \in S$, the expectation value of the number operator N^J in the state $\pi_J(W(f)|\xi\rangle$

turns out to be $||f||^2$. See Roepstorff 1970, 303 for details.) Thus satisfying the Fock condition, ω_J's GNS representation is an irreducible Fock space representation, in the sense of Def. 9.1. The cyclic GNS vector $|\xi\rangle$ is the vacuum state of this Fock space: $a(f)|\xi\rangle = 0$ for all $f \in S$. $\mathcal{H}_{\omega_J}$ itself is the symmetric Fock space $\mathfrak{F}(\mathcal{H}_J)$ over the single particle Hilbert space $\mathcal{H}_J$ defined by J and Ω.

The foregoing procedure is a special case of a more general, and mathematically rigorous, construction in whose terms results relevant to this chapter and the next will be stated. Starting with the real symplectic space (S, Ω) of solutions to the Klein–Gordon equation with compact support, one finds a real inner product $\mu(\varphi, \chi)$ on S—i.e. a bilinear map $\mu : S \times S \to \mathbb{R}$ which is symmetric and positive (i.e. $\mu(\varphi, \varphi) \geq 0$)—satisfying:

$$\mu(\varphi, \varphi) \geq \frac{1}{4} l.u.b._{\chi \neq 0} \frac{(\Omega(\varphi, \chi))^2}{\mu(\chi, \chi)} \qquad (9.23)$$

A subset of these maps correspond to Ω-compatible complex structures via:

$$\Omega(\varphi, \chi) := 2\mu(\varphi, J\chi) = \langle \varphi, J\chi \rangle \qquad (9.24)$$

In a sense made precise by Kay and Wald (1991), the choice of a real inner product μ satisyfing (9.23) on S is *equivalent* to the choice of a single particle structure for a Fock space representation of the Weyl relations.

A μ satisfying (9.23) fosters a single particle Hilbert space $\mathcal{H}_\mu$ underlying a symmetric Fock space $\mathfrak{F}(\mathcal{H}_\mu)$ carrying a representation of the Weyl relations over (S, Ω). This representation is the GNS representation of a state ω_μ on the Weyl algebra determined by μ:

$$\omega_\mu(W(\varphi)) = \exp[\frac{\mu(\varphi, \varphi)}{2}] \qquad (9.25)$$

States thus attainable are *quasi-free*.[13] When the inequality (9.23) is saturated, the GNS representation of ω_μ is irreducible, indeed, an irreducible Fock space representation. It follows from the representation's irreducibility that ω_μ is pure. The class of pure quasi-free states includes all vacuum states obtained by following the frequency-splitting procedure. Under the supposition that particle notions sustaining fundamental particle interpretations will be allied with irreducible Fock space representations, each pure quasi-free state circumscribes a candidate fundamental particle notion.[14]

---

[13] An n point function for a state ω is given by $\omega(W(\varphi_1)W(\varphi_2)\ldots W(\varphi_n))$ for $\varphi_1, \varphi_2, \ldots, \varphi_n \in S$. It can be taken to describe correlations between physically interpretable events, e.g. a particle in one state is created at one place while a particle in another state is destroyed at another (Bratteli and Robinson 1997, 39). In a quasi-free state, all one-point and all higher-order ($n > 2$) correlations vanish: n-point functions are sums and products of two-point functions.

[14] Complicating the candidacy is the circumstance, anticipated in note 8, that no account has been given about how to resolve the total number operator associated with an irreducible Fock space representation into a collection of variety number operators which form a complete set of operators (cf. Def. 9.2) with respect to that Fock space. The completeness of such a set is what we're taking to qualify a particle notion as fundamental in the physicist's sense. It seems plausible that, in the case that the classical solution space S

9.5.2 Unitary inequivalence and incommensurability

This section discusses the conditions under which Fock space representations are unitarily equivalent. It also describes how the failure of unitary equivalence constrains the application of a particle notion affiliated with one Fock space representation to states normal with respect to another.

Criteria of unitary equivalence for Fock space representations come in many forms. In lieu of exhibiting one adapted to the most general case of such representations (for which, see Wald 1994, §4.4), I'll discuss a simpler criterion adapted to a vivid special case.

Consider two irreducible Fock space representations, set in a stationary spacetime whose Cauchy surfaces are toroidal, so that general solutions to the Klein–Gordon equation can be expressed as discrete sums over Fourier modes $u_{\vec{k}}$.[15] Let us attempt to express the creation and annihilation operators of one representation ($\{a_{\vec{k}}^\dagger, a_{\vec{k}}\}$) in terms of those of the other ($\{a_{\vec{k}}'^\dagger, a_{\vec{k}}'\}$).[16] The simplest case of a *diagonal* transformation acts on each normal mode independently:

$$a_{\vec{k}}' = \alpha_k a_{\vec{k}} + \beta_k a_{\vec{k}}^\dagger \tag{9.26}$$

A straightforward calculation for this case reveals that, formally at least,

$$\langle 0|N'|0\rangle = \sum_k |\beta_k|^2 \tag{9.27}$$

Primed and unprimed representations are unitarily equivalent—equivalently, the transformations can be implemented by a unitary U such that for all k, $U^{-1}(a_{\vec{k}}') = \alpha_k a_{\vec{k}} + \beta_k a_{\vec{k}}^\dagger$—iff the quantity on the r.h.s. of (9.27) is finite (Fulling 1989, 143). And when it is finite, that quantity gives the expectation value the unprimed vacuum state assigns the primed total number operator. When it's infinite, the primed total number operator is undefined on the unprimed representation. (As Clifton and Halvorson, 2001b, emphasize, the expression $\langle 0|N'|0\rangle$ is nonsense, because N' isn't a well-defined operator on the Hilbert space $|0\rangle$ belongs to. Hence the hedge with which (9.27) was introduced above. But its import is clear enough, and I will continue to indulge in the suspect notation.)

---

admits a symplectic orthonormal system (cf. note 12) $\{f_i\}$, the variety number operators $N_{f_i}^j = a(f_i)a^\dagger(f_i)$, with $a(f_i)$ and $a^\dagger(f_i)$ given by (9.21) and (9.22) respectively, do the trick. Although I offer no general proof that an irreducible Fock space representation hosts a complete collection of variety number operators suitably related to its total number operator, the body of the text will proceed on the assumption that it does. This reflects my policy of considering circumstances maximally hospitable to particle interpretations.

[15] This discretization is essential to the present example. Non-trivial diagonal transformations on a continuous spectrum of modes are *never* unitarily implementable (Fulling 1989, 146).

[16] Here's a terminological complication: some authors (e.g. Birrell and Davies 1982, 109) are happy to call such re-expression "Bogoliubov transformations" whether they're mathematically well-defined or not. Others (e.g. Wald 1994, 69; Bratteli and Robinson 1997, 22) reserve the term for *unitary* transformations between the bases of two representations.

More generally, given a state ω and an irreducible Fock space representation π, the behavior under ω of the total π-particle number operator N^π furnishes a criterion of whether ω lies in π's folium.

Criterion. ω is π-normal iff $\langle N^\pi \rangle_\omega$ is finite.

(Chaiken 1967, thm. 4; Clifton and Halvorson 2001b, prop. 10; Bratteli and Robinson 1997, 26.) In words, states which are normal with respect to a Fock space representation π are exactly the states with finite π-particle content. And so, states with finite π-particle content are exactly the states that can be implemented by density operators on a Hilbert space with respect to which $\{N_k^\pi\}$ form a complete set of observables. The completeness of $\{N_k^\pi\}$ with respect to a Hilbert space whose density operator states contain all physical states is, §9.3.2 urged, a necessary condition for interpreting the theory by means of a fundamental π-particle concept.

We can use these results to begin to classify the relationship between particle notions associated with different (call them *primed* and *unprimed*) Fock space representations. I will distinguish three ways the primed and the unprimed Fock space representations can be related.

First Case (*coincident particle notions*): $\langle 0|N'|0\rangle = 0$. In this case, not only do the primed and unprimed theorists' Fock space representations share a common folium, the affiliated particle notions parse the particle contents of the common folium of those representations in the same way. There is no basis for rivalry between the particle notions.

Second Case (*non-coincident but unitarily equivalent particle notions*): $\langle 0|N'|0\rangle$ is non-zero but finite. In this case, the primed total number operator is defined on the unprimed representation. This time, although the primed and unprimed theorists recognize the same folium of states, they disagree about the particle contents of those states. For instance, the primed theorist will discern particles in the unprimed vacuum state. It has been suggested that these differences are sufficient to render the primed and unprimed particle notions "completely different" (Fulling 1972, 2850).[17]

But we shouldn't exaggerate the mismatch of particle notions falling under this Second Case. Notwithstanding the non-zero expectation value of the primed number operator in the unprimed vacuum, the two theorists wield unitarily equivalent Hilbert space representations of the Weyl algebra, between which harmony-preserving translations are possible. The primed representation contains a vector identifiable, via the unitary map implementing these translations, with the unprimed vacuum state.

---

[17] Fulling is speaking here of particle notions associated with the Rindler and Minkowski ("free field") quantizations (see §9.6), which fall not under the Second but the Third Case (a fact for which Fulling lacked a rigorous proof at the time). What's more, Fulling takes the facts that "the vacuum of the Rindler-space theory is not the ordinary vacuum of the free field, one-particle states in one theory are not one-particle states in the other, and so on" to suffice for the conclusion that "*the particles or quanta of the Rindler-Fock representation cannot be identified with the physical particles described by the usual quantum theory of the free field*" (1972, 2854). Likewise, Hajicek asserts that if "the transformation between two systems mixes positive and negative 'frequencies'," then the affiliated particle notions are "essentially different" (1976, 608).

(This vector will be a superposition of N' eigenvectors with even eigenvalues; see Fulling 1989, 146, for details.) Moreover, for each unprimed state that is an eigenvector (with finite eigenvalue) of N', the primed representation contains a vector identifiable with it.

Thus any state the unprimed theorist can describe in terms of the total number of (unprimed) particles it contains, the primed theorist can describe (albeit somewhat differently) in terms of the total number of (primed) particles it contains. The set of states to which the theorists apply their total particle notions coincide; when one theorist can distinguish between pair of states in this set in terms of their particle content (according to him), the other can distinguish between those states in terms of their particle content (according to her); they moreover agree about the relative possibility of (i.e. inner product between) each pair of states identified under the unitary map. In short, they agree about what states are in the extension of the particle concept, and about what discriminations, even modal ones, that concept makes possible. This suggests to me that the theorists disagree not about what particles are, but about how to enumerate them. Insofar as their enumerations are intertranslatable, their particle notions are notational variants on one another.

Of course, it is both true and noteworthy that the primed vector identifiable with $|0\rangle$ will in general be a superposition of eigenstates of the primed number operator. So (assuming that they both embrace the eigenvector–eigenvalue link) the primed and unprimed theorists will disagree about which states have determinate numbers of particles—a circumstance calling for more commentary than I furnish here (see Teller 1995, 110–13). Turning to our third case, we see that particle notions can differ more deeply than this.

Third Case (*incommensurable particle notions*): The formal sum that would define $\langle\psi|N'|\psi\rangle$ is divergent, for every state $|\psi\rangle$ in the unprimed Fock space. In this case the primed and the unprimed representations are disjoint. No state normal with respect to one representation is normal with respect to the other. Recall the criterion Def. 9.3 announces for a collection of observables to be fundamental. The variety number operators making up the primed particle notion are fundamental only if they admit a collection of "D-states" (i.e. simultaneous eigenstates whose members are *identified* by the eigenvalues they assign those number operators) and every other state recognized by the theory defines a probability distribution over D-states. That is to say, the primed particle notion is fundamental only if every state of the theory is a normal state of the primed representation, a state to which the primed total number operator N' applies. And *mutatis mutandis* for the unprimed total number operator: the set of variety number operators making it up satisfies Def. 9.3's criterion for fundamental status only if the space of physical states is confined to those normal with respect to the unprimed representation. In short, if one theorist's particle notion is fundamental, the other theorist's states can't be physical. In the case of unitarily equivalent particle notions, primed and unprimed fundamental particle interpreters are conducting QFT in a different idiom. In the case of disjoint particle notions, they're doing so in a different folium. Their particle concepts are domain-incommensurable.

When at least one representation is non-factorial, there is a fourth possibility: that the primed and unprimed representations are neither quasi-equivalent nor disjoint. In this case, one representation's total number operator could be well-defined only on some, but not on all, subrepresentations of the other representation. Because we are taking representations circumscribing particle notions to be irreducible and hence factorial, this chapter will have no more to say about this possibility.

For now, let us take the foregoing to suggest that when irreducible Fock space representations are unitarily inequivalent, the particle notion affiliated with one fails to characterize states characterizable by appeal to the particle notion affiliated with the other, and vice versa. (The next chapter will consider, and find wanting, a number of attempts to resist the suggestion.) Commitment to one of a disjoint pair of particle notions, commitment in the form of confidence that the notion will suffice to characterize all physical states, precludes commitment to the other. In this sense, disjoint particle notions are incommensurable. Such incommensurability would close the loophole in §9.4's argument against fundamental particle interpretation. The next section establishes that Jack's and Jill's particle notions are indeed incommensurable.

9.6 The incommensurability of Jack's and Jill's particle notions

9.6.1 Jill's spacetime

Let x, y, z, t be coordinates for Minkowski spacetime in which the line element assumes the familiar form:

$$ds^2 = dx^2 + dy^2 + dz^2 - dt^2 \tag{9.28}$$

The *right Rindler wedge* is the region $\mathcal{R} : x > |t|$. A coordinate system for $\mathcal{R}$ can be obtained from the standard coordinates by the transformation:

$$x = \xi \cosh \eta, \quad y = y, \quad z = z, \quad t = \xi \sinh \eta \tag{9.29}$$

These *Rindler coordinates* break down on the boundary $x = |t|$: because $\eta = \tanh^{-1}(t/x)$, $x = t$ corresponds to $\eta = +\infty$ and $x = -t$ corresponds to $\eta = -\infty$.[18] The region $\mathcal{L} : x < |t|$ of Minkowski spacetime, a.k.a. the *left Rindler wedge*, is covered by a coordinate system (ξ', y, z, η'), which can be obtained from standard Minkowski coordinates by sticking minus signs in the r.h.s.s of the first and last transformation equations in (9.29). Notice that whereas in the right Rindler wedge, the Rindler time

---

[18] Indeed, describing Minkowski spacetime in terms of Rindler coordinates inflicts a *coordinate singularity* upon it. See Arageorgis, Earman, and Ruetsche (2003).

coordinate η increases with increasing Minkowski time, in the left Rindler wedge, the time coordinate η' decreases as t increases.

Now the right Rindler wedge can be considered as a spacetime in its own right. Like Minkowski spacetime, this *Rindler spacetime* is globally hyperbolic: the planes $\eta = const$ are Cauchy surfaces. Like the surfaces of constant inertial time in Minkowski spacetime, these Cauchy surfaces are related by time-like isometries, namely Lorentz boosts. ($\eta_0 \mapsto \eta_0 + \eta$ corresponds to a Lorentz boost in the x-direction with speed $\tanh \eta$.) That is to say, Rindler spacetime is stationary. Because the hyperbolas $\xi^2 = x^2 - t^2 = const$ which are orbits of Lorentz boost isometries are hypersurface-orthogonal, it is moreover, like Minkowski spacetime, static: one can quantize the Klein–Gordon field by following the frequency-splitting heuristic with respect to Rindler time η. The frequency-splitting vacuum $|0_R\rangle$ thereby obtained is described in the next section. Since the hyperbolas $\xi^2 = x^2 - t^2 = const$ are the worldlines of uniformly accelerating observers (where the hyperbola $\xi^2 = const$ corresponds to constant proper acceleration $a = \xi^{-1}$), this vacuum is the natural one for a uniformly accelerating observer (like Jill) to adopt.

9.6.2 Jack's notion, and Jill's

Jill's spacetime is static; her frequency-splitting quantization yields an irreducible Fock space representation circumscribing a particle notion. Likewise for Jack. This section makes explicit the claim that Jack's particle notion and Jill's are incommensurable, in the sense of applying to disjoint sets of states. Jack and Jill disagree not about conventions for describing a common realm of particulate facts. Their disagreement is deeper: it's about what the realm of particulate fact *is* to begin with.

Let the C^* algebra $\mathfrak{A}(\mathcal{M})$ be the Weyl algebra over the symplectic vector space of solutions to Klein–Gordon equation with compact support on inertial time slices of Minkowski spacetime. Isotony requires that $\mathfrak{A}(\mathcal{M})$ contains as a proper subalgebra the right Rindler wedge algebra $\mathfrak{A}(\mathcal{R})$, which is the Weyl algebra over the space of solutions with compact support on the $\eta = const$ time slices of Rindler spacetime. On these algebras we can define states that are the algebraic counterparts of the Minkowski vacuum state $|0_M\rangle$ and the Rindler vacuum state $|0_R\rangle$.[19] Call these states ω_M and ω_R respectively.

Restricting the Minkowski vacuum state ω_M to the right Rindler wedge algebra defines on that algebra a state $\omega_{M|_R}$ which admits direct comparison with ω_R. $\omega_{M|_R}$ is one of the Type III factor states that gave such dramatic impetus to Chapters 7 and 8. The von Neumann algebra affiliated with the GNS representation of $\omega_{M|_R}$ on $\mathfrak{A}(\mathcal{R})$ is a Type III factor. Because each pure state on $\mathfrak{A}(\mathcal{R})$ is affiliated with a von Neumann

---

[19] The situation is more complicated than the foregoing suggests. The Minkowski case is perfectly tractable. But in the Rindler case, the stationary quantization procedure discussed in §9.5 breaks down for the vectors V^a whose orbits are Lorentz boost isometries, because the technical condition mentioned in note 10 is violated. Kay (1985) describes how to construct a Rindler vacuum state anyway.

algebra of Type I, it follows immediately that $\omega_{M|R}$ is *disjoint* from every pure state on $\mathfrak{A}(\mathcal{R})$. The disjointness of restricted Minkowski vacuum state $\omega_{M|R}$ and the Rindler vacuum state ω_R secures the incommensurability of the particle notions circumscribed by the Fock spaces in which those states serve as vacua. In the Rindler and Minkowski case, the incommensurability is particularly pronounced: in the Minkowski vacuum state, not only is the Rindler total number operator N^R undefined, so too are the all the Rindler variety number operators N_k^R, the ones that count monochromatic Rindler particles (Letaw and Pfautsch 1981, 1495).

But notice that the Rindler particle notion is not the only one estranged from the Minkowski particle notion. Recall that there are myriad pure quasi-free states ω_μ definable on the Rindler wedge algebra $\mathfrak{A}(\mathcal{R})$. With each is associated a particle notion via the Fock space structure of its GNS representation. As pure, each ω_μ is disjoint from $\omega_{M|R}$, and so each affiliated particle notion is incommensurable with the Minkowski particle notion.

Of course, among the particle notions available for Rindler spacetime, the Rindler particle notion can claim the privilege of being the particle notion constructed by appeal to the Lorentz boost isometries. Indeed, the Rindler vacuum can claim privilege of another sort. There are six classes of stationary coordinate systems covering all or part of Minkowski spacetime (see Letaw and Pfautsch 1982). Some of these, like the familiar system of inertial coordinates, are free of event horizons; others, like the Rindler coordinate system, have event horizons. Letaw and Pfautsch (1981) show that the stationary vacua associated with horizon-free coordinate systems are all unitarily equivalent to the Minkowski vacuum state ω_M. Every stationary vacuum associated with a coordinate system with horizons, on the other hand, is unitarily equivalent to the Rindler vacuum ω_R. So the Rindler vacuum is the stationary vacuum for portions of Minkowski spacetime with event horizons—regardless of whether their timelike isometries are Lorentz boost isometries or not!

9.7 Operationalizing the particle notion?

These stationary coordinate systems give me occasion to address an approach to particles in QFT that, superficially considered, promises to reconcile Jack's and Jill's particle notions. The approach is to *operationalize* the particle notion. Arageorgis (1995, 283) locates its origin in 1975 remarks of Bryce DeWitt concerning the debate then raging over the reality of Hawking radiation (formally analogous to Rindler quanta; see Wald 1994, chs. 5 and 7), a debate complicated by "the difficulty of defining 'particle' in the absence of a timelike Killing vector" (DeWitt 1975, 335). DeWitt comments, "If only to stop nonsense discussions, it seems fairly urgent to attempt to settle the issue by building explicit models of 'particle detectors' and computing how they perform under conditions of acceleration and free fall." The operationalist maneuver, then, is to defer questions about the presence of particles to a physical apparatus, whose responses are *constitutive* of their presence. For Davies "Any discussion about what is a 'real, physical vacuum,' must therefore be related to the behavior of real, physical measuring devices, in this case particle-number detectors" (1984, 69). In a slogan, particles are what particle detectors detect (Davies 1984, 75).

The promise of reconciliation comes from the observation that Jack and Jill deploy *different classes* of detectors. Jack's are in inertial motion; Jill's are accelerating uniformly. There is no logical necessity for the responses of detectors from different classes to coincide; there is the logical possibility that detectors from Jack's class register no particles when drifting through the Minkowski vacuum state, while detectors from Jill's register a thermal distribution of them. The promised reconciliation is that QFT is particle physics, where the particle notion has an unobvious complexity it takes particle detectors to unravel. Whereas the Lorentz transformations mediate Mary's and Martha's description of the electromagnetic field, whose unobvious complexity is that its decomposition into electric and magnetic fields depends on the state of motion of the observer, operationalism in collusion with particle detectors mediates Jack's and Jill's description of the particle content of the Minkowski vacuum, whose unobvious complexity is that it's registered by different classes of detectors in different ways.

I would join Andy Pickering (1995; 2010) in finding it intriguing that whereas Theory mediates Mary's and Martha's reconciliation, the reciprocally performative agency of material detectors mediates Jack's and Jill's. Except that I believe the promised reconciliation rings hollow. The problem is that there isn't in general sufficient coherence between detector response, detector trajectory, and particle concept for detector behaviors to mediate intelligibly between particle concepts associated with different trajectories.

Let π_a and π_b be disjoint irreducible Fock space representations corresponding to frequency-splitting quantizations naturally associated with trajectories *a* and *b* respectively. I propose that a detector can mediate between a pair of Fock space representations only if it is calibrated to at least one of them in the following sense:

[Calibration] A detector is calibrated to a representation π_a only if traveling trajectory *a* in the π_a vacuum state, it registers a null response.

Having assured ourselves that a detector is π_a-calibrated, we have a reason to take its response when traveling trajectory *b* to indicate something about π_b particle contents: its response when traveling trajectory *a* tells us something about π_a particle contents! In some circumstances, the exemplary DeWitt detector (explicated in Birrell and Davies 1989, §5.5; see Arageorgis 1995, 285 ff., for a discussion of its engineering presuppositions) exhibits calibration. In the vacuum state of the massless Klein–Gordon field on 4-dimensional Minkowski spacetime, a DeWitt detector in inertial motion registers no particles. In the same setting, a uniformly accelerated DeWitt detector registers a thermal spectrum of them (provided that it's both activated and uniformly accelerated forever: Bäuerle and Koning 1988). In light of Calibration, this suggests that a thermal spectrum of Rindler particles is present.

But in other circumstances, detectors behave bizarrely. Takagi (1986) shows that for massive scalar fields, or for odd-dimensional spacetimes, uniformly accelerating detectors don't register a thermal distribution of bosons. Incredibly, in the latter case, they register a *Fermi* distribution! Further evidence of the DeWitt detector's inability to mediate between the particle concepts of Fock spaces adapted to different stationary frames comes from Letaw and Pfautsch (1981), which considers a DeWitt detector travelling a helix in Minkowski spacetime. The frequency-splitting vacuum associated with the rotating (relative to inertial coordinates) coordinate system in which this detector is at rest is the Minkowski vacuum. A detector satisfying Calibration, a necessary condition for mediating successfully between the inertial and rotating observers' descriptions of the particle contents of the Minkowski vacuum, would register a null response. Yet the detector detects (a non-thermal distribution of) particles. These strange results dramatically temper the hope that operationalism might reconcile the unitarily inequivalent Fock space representations that come naturally to observers in different states of motion.

9.8 The Unruh effect without particles

The foregoing sections have established that the usual statement of the Unruh effect—according to which Jack assigns the Minkowski vacuum state a particle content at odds with the particle content Jill assigns that state—is misleading. Any state Jack can describe in terms of its particle contents is a state Jill cannot so describe, due to the incommensurability of their particle notions. It is the aim of this parenthetical section to describe the Unruh effect without appeal to particles.

In all its glory, the Unruh effect is not simply that an observer accelerated uniformly through Minkowski spacetime in its vacuum state detects particles, but also that "an observer with uniform acceleration a will feel that he is bathed by *a thermal distribution* of quanta of the field at temperature T given by $kT = \frac{\hbar a}{2\pi c}$" (Unruh and Wald 1984, 1047; italics mine). This is the traditional full-on Unruh effect: the uniformly accelerated observer finds herself immersed in a thermal bath. Or, as Unruh has it, "You could cook your steak by accelerating it" (Unruh 1990, 108–9).[20]

The restricted Minkowski vacuum state $\omega_{M|R}$ harbors no (non-trivial) implications about Rindler particles, yet it is supposed to contain a thermal distribution of them. As incoherent as this seems, it is not entirely false. For we needn't invoke *any* particle notion to understand $\omega_{M|R}$ as a thermal state. The key is the notion of equilibrium afforded by the KMS condition (§7.3).

Chapter 7 explained why KMS states are easy to come by in Minkowski spacetime. Any faithful normal state on a von Neumann algebra defines on that algebra a one-parameter group of automorphisms—its modular group—with respect to which

---

[20] "(if the minor problem, that a temperature of 300° C requires an acceleration of about 10^{24} cm/sec^2, didn't make the technique somewhat impractical)!" he continues.

it satisfies the KMS condition. When $\mathfrak{A}(\mathcal{R})$ is the subalgebra of $\mathfrak{A}(\mathcal{M})$ associated with a Rindler wedge, the restricted Minkowski vacuum state $\omega_M|_R$ is a KMS state at temperature $T = 1/2\pi$ with respect to the automorphism group α_η generated by the Rindler isometries.[21] In other words, the Minkowski vacuum state, when restricted to the Rindler wedge algebra, is a thermal state with respect to Lorentz boosts.

This result—which expresses the full-blown Unruh effect, insofar as that effect admits legitimate expression—is obtained in an unadulteratedly algebraic framework. $\omega_M|_R$'s status as thermal has nothing to do with concrete representations of the Weyl algebra or the particle notions they circumscribe. It has nothing do do with Rindler—or any other kind of—particle.

9.9 Conclusion: the case against, restated and extended

Section 9.4 involved the Unruh effect in an argument against a fundamental particle interpretation, which argument had an enormous loophole. That loophole was that the mismatch of Jack's and Jill's assessment of the particle contents of $|0_M\rangle$ could be due, not to the failure of that state to manifest particulate facts, but to an unobvious complexity in the particulate facts it did manifest. The discussion of §§9.5–9.6 sought to close the loophole, by exhibiting Jack's and Jill's particle notions as incommensurable. Jack's and Jill's particle assessments aren't different descriptions of the same set of facts (however complicated), but descriptions of disjoint sets of facts.

The argument of §9.4 does not emerge unscathed from the appeal to incommensurability to close its loophole. The incommensurability of Jack's and Jill's particle notions conflicts with the usual statement of the Unruh effect, which supplies the first premises of the argument of §9.4. But this circumstance is no reprieve for fundamental particle interpretations. Instead, it suggests the following, deeper argument against such interpretations. Let T be a QFT whose canonical Weyl algebra is $\mathfrak{A}$.

1. An irreducible Fock space representation π of $\mathfrak{A}$ circumscribes a particle notion, dubbed the π-*particle notion*, for T.
2. If π is an irreducible representation, there are physically relevant states on $\mathfrak{A}$ which are not π-normal states.
3. If a state ϕ is not π-normal, then the π-particle notion fails to characterize ϕ.
4. In order for the π-particle notion to be fundamental, it must characterize all physically relevant states.
5. No particle notion is fundamental to T.

---

[21] By rescaling the automorphism group, this temperature can be squared with the equilibrium temperature $T = a/2\pi$ more familiar from statements of the Unruh effect. See Arageorgis, Earman, and Ruetsche (2003).

The relevant sense of "characterize" is spelled out by Def. 9.3: states characterizable in terms of the π-particle notion are states that define probability distributions over simultaneous eigenstates of the variety numbers making up that particle notion.

Jack and Jill are special cases of a general possibility: the possibility of unitarily inequivalent, yet nevertheless physical, representations, evoked in the second premise. Such multiplicity, coupled with premise 3, is the bludgeon the restated argument uses against fundamental particle interpretations.

David Baker (2009) has argued persuasively that if this bludgeon beats the plausibility out of particle interpretations, a similar bludgeon does the same for field interpretations. The argument targets field interpretations which make use of wave-functional space, a Hilbert space whose elements describe superpositions of classical field configurations (see Huggett 2000 or Arageorgis 1995, §5.4.5, for elaboration). Just as a Fock space enables the particle interpreter to understand its states as superpositions of states with determinate particle contents, a wave-functional space enables the field interpreter to understand its states as superpositions of states with determinate field contents. Now, the field interpretation strategy succeeds only if a wave-functional space is a plausible setting for a QFT. Section 9.3.2 mentioned a compelling plausibility consideration: Fock space representations are plausible settings for QFT, and any Fock space representation is unitarily equivalent to a wave-functional representation. As Baker puts it, "where [field] interpretation succeeds, it does so precisely because wavefunctional space is equivalent... to the Fock space that represents states as particle configurations" (2009, 586). But if such a Fock space of states is overly limited in its representational resources—which is what the restated argument given above contends—so too is the wave-functional space underlying a field interpretation. If the restated argument does particle interpretations in, field interpretations succumb as well.

The next chapter considers two defenses a fundamental particle interpreter might mount to the restated argument. The first denies premise 3, and seeks to understand particle notions in such a way that their range of fundamental application outruns a single Fock space. In terms of the interpretive options outlined in Chapter 6, this defense of fundamental particle interpretation is a form of (tempered) Universalism. The second defense rejects premise 2. This averts the need to extend a favored particle notion beyond its environing Fock space, by arguing that all physically reasonable states are expressible as density operators on that Fock space. This makes the fundamental particle interpreter a natural ally of the Hilbert Space Conservative. The next chapter develops doubts that either alliance prevails.

10

Particles and the Void

A fundamental particle interpretation of a QFT understands every state of that QFT in terms of its particle contents. Chapter 9's challenge to fundamental particle interpretation conjoins three claims.

(i) In order to be fundamental, a particle notion must apply to all physically reasonable states.
(ii) The range of applicability of a particle notion is the folium of an irreducible Fock space representation.
(iii) The space of physically reasonable states is larger than the folium of an irreducible Fock space representation.

This chapter considers two classes of responses to the challenge. Section 10.1 covers denials of (ii), in the form of attempts to extend a particle notion beyond the folium of a single irreducible Fock space representation. It finds these attempts wanting, on the grounds that the extended particle notions in question lack key features that made a particle notion an attractive vehicle for interpreting QFT to begin with.

This leaves the fundamental particle interpreter with the option of resisting (iii). Section 10.2 develops resources underwriting such resistance. Writ large, the resources are the spacetime symmetries that could privilege the folium of an irreducible Fock space representation as *the* unique repository of physically reasonable states. Such a privileging strategy, however, threatens to intensify the challenge the Unruh effect poses to fundamental particle interpretations. Both the particles Jill sees and the particles Jack doesn't see are in the extension of particle notions constructed by appeal to timelike isometries. Yet their affiliated Fock space representations are unitarily inequivalent. The upsetting possibility for the fundamental particle interpreter is that not even the admirable symmetries of Minkowski spacetime serve to privilege a unitary equivalence class of irreducible Fock space representations. Section 10.3 examines this possibility to reveal its vulnerability to one more arrow in the fundamental particle interpreter's quiver: the Hadamard condition (see §1.2), which a state on the quasilocal algebra must satisfy in order to define an expectation value for the stress-energy tensor. In concert with other criteria of reasonableness, the Hadamard condition privileges Jack's Fock space over Jill's. Thus, I will contend, in highly symmetric spacetimes conscripted into the service of semi-classical quantum gravity, a fundamental particle interpretation of free bosonic field theory is available.

But is a fundamental particle interpretation desirable whenever it is available? Not always, §10.4 argues, for there are *prima facie* valuable applications of QFT that a fundamental particle interpretation, so pursued, would hobble. Section 10.4 develops this circumstance as a strike against the ideal of pristine interpretation: principles privileging an irreducible Fock space representation, applied in advance of and oblivious to particularities of theoretical application, issue interpretations more rigid than the empirical practice they aspire to explain.

10.1 Extended particle notions

I have argued that the real threat the Unruh effect poses to a fundamental particle interpretation lies in its suggestion that physically reasonable states outrun the folium of any single irreducible Fock space representation. One way to defuse this threat is to develop a particle notion whose range of applicability is not confined to such a folium. This section explores three extension strategies. Consider particle notions associated with disjoint (call them "primed" and "unprimed") representations. The first extension strategy, due to Clifton and Halvorson (2001b), identifies predictions about unprimed particle content afforded by states in the primed folium. The second strategy, suggested by Wald (1994), appeals to Fell's theorem to establish the presence in the primed folium of states with unprimed particle contents. The third strategy devises a "universalized" particle notion detached from any particular representation, and so applicable across a wide range of folia. This section reviews, and raises serious reservations about, each of these strategies.

10.1.1 An appeal to exfoliated predictions

Against the claim that disjoint particle notions are incommensurable, some have urged that there are, after all, ways to apply the unprimed particle notion to states in the primed folium. Fulling would say that each such state contains infinitely many unprimed particles (1989, 141). Clifton and Halvorson prefer the reformulation afforded by proposition 11 of their (2001b): each regular state outside the unprimed folium assigns every finite number of unprimed particles probability 0. And this, they suggest, tempers the troubling mismatch of disjoint particle notions.

The ground state of one Fock representation makes definite, if sometimes counterintuitive, predictions for the "differently complexified" degrees of freedom of other Fock representations... So long as a [primed] state prescribes a well-defined probability measure over the spectral projections of the unprimed theorist's total number operator—and all states in his *and* the folium of any primed theorist's representation *will* [that is the import of Prop. 11]—we fail to see the difficulty. (2001b, 458–9)[1]

The potential difficulty I see resides in the inability of the unprimed particle concept to *discriminate* between primed states. The exemplary sort of discrimination, whose features Def. 9.3 seeks to distill, is offered by a set of observables complete with respect to an irreducible Hilbert space representation whose normal states constitute the family of theoretically possible states. In this case, the space of possible pure states is spanned by a basis, each of whose elements is *identified* by the eigenvalues it assigns the operators in the complete set. Now, grant that the unprimed particle notion applies to primed states, and does so in the manner Clifton and Halvorson describe. It follows from Clifton and Halvorson's proposition 11 that *all* these states receive the same characterization in terms of the unprimed total number operator N: they assign every finite number of unprimed particles probability 0. The characterization of primed states N provides does not succeed in distinguishing distinct primed states from one another. *No* state in the primed folium can be *identified* by appeal to its total unprimed particle contents. Indeed, any pure state identifiable by the probability distribution it defines over eigenstates of N is arbitrarily far (as gauged by the algebraic transition probability $1 - \frac{1}{4}||\omega - \rho||$ between pure states ω and ρ) from every pure state in the primed folium.

Notice, incidentally, that Clifton and Halvorson's proposition 11 brings N into tension with §6.4.2's Harmony principle, which required of a physically significant observable that the expectation value it receives in a state be determined by the expectation values that state assigns observables—e.g. those representing the canonical commutation relations defining a theory—antecedently recognized as significant. Proposition 11 finesses the failure of states outside the unprimed folium to assign N an expectation value by redescribing the probabilities lurking in those states. But the redescribed probabilities pertinent to N commit the sin against which Harmony is vigilant: the sin of floating free of constraints imposed by the probabilities physically reasonable states assign obviously physically significant observables.

Still, N's discriminatory incapacity needn't call the fundamental status of the *set* of unprimed particle observables into question. After all, the fact that many distinct classical states have the same position coordinate does not cast aspersions on the fundamental status of the position observable. What distinguishes those states are the values they assign the (also) fundamental momentum observable.

The particle notion has conceptual anatomy just because it brings a number of magnitudes—not only the total number operator N but also the variety number operators N_k summed to obtain it—into inferential relationships. We should allow that this family of magnitudes *as a committee* circumscribes the unprimed particle notion.

---

[1] Potentially distracting shudder quotes omitted. The "differently complexified degrees of freedom" correspond to different choice of complex structure *J*. See §9.5.

The total unprimed number operator N is defined by an infinite sum of unprimed variety number operators N_k. Even if N is undefined on the primed representation, we can't rule out *ab initio* that some or all of the N_ks are well-defined there: an infinite sum can diverge even if each of its summands is finite. The discriminatory capacity qualifying a particle notion as fundamental would be preserved if unprimed *variety* number operators succeeded in distinguishing between a basis of distinct states N on its lonesome fails to distinguish.

Unfortunately for the fundamental particle interpreter, there are disjoint Fock space representations such that no observable in the committee making up one representation's particle notion is defined on states normal with respect to the other representation. Jack's and Jill's representations are an example. Letaw and Pfautsch (1981, 1495) establish this by using the expansion of positive Minkowski frequency modes in terms of positive Rinder frequency modes to express the Minkowski variety number operators as infinite sums of linear combinations of Rindler creation and annihilation operators. These infinite sums diverge. Undefined on the Rindler representation, the Minkowski variety number operators N_k are incapable of distinguishing between distinct states normal with respect to that representation. If states in the Rindler folium are physically significant, the Minkowski particle concept, even working through a committee of observables, cannot discharge its fundamental discriminatory task. There are physically significant pure states—states in the Rindler folium—that aren't superpositions of states identified by their Minkowski particle contents. If states in the Rindler folium are physically significant, the collection of Minkowski particle operators is not complete.

Even if *some* unprimed variety number operators are well-defined on the primed representation, extending the unprimed particle notion beyond its home folium comes at a price. The price is attenuating the conceptual ties that bind the different observables constituting a particle notion into the notion it is. In the primed folium, inferences from "For each k, the number of unprimed particles of variety k is 0" to "The total number of unprimed particles is 0" are blocked. And inferences from "There are finitely many particles of variety k ($\langle N_k \rangle < \infty$)" to "There are infinitely many particles of varieties other than k ($\sum_{i \neq k} \langle N_i \rangle = \infty$)" are enabled. These inferential novelties signal that when the unprimed particle notion is extendable beyond its home folium, its conceptual structure mutates. Having been stripped of localizability and genidentity, the particle observables associated with an irreducible Fock space representation at least exhibited enumerability in accordance with familiar rules. Stretching a particle notion beyond its home folium breaks even this analogy between the particles of QFT and their classical counterparts.

Finally, we can give a general argument that no "ex-foliated" π-particle concept is fundamental if states that aren't π-normal are physical. Suppose that $\{N_k^\pi\}$ is fundamental in the sense of Def. 9.3, and let ω be a state that's not π-normal. The "D-states" of the set $\{N_k^\pi\}$ are simultaneous eigenstates of its members; each is a π-normal vector state. Def. 9.3 requires every other possible state to define a probability distribution over these D-states. Thus ω must be expressed as a superposition of simultaenous eigenstates of $\{N_k^\pi\}$, or a mixture of states so expressible. This renders ω a π-normal state, contrary to our supposition. I conclude that exfoliated particle notions can't deliver the goods required of a fundamental particle notion.

10.1.2 An appeal to Fell's theorem

Chapter 9 complained about presentations of the Unruh effect which express the Minkowski vacuum state $|0\rangle_M$ as a superposition of Rindler modes. Owing to the disjointness of Rindler and Minkowski representations, this is something $|0\rangle_M$ cannot be, a circumstance that often passes without comment in expositions of the Unruh effect. Wald, however, explicitly recognizes the hiatus, and invokes Fell's theorem to bridge it (1994, 108). Because Wald's invocation reveals another sense in which a particle notion circumscribed by a Fock space representation could be taken to apply to states outside the folium of that representation, it merits closer examination.

$\omega_{M|R}$, the restriction of the Minkowski vacuum state to the Rindler wedge algebra $\mathfrak{A}(\mathcal{R})$, and ω_R, the Rindler vacuum state itself, are both states on the Rindler wedge algebra. It follows from Fell's theorem (see §6.4.1) that, given a set $\{A_i\}$ of observables in the Weyl algebra, and a set $\{\epsilon_i\}$ of experimental errors, there is a state ψ in ω_R's folium that reproduces $\omega_{M|R}$'s expectation value assignment to those observables within those experimental errors. Lying in the folium of the Rindler vacuum state, ψ *can* be characterized in terms of its Rindler particle content. This, Wald suggests, justifies expanding $\omega_{M|R}$ in terms of Rindler modes. Strictly false of $\omega_{M|R}$, the expansion is nevertheless strictly true of ψ, an approximator of $\omega_{M|R}$ guaranteed to exist by Fell's theorem.

I'm not persuaded by the appeal to Fell's theorem to warrant talk of the Rindler particle content of the Minkowski vacuum. For one thing, the equivalence claimed on the basis of Fell's theorem is equivalence with respect to purely algebraic observables. These exclude observables circumscribing particle notions: a total number operator N^π, if it exists at all for a representation π of the Weyl algebra $\mathfrak{A}$, resides in that representation's weak closure $\pi(\mathfrak{A})''$, not $\pi(\mathfrak{A})$ itself.[2] A symptom of this exclusion is that for *any* pure quasi-free state ω_μ on $\mathfrak{A}(R)$, Fell's theorem implies that for any set of algebraic observables and any set of experimental errors, there is a state in ω_μ's folium that reproduces $\omega_{M|R}$'s expectation value assignment to those observables within those experimental errors. Associated with ω_μ will be a Fock space $\mathfrak{F}(\mathcal{H}_\mu)$ circumscribing a particle concept; thus a state characterizable in terms of its π_μ particle content will approximate the restricted Minkowski vacuum state just as thoroughly as will a state characterizable in terms of its Rindler particle content. The invocation of weak equivalence to validate extending the *Rindler* particle notion into the Minkowski folium works for *every* particle notion—one suspects, by withholding significance from each of them.

---

[2] To be precise, N^π's spectral projections reside in $\pi(\mathfrak{A})''$. Because number operators are unbounded, this is the way to understand claims made here to the effect that a C^* algebra contains a number operator.

There is, I believe, a deeper difficulty with the appeal to weak equivalence to extend a particle notion beyond the folium that circumscribes it. This difficulty (which recalls a difficulty for apologetic Imperialism developed in §6.4.1) attends the empirical role played by state descriptions in terms of particle contents. Depict that role roughly as follows: by means of a pre-measurement procedure, we prepare a state, which we characterize in terms of its particle contents; we extract from this prepared state predictions about the outcome of subsequent measurement procedures, also described in terms of their particle contents. Adopting the framework of Fell's theorem, model both preparation and measurement procedures as fixing the expectation values of finitely many algebraic observables to finite precision: that is, let (A_i^P, ϵ_i^P) stand for the preparation procedure and let (A_i^M, ϵ_i^M) stand for the measurement procedure.

Now, the claim crucial to validating the application of a particle notion to states outside the folium of the Fock space representation circumscribing that notion is that for any algebraic state ψ there is a state ρ in ω_μ's folium (and so characterizable in terms of π_μ particle content) indistinguishable from ψ with respect to the preparation procedure (A_i^P, ϵ_i^P). The catch is that Fell's theorem offers absolutely no guarantee that this ρ will continue to ape ψ with respect to the measurement procedure (A_i^M, ϵ_i^M). So reverting via Fell's theorem to a characterization of an experimentally prepared state in terms of π_μ particle content does not underwrite the use of that state, *so characterized*, for future predictions. But then what's the point of having prepared it?

I take these considerations to strongly suggest that Fell's theorem does not underwrite the extension of a particle notion beyond the confines of its home folium.

10.1.3 Universalizing

The strategy I call "universalizing" responds to the challenge the Unruh effect poses for fundamental particle interpretation by contending that the particle notion fractures, into as many particle notions as there are Fock space representations hosting physically significant states.[3] For a suggestive analogy, suppose height were a property fundamental to some theory of the present condition of humankind. The committee of properties such as *height at eight years of age* and *height at nine years of age* would resolve the fundamental property *height* into properties with disjoint ranges, *without undermining the capacity of height [simpliciter] to characterize elements of humankind*. Just so, the fan of particle interpretation might contend, particle notions circumscribed by unitarily inequivalent Fock space representations, disjoint though their ranges might be, don't undermine the capacity of a particle notion [*simpliciter*] to characterize each and every state of physical interest.

I believe that the most promising way to pursue this strategy is as follows. Starting with the set of states giving rise to physically significant irreducible Fock space representations, pick a member from each unitary-equivalence class in this set to form

---

[3] I thank Arthur Fine for getting me to think about this, and Tracy Lupher and David Baker for suggesting more promising ways to implement the idea than occurred to me spontaneously.

the set S. Consider the representation $\tilde{\pi} := \oplus_{\omega \in S} \pi_\omega$ acting on the Hilbert space $\tilde{\mathcal{H}} = \oplus_{\omega \in S} \mathcal{H}_\omega$. Here's a helpful fact:

Fact 10.1 (Kadison and Ringrose 1997b, thm. 10.3.5, 738). Let $\{\pi_a\}$ be a set of representations of a C^* algebra $\mathfrak{A}$, and let $\pi := \oplus_a \pi_a$ be their direct sum. The set $\{\pi_a\}$ is pairwise disjoint iff:

$$\pi(\mathfrak{A})'' = \oplus_a \pi_a(\mathfrak{A})''.$$

The helpful fact implies that the von Neumann algebra affiliated with $\tilde{\pi}$ is the direct sum, over states in S, of their affiliated von Neumann algebras. When ω's GNS representation has an irreducible Fock space structure, the spectral resolution of the total number operator N^ω for that Fock space resides in the von Neumann algebra $\pi_\omega(\mathfrak{A})''$ affiliated with that GNS representation (but not in the C^* algebra $\pi_\omega(\mathfrak{A})$). So $\tilde{\pi}(\mathfrak{A})''$ will include:

$$\tilde{N} := \oplus_{\omega \in S} N^\omega \tag{10.1}$$

This $\tilde{N}$ is what the universal particle interpreter can regard as *the* number operator (i.e. the analog of *height simpliciter*), while relegating the parochial number operators N^ω to the status of a decomposing committee (i.e. the analogs of *height at n years of age*). She treats other components of the particle notion likewise: $\tilde{N}_k = \oplus_{\omega \in S} N_k^\omega$ and so on.

Chapter 9 alleged that a π-particle notion shirked its fundamental duty to characterize physically significant states that aren't π-normal. The notion circumscribed by $\tilde{N}$ and cognates is no shirker. Any state in the folium of some irreducible Fock space representation assigns expectation values to $\tilde{N}$ and cognates; different states in the same folium define different probability distributions over eigenvalues of those observables. No matter what subrepresentation $\tilde{N}$ is restricted to, it has the same functional dependence on that subrepresentation's canonical observables, and likewise for the universalized variety number operators $\tilde{N}_k^\omega$. That is, these operators promise to respect the Harmony principle. So it might seem that universalizing could hold the key to maintaining a fundamental particle interpretation in the face of the ballyhooed incommensurabilty of parochial particle notions.

I maintain that this key is miscast. The claim of $\tilde{N}$ (and cognates) to be fundamental is vitiated by at least two sins. First, universalized particle observables can be attributed identical collections of eigenvalues by states that are (by the lights of the universalizer) different states, thereby failing clause (ii) of Def. 9.3's criterion of fundamentality. Second, "particle detectors" as we construe them at present are dreadfully ill-suited to operationalize the universalized particle notion. The first sin exposes the universalized particle notion as incapable of articulating the content of a QFT to which it is applied—incapable, that is, of characterizing and distinguishing between worlds possible according to the theory. The second sin suggests that insofar as our grasp on what it is to be a particle is mediated by our practices of particle detection, the universalized particle notion eludes our grasp. (I can take the second sin to be a sin *without* making the claim criticized in §9.7, that a particle notion is *constituted* by the behavior of particle detectors.) These sins will turn out to be related.

To see that a particle concept contoured by the operators $\tilde{N}$, $\tilde{N}_k$, et al. won't distinguish between distinct states, suppose that ω, ω' are distinct elements of S and so disjoint states on $\mathfrak{A}$. We are supposing that the universal particle interpreter identifies the collection of physical observables with the algebra $\tilde{\pi}(\mathfrak{A})''$ that contains the operators implementing her universalized particle notion. The universalizing particle interpreter will admit that ω, ω' are distinct, because to be different states on $\mathfrak{A}$, ω and ω' must differ on elements of $\tilde{\pi}(\mathfrak{A}) \subset \tilde{\pi}(\mathfrak{A})''$. It's consistent with the foregoing that ω be implemented by the $\mathcal{H}_u$ vector $|1_k\rangle_\omega \oplus 0 \oplus 0...$, and that ω' be implemented by the $\mathcal{H}_u$ vector $0 \oplus |1_k\rangle_{\omega'} \oplus 0....$ The catch is that *these vectors assign the same eigenvalue to each universalized particle operator*. By the lights of the particle interpreter taking the universalized particle notion to be fundamental, ω and ω' are identical—even though the states they implement are not only distinct but also disjoint! The universalized particle notion isn't fundamental because the operators constituting it fail to discriminate between states—including simultaneous eigenstates of the observables making up the particle notion—that are *by the lights of the universalizing particle interpreter* distinct.

Of course, if $\tilde{N}$ and its cognates were joined by allies in the form of *parochial* observables such as N^ω, this enhanced set would suffice to distinguish between distinct states. But notice that we're back to taking seriously the "shirkers"—the parochial observables—it was the innovation of the genuinely universalized particle notion to demote from fundamental status. In the toy theory that motivated the universalizing move, height *simpliciter* was the observable fundamental to a theory of humankind. It happened to admit decomposition into a committee of height-at-age-n observables with disjoint domains, but *those* observables weren't imagined to be fundamental; that is, there weren't tasks of characterization and discrimination that height *simpliciter* needed their aid to perform. The point of the toy theory was to illustrate that a fundamental observable could be susceptible to a decomposition in terms of observables with disjoint ranges. The suggestion that the shirkers $\tilde{N}^\omega$ (and cognates) decomposing the range of $\tilde{N}$ (and cognates) be promoted to fundamental observables breaks the analogy with the toy theory that motivated the universalizing move.

The second problem with taking $\tilde{N}$ and its cognates to be fundamental has to do with how to understand the empirical impact of these observables, and follows from the first problem. While, as §9.7 divulged, the theory of particle detectors has hardly matured to completion, its best-articulated portions predict radically different behaviors for (say) detectors accelerating uniformly in the Minkowski and the Rindler vacuum state. But these are states that assign the same expectation values to $\tilde{N}$ and cognates. So even if advocates of a universalized particle notion could contend with worries about the capacity of that notion to discriminate between distinct states, they'd owe us an account of how to operationalize the particle notion circumscribed by those observables.

None of the foregoing considerations deals the universalized particle notion a death blow. But they do attach significant costs to sticking with that notion. My own sense of fiscal-conceptual responsibility is better served by the policy announced in §10.2: the policy of *situational* particle interpretations, adopted only in settings of sufficient symmetry (or other compelling grounds for privilege) to confine the space of possible states to the folium of a single irreducible Fock space representation.

10.2 Spacetime matters

This section develops, on behalf of the fundamental particle interpreter, a response to the anti-particle argument of §9.9. Briefly, the response is that, once we learn how to individuate QFTs, we see that there are instances of free bosonic field theories for which there are grounds for limiting physically significant states to the folium of an irreducible Fock space representation. Standing on these grounds, the fundamental particle interpreter needn't be dissuaded from a particle interpretation of *those* QFTs.

10.2.1 Killing particles and μ-born particles

Sections 9.3's and 9.5's accounts of quantization procedures suggest that we may distinguish between two sorts of Fock space construction by appeal to their precedents. The first sort of construction, set in a stationary spacetime, descends from the timelike isometries of that spacetime, which it exploits to follow a frequency splitting procedure. The resulting Fock space I'll call a *Killing Fock space* (hence: *Killing vacuum*, *Killing particles*, and so on). The second sort, set in a non-stationary spacetime, or in a stationary one whose symmetries it ignores, descends from a real inner product μ, by which it supplements the symplectic structure of the classical theory to obtain what I'll call a μ-*born Fock space* (hence: μ-*born vacuum*, μ-*born particles*, and so on). (Of course, in a stationary spacetime, μ can be chosen in such a way that $\mathcal{H}_\mu$ is unitarily equivalent to the single particle structure obtained by following the frequency-splitting procedure. I stipulate that the Fock space over such a $\mathcal{H}_\mu$ is a Killing Fock space.) Fock spaces of either sort circumscribe particle notions. But the particle notions they circumscribe differ significantly. Here I detail how.

The principle differences can be put briefly. First, the Killing vacuum is invariant under the timelike isometries of its spacetime. The μ-born vacuum cannot be, simply because either its spacetime lacks such isometries or it fails to respect them. Second, states containing a single Killing particle admit a principled and pleasing ascription of energy. Killing particles are, in this sense, and like all good substances, bearers of properties, not just any properties, but nomically significant ones. Because the principle at issue invokes timelike isometries, there is no analogously pleasing and principled way to ascribe energy to single μ-born particle states. μ-born particles therefore stake no analogous claim to substantiality. Finally, in a spacetime with timelike isometries, the Killing vacuum state constructed by appeal to those isometries is the unique vacuum state exhibiting certain desirable features. In any globally hyperbolic spacetime, μ-born vacua are legion; in the absence of timelike isometries, no μ-born vacuum is obviously preferable to any other. In light of this complex of considerable differences, I claim that Killing particles are *substantial* particles and that μ-born particles are merely *formal* particles. The remainder of this section attempts to clarify and support this claim.

Consider a globally hyperbolic spacetime $(\mathcal{M}, g_{ab})$, in which Klein–Gordon theory is quantized by finding a representation of the Weyl relations over (S, Ω). Suppose that $(\mathcal{M}, g_{ab})$ is stationary. Then there exists a one (real) parameter family of timelike isometries $\iota_t : \mathcal{M} \to \mathcal{M}$. This family of isometries induces a one (real) parameter family of transformations $A(t) : S \to S$ via:

$$A(t)\varphi(x) = \varphi(\iota_t[x]) \qquad (10.2)$$

for all $\varphi \in S$, $x \in \mathcal{M}$, and $t \in \mathbb{R}$. The symplectic structure Ω is invariant under $A(t)$, which exhibits the group property $A(t)A(s) = A(t+s)$.

A quantization of the classical theory furnishes a map from the classical solution space S to unitary operators $W(\varphi)$ satisfying the Weyl relations. These $W(\varphi)$ are generators of the Weyl algebra $\mathfrak{A}(\mathcal{M})$. Thus we can use the group $A(t)$ of symplectic tranformations on S to induce a one (real) parameter family $\alpha_\iota(t)$ of automorphisms of $\mathfrak{A}(\mathcal{M})$. Simply lift $A(t)$'s action on the classical solution space to $\alpha_\iota(t)$'s action on the algebra's generators:

$$\alpha_\iota(t)[W(\varphi)] = W(A(t)[\varphi]) \qquad (10.3)$$

for all $\varphi \in S$ and $t \in \mathbb{R}$.

Consider a candidate vacuum state in the form of pure quasi-free algebraic state constructed via (9.25) from a real inner product saturating (9.23). Call this state ω_μ. By means of its natural Fock space structure, the GNS representation of ω_μ, which realizes ω_μ as its cyclic vacuum state, will intimate a particle notion of some sort. Now suppose that the spacetime on which ω_μ serves as a state has timelike isometries ι_t. It is an axiom of axiomatic approaches to QFT in Minkowski spacetime that ω_μ should respect that spacetime's isometries, by being invariant under the action of the algebraic automorphism $\alpha_\iota(t)$ implementing them. Generalizations of the Minkowski axioms generalize the requirement: a QFT set in a spacetime with isometries should have a vacuum state invariant under those isometries. Such isometry-invariance requires:

$$\omega_\mu[\alpha_\iota(t)B] = \omega_\mu(B) \qquad (10.4)$$

for all $B \in \mathfrak{A}(\mathcal{M})$ and $t \in \mathbb{R}$. Insofar as the invariance demanded by (10.4) is desirable *because* $\alpha_\iota(t)$ implements timelike isometries, the demand makes sense only in stationary spacetimes.

Where $\mathcal{H}_\mu$ is the single-particle Hilbert space constructed from the real inner product μ, the invariance demanded by (10.4) implies that there act on $\mathcal{H}_\mu$ a unitary group $U(t)$ "intertwined" with the symplectic group $A(t)$ by a map $K : S \to \mathcal{H}_\mu$ implicated in the construction of the single-particle Hilbert space $\mathcal{H}_\mu$ from the symplectic space

(S, Ω) of classical solutions and a real inner product μ (see Kay and Wald 1991).[4] Kay and Wald show that, if a quasi-free state ω_μ on the Weyl algebra is invariant under an automorphism group $\alpha(t)$ arising from a group $A(t)$ of symplectic transformations, then ω_μ has a unique (up to unitary equivalence) *one-particle structure* $(K, \mathcal{H}_\mu, U(t))$, where $U(t)$ satisfies the intertwining condition of note 4. (Notice that a one-particle structure is a single-particle Hilbert space with a unitary dynamics supplied.) The force of the uniqueness claim is that no triple unitarily inequivalent to $(K', \mathcal{H}'_\mu, U'(t))$ realizes the invariance of ω_μ under $\alpha(t)$. From $U(t)$ on $\mathcal{H}_\mu$ is built a unitary operator $\tilde{U}(t)$ on the symmetric Fock space $\mathfrak{F}(\mathcal{H}_\mu)$ over $\mathcal{H}_\mu$.

A brief lexicon prefaces the next results. Consider a quasi-free state ω_μ invariant under $\alpha(t)$. If the cyclic vector $|\xi\rangle$ of ω_μ's GNS representation is invariant under the (unique) unitary group $\tilde{U}(t)$ implementing $\alpha(t)$, which unitary group is, moreover, strongly continuous, then ω_μ is a *ground state* for $\alpha(t)$ (see the Vacuum axiom of Chapter 5). The strong continuity of the group $\tilde{U}(t)$ implies that it has a positive generator H: $\tilde{U}(t) = \exp(-iHt)$. This H is the *quantum field Hamiltonian*. Thus ground states enable the recovery of standard quantum dynamics, unitarily implemented and generated by a positive operator. Such recovery is noteworthy because QFT abounds with states and automorphisms that are in some sense dynamical but that cannot be implemented unitarily in the GNS representation of those states (see Arageorgis, Earman, and Ruetsche 2002 for examples).

In a ground state, the unitary group $U(t)$ implementing $\alpha(t)$ on the single-particle Hilbert space $\mathcal{H}_\mu$ also has a positive "one-particle" generator h: $U(t) = \exp(-iht)$. h can be understood as the energy operator for single-particle states. Iff h has no "zero modes"—no eigenvectors with eigenvalue 0—the ground state ω_μ is *regular*. In regular ground state, energy is strictly positive.

Kay and Wald show that for any automorphism group $\alpha(t)$ generated (via e.g. (10.2)) from symplectic symmetries, there can exist at most one regular quasi-free ground state on the Weyl algebra arising from a linear scalar field theory (1991, prop. 3.2). A corollary is that such a state, if it exists, is pure. Examples of such states are Killing vacuums, the frequency-splitting vacuums constructable for stationary spacetimes.

In a stationary spacetime, it is axiomatic to demand the vacuum state of one's QFT construction be isometry-invariant. If one also demands that this vacuum be a quasi-free regular ground state—i.e. a state on which dynamics are unitarily implementable and generated by an energy operator which is strictly positive: a state, in short, whose energetics are "nice"—then this state is unique, if it exists at all. This state, if it exists at all, has an irreducible GNS representation whose Fock space structure circumscribes a particle notion.

The GNS representation of a quasi-free regular ground state exhibits another desirable feature. On its one-particle structure, the one-particle generator h of the quantum

---

[4] K intertwines $A(t)$ and $U(t)$ because for all $\varphi \in S$ and $t \in \mathbb{R}$, $U(t)(K\varphi) = K(A(t)\varphi)$.

field Hamiltonian supplies a similarly privileged notion of energy. Kay and Wald (1991; see also Ashtekar and Magnon 1975, 382) show that, where $G : S \to \mathbb{R}$ gives the *classical* energy associated with a solution,

$$G(\varphi) = \langle K\varphi, hK\varphi \rangle = \langle h \rangle_{K\varphi} \qquad (10.5)$$

This means that the expectation value of the quantum energy h in the single-particle state $|K\varphi\rangle$ coincides with the classical energy associated with φ. The particles of this Fock space—Killing particles—carry energy. Bearers of nomically significant properties, Killing particles are substantial particles. In virtue of the uniqueness of regular ground states in stationary spacetimes, they are the *unique* ι_t isometry-respecting energetically "nice" particles in whose terms we might parse the content of the quantum field.

Compare μ-born particles. A μ-born vacuum ω_μ in a non-stationary spacetime clearly can't respect the (non-existent) timelike isometries of that spacetime. Thus there is no one real parameter family of automorphisms of the Weyl algebra under the action of which it is reasonable to demand ω_μ be invariant. In the absence of the invariance demand, there is no way to identify a unique one-particle structure (K, $\mathcal{H}_\mu$, $U(t)$) satisfying the demand, and so no way to privilege, as the generator of $U(t)$, a one-particle Hamiltonian h in whose terms the energetics of the μ-born particles can be understood. Without a physically principled way to ascribe μ-born particles energy, there are no grounds to claim that such particles bear physical (as opposed to merely formal) properties. μ-born particles are not substantial particles. Moreover, again because of the absence of a gripping $\alpha(t)$ invariance requirement, infinitely many μ-born particle notions are available in general spacetimes, and no unitary equivalence class of them has any claim to priority.

The Kay and Wald result suggests a strategy for pursuing a fundamental particle interpretation. The strategy is to adopt the criterion of physical reasonableness:

[*Energy*] If a spacetime $\mathcal{M}$ admits timelike isometries, a state ω on the quasilocal algebra $\mathfrak{A}(\mathcal{M})$ for that spacetime is physically reasonable only if ω is normal with respect to some representation π of $\mathfrak{A}(\mathcal{M})$ that "respects those isometries."

The scare-quoted notion of "respecting isometries" is left vague. But suppose that such respect requires the presence in a representation of a regular quasi-free ground state. Then the Kay and Wald result implies that if $\mathcal{M}$ is stationary, physically reasonable states on $\mathfrak{A}(\mathcal{M})$, if they exist at all, lie in the folium of an irreducible Fock space representation of $\mathfrak{A}(\mathcal{M})$. The Killing particle notion circumscribed by this Fock space is thus complete with respect to the folium of physical states. Doing physics in terms set by that Killing particle notion, one is underwriting a fundamental particle interpretation.

10.2.2 Adulteration?

The states deemed physical by *Energy* lie in the folium of an irreducible Fock space representation π of $\mathfrak{A}(\mathcal{M})$, and the magnitudes constituting a particle notion lie in the von Neumann algebra $\pi(\mathfrak{A}(\mathcal{M}))''$ affiliated with π. That von Neumann algebra is just $\mathfrak{B}(\mathcal{H}_\pi)$; its normal states are just density operator states on $\mathcal{H}_\pi$. Thus a fundamental particle interpretation defended on the present grounds is a species of Hilbert Space Conservatism. But is it, this section asks, a Hilbert Space Conservatism adulterated? One suspected adulteration takes the form of making the *content* of the QFT depend, not on nomically endogenous matters such as its constitutive CCRs, but on the "exogenous" and contingent matter of the details of the spacetime in which it happens to be set. It's the symmetries of that spacetime, the worry goes, not the kinematics or dynamics of the QFT, that selects the folium of states the particle interpreter/Hilbert Space Conservative deems physical.

But *this* adulteration worry is misplaced. Mass m free bosonic field theory on curved spacetime is not a single theory, demanding a uniform interpretation, but many theories. Recall the point of departure for algebraic approaches to QFT: a QFT is identified in terms of an algebra $\mathfrak{A}$ *and* a map between subalgebras of $\mathfrak{A}$ and subregions of the spacetime $(\mathcal{M}, g_{ab})$ housing the QFT. Thus the spacetime setting $(\mathcal{M}, g_{ab})$ is part and parcel of the nomic structure of a QFT, and even the Algebraic Imperialist, who has more permissive criteria of theoretical identity than the Hilbert Space Conservative, would distinguish between free bosonic field theory set in one spacetime and free bosonic field theory set in another.

Given these criteria of individuation for mass m bosonic field theories, the particle interpreter cannot be expected to announce, in advance of learning the details of the spacetime in which such a theory is set, *the* particle notion in whose terms to interpret theory. There is no more reason to expect there to be such a particle notion than there is to expect there to be a single particle notion adequate to free bosonic field theories for different values of mass. The ideal of pristine interpretation does not require the interpreter to proceed in ignorance of the theory she is interpreting. It merely requires her to frame the content of that theory without appeal to accidents incidental to the nomic core of the theory.

Plausibly, *Energy* does express suitably general considerations. A case for *Energy* might be reconstructed from the following elements: (i) concern with spatiotemporal symmetries and their implications for objectivity and individuation, which have been the stock-in-trade of metaphysics for centuries; (ii) the similarly general desire to avert antinomies arising from the attainability of negative energies. In league with the Kay and Wald result, these concerns identify at most one representation of $\mathfrak{A}(\mathcal{M})$ per family of timelike isometries—one, moreover, with an irreducible Fock space structure. An interpretation in terms of this representation, having been identified by these considerations, promises to be pristine.

It's nevertheless worth noticing that the Killing route to a fundamental particle interpretation only works for certain spacetime settings. Supposing that the Killing route to a fundamental particle interpretation is also the best grounds for Hilbert Space Conservatism about a QFT, this suggests that Hilbert Space Conservatism isn't viable as an interpretive strategy for all QFTs, much less all theories of QM_∞. It's viable as an interpretive strategy only for certain special instances of QFT. Only in those instances do strategies for identifying worlds possible according to the QFT culminate in a set of worlds that can be understood in terms of their particle contents.

Although it might be expected that different theories demand different interpretive strategies, it could be surprising that mass m free bosonic field theories set in different spacetime settings are relevantly different enough to issue such a demand—that (with apologies to Lucretius) whether or not there are particles depends on what the void is shaped like. It's a surprise that casts doubt on the prospects for success of certain extremist pursuits of the ideal of pristine interpretation—for instance, those antecedently committed to a particle ontology, no matter what the spacetime setting.

My provisional conclusion is that the Hilbert Space Conservative can, without fear of adulteration, travel the Killing route to a fundamental particle interpretation, *when that route is available*. The next section concerns a bump in that route.

10.3 Matter matters

To recap the story thus far: a fundamental particle interpretation of a given QFT identifies states possible according to that QFT with those in the folium of an irreducible Fock space representation. That privileged representation circumscribes a particle notion in whose terms each possible state may be characterized. Such a characterization is a necessary component of a particle interpretation of that QFT. A spacetime with a family of timelike isometries furnishes grounds for privileging a Fock space representation of the Klein–Gordon field, grounds only Killing particles can stand on. If there is a Fock space representation whose vacuum is, with respect to that family of isometries, a regular quasi-free ground state, it is the unique such Fock space representation. Reasons to prefer representations with regular quasi-free ground states are reasons to privilege that Fock space representation.

But it may not be the only Fock space representation so privileged. If the spacetime harbors *other* families of timelike isometries, those isometries could breed Killing particle notions rival to the first. If there are many families of timelike isometries, there could be many Fock spaces privileged with respect to them, and no guarantee that those Fock spaces are unitarily equivalent.

Interpreters of QFT should take notice of Jack and Jill, not because Jill sees particles in the Minkowski vacuum state and Jack doesn't, but because when each follows an isometry-guided heuristic for quantizing the Klein–Gordon field, each obtains a representation whose vacuum is a regular, quasi-free ground state and . . . *those representations are disjoint*. Jill's spacetime is a subset of Jack's. If physical states on Jill's spacetime lie in both Jack's folium and Jill's, no single particle notion suffices for their articulation.

The incommensurability of Jack's and Jill's Killing particle notions precludes a particle interpretation of the free bosonic field theory on Minkowski spacetime only if their quantizations have equal claim to be physical. This section will announce principles for distinguishing invidiously between Jack's representation and Jill's—principles which favor Jack's. Then it will ask whether invoking those principles adulterates the interpretation they help identify.

10.3.1 For Jack

Jill's quantization has the "nice" energetics characteristic of the representation of a regular, quasi-free ground state. However, it also has a certain blemish. The eigenvalues of her one-particle Hamiltonian h are not bounded away from 0. That is to say, Jill's h lacks a mass gap. Jill's quantization is an *infrared* model in which states of finite energy can contain infinitely many quanta (Fulling 1972, 275; Arageorgis 1995, 304).

That shows Jill's quantization is weird. To castigate it as physically unacceptable, the fundamental particle interpreter can bring another consideration on board. Suppose one wanted to calculate the expectation value of the stress-energy tensor T_{ab} in a quantum field theoretic state—for instance, because one was pursuing semiclassical quantum gravity, which calculates the back reaction on the spacetime metric wrought by quantized matter fields by replacing the stress-energy tensor T_{ab} in Einstein's Field Equations with its quantum expectation value $\langle T_{ab} \rangle$ for those fields. One would immediately encounter the impediment that the Weyl algebra for the Klein–Gordon field does not in general contain an element corresponding to the stress-energy tensor. So an algebraic state ω does not automatically determine an expectation value $\langle T_{ab} \rangle_\omega$. If one tries to add T_{ab} to the catalog of quantum observables, by calculating the expectation value of:

$$T_{ab} := \nabla_a \phi \nabla_b \phi - \frac{1}{2} g_{ab} (\nabla_c \phi \nabla^c \phi + m^2 \phi^2) \tag{10.6}$$

for each state, one encounters another impediment. Impressionistically put, the impediment is that no QFT state can assign the last term on the r.h.s. a well-defined expectation value, because the product $\hat{\phi}(x)^2$ of a quantum field with itself at a spacetime point is singular. However, for *Hadamard states*—states for which $\langle \hat{\phi}(x)\hat{\phi}(x') \rangle_\omega$ possesses singularity structure of Hadamard form as the spacetime points x and x' approach one another—there is a "point-splitting" procedure for defining the renormalized expectation value $\langle T_{ab} \rangle_\omega$ anyway.[5] Hadamard states are those for which a stress-energy observable is well-defined. This suggests a criterion of physical reasonableness for states of the quantum field:

[*Hadamard*] A state ω on a quasilocal algebra $\mathfrak{A}(\mathcal{M})$ is physically reasonable only if it is Hadamard.

Wald (1994) proposes such a criterion, as do Kay and Wald (1991).

---

[5] To first approximation, the procedure consists in defining the expectation value of the stress-energy tensor in a state ω as the difference between $\langle T_{ab} \rangle_\omega$ and $\langle T_{ab} \rangle_{\omega_0}$ for some reference state ω_0 (where (10.6) is used to calculate each). If $\langle \hat{\phi}(x)\hat{\phi}(x') \rangle_\omega$'s singularity structure mirrors $\langle \hat{\phi}(x)\hat{\phi}(x') \rangle_{\omega_0}$'s as x and x' approach one another, the difference is well-defined. (Higher order approximation: what plays the role of the "reference state" isn't really a state on the entire quasi local algebra, but a collection of local bi-distributions. See Wald 1994, ch. 4.6, for elaboration.)

To bring this to bear on Jill's quantization, we need to observe that the spacetime whose timelike isometries Jill's quantization respects, Rindler spacetime, is *extendable*, that is, it's isometrically embeddable as a proper subset of another spacetime, Minkowski spacetime. The Isotony axiom requires that the Rindler wedge algebra $\mathfrak{A}(\mathcal{R})$, the algebra represented in Jill's quantization, be a proper subalgebra of the Minkowski algebra $\mathfrak{A}(\mathcal{M})$. Now, where $\mathcal{A}$ and $\mathcal{B}$ are regions of a spacetime in which a QFT is set, *suppose*:

[*Extendability*] If $\mathcal{A} \subset \mathcal{B}$, a state on $\mathfrak{A}(\mathcal{A})$ is physically reasonable only if it is extendable to a state on $\mathfrak{A}(\mathcal{B})$ that is physically reasonable.

I'm articulating the supposition to give the fundamental particle interpreter a fighting chance, but can imagine it defended by some sort of appeal to a principle of plenitude to the effect that a physical possibility not be (spatiotemporally) curtailed without reason.

This supposition enables the fundamental particle interpreter to administer the *coup de grâce*. Jill's vacuum state ω_R cannot be extended to a Hadamard state on Minkowski spacetime. Its behavior on the edges of the Rindler wedge are to blame: $\langle T_{ab}\rangle_{\omega_R}$ diverges there. Supposing that states must be Hadamard to be physical, it follows that Jill's vacuum state cannot be extended to a physically reasonable state on $\mathfrak{A}(\mathcal{M})$. Given the criteria announced in the previous paragraphs, this exposes Jill's quantization as unphysical. Jack's quantization, by contrast, passes these tests for physical reasonableness with flying colors. The Minkowski vacuum state ω_M is a Hadamard state *par excellence*; it is moreover a state on an algebra $\mathfrak{A}(\mathcal{M})$ for an inextendable spacetime. There is, after all, something to privilege Jack's quantization over Jill's.

Perhaps Jack is a special case. It could be that other particle notions, apparently privileged by their respect for a global family of timelike isometries, fail to be similarly robust. If the Fock space representations privileged by different families of *global* timelike isometries turned out to be unitarily inequivalent, no single folium could be isolated by its respect for those isometries. Consider a spacetime with global timelike Killing fields V^a and V'^a. Let states ω and ω' be the stationary vacua associated with these Killing fields. Chmielowski (1994) shows that if V^a and V'^a commute (as e.g. the Killing vector fields associated with the time translation isometries of different inertial observers in Minkowski spacetime do), then ω and ω' coincide.[6] Their affiliated particle notions are identical, and the basis for a particle interpretation of the theory—that it has a privileged Fock space representation—remains. I know of no example of a (physically reasonable) spacetime that has two non-commuting global timelike Killing fields.[7] In the absence of such an example, hope persists for a

---

[6] This is under the technical assumption (stated in note 10 of Chapter 9) necessary for the generalization of the frequency-splitting heuristic to stationary spacetimes.

[7] Chmielowski (1994) conjectures that, even if there were such an example, the frequency-splitting vacua associated with Killing fields would coincide.

particle interpretation of the quantized Klein–Gordon field when it is set in such a spacetime.

10.3.2 More adulteration?

The fundamental particle interpreter I've imagined exploits the timelike isometries of the spacetime in which his QFT is set to dismiss states disjoint from quasi-free regular ground states with respect to those isometries. If this maneuver still leaves disjoint irreducible Fock space representations and their associated incommensurable particle notions—e.g. Jack's and Jill's—standing, the fundamental particle interpreter appeals to *Extendability* and *Hadamard* to eliminate all but one of these. The question before us is whether such appeal adulterates the interpretation it determines and motivates.

By contrast to the appeal to *Extendability* or *Energy*, the appeal to the Hadamard criterion rests not on metaphysical considerations of long-standing, but on the details of a piece of aspirational physics. We set out to interpret a QFT on a curved spacetime $(\mathcal{M}, g_{ab})$, a QFT associated with the Weyl algebra $\mathfrak{A}(\mathcal{M})$. In narrowing the space of its physically reasonable states to the folium of a Hadamard quasi-free regular ground state on $\mathfrak{A}(\mathcal{M})$, the particle interpreter of the theory relies not only on mathematical and "metaphysical" considerations that are supposed to apply to all physically possible words, but also on apparently exogenous contingencies of various sorts: the theory of gravity reigning at present in this part of our universe, and the present schematic of our hopes for extending it. I think that the reliance is healthy. But it is also adulterating. I'll close this section by trying to explain both claims.

One way to cleanse a fundamental particle interpretation of this alleged adulteration is to articulate it without appeal to *Hadamard*, for instance, by adopting the pristine resolution:

[*Big Picture*] Criteria of physical reasonableness (such as *Energy* or *Hadamard*) apply only to states ω on algebras $\mathfrak{A}(\mathcal{M})$ such that $\mathcal{M}$ is *inextendable*.

Big Picture interpretations take "maximal" theories to be the basic units of significance. Thus a *Big Picture* interpreter plying the *Energy* criterion won't apply it *directly* to the Rindler vacuum state, a state on the algebra associated with an extendible Rindler wedge. A *Big Picture* interpreter will apply the *Energy* criterion to states on the quasilocal algebra associated with all of Minkowski spacetime—and discover that only states in the folium of the Minkowski representation pass the criterion. Footnote 7 mentions a conjecture that no physically reasonable globally hyperbolic spacetimes admit non-commuting families of global timelike isometries. If that's true, it follows from the Kay and Wald result that a *Big Picture* interpreter, plying the *Energy* criterion in a spacetime $\mathcal{M}$ with timelike isometries, will recognize only states in the folium of an irreducible Fock space representation of $\mathfrak{A}(\mathcal{M})$ as physically reasonable, if she recognizes any physically reasonable states at all. When $\mathcal{M}$ is Minkowski spacetime, the physical folium is the folium of the standard Minkowski representation. Rindler problem solved.

While recognizing how pristine a particle interpretation/Hilbert Space Conservatism descended from the Big Picture resolution and the Energy criterion is, I want to suggest that it's unhealthy. It's unhealthy because its insistence that the "maximal" theories are the basic units of physical significance is methodologically hobbling. *Without further review*, Big Picture denies *direct* physical significance to quantizations set in Schwartzschild spacetime (an extendible subset of Kruskal spacetime), as well as in the region of Schwartzschild spacetime exterior to the event horizon of the Schwartzschild black hole, to take some examples. By *Big Picture*'s lights, those quantizations must be exhibited as embeddable in "maximal" theories to earn their stripes. But practicing physicists break the Big Picture resolution as a matter of course. For Fulling, "the suggestion that a trustworthy field quantization can only be performed on 'the whole space' is especially frightening" (1972, 292).

Fulling does not dwell upon what elicits this fright, but it is not hard to guess. Non-maximal spacetimes may enjoy features crucial to quantization techniques—for instance, global hyperbolicity or timelike Killing fields—missing from the maximal spacetimes in which they're embeddable. Applying the techniques in question to non-maximal spacetimes can guide and illuminate physics. (Indeed, the examples of quantizations set in non-maximal spacetimes mentioned above help articulate the phenomena of Hawking radiation and black hole entropy.) Given the potential of non-maximal quantizations to be revealing of the future direction, or local shape, of physics, I'm inclined to agree with Fulling that it would be frightening to declare them unphysical until proven physical by embedding them in maximal quantizations.

The route to a fundamental particle interpretation mediated by the Hadamard criterion is healthier because an interpretation so mediated is responsive to the explanatory and developmental needs of physics. But the responsiveness which renders an interpretive strategy healthy also predisposes it to adulteration. A mark of the Hadamard strategy's adulteration is the existence of criteria *motivated in the same way* as *Hadamard* is motivated, that identify *different* sets of possible states for the same theory. A particular explanatory/developmental agenda—the agenda of conscripting QFT on curved spacetime into the services of semi-classical quantum gravity—motivates *Hadamard*. Other explanatory/developmental agendas motivate criteria in tension with *Hadamard*. For instance, an agenda for those quantizing Klein–Gordon theory in Minkowski spacetime might be to explain the response of a uniformly accelerating Unruh box detector in terms of *particles* it is detecting. A criterion for admissible states underwriting this agenda would, in bald defiance of *Hadamard*, declare states in Jill's quantization physical.

Even an agenda consonant with the one motivating *Hadamard* can militate for the physical significance of states violating that criterion. An example comes from a QFT set in an exotic spacetime. Candelas (1980) makes a case that the Unruh vacuum is the state on extended Schwartzschild spacetime that makes the best sense of the physics exterior to an evaporating black hole as we understand it. It's the state that predicts of observers in the distant future of such a black hole that they see a thermal flux of

Hawking radiation. That is, the Unruh vacuum is the state that saves the "phenomena" of Hawking radiation and black hole thermodynamics, which (even though they're not really phenomena, in the sense of residents of actual experimental records, at all) are among the closest things to an empirical constraint quantum gravity faces. And the Unruh vacuum, physically significant in the sense of significant to physics in the ways just described, is not a Hadamard state. In the Unruh vacuum, $\langle T_{ab} \rangle$ is singular on the past event horizon. As well as *Hadamard* supports some agendas for seeking a quantum theory of gravity, it frustrates others.

A suspicion may be dawning that the price of responsiveness is adulteration: that interpretations that make the best sense of physics as physics draw on circumstantial and situational resources to equip theories with content. This section prompted the suspicion in the context of criteria of physical reasonableness motivated by some explanatory agendas but violated by states promoting others (or even different pursuits of the same). The next section fans the suspicion by describing circumstances in which there is reason to override criteria of physical reasonableness motivated by the axioms of QFT themselves.

10.4 Coherent states

How classical can a quantum state be? Consider the standard quantization for a point particle moving in one linear dimension. The wave function $\psi(x, t)$ of this system, an element of $L^2(\mathbb{R})$, will determine position and momentum uncertainties $\Delta \hat{q}$ and $\Delta \hat{p}$, as well as expectation values $\langle \hat{q} \rangle$ and $\langle \hat{p} \rangle$ for those canonical observables. One way for $\psi(x, t)$ to behave as classically as possible is for $\Delta \hat{q} \Delta \hat{p}$ to be the minimum allowed by the Heisenberg uncertainty principle, with $\Delta \hat{q} = \Delta \hat{p}$, so that uncertainties are simultaneously minimized. Another way for $\psi(x, t)$ to behave as classically as possible is for its Schrödinger evolution to imply that over time the expectation values $\langle \hat{q} \rangle$ and $\langle \hat{p} \rangle$ reproduce a classical trajectory.

There are quantum states that behave this classically both ways at once. They're known as *coherent states*. Klauder and Sudarshan describe them: "Pictorially speaking, the state does not change its form in time but only its mean position and momentum, and these change in accord with the classical equations of motion" (1968, 107–8). Schrödinger introduced coherent states in 1926, in the service of articulating the correspondence principle. But a yen for the classical is not the only scientific desire these states satisfy. A laser is a source of electromagnetic waves that retain their shape as they propagate. Coherent states thus promise to capture the physics of lasers. Glauber (1963) gave voice to the promise. His 2005 Nobel prize recognizes the fundamental role coherent states have since played in quantum optics.

This section introduces coherent states, starting with the simple harmonic oscillator in one dimension and building up to the free boson field. It will transpire that the latter has coherent state representations disjoint from its standard Fock representation. While such coherent representations lack even the respect for Poincaré symmetries demanded by the covariance axiom (described in §5.2), they have physical applications. These circumstances bring favored principles for the interpretation of QFT into collision with the desideratum that an interpretation of a theory makes sense of its successes as physics.[8]

10.4.1 One degree of freedom

Let us revisit §2.3.4's account of the quantization of the simple harmonic oscillator, setting $\omega = m = \hbar = 1$ for convenience.[9] Hilbert space operators will wear hats in this subsection and the next, to make the definitions which follow easier to read. Where $\hat{p}$ and $\hat{q}$ satisfy the CCRs, the ladder operators $\hat{a} = \frac{1}{\sqrt{2}}(\hat{q} + i\hat{p})$ and $\hat{a}^\dagger = \frac{1}{\sqrt{2}}(\hat{q} - i\hat{p})$ facilitate a compact expression of the Hamiltonian: $\hat{H} = \hat{a}^\dagger \hat{a} + \frac{1}{2}$. Let $|0\rangle$ denote the ground state of this Hamiltonian, and let $|n\rangle$ denote its energy eigenstates (i.e. $\hat{H}|n\rangle = (n + \frac{1}{2})|n\rangle$).

Where z is a complex number, the "shift operator" $\hat{D}(z) = e^{z\hat{a}^\dagger - z^*\hat{a}}$ acts as follows on the ladder operators:

$$\hat{D}(z)^\dagger \hat{a} \hat{D}(z) = \hat{a} + z \qquad (10.7)$$

$$\hat{D}(z)^\dagger \hat{a}^\dagger \hat{D}(z) = \hat{a}^\dagger + z^* \qquad (10.8)$$

This definition of $\hat{D}(z)$ is hardly idle: $\hat{D}(z)$'s action on the ground state of the simple harmonic oscillator yields a ground state of the driven harmonic oscillator, whose Hamiltonian adds a potential term $f(t)\hat{q}$ to the free Hamiltonian given above. To see what these "shifts" in the ladder operators mean for the position and momentum observables, rewrite the complex number z in terms of real numbers P_0 and Q_0: $z = \frac{1}{\sqrt{2}}(Q_0 + iP_0)$. Then[10]:

$$\hat{D}(z)^\dagger \hat{p} \hat{D}(z) = \hat{p} + P_0 \qquad (10.9)$$

$$\hat{D}(z)^\dagger \hat{q} \hat{D}(z) = \hat{q} + Q_0 \qquad (10.10)$$

$\hat{D}(z)$ is an operator that induces a shift of P_0 and Q_0 in momentum and position respectively.

Now let us introduce the state $|z\rangle := \hat{D}(z)|0\rangle$, and investigate its behavior. Start with time evolution. It can be shown[11] that:

$$e^{-i\hat{H}t}|z\rangle = |ze^{-it}\rangle \qquad (10.11)$$

---

[8] I learned about coherent states from Baker (2009). See that essay for more on their foundational significance.

[9] The same convenience can be purchased by working with "dimensionless units"—see Gottfried and Yan (2003, 174), whose discussion I follow here, for an introduction.

[10] Applying the BCH theorem—see ibid., 182.

[11] Via a redefinition of the zero point energy; see ibid.

Re-expressed in terms of P_0 and Q_0, ze^{it} becomes:

$$ze^{it} = \frac{1}{\sqrt{2}}[(Q_0 \cos t + P_0 \sin t) + i(P_0 \cos t - Q_0 \sin t)] \tag{10.12}$$

Gottfried and Yan elaborate: "Hence at time t the [evolved state $|ze^{-it}\rangle$] is the undistorted ground state displaced to the position and momentum a classical oscillator would have at that time were it to have been set in motion at $t=0$ with the initial condition (Q_0, P_0)" (2003, 182). This is just the "as classical as possible" behavior distinguishing a coherent state. Thus we will call states $|z\rangle := \hat{D}(z)|0\rangle$ *coherent states*.

The coherent state $|z\rangle$ can be described in other ways. From (10.7):

$$\hat{a}\hat{D}(z) = \hat{D}(z)(\hat{a} + z) \tag{10.13}$$

Letting each side act on $|0\rangle$ and simplifying, we get:

$$\hat{a}|z\rangle = z|z\rangle \tag{10.14}$$

The coherent state $|z\rangle$ is an eigenstate of the annihilation operator $\hat{a}$.[12] $|z\rangle$ can be expanded in terms of energy eigenstates

$$|z\rangle = e^{-\frac{1}{2}|z|^2} \sum_{n=0}^{\infty} \frac{z^n}{\sqrt{n!}} |n\rangle \tag{10.15}$$

over which it thereby defines a Poisson distribution. From this, a particularly nice expression for the expectation value a coherent state assigns the number operator $\hat{N} = \hat{a}^\dagger \hat{a}$ follows:

$$\langle z|\hat{N}|z\rangle = |z|^2 \tag{10.16}$$

10.4.2 n degrees of freedom

The foregoing generalizes straightforwardly to n degrees of freedom. This time, $\mathbf{z}$ is an n-tuple of complex numbers z_k, each with real and imaginary components Q_k and P_k. The shift operator $\hat{D}(\mathbf{z})$ acts as follows on canonical observables:

$$\hat{D}^\dagger(\mathbf{z})\hat{p}_k\hat{D}(\mathbf{z}) = \hat{p}_k + P_k \qquad \hat{D}^\dagger(\mathbf{z})\hat{q}_k\hat{D}(\mathbf{z}) = \hat{q}_k + Q_k \tag{10.17}$$

$$\hat{D}^\dagger(\mathbf{z})\hat{a}_k\hat{D}(\mathbf{z}) = \hat{a}_k + z_k \qquad \hat{D}^\dagger(\mathbf{z})\hat{a}_k^\dagger\hat{D}(\mathbf{z}) = \hat{a}_k^\dagger + z_k^*$$

Clearly, the collection of operators $\hat{q}'_k := \hat{q}_k + Q_k$ and $\hat{p}'_k := \hat{p}_k + P_k$ satisfy the CCRs if the $\hat{q}_k$s and $\hat{p}_k$s do. The *coherent transformations* defined by (10.17) form the lowest rung in a ladder of transformations that map one representation of the CCRs to another. The nth rung of the ladder sets:

---

[12] This turns out to imply another respect in which coherent states are as classical as possible: they minimize the products of the uncertainties $\Delta \hat{p}$, $\Delta \hat{q}$. See Klauder and Sudarshan (1968, 108–9) for a simple argument.

$$q_k \mapsto \hat{q}_k \qquad (10.18)$$

$$\hat{p}_k \mapsto \hat{p}_k \sum_{k_1,\ldots,k_n} \lambda_{kk_1\ldots k_n} : \hat{q}_{k_1} \cdots \hat{q}_{k_n} :$$

where : : denotes normal ordering and the complex coefficients $\lambda_{kk_1\ldots k_n}$ are symmetric in their indices (see Florig and Summers 2000 for further discussion). Where $|0\rangle$ is the vector s.t. $a_k|0\rangle = 0$ for all k, the coherent state $|\mathbf{z}\rangle := \hat{D}(\mathbf{z})|0\rangle$ assigns the total number operator $\hat{N} = \sum_{k=1}^n \hat{a}^\dagger \hat{a}$ the expectation value:

$$\langle \mathbf{z}|\hat{N}|\mathbf{z}\rangle = \sum_{k=1}^n |z_k|^2 \qquad (10.19)$$

Letting $n \to \infty$, we can let $\mathbf{z}$ be an *infinite* sequence of complex numbers z_k; define the shift operator $\hat{D}(\mathbf{z})$ in terms of canonical observables providing a representation of the CCRs on the model of (10.17); and use the ground state $|0\rangle$ of that representation and $\hat{D}(\mathbf{z})$ to define the coherent state $|\mathbf{z}\rangle := \hat{D}(\mathbf{z})|0\rangle$. Here we encounter a snag. $|\mathbf{z}\rangle$ is a vector in the representation with which we started only if $\sum_{k=1}^n |z_k|^2 < \infty$. In this case, $|\mathbf{z}\rangle$ assigns the number operator affiliated with the representation an expectation value $\langle \mathbf{z}|\hat{N}|\mathbf{z}\rangle = \sum_{k=1}^n |z_k|^2 < \infty$ — the condition familiar from Chapter 9 for the state $|\mathbf{z}\rangle$ to lie in the folium proper to the number operator $\hat{N}$.

This should prepare the ground for the moral of the next section, which will be expressed in operator-algebraic terms: In QFT, coherent states of a field with a natural Fock vacuum state ω can lie outside ω's folium.

10.4.3 QFT

We will, as usual, work with the example of the free boson field. Let (S, Ω) be the symplectic vector space of solutions to the classical Klein-Gordon equation. A quantization of this classical field theory is a representation of the Weyl relations over (S, Ω): that is, a map W from S to unitary operators on a complex Hilbert space satisfying the Weyl relations (see §9.5).

We can take the n-dimensional equations (10.17) to describe a transformation between a *ground state representation* of the CCRs (where the likes of $\hat{a}_k$ and $\hat{a}_k^\dagger$ play the role of canonical variables) and *coherent representation* of the CCRs (where $\hat{a}'_k := \hat{a}_k + z_k$ and $\hat{a}'^\dagger_k := \hat{a}_k^\dagger + z_k^*$ play those roles). Because n is finite, the transformation is unitarily implementable, so we don't get much mileage out of the notion of a coherent representation: it's the same (up to unitary equivalence) as an ordinary ground state representation.

But in QFT, the distinction has bite. Focus on the massless Klein–Gordon field, appropriate for the description of photons. Consider a Poincaré-invariant Fock space representation of the Weyl relations over (S, Ω), in which field operators $\Phi(f)$ satisfy (9.20). On the model of equations (10.17), to obtain a *coherent representation*, we simply shift each $\Phi(f)$ by a real number. To each real linear function $\ell : S \to \mathbb{R}$ on S there corresponds the *coherent transformation*[13]

$$\Phi_\ell(f) \mapsto \Phi(f) + \ell(f)\mathbb{I} \qquad (10.20)$$

The field operators $\Phi_\ell(f)$ furnish a *coherent representation* of the Weyl relations over (S, Ω).

We can put the foregoing more abstractly. Let $\mathfrak{A}$ be the Weyl algebra obtained as the uniform closure of the unitaries $W(f)$ conspiring in a representation of the Weyl relations. $\mathfrak{A}$ is a representation-independent C^* algebra. Suppose that there is a privileged (say, by spacetime symmetries and the *Energy* principle) Fock space representation π_0 of $\mathfrak{A}$, the GNS representation of the Fock vacuum state ω_0 on that algebra. Then the following defines a coherent representation π_ℓ of $\mathfrak{A}$:

$$\pi_\ell(W(f)) = \pi_0(W(f))e^{i\ell(f)} \qquad (10.21)$$

π_ℓ is the GNS representation of the coherent state ω_ℓ on $\mathfrak{W}$. ω_ℓ is related to the Fock vacuum state by:

$$\omega_\ell(W(f)) = \omega_0(W(f))e^{i\ell(f)} \qquad (10.22)$$

From (10.20) and (10.21) it should be evident that in the Fock representation π_0, the field operators $\Phi(f)$ satisfy the CCRs (9.20), while in the coherent representation π_ℓ, the field operators $\Phi_\ell(f)$ satisfy them.

10.4.4 Some properties of coherent representations

We will of course be interested in whether the coherent representation π_ℓ is unitarily equivalent to the reference representation π_0 from which it is obtained. The rough answer is that π_ℓ and π_0 are unitarily equivalent iff ℓ is bounded.[14] This boundedness serves as well as the criterion for the unitary equivalence of the reference vacuum ω_0 and the coherent state ω_ℓ. When the boundedness condition fails, the number operator N affiliated with the reference representation is not well-defined on the folium of the coherent representation. Informally, $\omega_\ell(N) = \infty$.

A coherent state ω_ℓ not unitarily equivalent to ω_0 nevertheless shares with that reference vacuum the privilege of being a *pure, quasi-free* state (Florig and Summers 2000, 470). This has several consequences. First, unitarily inequivalent ω_0 and ω_ℓ are disjoint. Second, the GNS representation of a coherent state ω_ℓ disjoint from ω_0 is an irreducible Fock space representation, on which is defined a number operator N^{π_ℓ} whose expectation value diverges for states in ω_0's folium, and *mutatis mutandis*. Because in

---

[13] For the mass 0 Klein–Gordon field, these are symmetry transformations which need not be unitarily implementable in the privileged Fock space representation. Ch. 14 will return to this point.

[14] The smooth answer concerns a complex linear functional Λ_ℓ, determined by ℓ, on the single-particle Hilbert space of the reference representation. Where J is the complex structure on S associated with the reference representation, $\Lambda_\ell(f) := \ell(f) - i\ell(Jf)$. π_ℓ and π_0 are unitarily equivalent iff Λ_ℓ is bounded (Reents and Summers 1994, thm. 3.1 (184)).

the mass 0 case, the total Hamiltonian can be well-defined on ω_ℓ's folium even if the reference number operator N^{π_0} isn't (Strocchi 2008, 79), ω_ℓ still has a well-defined energy. The "physical" account of ω_ℓ is that it assigns N^{π_0} an infinite expectation value because it describes infinitely many photons. This is supposed to be OK because the energies of these photons are so small—so far into the "infrared"—that the total energy is finite anyway. Coherent representations π_ℓ unitarily inequivalent to the reference representation π_0 are known as *infrared* representations.

The behavior under symmetry transformations of "infrared" coherent states is provoking. Roepstorff (1970) shows that rotations cannot be implemented unitarily in the GNS representation of such a state. But rotation is a Poincaré symmetry, and the Covariance axiom of axiomatic QFT demands that its Poincaré symmetries be implemented unitarily. If we require physically reasonable representations to satisfy the axioms of axiomatic QFT, infrared representations are not physical. *Mutatis mutandis*, if ω_0 is the vacuum state of irreducible Fock space representation satisfying *Energy* for a QFT on a spacetime where that principle has bite, infrared coherent states ω_ℓ disjoint from ω_0 are declared unphysical by that principle.

Their evidence unphysicality notwithstanding, coherent representations can be useful. The next section documents how.

10.4.5 Coherence: what is it good for?

We've already observed that coherent representations play a central role in quantum optics. The coherent state representations Glauber pressed into the service of laser physics were typically *not* infrared representations. But a standard account of their amenability to such service makes covert reference to infrared representations.

Laser light is "coherent, in the sense of being monochromatic with a sharply-defined phase angle" (Sewell 1986, 189). Contrast the phase angle of a normal light beam, which is not sharply defined: correlations between the electric field describing a normal light beam at different spacetime points are wiped out by phase fluctuations. A standard account of the suitability of coherent states to represent laser light embarks from the observation that a coherent state $|z\rangle$ is an eigenstate of the annihilation operator: $\hat{a}|z\rangle = z|z\rangle$, where z is a complex number and we're working with one mode to keep life simple and hats (in this paragraph) to make things vivid. As a complex number, $z = |z|e^{i\theta}$. Heuristically, the annihilation operator admits a polar decomposition in terms of a "phase operator" $\hat{\theta}$: $\hat{a} = |\hat{a}|e^{i\hat{\theta}}$. It follows that the coherent state $|z\rangle$ is an eigenstate of the phase operator $\hat{\theta}$ with eigenvalue θ. Thus would coherent states explain the coherence of laser light.

This explanation has a hitch. A phase operator $\hat{\theta}$ is supposed to join a total number operator N to satisfy the canonical commutation relation[15]

$$[N, \hat{\theta}] = i \tag{10.23}$$

---

[15] See Teller (1995, 107–8) for an argument that a state of the electromagnetic field predicting a low dispersion in phase angle must be one that assigns a large uncertainty to the number of quanta.

Now $|z\rangle$ is in the folium of the standard Poincaré-invariant Fock representation π_0 and thus in the domain of the associated number operator N^{π_0}. The hitch is that no self-adjoint "phase operator" exists on the representation Hilbert space $\mathcal{H}_0$ such that (10.23) holds. The heuristic explanation of the suitability of non-infrared π_0-normal coherent states to laser physics has a gap where its phase operators should be.

Phase operators aren't chimerae, however. There is a sense in which they exist on infrared coherent representations (Provost et al. 1975). Thus, there is a sense in which the physical applicability of even π_0-normal coherent states is theoretically understood only by appeal to infrared coherent states. If it is a virtue of QFT for its explanation of phase coherence to rest on physically significant magnitudes, QFT exhibits that virtue only under interpretations extending physical significance to infrared coherent representations and observables parochial to them. It exhibits that virtue, then, only under interpretations violating *Energy*.

Infrared coherent representations also have a host of direct physical applications (see Provost et al. 1975; Klauder and Skagerstram 1985; Honegger and Rieckers 1990). They are thought to apply to the physics of condensates, and optical squeezing, as well as to describe experimental situations involving high photon densities. And infrared representations have been repeatedly prescribed in the treatment of scattering problems in QFT.

Perhaps the most significant is the so-called infrared problem of QED (see Peskin and Schroeder 1995, §§6.3–6.5): the problem that key calculations are divergent, owing to the contribution of low-frequency (infrared) terms. An "infrared cutoff"—basically, a limit of integration excluding contributions from frequencies below some reference frequency—is a stopgap solution. A more principled solution is mediated by infrared coherent states. Carried out in terms of these states, calculations of scattering amplitudes give finite results, *even when the infrared cutoff is removed* (see Morchio and Strocchi 1986 and references therein). The 1937 Bloch–Nordsieck model gives a taste of how this works.

In the model a quantum radiation field interacts with a classical source associated with a real scalar field $j(\vec{x},t)$. A QFT obtained by quantizing the source-free mass 0 Klein–Gordon equation for a vector valued field $\vec{A}(\vec{x},t)$:

$$\Box \vec{A}(\vec{x},t) = 0 \tag{10.24}$$

describes the quantum radiation, in the absence of interaction. The quantum theory has a natural Fock space representation, call it π, and let ω_0 be its vacuum state. A QFT obtained by quantizing the same equation, but with $j(\vec{x},t)$ playing the role of the source:

$$\Box \vec{A}(\vec{x},t) = j(\vec{x},t) \tag{10.25}$$

describes the interaction. This QFT also has a Fock space representation π_j. Its pure normal states are "scattering states": they'll be used to calculate probability amplitudes for events described by means of operators from the free representation, the representation supposedly salient to the distant aftermath of the interaction.

Now, the creation and annihilation operators of the representations are related to one another by a non-unitarily implementable coherent transformation determined by the classical field. Taking π to be the reference Fock representation, π_j is an infrared coherent representation, and a π_j-normal state ψ_j—a scattering state—is an infrared coherent state. ψ_j therefore assigns the number operator N^π an infinite expectation value. ψ_j, however, manages to assign the free Hamiltonian a finite expectation value. Strocchi comments: "The physical meaning of the above result is rather basic; in a scattering process of a charged particle the emitted radiation has a finite energy but an infinite number of 'soft' photons, in the sense that for any finite ϵ, the number of emitted photons with energy greater than ϵ is finite, but the total number of emitted photons is infinite" (2005, 87).

So-called "soft photons," residents of infrared coherent representations, are supposed to hold the key to solutions to the infrared problem in QED.

So there is call to put infrared coherent representations to explanatory use. But answering this call requires abandoning principled restrictions on physically reasonable states—even ones arising from the apparently unadulterated covariance axiom, and even in the setting of Minkowski spacetime, which appears maximally suited to privileging strategies that are both motivated and biting. Conversely, a pristine commitment to principled criteria of physical reasonableness, criteria announced in advance of trying to apply the theory to disparate and particular situations and inflexibly adhered to, disables some of the explanatory capacity of QFT.

10.5 Conclusion

Suppose it is healthy to contour the content of a physical theory in a way that supports the explanatory and developmental aspirations of the physicists who work with it. Such a contouring constitutes an interpretation of that theory. In the case that explanatory and developmental aspirations are sufficiently diverse, *different* interpretations will be indexed to—and adulterated by—different aspirations. The aspirations adulterate because they arise *along with* particular applications of the theory. The considerations they underwrite lack the generality of metaphysics.

One response to this might be to try to ride herd on explanatory/developmental aspirations. Perhaps the set of "acceptable" ones could be narrowed to the point that they underwrite a single and univocal interpretation of the theory to which they apply. Such a narrowing is certainly a conceptual possibility. The foregoing examples, in which individually respectable aspirations generate a bevy of disparately adulterated interpretations, suggest that it's a methodologically distasteful one.

11

Phenomenological Particle Notions

This chapter changes gears to consider a different sort of impetus to interpret quantum field theory (QFT) in terms of particles: the existence of technical and experimental practices, as well as explanatory strategies, that appeal to particle notions. Invoking Hempel, §1.4 announced a criterion of adequacy for an interpretation of a physical theory. Vaguely put, the criterion demands an interpretation to support a theory's explanatory aspirations. Put somewhat less vaguely, the criterion demands that the set of worlds the interpretation declares possible be a set of worlds manifesting the properties or structures exploited by prominent theoretical explanations. Put either way, the criterion appears to be violated by any interpretation of QFT that forswears a particle notion. Particle physics is called "particle physics" for a reason. Its iconic phenomena—a cloud or bubble chamber photograph depicting "the three body decay of a long-lived meson" or "the decay of the Ω^-" or "electron-positron scattering"—appear for all the world to record the trail left by a well-localized and enduring particle as it travels through a detection device. What's more, the theory saved by and saving these phenomena, the Standard Model, customarily surrenders predictions when prodded with the calculational technique of Feynman diagrams (Fig. 11.1 below), which for all the world appear to depict particles, virtual and otherwise, in interaction. To make sense of both its iconic phenomena and the success of its iconic techniques of calculation, one suspects, a QFT must be interpreted in terms of particles.

There are in QM_∞ other explanations whose currency is a particle notion. "The basic physical effect [of QFT on curved spacetime]," writes the mathematical physicist Bernard Kay, "is that a time-dependent gravitational field can create pairs of particles out of the vacuum" (1988, 375). The stress energy associated with particles created in an expanding universe has been invoked to explain its dynamics (see Parker 1969; 1971). Such cosmological particle creation has been cited as a possible origin of all the matter in the universe (Grib, Mamayev, and Mostepanenko 1976, 536), as well as a seed mechanism for galaxies (Parker 1971, 353). An interpretation that withheld particles from the ontology of QFT on curved spacetime would hamstring explanations predicated on particles created by cosmological expansion.

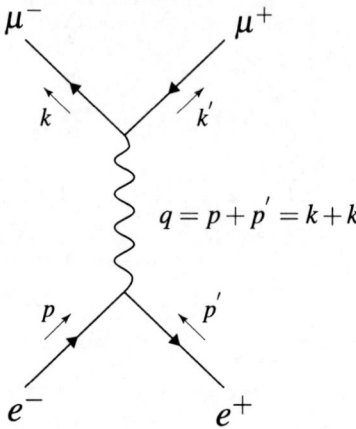

Figure 11.1. Feynman diagram for electron-positron annihilation

To identify precisely what particle notion or notions inform such explanations would be a worthy endeavor. But it is not the aim of this chapter, which aspires instead to argue that *whatever those particle notions are,* they're not fundamental particle notions in the sense of Chapters 9 and 10. Those chapters dealt with particle notions fundamental in the "physicist's" sense. Such a notion applies to a set of physical states confined to an irreducible Fock space representation. With respect to this set of states, the set of particle observables affiliated with that Fock space is complete: any pure state in the space can be approximated arbitrarily well by a superposition of states *identified* by the eigenvalues they assign those particle observables. The particle notion circumscribed by those particle observables suffices to characterize the pure states of the theory, and so promises to guide the interpretation of theoretical states in terms of their particle contents. If the particle notions informing the explanations and techniques discussed in this chapter don't adopt the guise of a fundamental particle interpretation, then appeal to the Hempelian criterion can't restore a fundamental particle interpretation to settings of insufficient symmetry to support it in the first place.

Thus this chapter will blur two questions hitherto kept distinct. First is the question of how to identify the content of—including the set of states possible according to—a particle interpretation. It is with respect to how an interpretation handles this question that it is measured against the ideal of pristine interpretation. The second question arises after an interpretation has been articulated, and concerns not the resources for constructing the interpretation but the reasons for adopting it. It's the question of whether to accept an interpretation. The two questions are blurred here because I'm characterizing a phenomenological particle interpretation in terms of reasons for adopting it: a phenomenological particle interpretation is (supposed to be) that interpretation of QFT that makes sense of explanatory and experimental practices marshaling the particle notion.

One impediment to explanations of particle physics phenomenology framed by a fundamental particle notion is Haag's theorem, which implies that incommensurable particle notions—particle notions circumscribed by observables parochial to unitarily inequivalent irreducible Fock space representations—pertain to different stages of a particle physics experiment. (Section 11.1 will develop and defend this implication more carefully.) It follows that no single fundamental particle notion is explanatorily adequate to the entire history of a particle physics experiment. Thus no fundamental particle notion is also a phenomenological particle notion. What impedes cosmological explanations framed by a fundamental particle notion is that incommensurable particle notions can pertain to different epochs of a cosmos in which there is particle creation. There are accounts of cosmological particle creation no single fundamental particle notion can underwrite. Section 11.2 will develop and defend this claim more carefully.

11.1 Particle physics, redux

Consider first a textbook account of how the detectors employed by particle physicists work: "The detection of charged particles and photons is based on the fact that these particles produce ionization when traversing charged matter.... [in cloud and bubble chambers] the path of the particle can be determined by making the ionization observable, so that it can be seen or photographed" (Tipler 1969, 418). The most straightforward explanation of the appearance of cloud and bubble chamber tracks is this: they look like particle trajectories because *they were made by passing particles*. Any interpretation of QFT as unparticulate undermines this explanation.

Consider next a textbook account of Feynman diagrams, "the basic calculational method of quantum field theory" (Peskin and Schroeder 1995, 3). The "simplest of QED processes" discussed in the passage below is pair production by electron-positron annihilation (see Fig. 11.1).

The bad news is that even for this simplest of QED processes, the exact expression for $\mathcal{M}$ [the probability amplitude for the process] is not known... The best we can do is obtain a formal expression for $\mathcal{M}$ as a perturbation series in the strength of the electromagnetic interaction, and evaluate the first few terms in this series.

The good news is that Feynman has invented a beautiful way to organize and visualize the perturbation series: the method of *Feynman diagrams*. Roughly speaking, the diagrams display the flow of electrons and photons during the scattering process. For our particular calculation, the lowest-order term in the perturbations series can be represented by a single diagram... the diagram is made up of three types of components: external lines (representing the four incoming and outgoing particles), internal lines (representing "virtual" particles, in this case one virtual photon), and vertices....

According to the *Feynman rules*, each diagram can be translated directly into a contribution to $\mathcal{M}$. The rules assign a short algebraic factor to each element of the diagram, and the product of these factors gives the value of the corresponding term in the perturbation series. (Peskin and Schroeder 1995, 5)

The Feynman diagram depicted in Fig. 11.1 corresponds to the lowest-order term in the perturbative expansion of the probability amplitude for the simple electron-positron annihilation interaction. Heuristically, it depicts an incoming positron of momentum p' encountering an incoming electron of momentum p to create a virtual photon (the squiggly line) of momentum $p + p'$, which decays into a pair of muons. This Feynman diagram is conscripted into the service of calculating the quantum mechanical probability amplitude for the process.

Calculations mediated by Feynman diagrams and Feynman rules are mind-bogglingly successful, agreeing with observation over many orders of magnitude and to many significant figures. There is a tempting and straightforward explanation of this success. The cross-sections predicted by the method of Feynman diagrams are upheld by experiment because the particle interactions those diagrams depict (and by depicting, guide the calculation) occur in nature. Once again, any interpretation of QFT as unparticulate undermines this explanation.

To look at bubble chamber photographs or Feynman diagrams and see particles is so natural as to be almost irresistible. Explanations of a certain sort are implicit in this way of seeing, particle-invoking explanations. The legitimacy and exact nature of these explanations will not be my focus here. Instead, I will suppose that whatever other features these explanations exhibit, the particle notion they invoke is meant to *apply over the entire microhistory of a particle physics experiment*: to the particles entering the apparatus, to the interacting particles evidently depicted in Feynman diagrams, to the particles exiting the apparatus whose ionization trails are recorded in cloud and bubble chamber photographs. In the first part of this section I will argue that Haag's theorem shows that no fundamental particle notion is up to the task italicized above. For it follows from Haag's theorem that the micro history of a scattering experiment (at least one compliant with predictions obtained via Feynman diagrams) is not confined to the folium of single Fock space representation. No single fundamental particle notion embraces the ingoing and outgoing particles encountered in the iconic phenomena of particle physics, as well as the interacting particles portrayed in Feynman diagrams.

Haag–Ruelle scattering theory and the $(\phi^4)_2$ model of interacting QFT promise some solace for the fundamental particle interpreter. The second part of this section describes the solace, but questions its depth. I will suggest that the explanatory connection between particle physics phenomenology and fundamental particle notions allied with Haag–Ruelle scattering theory or the $(\phi^4)_2$ model is decidedly attenuated. The capacity to explain particle physics phenomenology is constitutive of a phenomenological particle notion. Hence the particle notions of Haag–Ruelle scattering theory or the $(\phi^4)_2$ model aren't the phenomenological particle notions we're after.

11.1.1 The interaction picture and Haag's theorem

Feynman diagrams oversee the calculation of scattering QED cross-sections on behalf of the interaction picture, which at first approximation works like this. What we want is a scattering amplitude between incoming states $|\psi_{in}\rangle$ (e.g. the state of an electron-positron pair) and outgoing states $|\phi_{out}\rangle$ (e.g. the state of the pair of muons produced by electron-positron annihilation). The gadget that gives us this scattering amplitude is the S-matrix S; that is:

$$\langle \psi_{in}|S|\phi_{out}\rangle \qquad (11.1)$$

gives us the quantum mechanical amplitude that a system enters the detector in the state $|\psi_{in}\rangle$ and leaves it in the state $|\phi_{out}\rangle$. Insofar as both $|\psi_{in}\rangle$ and $|\phi_{out}\rangle$ admit an interpretation in terms of particle contents, the event for which (11.1) gives the amplitude can be understood as a particle-scattering event. The in states $|\psi_{in}\rangle$ are supposed to span a Hilbert space $\mathcal{H}_{in}$ while the out states $|\phi_{out}\rangle$ are supposed to span a Hilbert space $\mathcal{H}_{out}$. The S-matrix effects a map between these Hilbert spaces. Let us suppose that the Hilbert spaces carry representations (π_{in} and π_{out}, respectively) of the relevant canonical algebra (whatever it may be), and let us suppose further that each is an irreducible Fock space representation. (It will soon become apparent that neither of these suppositions is non-trivial.) Previous chapters have argued that the particle notion affiliated with an irreducible Fock space representation is fundamental to a theory only if that theory's state space is confined to the folium of that Fock space representation.

Here's how, on the interaction picture, we obtain scattering amplitudes. Assume that at early and late times—i.e. $t = \pm\infty$—the system is governed by a free Hamiltonian H. What would underwrite a fundamental particle interpretation embracing both in and out states $|\psi_{in}\rangle$ and $|\phi_{out}\rangle$ is an irreducible Fock space representation of this free-field theory set in a Hilbert space encompassing those states. But what generates interesting scattering amplitudes is the imposition of a non-zero interaction Hamiltonian H_{int} at intermediate times.

By letting H_{int} govern state evolution while fields evolve freely, the interaction picture provides a recipe for expressing S-matrix elements such as $\langle\psi_{in}|S|\phi_{out}\rangle$ in terms of the free and interacting Hamiltonians. This expressions admit a perturbative expansions, and Feynman diagrams, in concert with the Feynman rules, direct the calculation of the first few terms of this expansion. Extending a fundamental particle interpretation of incoming and outgoing particles to particles at intermediate times $-\infty < t < \infty$ (the interacting particles presumably depicted in Feynman diagrams) we assume that a single irreducible Fock space representation embraces them all.

Haag's theorem reveals this assumption to be false, at least if H_{int} is different from 0. To simplify somewhat (see Earman and Fraser 2006 for further discussion), Haag's theorem shows that *if certain assumptions* (documented below) *about the interacting theory are satisfied*, then that theory is unitarily equivalent to the free theory (i.e. $H_{int} = 0$), and so not an interacting theory at all. A consequence is that any non-trivial (i.e. $H_{int} \neq 0$) interacting theory satisfying the assumptions is unitarily *inequivalent* to the free theory, and thus in the ambit of a particle notion incommensurable with the free particle notion, if it's in the ambit of any particle notion at all. The incommensurability of these particle notions precludes extending a single fundamental particle notion over the entire microhistory of a scattering experiment.

The first assumption is that at intermediate times, the interacting theory has a Fock space representation. That is, there are interacting field operators $\phi_I(\vec{x}, t)$ and conjugate momentum operators $\pi_I(\vec{x}, t)$ giving an irreducible Hilbert space representation of equal time commutation relations (9.4). Creation and annihilation operators $a_I^\dagger(\vec{k}, t)$ and $a_I(\vec{k}, t)$ can be defined in terms of $\phi_I(\vec{x}, t)$ and $\pi_I(\vec{x}, t)$ in the usual way. The assumption that such a representation is available at intermediate times allies a particle notion—the particle notion circumscribed by the collection of variety interacting number operators $N_k^I := a_I^\dagger(\vec{k}, t) a_I(\vec{k}, t)$—with those times.

The second set of assumptions requires this representation to satisfy key axioms of QFT (see §5.2): the Hilbert space on which these operators act has a unique normalizable state $|0_I\rangle$ invariant under Euclidean transformations,[1] which are therefore unitarily implemented on the representing Hilbert space; the infinitesimal generator of time translations is positive, so that no states of negative energy exist.

If these assumptions hold, the particle notion applicable to intermediate times is commensurable with the $t = \pm\infty$ particle notion only if the interaction is trivial. This follows from a refinement of Haag's work due to Hall and Wightman (1957). They show that if there is a non-negative energy free field ($H_{int} = 0$) representation $\phi(\vec{x}, t)$ and $\pi(\vec{x}, t)$ of the equal time CCRs that satisfies the demands of Euclidean invariance, and if the free and interacting fields are related at time t by unitary transformations $U(t)$ such that:

$$\phi(\vec{x}, t) = U(t)\phi_I(\vec{x}, t)U^{-1}(t) \qquad \pi(\vec{x}, t) = U(t)\pi_I(\vec{x}, t)U^{-1}(t) \qquad (11.2)$$

then the free field and the interacting field are the same. In other words, either there is no interaction (and so no interacting particle to which to extend the fundamental particle interpretation underwritten by the free field theory), or the interacting and free field theories are unitarily inequivalent (so that whatever particle notion applies to the interacting theory is incommensurable with the particle notion underwritten by the free field theory). In either case, a fundamental particle interpretation of the entire history of an interaction—an interpretation that embraced both the particles evinced by Feynman diagrams and those implicated in cloud and bubble chamber photographs—is untenable.

To avoid this consequence, the fundamental particle interpreter can deny the theorem's assumptions. And for each assumption, there is a reason to do so. The catch is that in each case the reason also undermines a fundamental particle interpretation. This leaves the fundamental particle interpreter with a dilemma: *either the assumptions of Haag's theorem do not hold, in which case there is no particle notion applicable to a scattering experiment at intermediate times, or they do, in which case the particle notion applicable at intermediate times is incommensurable with the ingoing/outgoing particle notions, if the interaction is non-trivial.*

The first assumption is that at intermediate times, the interacting theory has a Fock space representation. The reason to deny it is that interacting fields are typically

---

[1] i.e. spatial rotations and translations; a subgroup of Poincaré transformations.

nonlinear. Given the crucial role linearity plays in the explicit construction of Fock space representations of the CCRs for free field theories (see §9.3), it is hardly obvious that, in the interacting case, any such representation is available. (Indeed, it is hardly obvious *what* the relevant CCRs are. In the non-interacting case, the CCRs were identified by appeal to the Weyl relations for the symplectic vector space (S, Ω) determined by the—linear!—classical theory. The interacting case offers no analogous route to its CCRs.) Conjuring this worry gives the fundamental particle interpreter a way out of Haag's theorem. But if the worry is valid, no particle interpretation of the interacting theory at intermediate times, much less one commensurable with the particle interpretation of incoming and outgoing states, is possible. So a fundamental particle interpretation of the entirety of a scattering interaction presupposes that the first assumption holds.

A similar remark applies to desperate complaints about the second assumption, and in particular about Euclidean invariance. Impressionistically put, the complaint is that on the interaction picture, there's some region of spacetime—Fermilab in 2007, say—where the interaction is turned on, and other regions of spacetime—the remote past, remote future, and remote elsewhere—where it isn't. So doesn't Euclidean invariance fail? (The complaint is desperate because the QFT can be Euclidean-invariant at the same time particular states of the theory violate Euclidean invariance.) The remark is that if Euclidean invariance fails, so much the worse for running a particle interpretation at intermediate times of the scattering experiments, because symmetries are essential to singling out, from the myriad of formally qualified Fock spaces available, *the* physical (i.e. symmetry-respecting) Fock space circumscribing that particle interpretation.

I conclude that within the confines of the interaction picture, no fundamental particle interpretation can frame explanations of particle physics phenomenology and its compliance with calculations mediated by Feynman diagrams. If the assumptions of Haag's theorem fail, there is no particle notion applicable to intermediate time phenomena; if they hold and the interaction is non-trivial, the applicable particle notion is incommensurable with the particle notion applicable at early and late times. There is no single Fock space folium in terms of whose states one might recount the history of a particle physics experiment. A fundamental particle interpretation of that history is available only if there is such a folium. Therefore, no fundamental particle interpretation is available.

The story cannot end there. For there are accounts of scattering and interactions that leave the confines of the interaction picture. Next we examine these for evidence of hospitality to a fundamental particle notion that does the explanatory work constitutive of a phenomenological particle notion.

11.1.2 *Hope: Haag–Ruelle?*

The interaction picture is not the only approach to scattering theory. For our purposes, an intriguing alternative is Haag–Ruelle scattering theory. It is intriguing because it circumvents Haag's theorem, the *prima facie* impediment to embracing the history of a scattering physics experiment by a single fundamental particle notion.

Haag's theorem characterizes conditions under which representations of the CCRs for free and interacting fields are unitarily inequivalent. Haag–Ruelle scattering theory rescues unitarity from the maw of this result by constructing a unitary S-matrix directly relating states of an interacting field theory *postulated* to obey the Wightman axioms. The states of the interacting theory related by the S-matrix are supposed to be surrogates of the *in* and *out* states of the free field theory.

Reed and Simon's exposition of Haag–Ruelle scattering theory begins with a question: "How can we hope to construct a scattering theory for a general quantum field theory . . . ? The answer is simple. Our construction is axiomatic. We shall assume that the Gårding-Wightman axioms hold; in particular, we assume the existence of a unitary interacting dynamics" (1980b, 317). They hasten to acknowledge that "the first conceptual problem that must be faced is that there is no natural 'free' dynamics for the interacting dynamics to approach as $t \to \pm\infty$" (ibid.). An otherwise exemplary candidate for free dynamics—those defined by the free Hamiltonian H_0—is defeated by Haag's theorem and the Gårding–Wightman demand for a unitary S-matrix. The free dynamics H_0 defines acts on a Hilbert space representation of the CCRs unitarily inequivalent to that postulated for the interacting dynamics. Haag–Ruelle scattering theory does an end run around Haag's theorem by identifying counterparts in the interacting theory for free fields. The identification proceeds in the scope of further physical assumptions—that the interacting theory has a mass gap, and that its vacuum is coupled to its one-particle states—and the counterpart relation is rather involved. Fortunately, we don't need to get into the details (for which see Reed and Simon 1980b, ch. 9). For our purposes, the crude sketch developed so far will suffice.

Let $\mathcal{H}$ be the Hilbert space in which the interacting theory is set. Call the subspace of $\mathcal{H}$ spanned by the counterparts of incoming free field states $\mathcal{H}_{in}$; call the subspace of $\mathcal{H}$ spanned by the counterparts of outgoing free field states $\mathcal{H}_{out}$. *Asymptotic completeness* holds just in case $\mathcal{H}_{in} = \mathcal{H}_{out} = \mathcal{H}$. And if asymptotic completeness holds, in states and out states are related by a *unitary* S-matrix. Haag's theorem notwithstanding, scattering theory can be conducted in the folium of the Hilbert space representation of the interacting field theory.

Could this representation circumscribe a fundamental particle notion applicable to the whole history of a particle physics experiment? Several circumstances might give the fundamental particle interpreter pause. The first is the inexplicitness of the representation. Haag–Ruelle scattering theory postulates, rather than constructively demonstrates, its existence. In the absence of explicit examples of Hilbert space representations of interacting field theories, there is room to wonder whether the postulate is correct. And even if the postulate is correct, lacking examples of the representation postulated, we also lack evidence of its character. Does it have a Fock space structure at all, and if it does, *which* Fock space structure? Even if the folium of the representation of the CCRs for an interacting field theory postulated by Haag–Ruelle comprehends the microhistory of a scattering experiment, the ineffability of the representation postulated interferes with *our* comprehension of the particle notion circumscribed by that representation.

Not only is the particle notion circumscribed by Haag–Ruelle scattering theory inexplicit, it is also missing from the explanatory nexus—the nexus embracing particle physics phenomenology and the calculational techniques underwriting the predictions it saves—invoked in the Hempelian route to a particle interpretation of QFT. The scattering amplitudes implicit in Haag–Ruelle's (unitary) S-matrix aren't calculated by applying Feynman diagrams and Feynman rules to a perturbative expansion. When they're calculated at all, the mathematical intermediaries are much less suggestive. For Klaus Hepp, "The great advantage of the Haag-Ruelle construction...lies in the clarity of the underlying physical concepts and in the mathematical rigor of the construction—the main drawback on the other side is the very involved way, by which the measurable quantities as S-matrix elements are expressed by the local interpolating fields" (1963, 92). This attenuates the explanatory connection between, on the one hand, the particles apparently depicted in Feynman diagrams and apparently detected in experiments confirming Feynman diagram-mediated calculations, and on the other, any particle notion circumscribed by Haag–Ruelle scattering theory. Whereas the interaction picture and the practices surrounding it furnish the raw materials for an explanationist defense of a particle interpretation that can't for reasons of mathematical physics be fundamental, Haag–Ruelle scattering theory is a piece of mathematical physics framing (modulo concerns about its inexplicitness) a fundamental particle interpretation, but which is not surrounded by practices supporting an explanationist defense of that interpretation. *If* Haag–Ruelle scattering theory circumscribes a particle notion, it's not a particle notion that particle physics phenomenology gives us an explanationist reason to adopt.

11.1.3 The $(\phi^4)_2$ theory

To silence complaints raised in the foregoing section about the ineffability of the interacting QFT postulated by Haag–Ruelle scattering theory, the fundamental particle interpreter might summon the example of the $(\phi^4)_2$ theory, an interacting QFT set in an explicit Hilbert space. The $(\phi^4)_2$ theory describes a self-interacting scalar field (the self-interaction is implemented by a ϕ^4 term in the interaction Hamiltonian) in two spacetime dimensions. The field obeys the non-linear equation of motion:

$$(\Box + m^2)\phi(x, t) = -4\lambda\phi^3(x, t) \tag{11.3}$$

Its quantization can be modeled as Schrödinger evolution generated by a positive self-adjoint Hamiltonian acting on a Hilbert space $\mathcal{H}_{ren}$ of "physical" states. Here I'll sketch the nifty maneuver by which this Hilbert space representation is obtained (for a further overview, see Glimm 1969), then explain why, explicit though it is, the $(\phi^4)_2$

theory doesn't ratchet particle physics phenomenology into a Hempelian argument for a fundamental particle interpretation of QFT.

The interaction H_{int} of the $(\phi^4)_2$ theory is[2]

$$H_{int} = \lambda \int :\phi^4(x): dx - E \tag{11.4}$$

As Haag's theorem would lead us to expect, H_{int} is not an operator on Fock space of the free field. But H_n defined in terms of $(\phi^4)_2$ and a "spatial cutoff" is.[3]

$$H_n = \lambda \int :\phi^4(x): g(x/n) dx - E \tag{11.5}$$

The function $g(x)$ is set to 1 for $|x| < 1$ and 0 for $|x| \gg 1$. $g(x)$ "turns off" the spatially curtailed interaction Hamiltonian H_n for $|x| < n$, but enables H_n to mimic H_{int} for $|x| < n$.

To construct $\mathcal{H}_{ren}$ for the full interacting theory, Glimm and Jaffe proceed as follows. Each well-defined H_n has a vacuum vector $|\Omega_n\rangle$. Each $|\Omega_n\rangle$ assigns expectation values to elements of the quasilocal algebra $\mathfrak{A}$ for the free field theory via:

$$\omega_n(A) := \langle \Omega_n | A | \Omega_n \rangle \tag{11.6}$$

The key result is that as $n \to \infty$, $\omega_n(A)$ converges to a limit $\omega(A)$ for each $A \in \mathfrak{A}$. These limits thereby define a state ω on $\mathfrak{A}$. *The GNS representation of the state ω gives the physical Hilbert space $\mathcal{H}_{ren}$.* The cyclic vector implementing ω is the "dressed" vacuum. And the interaction Hamiltonian can be defined as an operator on $\mathcal{H}_{ren}$. Supposing this GNS representation has a Fock space structure, it circumscribes an explicit particle notion associated with an interacting QFT.

The catch for anyone mounting a Hempelian defense of a fundamental particle interpretation is that the $(\phi^4)_2$ theory—itself one of the few "successes" of constructive field theory—is a toy model. It's not a member of the family of interacting QFTs constituting the Standard Model and implicated in Feynman diagrams. Explicit as the particle notion associated with the $(\phi^4)_2$ theory may be, it gets no explanatory oomph from particle physics phenomenology. As with Haag–Ruelle scattering theory, the would-be particle interpreter puts a fundamental particle interpretation on firm conceptual footing at the price of alienating it from the phenomena it might demonstrate its ontological *bona fides* by explaining.

11.2 Cosmological particle creation

In cosmological particle creation, a time-dependent gravitational field—the gravitational field of an expanding universe, for example—creates pairs of particles.

---

[2] Recall that ": :" denotes normal ordering. E is a constant chosen to zero out the ground state energy.
[3] The heuristic explanation of this is that the spatial cutoff reinstates the Stone–von Neumann theorem by precluding "ultraviolet" modes.

Schrödinger is credited with the insight that cosmological particle creation could occur. Subsequent physicists have invoked the particles created in a host of explanations. Calculating that "the reaction of the particle creation (or annihilation) back on the gravitational field will modify the expansion in such a way as to reduce the creation rate" (Parker 1969, 1066), Parker argues heuristically *from* this minimization principle *to* the special case of the Einstein Field Equations. Cosmological particle creation has been imagined as the mechanism by which not only matter (Grib et al. 1976) but also galaxies (Parker 1971, 353) have made their cosmic debuts; it also has a provocative connection to spin and statistics (see Fulling 1989, 149).

Once again, my primary concern here is not to articulate and evaluate the explanations in which cosmologically created particles feature. It is rather to argue that these explanations cannot anchor a Hempelian case for a *fundamental* particle notion. A fundamental particle interpretation requires physically reasonable states to be confined to the folium of a single, irreducible Fock space representation, but the states these explanations invoke are not so confined. The key to the argument is that at different epochs of a universe capable of cosmological particle creation, incommensurable particle notions can apply.

Here I sketch cosmological particle creation in only enough detail to discern the incommensurable particle notions blocking the Hempelian route to a fundamental particle interpretation. (Fulling 1989 offers a more complete exposition.)

The line element of a universe with spacelike hypersurfaces which are Euclidean but expanding/contracting is given by:

$$ds^2 = -dt^2 + R(t)(dx^2 + dy^2 + dz^2) \qquad (11.7)$$

The factor $R(t)$—the coefficient of the metric for a three-dimensional Euclidean space—gives the expansion rate. Friedman–Robertson–Walker solutions to the Einstein Field Equations boast metrics of this sort. $R(t) \propto t^{\frac{1}{2}}$ corresponds to a Friedman universe dominated by radiation; $R(t) \propto t^{\frac{2}{3}}$ corresponds to a Friedman universe dominated by matter.

The metric (11.7) is isotropic. Associating different expansion rates with different spatial directions, the metric of a Bianchi Type I universe,

$$ds^2 = -dt^2 + a_x(t)dx^2 + a_y(t)dy^2 + a_z(t)dz^2 \qquad (11.8)$$

is not.

Spacetimes with metrics (11.7) and (11.8) are not stationary (although the former are conformally flat and the latter can be resolved into spacelike hypersurfaces with Euclidean metrics). In these spacetimes, just as in Minkowski spacetime, the covariant generalization of the $m \neq 0$ Klein–Gordon equation can be solved by separation of variables, i.e. by assuming that a general solution $\phi(\vec{x}, t) = \psi_j(\vec{x})\phi_j(t)$ (see Fulling 1989, ch. 7). But unlike Minkowski spacetime, the $\phi_j(t)$s are not given by the plane waves $e^{\pm \omega_j t}$ but by solutions to ordinary differential equations with time-dependent coefficients.

To finesse questions about how to associate a particle notion with a QFT set in an expanding universe,[4] let us sandwich flat early and late epochs around an expansion described by (11.7) at intermediate times. That is, let us suppose that $R(t) = constant$ for early $t < T_-$ and late $t > T_+$ times. Taking these early and late regions as (static) spacetimes in their own right, we can perform standard frequency-splitting QFT constructions in each. Call the Fock spaces obtained $\mathfrak{F}(\mathcal{H}_{in})$ and $\mathfrak{F}(\mathcal{H}_{out})$ respectively. Fulling argues that particle notions associated with these Fock spaces are physically appropriate to their spacetime setting:

> Our space-time is flat. This region contains entire Cauchy surfaces. The dynamics of the field there is that of the free field, and the physics of any experimental operation is surely the same as in the globally flat space of nongravitational physics. Therefore $\phi^{in}(k)e^{i\vec{k}\cdot\vec{x}}$ [a one-particle state in $\mathfrak{F}(\mathcal{H}_{in})$] is surely the wave function of a real particle, as far as that epoch of time is concerned. (Fulling 1989, 139)

And *mutatis mutandis* for $\mathfrak{F}(\mathcal{H}_{out})$. Supposing we accept the argument, we can pose the following question about particle creation: when the initial ($t < T_-$) state is the vacuum $|0_{in}\rangle$, how many particles are produced by expansion in the interval $T_- < t < T_+$? Because $\mathfrak{F}(\mathcal{H}_{out})$ circumscribes the particle notion appropriate to post-expansion times $t > T_+$, the answer will be the "expectation value" the in-vacuum assigns the number operator N^{out}.

When a universe whose Cauchy surfaces are compact (a.k.a. a "closed" universe) undergoes sudden expansion, or a universe whose Cauchy surfaces are $\mathbb{R}^3$ (a.k.a. an "open" universe) experiences any expansion at all, $\langle 0_{in}|N^{out}|0_{in}\rangle = \infty$, signaling the unitary inequivalence of the in and out Fock space representations, and the incommensurability of their associated particle notions. Fulling is not alarmed by this circumstance. Concerning infinite particle creation in open universes, he writes, "$\langle N_k^{out}\rangle = \infty$, even for the smoothest and gentlest expansion of the universe. The physical meaning of this divergence is quite clear and nonfrightening, however. A particle-creation process is taking place throughout a homogeneous, infinite space. The total number of particles produced, if nonzero, has to be infinite" (Fulling 1989, 141). Parker agrees:

> The state $|0_{in}\rangle$ cannot be expressed as a superposition of the eigenstates of $a_k^{out\dagger}$, a_k^{out} in the limit as $L \to \infty$ [L the volume of the universe]. Thus the Hilbert space becomes nonseparable in the limit of infinite volume. This is not surprising since any finite particle density implies an infinite number of particles in an infinite volume. No difficulties are encountered if one works with L finite and takes the limit $L \to \infty$ after the physically significant quantities have been deduced. (Parker 1969, 133; notation changed to match Fulling's conventions)

---

[4] For more on the fraught issue of associating a particle notion with times *during* expansion, see Fulling (1989, 152 ff.); Grib et al. (1976).

(The nonseparability Parker remarks is the signature of the unitary inequivalence of the in and out representations.)

The fundamental particle interpreter should not be so sanguine. The $\mathfrak{F}(\mathcal{H}_{out})$ particle notion invoked to discern created particles is incommensurable with the $\mathfrak{F}(\mathcal{H}_{in})$ particle notion invoked to describe the vacuum from which they are created. There is no irreducible Fock space representation that comprehends both the void $|0_{in}\rangle$ and the particles sprung therefrom by expansion. A fundamental particle interpretation is available only when there is an irreducible Fock space representation comprehending all physically possible states. The particles concerned in cosmological particle creation are not the particles of a fundamental particle interpretation, and explanations that appeal to created particles are not grounds for adopting a fundamental particle interpretation. To the contrary, a fundamental particle interpretation *precludes* explanations that appeal to cosmological particle creation, because any irreducible Fock space representation, to which a fundamental particle interpretation would confine physically significant states, must exclude either $|0_{in}\rangle$ or the entire folium of states to which the created particle notion applies.

At this point, the opponent of the fundamental particle interpreter, or of Hilbert Space Conservatism in general, can turn the Hempelian tables. *Supposing* that there are explanations worth sustaining that invoke cosmological particle creation, no fundamental particle interpretation—indeed, no Hilbert Space Conservatism—can sustain them. For the fundamental particle interpreter and the Hilbert Space Conservative share the assumption that the folium of a single Fock space representation comprehends all physically reasonable states. And this assumption stymies explanatory aspirations appealing to the creation from $|0_{in}\rangle$ of particles described by N^{out}.

But can interpretations forswearing fundamental particles do any better? Perhaps they can. One strategy would be to argue that non-fundamental *epiparticles* satisfy the explanatory needs expressed in appeals to cosmological particle creation. Another, possibly related strategy would be to rephrase explanations taking created *particles* at face value as explanations in terms of other commodities, ones that don't privilege particular Fock space representations. For instance, one could rephrase an appeal to the stress energy of *created particles* to explain what puts the brakes on cosmic expansion as an appeal to the expectation value $\langle T_{ab}\rangle_\omega$ of a stress-energy observable T_{ab} in the state ω. The appeal can regard ω as the (Hadamard) state of the cosmos, whose particle contents are left unarticulated. Assessing either strategy requires a closer examination of particle-invoking explanations than I will mount here.

We've examined cosmological particle creation closely enough to conclude that the incommensurability of created particles with the voids from which they were created foils explanationist defenses, based in cosmological particle creation, of a fundamental particle interpretation. The next section attempts to summarize my prolonged discussion of the particle concept in QFT.

11.3 Conclusion

So, is particle physics particle physics? The answer I've tried to support is: sometimes it might be. To offer a *fundamental particle interpretation* of a QFT, in Chapter 9's sense, is to attribute to that theory a dense set of pure states that can be characterized and distinguished by appeal to their particle contents. Arguing that irreducible Fock space representations circumscribe particle notions, and that unitarily inequivalent representations circumscribe *incommensurable* particle notions, Chapter 9 claimed that to give a fundamental particle interpretation of a QFT is to confine its possible states to the folium of an irreducible Fock space representation. Invoking a result due to Kay and Wald, Chapter 10 sketched a strategy for confinement available in spacetimes with timelike isometries. A QFT might be particle physics when it is set in such a spacetime.

The Unruh effect threatens to spoil the strategy. For it alerts us that QFT constructions guided by *different* families of timelike isometries in the same spacetime can culminate in unitarily inequivalent Fock space representations: Jack's Fock space, privileged by appeal to time translation isometries of Minkowski spacetime, and Jill's Fock space, privileged by appeal to the Lorentz boost isometries of the right Rindler wedge, *have no states in common*. No single particle notion comprehends both Fock spaces. If Jack's quantization and Jill's quantization are both physical, not even the generous symmetries of Minkowksi spacetime underwrite a fundamental particle interpretation of the quantized Klein–Gordon field.

Section 10.3 gave instructions for defusing this threat, in the form of further principles of privilege that distinguish invidiously between Jill's quantization and Jack's. The principle I gave the most attention was the *Hadamard* principle that, in order to be physical, a state of the quasilocal algebra had to support a regularization procedure for assigning the stress-energy tensor an expectation value. Invoking the Hadamard principle helps *single out* Jack's folium as the folium of physical states, and thereby restores Jack's particle notion to fundamental status. Invoking the Hadamard principle identifies the content of quantized Klein–Gordon theory on Minkowski spacetime. Chapter 10 closed with the suggestion that the fundamental particle interpretation thus obtained was adulterated by the entanglement of the Hadamard condition in the unprincipled (or not "metaphysically" principled) business of creative physics. Chapter 10 also applauded this adulteration. "Metaphysical" principles for isolating theoretical content, once and for all, run the risk of stymying attempts to harness the theory they interpret in fruitful projects of explanation and expansion. (Or at least they would if physicists took them too seriously.) Sometimes particle physics is, adulteratedly, particle physics, and that's a good thing.

The next chapter carries these themes to the thermodynamic limit of quantum statistical mechanics, many of whose purported explanations (I claim) would be hobbled by interpretive extremism.

12
A Matter of Degree
Making Sense of Phase Structure

What is the content of a quantum theory? Chapter 6 described and motivated a variety of "extremist" answers to these questions. Each extremist espouses a blanket policy for interpreting a quantum theory of canonical type $\mathfrak{A}$—i.e. a quantum theory whose characteristic (anti)commutation relations generate the C^* algebra $\mathfrak{A}$.

- The *Hilbert Space Conservative* identifies quantum observables with the self-adjoint elements of the concrete von Neumann algebra $\mathfrak{B}(\mathcal{H}_\pi) = \pi(\mathfrak{A})''$ affiliated with an irreducible representation $\pi : \mathfrak{A} \to \mathfrak{B}(\mathcal{H}_\pi)$ of $\mathfrak{A}$, and identifies quantum states with countably additive probability measures over the closed subspaces of $\mathcal{H}_\pi$. Thus possible states of a theory of canonical type $\mathfrak{A}$ lie in the folium of an irreducible representation π of $\mathfrak{A}$; they're exactly the states that are *normal* with respect to that representation, that is, its density operator states.
- The *Algebraic Imperialist* identifies quantum observables with self-adjoint elements of the abstract C^* algebra $\mathfrak{A}$, and quantum states with the set $S_\mathfrak{A}$ of normed, positive linear functionals over $\mathfrak{A}$.
- The *Universalist* specifies the content of a theory of canonical type $\mathfrak{A}$ in terms of $\mathfrak{A}$'s universal enveloping von Neumann algebra $\pi_U(\mathfrak{A})''$.[1] The Universalist identifies quantum observables with self-adjoint elements of $\pi_U(\mathfrak{A})''$ and quantum states with normal states of $\pi_U(\mathfrak{A})''$.

In the presence of unitarily inequivalent representations of $\mathfrak{A}$, extremist accounts of quantum content become rivals. If $\mathfrak{A}$ has unitarily inequivalent representations, the full set $S_\mathfrak{A}$ of states on $\mathfrak{A}$ is larger than the folium of any irreducible representation of that algebra, and the Imperialist disputes the Conservative's account of which states are possible. Because normal states on $\pi_U(\mathfrak{A})''$ coincide with $S_\mathfrak{A}$, the Universalist sides with the Imperialist in this dispute. But she has her own bone to pick. When $\mathfrak{A}$ is

---

[1] Recall that the universal enveloping von Neumann algebra $\pi_U(\mathfrak{A})''$ of a C^* algebra $\mathfrak{A}$ is the von Neumann algebra affiliated with $\mathfrak{A}$'s *universal representation* π_U, the representation obtained by taking the direct sum, over states on $\mathfrak{A}$, of their GNS representations. See §6.6.

infinite dimensional—as it will be for QM$_\infty$—$\pi_U(\mathfrak{A})$, which is isomorphic to the algebra of observables the Imperialist recognizes, is a proper subset of $\pi_U(\mathfrak{A})''$, the algebra of observables the Universalist recognizes. The Universalist attributes physical significance to observables the Imperialist disdains as unphysical.

I call these positions "extremist" because they announce strategies for assigning content to quantum theories that are insensitive to peculiarities of particular applications of those theories. They're extremist because they do not adapt themselves to the exigencies in which theories find themselves. Immune to adulterations arising from these exigencies, extremist interpretations are liable to be pristine. They are just as liable to hamstring the explanatory capacity of the theories they interpret. That, at least, is what this chapter will claim about extremist interpretations of the thermodynamic limit of quantum statistical mechanics.

Statistical mechanics models a macroscopic system, such as an iron bar, as an aggregate of microsystems, such as a lattice of electrons. It aims to explicate bulk properties of the macrosystem—its energy, its temperature, its persistence in equilibrium, and the like—in terms of statistical aspects of the physics of its microconstituents. *Quantum* statistical mechanics takes the underlying microphysics to be quantum. To take the thermodynamic limit of (quantum) statistical mechanics is to let the number N of microsystems and the volume V they occupy go to infinity, while their density $\frac{N}{V}$ remains finite. Taken to the thermodynamic limit, quantum statistical mechanics falls outside the scope of the Stone–von Neumann and Jordan–Wigner theorems. Unitarily inequivalent representations of the quantum microphysics become available.

This chapter describes some ways in which the thermodynamic limit of QSM presses unitarily inequivalent representations into explanatory service. Section 12.1 canvasses reasons to go to the thermodynamic limit, in the form of explanatory aspirations that are stymied by ordinary quantum statistical mechanics. One aspiration is to explain phase structure, the putative coexistence, at certain temperatures, of macroscopically distinct equilibrium phases—solid, liquid; paramagnetic, ferromagnet; gaseous, condensed. Section 12.2 sketches an account of phase structure made available in the thermodynamic limit, and §12.3 argues that none of our extremist positions sustain the account. Section 12.4 explores what this might imply for how we understand the contents of physical theories. The next chapter continues the exploration in the context of another phenomenon, a phenomenon QFT and the thermodynamic limit of QSM alike seek to accommodate: broken symmetry.

12.1 The thermodynamic limit: why go there?

"The miracle of thermodynamics," Geoffrey Sewell writes, "is that the equilibrium states of systems of enormously many particles can be classified in terms of just a few macroscopic variables" (2002, 144). Consider a gas, a system composed of enormously many ($\sim 10^{23}$) gas molecules, where the classical characterization of *each one* of these

12.1 THE THERMODYNAMIC LIMIT: WHY GO THERE?

molecules requires the specification of its (three component) position and its (three component) momentum. Thus, to specify the state of the gas by way of specifying the states of its microconstituents, one must supply on the order of 6×10^{23} numbers. Nevertheless, at equilibrium, the gas as a collective is characterized by a few bulk features—its energy, its pressure, its temperature, its volume, and so on—which remain more or less constant in time, and which manifest regularities captured by the laws of thermodynamics.

Statistical mechanics aims at a microphysical account of the bulk behavior of macrosystems, an account meant to explain thermodynamic laws its bulk properties obey. From the vantage of statistical mechanics, the temperature of a macrosystem corresponds to the mean kinetic energy of its microconstituents; its equilibrium state at this temperature corresponds to a particular probability distribution over microconfigurations of the system's $\sim 10^{23}$ microconstituents that are consistent with the bulk features of the system. Work on the foundations of statistical mechanics has focused on the nature and adequacy of the explanations and/or theoretical reductions it offers (see Sklar 1993, Albert 2000, or Batterman 2002 for a sample). For now, we will not engage such issues. Instead, this section will canvass explanatory aspirations practitioners of statistical mechanics actually adopt, and show how ordinary QSM stymies them. For the purposes of the demonstration, I will understand ordinary quantum statistical mechanics to be as follows:

Ordinary QSM. Ordinary QSM addresses systems of finitely many microconstituents. It identifies the observables pertaining to a system with the self-adjoint elements of a Type I von Neumann factor $\mathfrak{B}(\mathcal{H})$; identifies possible states of the system with density operators in $\mathfrak{B}(\mathcal{H})$; and supposes the evolution of the system to be implemented by a one (real) parameter family of unitaries $U(t) = e^{-iHt}$ generated by the Hamiltonian observable H.

The equilibrium state of a system with Hamiltonian H at inverse temperature $\beta = \frac{1}{kT}$, if it exists, is the *Gibbs state* given by the density operator

$$\rho = \exp(-\beta H)/\text{Tr}[\exp(-\beta H)] \qquad (12.1)$$

This equilibrium state is is well-defined only if the spectrum of $\hat{H}$ is pure discrete (see § 7.3). ♠

You can do a lot within the confines of ordinary QSM. Among the things you can't do are accommodate ergodicity, phase structure, and broken symmetry.

12.1.1 Ergodicity

Ergodicity classically . . . Ergodicity is a daunting topic. Fortunately, I need only scratch its surface to exhibit impediments ordinary QSM throws in its way.[2] In a slogan, an

---

[2] The following presentation is quite informal. For more extensive treatments, see Sklar (1993, ch. 5), Uffink (2007), and Emch (2007).

ergodic system is one for which phase averages (i.e. averages of observable values over accessible regions of phase space) equal time averages (i.e. averages of observable values over allowed dynamical trajectories).

To explicate the slogan, consider a classical mechanical system whose phase space M is equipped with deterministic dynamical trajectories. Let G be a hypersurface of constant energy in that phase space. Take a region R of that hypersurface, and a dynamical trajectory through phase space. The trajectory is ergodic iff in the limit as time $t \to \infty$, the proportion of the time the trajectory spends in R is the same as the proportion of G's area taken up by R. (This latter proportion is calculated by appeal to the standard equilibrium measure on phase space; the time-invariance of that measure is essential to the meaningfulness of the calculation.) We are considering deterministic systems. Hence each point of phase space lies on at most one dynamical trajectory. A point of phase space is defined to be ergodic iff the trajectory on which it lies is. And so a system is ergodic iff all points (except possibly for a set of points of measure 0 in the standard equilibrium measure) of its phase space are.

The von Neumann–Birkhoff ergodic theorem identifies a condition necessary and sufficient for a system to be such that the time averages and phase averages for suitable observables coincide along "most" trajectories. The condition is *metric transitivity*, which forbids the decomposition of a hypersurface G of constant energy into a pair of regions with positive measure such that no trajectory starting in one region ever reaches the other. As Uffink observes, "metric transitivity captures in a measure-theoretic sense the idea that trajectories wander wildly across the energy hypersurface" (2007, 1008)—the idea animating Boltzmann's original ergodic hypothesis: that an arbitrary trajectory will eventually visit every point in phase space.[3]

In an ergodic system, the infinite time average of an observable (understood as a phase function, i.e. a map from elements of phase space to $\mathbb{R}$) along a trajectory exists and is independent of the trajectory's starting point (except possibly for a set of points of measure 0 in the standard equilibrium measure). Moreover, the only invariant probability distribution that assigns probability 0 to the same regions to which the equilibrium probability distribution assigns probability 0 is the equilibrium probability distribution.

Ergodicity is thought by some—and adamantly denied by others—to be crucial to the explanation of irreversible behavior. And ergodicity is invoked by some—and adamantly abjured by others—to justify the use of the standard equilibrium measure. Without evaluating explanatory aspirations which rest on extending ergodicity into the quantum setting, I here examine what such an extension would require.

---

[3] The hypothesis fails if G is a manifold of more than one dimension. Ehrenfest's "quasi-ergodic hypothesis" is that an arbitrary trajectory will come arbitrarily close to every point in G. For a discussion of Ehrenfest's argument that the quasi-ergodic hypothesis entails the equality of time and phase averages, see Sklar (1993, 160).

12.1 THE THERMODYNAMIC LIMIT: WHY GO THERE?

...but (not) in ordinary QSM There's a quick and dirty argument that the notion of ergodicity—explicated just now in the classical setting—does not generalize to ordinary QSM. The explication of classical ergodicity appeals to the standard equilibrium measure over accessible microstates. In the interest of keeping the argument quick and dirty, let us, without venturing further into the details of the explication of quantum ergodicity, suppose that those details involve probabilities assigned by the Gibbs equilibrium state (12.1). Now the argument: the Gibbs equilibrium state is well-defined only if the spectrum of the Hamiltonian H is pure discrete. A quantum evolution generated by a Hamiltonian with a discrete spectrum is periodic.[4] But ergodic evolution cannot be. Intuitively, an ergodic trajectory "wanders wildly," but a periodic one does not. Returning to its starting point in a finite time, a periodic evolution can't possibly visit, or drop into a near neighborhood of, each of the infinitely many states possible for a system. Therefore any Hamiltonian that supports equilibrium in ordinary QSM foils ergodicity.

The aspirant to ergodicity within the confines of ordinary QM need not give up hope. For it could be that the state ϕ furnishing the time-invariant probability measures pertaining to an ergodic system is not a Gibbs equilibrium state. Freed of the demand to conspire in equilibrium as Eq. (12.1) understands it, the Hamiltonian H of a would-be ergodic system needn't have a discrete spectrum. Allowing part of H's spectrum to be continuous, we avert the quick and dirty argument from a discrete Hamiltonian to periodic (hence non-ergodic) evolution. And the following fact, establishing as it does the existence of a family of observables whose infinite time averages *do* exist, extends hope for an account of ergodicity in ordinary QM:

Fact 12.1 (ergodic limits...). Let $\mathfrak{C} \subset \mathfrak{B}(\mathcal{H})$ be the algebra of compact operators on $\mathfrak{C}$.[5,g] Let $\rho(t) = U(t)\rho U(-t)$ for $U(t) = e^{-iHt}$, where ρ is an arbitrary density operator in $\mathfrak{B}(\mathcal{H})$. Then for each $A \in \mathfrak{C}$, the limit:

$$\lim_{T \to \infty} \frac{1}{T} \int_0^T Tr(\rho(t)A) dt \tag{12.2}$$

exists. The set of such limits defines a positive linear functional ρ_∞ on $\mathfrak{C}$ (Emch 2007, 1098).

Eq. (12.2) is naturally interpreted as specifying the infinite time average of the observable A: just the sort of average that must exist for a system to be ergodic. The linear functional ρ_∞ is a state on $\mathfrak{C}$ encapsulating these infinite time averages.

The catch, for ordinary QSM, is the nature of any state on the observable algebra $\mathfrak{B}(\mathcal{H})$ that induces an ergodic-average-encapusulating state ρ_∞ on $\mathfrak{C}$.

---

[4] Emch (2007, 1096–8) explicates the connection.
[5] A is compact if ψ_n converges weakly to ψ implies $A\psi_n$ converges strongly to $A\psi$. Density operators are an example. $\mathfrak{C}$ is a C^* algebra.

Fact 12.2 (... and normal states). If the spectrum of H is continuous, then ρ_∞ can't be extended to a countably additive state on $\mathfrak{B}(\mathcal{H})$ (Emch 2007, 1098).

In ordinary QM, states must be countably additive to be physically admissible. When H has a purely continuous spectrum (or even a spectrum with a continuous interval (Emch 2007, 1098)), the states defined by ergodic averages aren't countably additive. This leaves the aspirant to ergodicity in ordinary QM with a dilemma: either H is pure discrete and the evolution it generates, as periodic, is non-ergodic; or H is not pure discrete, and the ergodic averages it underwrites define unphysical states.

Impediments to quantum ergodicity can be explicated in a way that reveals them to root in the very structure of ordinary QM. In the interest of full generality, let us characterize the quantum setting by what I'll call a C^* *statistical dynamical system*.

Definition 12.3 (C^* statistical dynamical system). A C^* statistical dynamical system is a triple $(\mathfrak{A}, \alpha_t, \phi)$. $\mathfrak{A}$ is a C^* algebra whose self-adjoint elements represent observables. α_t is an automorphism group which encodes the dynamics, in the sense that for all $A \in \mathfrak{A}$, $\alpha_t(A)$ represents its evolution through a time t. Finally, ϕ is a α_t-invariant state.

ϕ might, for instance, be an equilibrium state assigning probabilities in whose terms we might hope to explicate ergodicity. Thus, a C^* statistical dynamical system is just a C^* dynamical system $(\mathfrak{A}, \alpha_t)$ (see §7.3) equipped with an α_t-invariant state ϕ on $\mathfrak{A}$ from which statistics might be extracted.

Turning to ordinary QM, where $\mathfrak{A}$ is a von Neumann algebra $\mathfrak{B}(\mathcal{H})$ and α_t is implemented by a continuous family of unitary operators $U_t \in \mathcal{H}$, we can dash this hope in two steps. First, we'll announce (but fall far short of thoroughly motivating; for this, see Benatti 1993, ch. 4.3) a criterion for the ergodicity of the C^* dynamical system $(\mathfrak{A}, \alpha_t, \phi)$, and then we'll show that any theory of ordinary QM must violate the criterion.

The criterion of ergodicity for a quantum dynamical system is developed by analogy with the account of classical ergodicity, cashed out in terms of a C^* dynamical system whose algebra $\mathfrak{A}$ is abelian (i.e. commutative). An immediate impediment to the analogy is the failure of quantum observable algebras to be abelian. Fortunately, there is a sense in which an arbitrary observable algebra might, with the aid of the right sort of dynamics, be "abelian."

Definition 12.4. [asymptotically abelian]. A C^* dynamical system $(\mathfrak{A}, \alpha_t, \phi)$ is (norm) *asymptotically abelian* if and only if $\lim_{t \to \infty}[B, \alpha_t C]$ converges in C^* norm to 0 for all $B, C \in \mathfrak{A}$.

In the limit of large t, and independent of the system's starting state, an asymptotically abelian evolution α_t eradicates correlations between observables pertaining to the

12.1 THE THERMODYNAMIC LIMIT: WHY GO THERE?

system at present (B) and in future ($\alpha_t(C)$). Put loosely, the present configuration of an asymptotically abelian system tells us little about its distant future. Such asymptotic statistical independence is a hallmark of classical ergodicity.

Extending the analogy with the classical account (see Benattti 1993, ch. 4.3) leads to the following definition:

Definition 12.5 (ergodic system). A C^* dynamical system $(\mathfrak{A}, \alpha_t, \phi)$ is *ergodic* if it is asymptotically abelian, and ϕ is extremal among the set of α_t-invariant states.

(Notice that such a ϕ is also an ergodic state in Def. 7.28's sense of being an extremal member of the set of α_t-invariant states.) In lieu of attempting to motivate the definition, I'll proceed directly to showing that systems satisfying it aren't to be had in ordinary QM. The reason is simple. If α_t is *inner*, i.e. if $\alpha_t(A) = U_t A U_t^*$ for some unitary element U_t of $\mathfrak{A}$, then its action cannot be both asymptotically abelian and non-trivial (Benatti 1993, ex. 4.45). One way to see this is to let C in Def. 12.4 above be U_s for a finite $s \in \mathbb{R}$. Then the assumption that α_t is inner coupled with Def. 12.4 implies that $\lim_{t\to\infty}[B, U_t U_s U_t^*] = [B, U_s] = 0$ for all $B \in \mathfrak{A}$. This implies that U_s is the identity for all $s \in \mathbb{R}$, which implies in turn that the evolution is trivial.[6] "This case is of very little interest for a quantum statistical mechanics at finite temperature," Emch comments (1984, 450).

In ordinary QM, dynamics are implemented by a family of unitaries in the von Neumann algebra $\mathfrak{B}(\mathcal{H})$ of quantum observables. That is, dynamics are inner. Thus ordinary QM precludes dynamical systems that are asymptotically abelian, and therefore precludes dynamical systems that are ergodic.

Supposing the α_t-invariant state ϕ of a would-be ergodic system $(\mathfrak{A}, \alpha_t, \phi)$ is an α_t-KMS state, we can draw a moral slightly more general. Let $\mathfrak{A}$ be a von Neumann factor algebra. If ϕ is an α_t-KMS state, then α_t is the modular group ϕ defines for $\mathfrak{A}$. This modular group is inner iff $\mathfrak{A}$ is semifinite—i.e. Type I (the case of ordinary QM) or Type II$_1$ (see Fact 7.30). Thus if $\mathfrak{A}$ is a semifinite factor von Neumann algebra and ϕ is an α_t-KMS state, the statistical dynamical system $(\mathfrak{A}, \alpha_t, \phi)$ cannot be ergodic.

The ergodicity of equilibrium states is unattainable in settings where the Hilbert space representation of the quantum physics is unique up to unitary equivalence. In his 1984 book, Emch concludes that "the framework provided by the von Neumann postulates [i.e. the framework of ordinary QM] is too restrictive for formulation of quantum statistical mechanics with good ergodic properties" (1984, 432). Some 20 years later he remarks: "one can hardly resist the conclusion that quantum ergodic theory is now a mature mathematical theory in search of further physical applications"

---

[6] Replacing the demand for norm-convergence in Def. 12.4 with demands for weak or strong convergence in ϕ's GNS representation lends no aid to the pursuit of ergodicity in the context of ordinary QM. Short of the thermodynamic limit, no group actions are asymptotically abelian on the GNS representations of canonical equilibrium states for finite systems (Emch 1984, ch. 10, schol. 5).

(Emch 2007, 114). This may take some of the sting out of the failure of ordinary QM to accommodate ergodicity. The balance of this section, conjuring further physical applications beyond the scope of ordinary QM, puts the sting back in.

12.1.2 Phase structure

Iron is *ferromagnetic*: in the absence of an external magnetic field, above the Curie temperature 771° Celsius an iron bar at equilibrium exhibits no net magnetization (paramagnetic phase); below 771°, an iron bar at equilibrium exhibits spontaneous magnetization (ferromagnetic phase); at 771°, both phases, ferromagnetic and paramagnetic, are possible for it. Fig. 12.1 gives the phase diagram for a ferromagnet. It plots the magnitude of the magnet's spontaneous (because the external magnetic field B is assumed to be 0) magnetization **m** as a function of temperature. An abrupt transition—from multiple, non-zero spontaneous magnetizations to no spontaneous magnetization—occurs at the temperature T_C. Fig. 12.2 depicts "isotherms" for the same ferromagnet, constant-temperature plots of its net magnetization **m** as a function of an external magnetic field B for temperatures at and below T_C. (B and **m** are assumed to lie along the same axis.) Again, an abrupt transition—from $T \geq T_C$ isotherms with well-defined first derivatives to $T < T_C$ isotherms without—registers at T_C.

Phase coexistence of the sort manifested by the iron bar is pandemic in the natural world. At normal atmospheric pressure, water in equilibrium at 0° centigrade can be solid or liquid; at 100° it can be liquid or vapor. In equilibrium at 2.17° Kelvin, ^{4}He can occupy either a fluid or a superfluid phase. In the latter, a form of Bose–Einstein condensation, the vast majority of the substance's atoms occupy the quantum ground state, endowing the substance with surprising properties, such as vanishing viscosity. In equilibrium at 3.2° Kelvin, mercury enjoys both a solid and a *superconducting* phase; in the latter, also understood (via the BCS model) as a macroscopic quantum effect,

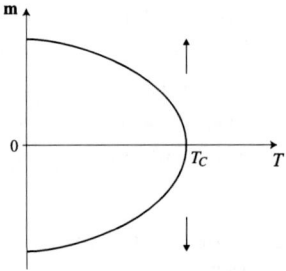

Figure 12.1. Phase diagram of a ferromagnet

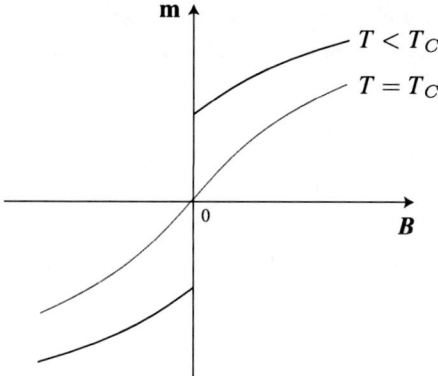

Figure 12.2. Isotherms for a ferromagnet

mercury's electrical resistance is 0. These examples illustrate *phase structure*, the putative availability, to certain systems in equilibrium at certain temperatures, of a variety of phases, differentiated from one another by drastically different bulk features.

The abrupt transitions in bulk features which differentiate phases are often dramatic changes in the value of an *order parameter*. For example, the ferromagnetic phase, which requires the spins of most of the (outer shell) electrons in an iron bar to line up, is more highly ordered than the paramagnetic phase, in which the spins of constituent electrons, distributed every which way, tend to cancel one another out, rendering the net magnetization **m** a gauge of how ordered the bar's constituents are. The abrupt change in the value of this order parameter **m** at the critical temperature T_c is a *critical phenomenon*. Its essence lies in how **m** varies with the temperature T of the bar as the critical temperature T_c is approached. Described by a function $(T_c - T)^\beta$, **m**'s critical behavior is captured by the *critical exponent* β. The very same critical exponent suffices as well to characterize the critical behavior of a wide range of substances—an example of what's known to physicists as "universality," and studied by the technique of renormalization group theory, which a popular QFT text deems "the most important conceptual advance in quantum field theory over the last three or four decades" (Zee 2010, 356). These are *prima facie* reasons to take phase structure and its investigation seriously as projects in living physics.

If phase structure is the availability, to a system at equilibrium at certain temperatures, of macroscopically distinct thermodynamic phases, then one might aspire to model phase structure by establishing the existence, at these temperatures, of *multiple distinct* equilibrium states. Supposing different states in this equilibrium set to correspond to different thermodynamic phases, we have the state structure requisite for representing the different thermodynamic phases accessible to a system at equilibrium. (A further explanatory aspiration, to account for the *dynamics* of phase transitions, I will not discuss here.)

As with explanatory aspirations grounded in ergodicity, the aspiration to model phase structure is dashed by ordinary QSM. The reason is that in ordinary QSM, the Gibbs state for a finite system at a given temperature and with respect to a given dynamics is, if well-defined, unique. Thus in ordinary QSM, there simply isn't the sort of multiplicity in equilibrium states at a fixed temperature the desired explanation of phase structure requires. This reinforces what may be a more familiar way of making a similar point: short of the thermodynamic limit, when the number of microsystems under consideration is supposed to be finite, the discontinuities in thermodynamic functionals signaling phase transitions simply don't occur.[7]

12.1.3 Broken symmetry

The time evolution of physical systems is governed by their dynamical laws. A theory's dynamical laws serve to distinguish histories of a system which are possible, according to that theory, from those which are impossible. Consider, for instance, a pair of masses in otherwise empty space. Some sequences of configurations of this two body system—e.g. ones where one mass traverses an elliptical orbit with the other mass at one focus—obey Newtonian mechanics and Newton's law of universal gravitation; other sequences of configurations—e.g. ones where one mass traverses a rectangular orbit about the other—do not.

Call a sequence of configurations satisfying the dynamical laws of Newtonian celestial mechanics a *history*. Here is a noteworthy feature of the set of histories: if you take any history in the set and "rotate it" (e.g. in the elliptical orbit sequences, tip over the plane containing the major and minor axes of the ellipse), you'll wind up with another history in the set. This indicates that the dynamical laws are *rotationally symmetric*; they favor no direction in space over any other. Rotation, an operation one can perform on configurations (or states) of the two body system, is a transformation implementing this symmetry. In general, and to first approximation, let us understand a theory's *dynamical symmetries* as transformations on the theory's state space under which the theory's set of dynamically allowed histories is stable.[8]

Now consider the aspiration to explain *broken symmetry*. For now (Chapter 13 will consider the question more thoroughly), let us say that a symmetry is broken when a physically significant state of a system—its ground state, say, or an equilibrium state—fails to be invariant under the system's dynamical symmetries. Nature appears to

---

[7] For a treatment in the context of the ferromagnetic-paramagnetic phase transition, see Emch and Liu (2002, 397).

[8] See Earman (1989, ch. 3) if you want this spelled out. We might legitimately want more from a symmetry than the stability under its action of a set of solutions. We may e.g. think the space of solutions itself has structures which any self-respecting symmetry would preserve. For instance, Wigner's venerable argument that quantum symmetries are implemented unitarily or anti-unitarily requires those symmetries to preserve transition probabilities, and we might expect symmetries of a theory of canonical type $\mathfrak{A}$ to preserve the algebraic structure of $\mathfrak{A}$ or its self-adjoint part (see Emch 1972, 150–59, for a discussion). The next chapter will develop this point.

abound with equilibrium states that break dynamical symmetries. Ferromagnetism furnishes one example. Without delving into the dynamics of ferromagnetic substances, let us suppose them to be rotationally invariant. Ferromagnetic equilibrium states break this symmetry—spontaneously magnetized, an iron bar has a net magnetization lying in a particular direction; neither this direction nor the state of the iron bar are invariant under arbitrary rotations. Crystals are another example—crystalline states generally break Euclidean symmetries but abide by those of a crystallographic subgroup of the Euclidean group.

Ordinary QSM quashes any hope of accommodating symmetry-breaking equilibrium states. A theory of ordinary QM equates possible states with density operators on a fixed Hilbert space $\mathcal{H}$ and implements dynamics by a family of unitary operators on $\mathfrak{H}$ generated by a self-adjoint Hamiltonian $H \in \mathfrak{B}(\mathcal{H})$. A symmetry of these dynamics corresponds to a unitary operator V that commutes with the Hamiltonian. (V's unitarity makes it a symmetry; its commuting with H makes it dynamical.) The Gibbs state ρ given by Eq. (12.1) is the equilibrium state with respect to these dynamics. But a simple argument establishes that ρ is invariant under any unitary V that commutes with H:

$$V \rho V^{-1} = V \frac{e^{-\beta H}}{Tr(e^{-\beta H})} V^{-1} = V V^{-1} \frac{e^{-\beta H}}{Tr(e^{-\beta H})} = \frac{e^{-\beta H}}{Tr(e^{-\beta H})} = \rho \qquad (12.3)$$

where V's commutativity is used in the second step, and its unitarity in the third. Within the confines of ordinary QSM, where the Gibbs criterion defines equilibrium, symmetry-breaking equilibrium states are not to be had.

Here is an argument that derives the same conclusion from the uniqueness of the Gibbs state in ordinary QSM. It is desirable that the *set* of equilibrium states for a given system at a given temperature be invariant under the dynamical symmetries of the theory governing that system. The idea is that, insofar as equilibrium is characterized *dynamically*, in terms of *stability* of bulk properties, you shouldn't be able to change something that is an equilibrium state into something that isn't by performing a transformation that's a symmetry of the dynamics. Of course, simply requiring the set of equilibrium states at a given temperature to be invariant under the dynamical symmetries of the system for which they're equilibrium states does not preclude the existence of equilibrium states that *break* the symmetry. (Think of unit vectors in $\mathbb{R}^2$ and rotations. An individual unit vector is not invariant under rotation, but the whole collection of them is.) If, however, the set of equilibrium states in question contains only one member, that set is invariant under dynamical symmetry transformations iff those symmetry transformations leave its sole member unchanged.

The uniqueness of the Gibbs equilibrium state in ordinary QSM is what inhibited a microanalysis of phase structure there. We are now in a position to see that it also inhibits a microanalysis of equilibrium states that break dynamical symmetries. For, according to the considerations above, in ordinary QM, the set of equilibrium states at a finite temperature for a system assigned a specific Hamiltonian has at most one member, which must therefore be invariant under that system's dynamical symmetries.

The story thus far: ordinary QSM sets terms of engagement that render accounts of ergodicity, phase structure, and broken symmetry unattainable. Homing in on phase structure, the next section sketches a model of the phenomenon of phase structure made available by taking QSM out of the scope of the Stone–von Neumann and Jordan–Wigner theorems and to the thermodynamic limit.

12.2 Phase structure: a closer look

Short of the thermodynamic limit, the aspiration to model phase coexistence is scuppered by the uniqueness of the Gibbs state. But at the thermodynamic limit, explanatory hope is restored. This section sketches how. It begins with a sharper characterization of the *pure thermodynamic phases* that coexist at critical temperatures. Then it explains why the exposition of equilibrium appropriate to the thermodynamic limit accommodates the possibility of temperatures at which distinct equilibrium states, *corresponding to distinct pure phases*, occur.

This section will repeatedly make an example of the toy model of the infinite spin chain. Here is a review of the setup:

[Review: the infinite spin chain.] To model an infinite chain of spin $\frac{1}{2}$ systems, let the positive and negative integers $\mathbb{Z}$ index a lattice, with a spin $\frac{1}{2}$ system resident at each lattice point $k \in \mathbb{Z}$. Reflect this residence by assigning each k the algebra $\mathfrak{M}(2)_k$, a copy of the algebra of 2×2 complex matrices generated by the Pauli spins $\sigma_x, \sigma_y, \sigma_z$. Algebras associated with different lattice sites commute. Assign each finite subset $K \subset \mathbb{Z}$ the algebra $\mathfrak{A}_K = \oplus_{k \in K} \mathfrak{M}(2)_k$: the tensor product of the algebras assigned elements of that subset. (For notational convenience, limit attention to subsets K that range from $-k$ to k for some integer k; K will approach $\mathbb{Z}$ in the limit $k \to \infty$.) Notice that these algebras satisfy Isotony: for $L \subset K$, $\mathfrak{A}_L \subset \mathfrak{A}_K$. The *quasilocal algebra* $\mathfrak{A}$ for this infinite spin chain is the C^* algebra obtained as the norm closure of $\cup_{K \subset \mathbb{Z}} \mathfrak{A}_K$. The infinite spin chain is symmetric under discrete translations that map lattice points to lattice points; this symmetry is implemented by the automorphism group $\tau(a)$ ($a \in \mathbb{Z}$) of $\mathfrak{A}$, with $\tau(b)$ understood as the automorphism effecting a shift of b steps along the lattice. The reader may want to revisit Chapter 5 to verify that $\mathfrak{A}_{CAR}$ so-configured satisfies the axioms of algebraic QSM. ♠

12.2.1 Pure phases

Pure phases are distinguished from one another by macroproperties, properties of the bulk thermodynamic system considered as a whole: the *global density* of water is lower than that of ice; the *global magnetization* of an iron bar in its ferromagnetic phase is positive; in its paramagnetic phase, it's 0, and so on. What's more, in a pure thermodynamic phase, these distinguishing macroproperties are dispersion-free: the pure liquid phase of water assigns the global density observable the *liquid* value with

12.2 PHASE STRUCTURE: A CLOSER LOOK

certainty: water, water everywhere.[9] Mixed phases, by contrast, exhibit dispersions in distinguishing macroproperties: a statistical mixture of liquid and solid phases predicts a dispersion in the global density of the substance; this dispersion can be interpreted as emblematic of phase coexistence: there is water here and ice there.

These distinguishing macroproperties are exactly the properties whose abrupt changes signal phase transitions. The aim of the present section is twofold. First, to explicate an account of macroproperties afforded by the thermodynamic limit of QSM.[10] Second, to register a technical claim crucial to the thermodynamic limit's accommodation of phase structure: the claim that *like pure thermodynamic phases, factor states assign dispersion-free values to macroproperties.*

Candidate macroproperties of an infinitely extended quantum system are observables that pertain to the system as a whole: the infinite volume limit of a space average of local observables, for example. Let us recall the infinite spin chain to illustrate. Elements of the algebras $\mathfrak{M}(2)_k$ associated with particular sites on the lattice of spins are *local observables*. So are elements of the algebras $\mathfrak{A}_K$ associated with finite collections $K = [-k, k]$ of sites. These algebras, which as finite tensor products of the two dimensional matricial algebra $\mathfrak{M}(2)$ are von Neumann algebras, pertain to microconstituents of the bulk system and finite collections thereof. The norm closure of their union is the quasilocal C^* algebra $\mathfrak{A}$, which pertains to the bulk system in its entirety and specifies its canonical type. Now consider the local observable $A_k \in \mathfrak{M}(2)_k$, an instance of the generic local observable A. A's average over a finite subset K of $\mathbb{Z}$ is also a local observable, the element of $\mathfrak{A}_K$ given by:

$$A_K = \frac{1}{2k+1} \sum_{i=-k}^{k} A_i \qquad (12.4)$$

The lattice has the symmetries $\tau(a)$ because its points are evenly distributed in space. Thus dividing $\sum_{i=-k}^{k} A_i$ by $\frac{1}{2k+1}$ on the r.h.s. of (12.4) amounts to taking a *spatial average* of the local quantity A. A_K, as defined by (12.4), is a *spatially averaged observable*.

The paradigmatic macroscopic observable is an "infinite volume" limit of spatially averaged local observables such as A_K.[11] Such an infinite volume limit, capturing a bulk feature of the system in its entirety, expresses a macroscopic observable, call it **A** (note the convention that macroscopic observables are bold-faced). **A** would be defined as a member of $\mathfrak{A}$ by the expression:

---

[9] Indeed, classically a pure phase is characterized as one in which the global densities of extensive conserved quantities, such as energy and specific heat, are dispersion-free.

[10] My primary sketch follows Sewell (2002; 1996), which follow Hepp's (1972) account of macroscopic observables. Emch (1984, 386–402) describes a related approach, due to Lanford and Ruelle, in terms of "observables at ∞." The approaches share the features relevant for my purposes, in particular, the representation-dependence of the macroscopic observables.

[11] Hepp's (1972) account identifies macroscopic observables with infinite volume limits of arithmetic means of uniformly bounded sequences of local observables. I focus on the simplest case of spatial averages because the case includes the macroscopic observables relevant to models of phase structure discussed here.

$$\mathbf{A} = \lim_{n \to \infty} \frac{1}{2n+1} \sum_{k=-n}^{n} A_k \qquad (12.5)$$

if only the limit on the r.h.s. were to be had in the topology defined by $\mathfrak{A}$'s C^* norm, the only topology available at present that's applicable to the system as a whole. (Because the von Neumann algebras $\mathfrak{A}_K$ pertain to *finite* sets of lattice points, $\mathbf{A}$ won't occur in them.) But in general, the sequences of sums defining macroscopic observables won't have norm limits.

Recall, for example, $\mathbf{m}_z$, the (z-component of the) global polarization of the infinite spin chain, defined on the model of equation (12.5), with $\sigma(z)_k$ in the place of A_k. $\mathbf{m}_z$, the infinite volume spatial average of the z components of the spins constituting the chain, gives the chain's net magnetization along the z-axis (see Ex. 4.15). Operationalized by such bulk behavior as the capacity to move iron filings, this net magnetization $\mathbf{m}_z$ is a macroscopic observable. The sequence of partial sums that would define $\mathbf{m}_z$ can fail to converge in the uniform topology of a representation of $\mathfrak{A}$. For instance, $\mathbf{m}_z$'s defining sequence of sums fails to converge uniformly in a representation π_x built on the models of π_+ and π_- but whose base state sandwiches a +eigenstate of $\sigma(z)$ between 4 −eigenstates, which are sandwiched in turn between 16 +eigenstates, ...But, because representations must be norm continuous, this implies that the sequence of partial sums that would define $\mathbf{m}_z$ fails to converge in $\mathfrak{A}$'s C^* norm. Not defined as a norm limit, the global polarization is not an element of $\mathfrak{A}$.

The lesson generalizes to other situations where a quasilocal C^* algebra $\mathfrak{A}$ supplies the thermodynamic ($V \to \infty$) limit of a net of local algebras A_V associated with systems occupying finite volumes. These can be continuous systems, such as the Bose gas (Araki and Woods 1963), as well as lattice ones. Macroscopic observables, understood as infinite volume limits of spatial averages of local observables, are typically absent from $\mathfrak{A}$, due to the failure of the expressions defining those averages, thence those observables, to converge in $\mathfrak{A}$'s norm topology.

However, and returning to the case of the infinite spin chain, given a particular state ϕ on $\mathfrak{A}$, the quantity:

$$\mathbf{A}(\phi) = \lim_{n \to \infty} \frac{1}{2n+1} \sum_{k=-n}^{n} \phi(A_k) \qquad (12.6)$$

will exist if the sequence of partial sums of *real numbers* $\frac{1}{2n+1} \sum_{k=-n}^{n} \phi(A_k)$ converges. That sequence can converge, even if the r.h.s. of Eq. (12.5), a sequence of partial sums of *operators*, fails to converge in norm. Not an element of $\mathfrak{A}$, a macroscopic observable $\mathbf{A}$ may nevertheless be well-defined *for a state ϕ on $\mathfrak{A}$*, in the sense that the r.h.s. of Eq. (12.6) converges for that state.

This prompts a picture of the macroscopic observable $\mathbf{A}$ as a *state-dependent functional*, a map from states ϕ to real numbers $\mathbf{A}(\phi)$, understood as expectation values of $\mathbf{A}$. In general, the domain of a macroscopic observable so understood is not the full collection of states on the quasilocal algebra $\mathfrak{A}$. For that collection could include states foiling the convergence of the r.h.s. of Eq. (12.6). The domain of the macroscopic observable **A** consists of those states on $\mathfrak{A}$ for which the r.h.s. of Eq. (12.6) converges. Notice that the domain of a macroscopic observable that's a *spatial* average over the infinite volume occupied by a system at the thermodynamic limit will include *translationally invariant* states, simply because in a translationally invariant state, local averages—averages over an index set of finite cardinality—and global averages of local observables will coincide. As the former are well-defined, so will be the latter.

For an alternate characterization of the domain of a macroscopic observable, observe that the failure of its defining limit $\lim_{n\to\infty} \frac{1}{2n+1} \sum_{k=-n}^{n} A_k$ to converge in $\mathfrak{A}$'s C^* norm needn't imply the failure of:

$$\lim_{n\to\infty} \frac{1}{2n+1} \sum_{k=-n}^{n} \pi(A_k) \qquad (12.7)$$

to converge in the weak or strong topologies associated with some concrete representation π of $\mathfrak{A}$. When (12.7) does converge weakly, it defines an observable, let's call it $\mathbf{A}^\pi$, in the von Neumann algebra $\pi(\mathfrak{A})''$. Let ϕ be a π-normal state. The weak convergence of (12.7) will be sufficient for ϕ to belong to the domain of the macroscopic observable **A** (conceived as state-dependent functional). For that weak convergence implies the convergence of the r.h.s. of Eq. (12.6) *for every state in ϕ's folium*. (Hence it's a stronger condition than the criterion for ϕ falling in **A**'s domain that the r.h.s. of Eq. (12.6) converge for ϕ.) $\mathbf{A}^\pi$ is the operator-algebraic implementation of the macroscopic observable **A** with respect to the representation π.

Let us turn to the infinite spin chain for illustration.

Example 12.6 (macroscopic observables for the infinite spin chain). Section 3.3.1 exhibited two concrete Hilbert space representations of the quasilocal algebra $\mathfrak{A}$ of the infinite spin chain: the representations π_+ and π_-. In the base state ϕ^+ of the former, every spin in the chain occupies the $\sigma(z)^+ (:= \pi_+(\sigma_z))$ eigenstate $|+\rangle$; the representing Hilbert space $\mathcal{H}^+$ is spanned by states that can be obtained from ϕ^+ by a finite number of spin flips. In the base state ϕ^- of the latter, every spin in the chain occupies the $\sigma(z)^- (:= \pi_-(\sigma_z))$ eigenstate $|-\rangle$; the representing Hilbert space $\mathcal{H}^-$ is spanned by states that can be obtained from ϕ^- by a finite number of spin flips. It should be plausible that both ϕ^+ and ϕ^- are translationally invariant.

The von Neumann algebra $\pi_+(\mathfrak{A})''$ affiliated with π_+ included a global polarization observable $\mathbf{m}^+$ that records the net magnetization of this system; the von Neumann algebra $\pi_-(\mathfrak{A})''$ affiliated with π_- includes a global polarization observable $\mathbf{m}^-$. These macroscopic magnetization observables aren't members of their respective representations $\pi_\pm(\mathfrak{A})$. Rather, they're obtained as weak limits of sequences of such members. For example, where $\sigma(z)_k^+$ is the representative under π_+ of the z-component of the Pauli spin of the electron at site k on the chain, the z-component of global polarization is defined by the limit:

$$\mathbf{m}_z := \lim_{n \to \infty} \frac{1}{2n+1} \sum_{k=-n}^{n} \sigma(z)_k^+ \qquad (12.8)$$

which, Ex. (4.15) argued, converges in $\mathcal{H}^+$'s weak topology. ♠

A crucial fact about macro-observables is:

Fact 12.7. If ϕ on $\mathfrak{A}$ is a factor state in the domain of a macroscopic observable $\mathbf{A}$, then $\mathbf{A}$ has the same value for every state in ϕ's folium (Sewell 2002, 43).

Fact 12.7 has the simplifying consequence that the convergence of the r.h.s. of Eq. (12.6) gaining a factor state ϕ's entry into the domain of the macroscopic observable $\mathbf{A}$ implies that Eq. (12.7) has a weak limit in ϕ's GNS representation. This weak limit identifies the element $\mathbf{A}^{\pi_\phi}$ of $\pi_\phi(\mathfrak{A})''$ that is the avatar of $\mathbf{A}$ on that representation. Let's belabor why. Let the vector $|\psi\rangle \in \mathcal{H}_\phi$ implement the state $\psi : \psi(B) := \langle \psi | \pi_\phi(B) | \psi \rangle$ for all $B \in \mathfrak{A}$. Evidently, ψ belongs to ϕ's folium. Fact 12.7 requires that:

$$\mathbf{A}(\phi) = \mathbf{A}(\psi) = \lim_{n \to \infty} \frac{1}{2n+1} \sum_{k=-n}^{n} \psi(A_k) = \lim_{n \to \infty} \frac{1}{2n+1} \sum_{k=-n}^{n} \langle \psi | \pi_\phi(A_k) | \psi \rangle \qquad (12.9)$$

But if Eq. (12.9) holds for an arbitrary vector $|\psi\rangle \in \mathcal{H}_\phi$, Eq. (12.7) converges in the weak topology of ϕ's GNS representation, defining the macroscopic observable $\mathbf{A}^{\pi_\phi}$ relative to ϕ. Indeed, it converges to the observable $\mathbf{A}(\phi)I$. Because $\mathbf{A}$ is a scalar, this observable is just a multiple of the identity operator. $\mathbf{A}^{\pi_\phi}$ lies in the (trivial) center of the factor von Neumann algebra $\pi_\phi(\mathfrak{A})''$.

This is no coincidence. In general, macroscopic observables relative to ϕ lie in the center of $\pi_\phi(\mathfrak{A})''$ (Emch 1984, 394).[12] When ϕ is a factor state, the center of its affiliated von Neumann algebra is trivial, and those macroscopic observables are multiplies of the identity. Fact 12.7 has the reassuring consequence that if $\mathbf{A}^{\pi_\phi}$ and $\mathbf{B}^{\pi_\phi}$ are both macroscopic observables relative to a factor state ϕ, then they commute with one another. Macroscopic observables that are space averages are moreover, and like the global intensive quantities characterizing pure thermodynamic phases, *dispersion-free* on factor states (Sewell 2002, 44).

We've forged this much of a connection between factor states and pure thermodynamic phases: the former exhibit dispersion-free values of macroscopic observables typical of the latter. But we want more. We want *different* factor states to behave like distinct thermodynamic phases. We already know from Fact 12.7 that distinct but quasi-equivalent factor states ϕ and ψ aren't different enough: quasi-equivalent states

---

[12] Emch's technical explication of "macroscopic observable" is provided by "observables at ∞." Crudely, the observables at ∞ relative to a state ϕ on $\mathfrak{A}$ reside in the intersection, over finite regions V, of the von Neumann algebras $\pi_\phi(\mathfrak{A}(\bar{V}))''$ affiliated with V's spatial complement $\bar{V}$. They're the observables associated with no finite region. Here's an intuitive argument that these include the spatial averages we've been treating as paradigmatic macroscopic observables: for any finite k, the expectation value ϕ defines for a local average over k sites *places no constraints on* the expectation value ϕ assigns the global average.

lie in one another's folia; quasi-equivalent factor states therefore agree with respect to all macroscopic observables in whose domains they fall. Factor states are either quasi-equivalent or disjoint (Fact 4.61). So our last hope is that disjoint factor states are macroscopically distinguishable. Fortunately

Fact 12.8. If ρ and ϕ are disjoint factor states, there exists a macroscopic observable **A** s.t. $\mathbf{A}(\rho) \neq \mathbf{A}(\phi)$ (Sewell 2002, 43).

Given Fact 12.7, this means that there's a macroscopic observable **A** whose value $\mathbf{A}(\rho)$ on every state in ρ's folium is different from its value $\mathbf{A}(\phi)$ on every state in ϕ's folium. The example of the infinite spin chain consolidates much of the foregoing.

Example 12.9 (disjoint factor states on the infinite spin chain). States ϕ^+ and ϕ^- on $\mathfrak{A}$ are both translationally invariant pure (ergo Type I factor) states. Each lies in the domain of the macroscopic global polarization observable **m**; thus $\mathbf{m}^+ \in \pi_+(\mathfrak{A})''$ and $\mathbf{m}^- \in \pi_-(\mathfrak{A})''$ are macroscopic observables relative to ϕ^+ and ϕ^-, respectively. Every π_+-normal state is an eigenstate of $\mathbf{m}^+$, with eigenvalue 1 along the positive z axis. That is, $\mathbf{m}^+ = \mathbf{m}(\phi^+)I$. Every π_--normal state is an eigenstate of $\mathbf{m}^-$, with eigenvalue -1 along the same axis. That is, $\mathbf{m}^- = \mathbf{m}(\phi^-)I$. Within the folium of each factor state, the macroscopic observable **m** takes the same value everywhere. ϕ^+ and ϕ^- are disjoint (see §3.3.1). The value of the macroscopic observable **m** on every state in ϕ^+'s folium is different from its value on every state in ϕ^-'s folium. Taking different values for **m**, those states, as well as the local modifications of those states collected into their folia, are macroscopically distinct. ♠

The foregoing suggests that factor states behave with respect to macroproperties in a manner worthy of pure thermodynamic phases: each factor state is dispersion-free with respect to macroscopic observables; disjoint factor states are "macroscopically distinguishable," in the sense that there is some macroscopic observable they assign distinct values. Next we harness this suggestion to evocative features of collections of equilibrium states to motivate an account of phase structure.

12.2.2 The set of equilibrium states

The Gibbs state is not in general well-defined for infinite systems. Nor is the Gibbs criterion for equilibrium applicable to the abstract C^* algebras in whose terms one might describe the physics of such systems. To pursue QSM in the thermodynamic limit, then, we must extend the notion of equilibrium afforded by the Gibbs state. The *KMS condition*, first encountered in §7.3, gives us a way to do so. Recall that a state ω on a C^* algebra $\mathfrak{A}$ is a *KMS state* with respect to the dynamical automorphism group α_t at inverse temperature β iff the formal expression:

$$\omega[A\alpha_{i\beta}(B)] = \omega(BA) \tag{12.10}$$

holds for all A, B in a dense subalgebra of $\mathfrak{A}$. We will say that ω is an (α_t, β)-KMS state. Section 7.3 catalogs reasons to regard (α_t, β)-KMS states as equilibrium states with respect to the dynamics α_t at inverse temperature β.

Suppose ω is an (α_t, β)-KMS state on $\mathfrak{A}$ for a finite inverse temperature β. Then it is typically the case at the thermodynamic limit that:

Fact 12.10. The von Neumann algebra $\pi_\omega(\mathfrak{A})''$ is a Type III factor (see §7.5).

The states ϕ^+ and ϕ^- of the infinite spin chain evince features of macroscopically distinct pure thermodynamic phases, features exploited in the account of phase coexistence. But that account will be in terms of equilibrium, i.e. KMS states, and we know that ϕ^+ and ϕ^- aren't (finite-temperature) KMS states. We know this because as pure states they're Type I factor states, whereas (according to Fact 12.10) finite temperature KMS states are typically Type III factor states. Example 12.11 shows that KMS states of the infinite spin chain adhere to this pattern.

Example 12.11 (KMS states of the infinite spin chain). We haven't yet said anything about the dynamics of the infinite spin chain, but let us suppose that they are finite in range and translation-invariant. (The Heisenberg model of §12.3.2 will be an example of such a dynamics: each spin interacts with and only with its nearest neighbors.) Every finite subset $[-k, k] \subset \mathbb{Z}$ falls within the scope of ordinary QSM. $\mathfrak{A}_K$ is just the concrete Type I von Neumann algebra $\otimes_{i=-k}^{k} \mathfrak{M}(2)_i$. The local dynamics define an automorphism group $\alpha_K(t)$ for $\mathfrak{A}_K$ that's implemented by unitary elements of $\mathfrak{A}_K$. With respect to these dynamics and at each inverse temperature β, the Gibbs recipe Eq. (12.1) defines an equilibrium state ρ_K.

Araki (1969) showed that in the thermodynamic limit—the limit as $k \to \infty$, that is, as K approaches the entirety of the spin chain—the local automorphism groups $\alpha_K(t)$ converge in norm to an automorphism group α_t of the quasilocal algebra $\mathfrak{A}$. In the same limit, the local states ρ_K uniquely determine a state ρ on $\mathfrak{A}$. What's more, ρ is an (α_t, β)-KMS state, indeed, a Type III factor state. ♠

We are at long last in a position to bring this discussion of phase structure to bear on a question of interpretation. Returning to the general case, we can draw out an unwelcome consequence of applying the interpretive policy of Hilbert Space Conservatism to QM_∞. If the von Neumann algebra $\pi_\omega(\mathfrak{A})''$ affiliated with a KMS state is a Type III factor, there is no irreducible representation of $\mathfrak{A}$ whose folium contains ω. Requiring physically possible states to reside in the folium of a an irreducible representation of $\mathfrak{A}$, the Hilbert Space Conservative denies that ω is physically significant. Refusing to extend physical significance to equilibrium states is not a promising way to interpret QSM.

An obvious way for the Hilbert Space Conservative to make room for equilibrium is to relax her requirement that the representation of $\mathfrak{A}$ whose folium contains all and only the physical states be an irreducible one. Relaxing this requirement implies dropping the expectation that the observable algebra takes the form $\mathfrak{B}(\mathcal{H}) (= \pi(\mathfrak{A})''$

for irreducible π). If the Hilbert Space Conservative relaxes too far—to the point that, say, the folium of the universal representation π_U becomes an acceptable receptacle for physically possible states—the distinction between her position and its rivals evaporates. But there are stable intermediate points that both preserve the distinctness of Hilbert Space Conservatism and enable it to count equilibrium states among the physical possibilities. For instance, the Hilbert Space Conservative could confine the space of physical possibilities to the folium of a *factor state* on $\mathfrak{A}$. She might cite as her motivation for doing so that insofar as pure states *are* factor states, this is a natural generalization of her original requirement.[13] Thus the relaxed Hilbert Space Conservative could count Type III factor equilibrium states as physically possible.

But this would bring her to a second awkward fact about KMS states. Let ω_1 and ω_2 be α_t-KMS states for inverse temperatures β_1 and $\beta_2 \neq \beta_1$ respectively.

Fact 12.12. If ω_1 and ω_2 are both Type III factor states, then they are are disjoint (Bratteli and Robinson 1997, thm. 5.3.35 (128–9)).

Fact 12.10 ensures that the antecedent will typically hold at the thermodynamic limit. So Fact 12.12 implies, of an infinite quantum dynamical system $(\mathfrak{A}, \alpha_t)$, that there is no single factor representation whose folium includes its equilibrium states at *different* temperatures. Not even the relaxed Hilbert Space Conservative can admit equilibrium states at different temperatures into her space of physically possible states.

In the setting of QSM's thermodynamic limit, Hilbert Space Conservatism commits us to some startling assessments of physical possibility. Could a sufficiently zealous supporter of the position, contending that only one temperature is ever actual, live with these assessments? Not, I'll try to show below, if she wants to sustain the aspiration to model phase structure.

Let $\mathfrak{A}$ be a C^* algebra equipped with a dynamical automorphism group α_t, and let K_β denote the set of (α_t, β)-KMS states (where $\beta \in \mathbb{R}$) on $\mathfrak{A}$. Here are a collection of suggestive truths about the structure of these sets K_β (see Bratteli and Robinson 1997, thm. 5.3.30 (116–17)). First,

Fact 12.13. K_β is convex.

We're investigating the aspiration to accommodate equilibrium temperatures at which multiple, distinct pure phases occur. This is just to accommodate the phenomenon of *mixed phases*, understood as the possibility for (or the presence to) a system in equilibrium of distinct pure thermodynamic phases, as when your hot coffee steams. The convexity of K_β suggests an accommodation strategy: regard extremal elements of K_β—those that can't be expressed as non-trivial convex combinations of other

---

[13] Further motivation emerges presently: KMS states form a convex set the extremal elements of which are factor states. So factor states play the role with respect to equilibrium that pure states play in general. Emch (1997) issues a plea for treating factor, rather than pure, states as the building blocks of quantum physics.

elements—as pure thermodynamic phases; regard non-extremal elements as mixed phases. The next fact suggests that we're on the right track:

Fact 12.14. $\omega \in \mathcal{K}_\beta$ is extremal iff it's a *factor* state.

Extremal KMS states, as factor states, bear the hallmark of pure thermodynamic phases: they assign dispersion-free values to the macrovariables in whose domain they lie. (For a roster of further reasons to identify extremal KMS states with pure thermodynamic phases, see Emch 2007, §5.7.)

It's a further virtue of the explanation we envision that $\mathcal{K}_\beta$ is a special sort of convex set.

Fact 12.15. Every $\omega \in \mathcal{K}_\beta$ can be represented as a *unique* convex combination of extremal elements of $\mathcal{K}_\beta$. That is to say, $\mathcal{K}_\beta$ is a *simplex*.

Insofar as the non-unique decomposibility of mixtures is an archetypal feature of quantum state spaces, this uniqueness is noteworthy. The fact that an arbitrary density operator $W \in \mathfrak{B}(\mathcal{H})$ does not legislate a preferred decomposition into pure states on $\mathfrak{B}(\mathcal{H})$ is what gives impetus to interpretations of ordinary QM which proceed by isolating a privileged set of pure states over which W might be understood to assign probabilities admitting an ignorance interpretation. (Some examples of such interpretations are discussed in Chapter 8 and include the collapse, modal, and "Bohm" interpretations.) By contrast, an arbitrary element of $\mathcal{K}_\beta$ carries its own interpretation in terms of a privileged decomposition into extremal elements of $\mathcal{K}_\beta$. Formally, this reflects the uniqueness of the central measure associated with a state $\omega \in \mathcal{K}_\beta$ (see Emch 1997 for elaboration). Physically it means that a KMS state that is not itself a pure thermodynamic phase admit a unique construal as a statistical mixture of pure thermodynamic phases.

The account of phase coexistence hinges on sets $\mathcal{K}_\beta$ of (α_t, β)-KMS states with multiple members. Extremal members of this set correspond to pure thermodynamic phases, different extremal members to macroscopically distinct such phases. So when $\mathcal{K}_\beta$ has multiple members, distinct pure thermodynamic phases are possible for a system in equilibrium at inverse temperature β. The next fact reveals why we had to go to the thermodynamic limit to construct this model.

Fact 12.16. If ω_1 and ω_2 are extremal elements of $\mathcal{K}_\beta$, then either they're equal or they're disjoint.

It follows that if $\mathcal{K}_\beta$ has multiple elements, and so "enough" states to accommodate phase structure, it has disjoint elements. But short of the thermodynamic limit and in the scope of the Stone–von Neumann and Jordan–Wigner theorems, an algebra of quantum observables *does not admit disjoint states*. Short of the thermodynamic limit, any non-empty set of (α_t, β)-KMS states contains as its only element the unique Gibbs states for those conditions. Thus we re-encounter the result that short of the thermodynamic limit, a set of equilibrium states at a fixed temperature lacks the multiplicity required to model phase coexistence.

We haven't said enough about the dynamics of the infinite spin chain to determine whether it admits sets $\mathcal{K}_\beta$ with multiple members. In fact, for the models in the scope of the Araki (1969) result—models with interaction that die off "fast enough" with distance—there are no phase transitions in one dimension. But the general point remains: the availability of unitarily inequivalent representations of the quasilocal algebra $\mathfrak{A}$ characterizing a theory at the thermodynamic limit at least opens up the possibility, foreclosed by ordinary QM, of such sets $\mathcal{K}_\beta$. And there are models, for instance, the anisotropic Heisenberg model (see Bratteli and Robinson 1997, ch. 6.2), for which the possibility is realized.

Now we may assemble the pieces laid out in this section into a schematic picture of phase coexistence. Phase coexistence occurs at those inverse temperatures β for which the set $\mathcal{K}_\beta$ of (α_t, β)-KMS states has more than one member, and in those states $\omega \in \mathcal{K}_\beta$ which are not extremal. Such an ω is a unique convex combination of extremal states. Each extremal state in this decomposition corresponds to a pure thermodynamic phase. Different extremal states, answering to different phases, are disjoint. Macroscopic observables, state-dependent functionals that lie in the centers of von Neumann algebras affiliated with extremal KMS states, effect the differentiation. The decomposition corresponds to the statistical separation of a system at equilibrium into pure thermodynamic phases. *This* is how to model phase structure at the thermodynamic limit.

12.3 Extremist obstructions to explanation

Chapter 1 announced a quasi-Hempelian criterion of adequacy for interpretations of physical theories. An adequate interpretation ought to attribute to a theory at least the content it requires to discharge its explanatory duties. Focusing on the explanation of phase coexistence, this section measures extremist interpretations of QM_∞ against this quasi-Hempelian standard, and argues that they fall short. Section 12.3.1 presents the *phase argument*, lodged at the same schematic level as the account of phase coexistence just sketched. According to the phase argument, none of our extremists can support accounts of that general form. Section 12.3.2 presents the W^* *dynamics argument*, lodged at the level of a concrete(ish) model of the ferromagnetic phase transition, due to Emch and Knops. According to the W^* dynamics argument, this model reveals the presence in QM_∞ of explanatory practices that bring each of our extremists into collision with a principled constraint on the space of physical possibilities we'd expect them to respect.

12.3.1 The phase argument

It is easy to see that §12.2's account, offered at the thermodynamic limit of QSM, of the structure and *differentia* of states at play in phase transitions is an account that neither Hilbert Space Conservatism nor Algebraic Imperialism can sustain. The account takes different thermodynamic phases to be possible for a C^* dynamical system $(\mathfrak{A}, \alpha_t)$ in equilibrium at an inverse temperature β such that $\mathcal{K}_\beta$ has more than one member, and identifies the different pure thermodynamic phases accessible to such a system with different extremal elements of $\mathcal{K}_\beta$, i.e. disjoint factor states on $\mathfrak{A}$. The account thereby presupposes

(**PA 1**) Disjoint states ϕ and ψ, corresponding to distinct pure phases, are physically possible.

Not even the *relaxed* Hilbert space Conservative (who confines possible states to the folium of a factor representation of $\mathfrak{A}$) can accept **PA 1**. Interpreted by the Hilbert Space Conservative, QSM in its thermodynamic limit recognizes too few states to support §12.2's model of phase structure.

Every member of $\mathcal{K}_\beta$ is a possible state of a system of canonical type $\mathfrak{A}$, according to the Algebraic Imperialist. So Algebraic Imperialism recognizes enough states to sustain §12.2's account of phase structure. The problem is, Algebraic Imperialism doesn't recognize enough observables. The account identifies pure thermodynamic phases *as* pure thermodynamic phases by appeal to their dispersion-free assignment of expectation values to *macroscopic observables*; it identifies distinct pure thermodynamic phases *as* distinct on the basis of the *different* values they assign macroscopic observables. Merely *characterizing* the phenomenon of phase coexistence targeted by §12.2's account presupposes:

(**PA 2**) Distinct pure phases ϕ and ψ are distinguished by a macroscopic *phase observable* **M** such that $\mathbf{M}(\phi) \neq \mathbf{M}(\psi)$.

Recall that **M** is dispersion-free on both ψ and ϕ, and that if ω is a π_ϕ-normal state, $\mathbf{M}(\omega)$ is well-defined and equal to $\mathbf{M}(\phi)$ (and *mutatis mutandis* for ψ). It follows, for the Imperialist, that **M** lacks physical significance, because it corresponds to no element of the quasilocal algebra $\mathfrak{A}$ of the QSM system exhibiting phase structure. Suppose, for reductio, that there is such an $M \in \mathfrak{A}$. The GNS representations π_ϕ and π_ψ are *weakly equivalent* (see §6.4.1 for a definition and discussion). This implies that there is some π_ψ-normal state χ such that $\chi(M)$ is within an arbitrarily small ϵ of $\phi(M)$. But if M implements the macro-observable **M**, $\chi(M) \neq \psi(M)$, and $\psi(M)$ and $\phi(M)$ are macroscopically distinct. It follows that no element of $\mathfrak{A}$ can do the work (**PA 2**) demands of a phase observable.

Explanatory aspirations entertained in the thermodynamic limit rely on physical discriminations that cannot be made with purely algebraic resources. The Algebraic Imperialist sustains those aspirations no better than does the Hilbert Space Conservative. To support the aspirations, an account of the content of a quantum theory needs to take seriously more observables than the Imperialist can, but more states than the Conservative can.

Universalism promises to do exactly this. Given a theory of canonical type $\mathfrak{A}$, let $\mathcal{S}_\mathfrak{A}$ be the set of states in the algebraic sense on $\mathfrak{A}$. The *universal representation* π_U is the representation obtained as the direct sum, over these states, of their GNS representations:

12.3 EXTREMIST OBSTRUCTIONS TO EXPLANATION

$$\pi_U = \oplus_{\omega \in S_\mathfrak{A}} \pi_\omega \tag{12.11}$$

Notice that the representative $\pi_U(A)$ of an element $A \in \mathfrak{A}$ takes the form $\pi_{\omega_1}(A) \oplus \pi_{\omega_2}(A) \oplus \pi_{\omega_3}(A) \oplus \ldots$. The *universal enveloping von Neumann algebra* is just the von Neumann algebra $\pi_U(\mathfrak{A})''$ affiliated with the universal representation. Let us suppose that our Universalist identifies the observables proper to a theory of canonical type $\mathfrak{A}$ with the self-adjoint elements of $\pi_U(\mathfrak{A})''$, and the possible states of the theory with normal states on $\pi_U(\mathfrak{A})''$ (these turn out to coincide with $S_\mathfrak{A}$).[14] It follows that Universalism removes the impediment presented by Hilbert Space Conservatism to §12.2's account of phase structure. Universalism endorses (**PA 1**).

What about (**PA 2**)? I will argue that, given another generic feature of phase observables, Universalism must deny (**PA 2**). The other feature is:

(**PA 3**)   The domain of a phase observable **M** is a proper subset of $S_\mathfrak{A}$.

The z-component of the global polarization of the infinite spin chain, $\mathbf{m}_z$, exemplifies (**PA 3**).

It follows from (**PA 3**) that no surrogate of the phase observable **M** occurs in the universal enveloping von Neumann algebra $\pi_U(\mathfrak{A})''$ the Universalist takes to encompass physical observables. Let $M_i \in \mathfrak{A}$ be a sequence such that, where π is the GNS representation of a state in **M**'s domain, $\pi(M_i)$ converges in that representation's weak topology to an observable we've been calling M^π. Assume that if an element $M \in \pi_U(\mathfrak{A})''$ implements the phase observable **M**, M must be the weak limit, in the universal representation's weak topology, of the sequence $\pi_U(M_i)$. (Otherwise something has gone awry with the interweaving of magnitudes constitutive of the phase observable.) It is clear that, if $\pi_U(M_i)$ converges weakly, then for each algebraic state ω, $\omega(M_i)$ converges. For it is a property of the universal representation that for each algebraic state, there's a vector in the Hilbert space $\mathcal{H}_U$, the Hilbert space on which π_U acts, implementing that algebraic state. Let ξ_ω be the vector implementing ω. If the sequence $\omega(M_i)$ fails to converge, then the sequence $\langle \xi_\omega | \pi_U(M_i) | \xi_\omega \rangle$ perforce fails to converge as well. But the failure of the latter sequence to converge spoils $\pi_U(M_i)$'s convergence in the weak topology, and foils the inclusion in $\pi_U(\mathfrak{A})''$ of a phase observable. Now, if (**PA 3**) holds, there is some ω such that the sequence $\omega(M_i)$ fails to converge. If (**PA 3**) holds, there is no phase observable in the universal enveloping von Neumann algebra, and Universalism disables the account sketched in §12.2 of phase coexistence.

A natural response to this complaint would be for the Universalist to offer the following reconstrual of the would-be phase observable **M**. Let $D(\mathbf{M})$ be the set of states ϕ such that $\pi_\phi(M_i)$ converges weakly in their GNS representations. For such a ϕ, let $M^\phi \in \pi_\phi(\mathfrak{A})''$ denote the weak limit of this sequence. Then the observable:

$$\mathbf{A}^U = \oplus_{\phi \in D(\mathbf{M})} M^\phi \tag{12.12}$$

---

[14] See Kronz and Lupher (2005) for one version of this proposal, which they attribute to Müller–Herold. Independently, Rob Clifton has made the proposal in conversation.

acting on the Hilbert space $\oplus_{\phi \in D(\mathbf{M})} \mathcal{H}_\phi$ reproduces the behavior of the phase observable within that observable's range.

The customization (12.12) appears to draw distinctions between states—e.g. between those within and those outside the domains of interesting phase observables—the Universalist officially forswears. Next I'll muster some concrete models of phase coexistence to argue that each of our extremists places on all fours states and/or folia between which the physics distinguishes.

12.3.2 The W^* argument

Start with the question, hitherto mostly suppressed, of where the dynamical automorphism group α_t on a quasilocal algebra $\mathfrak{A}$ comes from. In the sorts of models we've been discussing, dynamics are supplied locally, by specifying a Hamiltonian for a finite region. The Heisenberg model of ferromagnetism, for instance, supposes neighboring spins in the infinite chain to interact via the Hamiltonian:

$$H_K = - \sum_{\langle i,j \rangle \in K} J \sigma_i \cdot \sigma_j \tag{12.13}$$

where $\langle i,j \rangle \in K$ denotes summation over nearest neighbors in $K = [-k, k]$, J is a positive real number, and σ_k is the Pauli spin for site k. The Hamiltonian (12.13) aims to recover ferromagnetism by rendering states of the system in which spins are aligned energetically favorable.

Let $\mathfrak{A}_K$ be the algebra generated by subjecting spins at lattice sites in K to the CARs. When K is finite, the Jordan–Wigner theorem assures us that representations of $\mathfrak{A}_K$ are unique up to unitary equivalence. $\mathfrak{A}_K$'s dynamical automorphisms are given by the unitary group $\alpha_t^K(A) := e^{-iH_K t} A e^{iH_K t}$ for all $A \in \mathfrak{A}_K$.

Now take the thermodynamic limit by letting $k \to \infty$ and forming the quasilocal algebra $\mathfrak{A}$ as the closure in C^* norm of $\cup_K \mathfrak{A}_K$. $\mathfrak{A}$ is a C^* algebra. Should $\lim_{k \to \infty} \alpha_t^K$ converge in the norm topology, it will define an automorphism group α_t on the quasilocal algebra $\mathfrak{A}$. Norm convergence is known to obtain for models, like the Heisenberg model, positing short-range interactions (Streater 1967), or even long-range interactions that fade fast enough as the distance between lattice sites grows (Robinson 1968). But many interesting physical models—e.g. the Bardeen–Cooper–Schrieffer model of superconductivity, the model of Bose–Einstein condensate as an ideal bose gas (both reviewed in Emch 2007)—posit long-range interactions that spoil the norm convergence of the local dynamics in the thermodynamic limit. This leaves the quasilocal algebra $\mathfrak{A}$ without a generally defined dynamics.

It does not, however, follow that such models are bereft of all dynamics. For even if $\lim_{K \to \infty} \alpha_t^K$ fails to converge in $\mathfrak{A}$'s norm topology, there may be states ω such that $\lim_{K \to \infty} \pi_\omega(\alpha_t^K)$ converges in the weak topology of their GNS representations. In such cases, dynamics are available on a state-by-state basis, in the form of an automorphism of the concrete von Neumann (aka "W^*") algebra $\pi_\omega(\mathfrak{A})''$ affiliated with a particular representation, *even though no dynamics for the quasilocal C* algebra $\mathfrak{A}$ exists*. Following Sewell, call dynamics obtainable on these terms "W^* dynamics."

Some noteworthy features of the sort of W^* dynamics that have been studied (see Morchio and Strocchi 1985; Strocchi 1988) include: when $\pi_\omega(\alpha_t^V)$ converges in $\mathcal{H}_\omega$'s weak topology to an automorphism A_t^ω acting on elements of $\pi_\omega(\mathfrak{A})''$, A_t^ω shares the dynamical symmetries (which typically continue to be implementable by automorphisms of $\mathfrak{A}$, even when the overall dynamics aren't) of the α_t^V. What's more, the set of states on $\mathfrak{A}$ whose GNS representations sustain W^* dynamics is stable under the action of symmetries of the finite-volume dynamics, and contains states of physical interest, such as equilibrium and symmetry-breaking ones. W^* dynamics not only maintain features, such as dynamical symmetries, in which we're interested, but also engage a suitably broad and significant sets of states to make the pursuit of that interest potentially rewarding.

The Emch and Knops model of ferromagnetism for the infinite spin chain (1970; see also Emch 2007, 1148 ff.) is an exactly solvable example of W^* dynamics. The model replaces the nearest neighbor interaction of the Heisenberg ferromagnet with the supposition that each spin in a finite region K interacts via its z component with *the average magnetic field due to the z components of other spins in K*. (Models adopting this strategy are known as "mean field models.") That is,[15]

$$H_K = -\sum_{n=-k}^{k} Jm^{K/n}\sigma(z)_n \qquad (12.14)$$

where J is a real number encoding a coupling constant and:

$$m^{K/n} = \frac{1}{2k+1}\sum_{n\neq j=-k}^{k} \sigma(z)_j \qquad (12.15)$$

Notice that $m^{K/n}$ is defined on the model of a polarization observable, as a spatial average of local observables. The existence of a macroscopic global polarization observable, in the form of an infinite volume limit of such averages, is representation-dependent. So too the existence of the infinite volume limit of the Emch–Knops dynamics is representation-dependent. Although the automorphisms α_K determined by the Hamiltonian (12.14) lack a limit in the norm topology of $\mathfrak{A}$ as $k \to \infty$, $\pi_\omega(\alpha_t^V)$ does converge in the weak topology of some representations π_ω of $\mathfrak{A}$. The global polarization observable takes different values in different representations. So too the Emch–Knops W^* dynamics assumes different forms in different representations to which it applies. In each such representation, the W^* dynamics is implemented unitarily, and "lifts" to a C^* automorphism group α_t^ω of $\mathfrak{A}$. Because in the thermodynamic

---

[15] The full-blown Emch–Knops model is more complicated. One simplification made below, because *spontaneous* magnetization is the phenomenon under discussion, is to set the magnetic field external to the system to 0.

limit the W^* dynamics depend on the global polarization, which takes a different value in different states, these C^* automorphisms vary depending on which GNS representation they were lifted from.

The states ω whose GNS representations support Emch–Knops dynamics in W^* form include KMS states with respect to those dynamics, obtained (*à la* Araki) as infinite-volume limits of Gibbs equilibrium states on finite subalgebras $\mathfrak{A}_K$. And these KMS states have a structure suggestive of phase coexistence: in the thermodynamic limit, there is a temperature determined by the strength of the coupling constant J at and below which two extremal KMS states appear. What's more, these extremal KMS states are disjoint factor states, distinguished by the dispersion-free values they assign a macroscopic observable, an order parameter in the form of the z-component of global polarization.

The Emch–Knops model departs from the general strategy §12.2 sketches for accommodating phase coexistence in this detail: there is no automorphism group α_t on $\mathfrak{A}$ with respect to which the KMS states in which it traffics are equilbrium states. The Emch–Knops model nevertheless recapitulates much of the strategy §12.2 outlines. It puts disjoint states to explanatory work, and takes differences limned by observables not appearing in the quasilocal algebra $\mathfrak{A}$ seriously as physical differences. That is, the Emch–Knop model endorses all the premises of the phase argument.

It also suggests a different sort of argument against extremisms. In the light of explanatory practices deploying W^* dynamics, extremisms stand in tension with a long-standing account of what it is to be physical—an account we might attribute to Aristotle, who (as §6.4.2 mentioned) opens *Physics, II.i* by declaring, "The obvious difference between all these [natural] things and things which are not natural is that each of the natural ones contains within itself a source of change and of stability, in respect of either movement or increase and decrease or alteration." Sewell echoes the sentiment:

> the theory is constructed within the C^*-algebraic formalism and is based on the postulate that the *physical* states of a system constitute the maximal folium of its locally normal states which can support a one-parameter group of affine transformations that corresponds to a certain infinite volume limiting form of the time-translation group for a finite system of particles for the same species. (1973, 44)

On their behalf, consider the criterion:

[ARIS] To be physical, a state must admit of dynamical development.

An idea common to the Algebraic Imperialist and the Universalist is that every state ω on a canonical algebra $\mathfrak{A}$ is a physically possible state. In the presence of a theory of canonical type $\mathfrak{A}$ whose applications marshall W^* dynamics, [ARIS] condemns this idea. Placing every algebraic state on a modal par, the idea is more democratic than the physics, which distinguishes invidiously between algebraic states on the basis of their capacity to sustain dynamics. Supposing that genuine physical possibilities are those that admit of time development, algebraic states incapable of sustaining dynamics are on that account less genuinely physical than those that are capable. (The capable ones incidentally have further claims to physical significance; for instance, they're equilibrium states.) In the presence of W^* dynamics, there's a sense in which some states are more possible than others. Which states those are depend on contingent features of local dynamics, features that aren't given alongside the canonical algebra $\mathfrak{A}$, but specified in the context of gearing models to particular applications.

Contrast the situation in classical mechanics. There the phase space M of a theory is equipped with a dynamical flow when a Hamiltonian observable is announced. But M is not (or not typically) *reshaped* by that announcement. The Hamiltonian defines a flow through *every* point in M; which flow it defines has no impact on what points correspond to physically possible states. Put another way, dynamical details don't alter what situations are possible according to a classical Hamiltonian theory. They just alter which possible situations are dynamical developments of which others. [ARIS] has no bite.

Exceptions to this general state of affairs occur. Non-Lipshitz potentials that introduce singularities into Hamilton's equations at certain unlucky points in M are a recently discussed example (Norton 2008). But the volume and tone of foundational attention, often attention in the form of inquiries into whether the exceptions are *physical*, underscores the claim that, as a rule in classical mechanics, imposing concrete dynamics along with the criterion [ARIS] does not alter the space of states deemed physically possible by a theory.

The case is even clearer for ordinary QM. A theory whose states correspond to density operators on a Hilbert space $\mathcal{H}$ admits dynamics in the form of unitary operators on $\mathcal{H}$. It follows that every possible state lies in the domain of every possible dynamics. [ARIS] can never interact with a particular dynamics to recontour the space of physical possibilities.

In QM_∞, particularly in accounts of phase coexistence, phase structure and (the next chapter will reveal) broken symmetry, the story is different. The set of states of a given theory that are, by [ARIS]'s lights, fully physically possible can vary along with the dynamical details of specific applications of the theory. This section has focused the simplified Weiss–Ising model of ferromagnetism due to Emch and Knops. But more sophisticated applications of QM_∞—to superconductivity (discussed in the next chapter), to Bose–Einstein condensation, perhaps even to the electroweak theory—exemplify the moral.

It's not a moral that Universalism or Imperialism, holding as they do that all representations are equally valid, can take to heart. Nor is it one the Conservative can embrace, so long as the dynamically distinguished folia fall outside the embrace of a factor state. None of the extremisms announced at the outset of this chapter adequately reflect the play of possibilities directed by QM_∞ accounts of phase coexistence.

12.4 Complicating content

Can we construe the content of quantum theories in a way that accommodates the form and texture of explanations offered in the thermodynamic limit of QSM? I'll close with a suggestion that rests on a diagnosis of why the extremisms fail to do so. Extremist interpretations proceed as though a physical theory's content is to be specified by a simple sort of logical possibility into one of two disjoint and exhaustive categories: the physically possible and the physically impossible. Different extremisms offer different sorting mechanisms: a Hilbert space structure of observables, an abstract algebraic structure, a universal representation structure, and so on. But each of them constructs their sorting mechanisms on the basis of principles of the sort I've labelled "metaphysical"—principles, such as the Imperialist's operationalism or the Universalist's tolerance, that apply indiscriminately to all physical situations, principles that equip a theory, once and for all, with a content that's invariant across those situations. What makes extremisms extreme is the commitment to configure theoretical content in advance of theoretical application.

In Chapter 6, another approach to interpreting physical theories, dubbed the *coalescence approach*, made a brief appearance. The idea of the coalescence approach is that the uses to which theories are put have a hand in determining the content of those theories. Thus the coalescence approach refuses to specify the content of a physical theory in one fell swoop, separating states simply possible according to it from states simply impossible. Allowing that the content a theory has could vary with the scientific conditions in which that theory finds itself, the coalescence approach is consistent with a more kaleidoscopic notion of physical possibility. On the coalescence approach, "possible according to T" can be vague, with the set of worlds providing its extension changing from circumstance to circumstance. Rather than acquiring content in a single block pristinely identified, a theory interpreted on the coalescence approach can accrete content, or shed it, as circumstances allow. And this accretion can happen in stages, corresponding to different grades of physical possibility recognized by the theory. For instance, a sort of primordial stage might specify the theory's content, as it were, *a priori*, before taking the contingencies of physical applications into account. For a theory of canonical type $\mathfrak{A}$, a reasonable account of primordial content identifies it with the worlds possible according to the theory in the broadest sense, e.g. the space $S_{\mathfrak{A}}$ of algebraic states. Self-adjoint elements of $\mathfrak{A}$ correspond to the theory's most basic physical magnitudes, the ones that apply as it were *ab initio* to all systems in the scope of the theory.

A further stage of physical content specification could take contingencies into account, for instance, by supplementing the basic observable algebra $\mathfrak{A}$ with elements coalesced from features specific to particular applications. The discussion of phase structure gives examples of coalesced second-stage observables: state-dependent functionals defining macroscopic observables fostering thermodynamical applications, a Hamiltonian specific to a representation that serves as the generator of dynamics in the folium of that representation. QFT also furnishes candidate coalesced second-stage observables: particle observables, for example, or the magnitudes provided by the point-splitting procedures defining stress-energy observables conspiring in the project of semi-classical quantum gravity. Sometimes these coalesced observables will be elements of a von Neumann algebra $\pi(\mathfrak{A})''$ affiliated with a representation π of $\mathfrak{A}$ most relevant to the application at hand. Again, observables confined to von Neumann algebras parochial to less relevant representations aren't once and for all *unphysical*. They're just inadequate to the sorts of discriminations demanded by the application at hand. There may be other applications to which they're well-suited, applications that might be hindered by insisting on the physical significance of every element of $\pi(\mathfrak{A})''$—as certain applications are disabled by insisting that physical states be Hadamard, or have standard Minkowski particle contents.

Enlarging the constituency of observables during this coalescence stage could interact with ideologies of state—for example the requirement [ARIS] or the demand that states be countably additive—to shrink the space of possible states. Again, a state ω on $\mathfrak{A}$ outside the shrunken space is not outright unphysical. It's just, given the contingent application at hand, less physical than states within the shrunken set.

The point of shrinking the set, rather than simply leaving the less applicable states in it, is sustaining the "laws" invoked in applications: semi-classical quantum gravity sets up the Einstein field equations, with an *expectation value* for the stress-energy tensor on the r.h.s., as a law. It thereby coalesces a set of possible worlds—those corresponding to Hadamard states—that instantiate the law. With respect to this coalesced possibility set, the law has the force of nomic necessity, and all the attainments pertaining thereto.

Universality is a reason to supplement the magnitudes recognized by a theory applied to phase transitions to include the coalesced magnitudes that mark phase transitions. The sudden change in the value of a global property of a system (e.g. the net magnetization of an iron bar) as a relevant parameter (e.g. temperature) passes through a critical value is the hallmark of a *critical phenomenon*. The behavior of such bulk features at critical points tends to exhibit *universality*: the way the global magnetization **m** of iron varies with temperature near 771°C is characterized by a few critical exponents, the very same critical exponents which characterize a host of other critical phenomena, instantiated not only by other ferromagnetic substances but also by (among other things) a classical bead free to roam on a spinning hoop. The study of universality, particularly by the technique of renormalization group theory, has proven to be of enormous fecundity (Batterman 2002 is a discussion aimed at philosophers). This is inducement to regard universality as a well-founded physical phenomenon, and the behavior of critical exponents uncovered by renormalization group theory as law-like.

Critical exponents govern the behavior of coalesced magnitudes such as magnetization, and in particular concern abrupt transitions in the values of those magnitudes at critical temperatures. If those magnitudes lack physical significance, their governance by critical exponents lacks the character of physical law. In order to see the phase diagram (Fig. 12.1) and its ilk as manifesting something nomic, we need to understand the magnitudes they link as genuinely physical. And to do that, I've argued, we need to admit more magnitudes into the fold than we get with the canonical quasilocal C^* algebra encoding the primordial observables. This expansion has a collateral consequence for the possible states of the theory. That subset of states sustaining the definition of these coalesced magnitudes acquires a physical significance which the collection of states foiling their definition lacks.

Identifying physical possibility with primordial possibility, period, we undermine the nomic force of many explanations encountered in physical practice—explanations mediated by laws which hold only of a proper subset of primordial possibilities. On the other hand, identifying physical possibility *only* with a particular such proper subset, we deprive the space of primordial possibilities of its role as a reservoir from which different physical possibility spaces might be coalesced, to meet different practical or explanatory challenges.

In a sense, all extremists obscure the role of coalescence in content specification. The Algebraic Imperialist suppresses it entirely. The Hilbert Space Conservative shackles it to a single concrete Hilbert space, which is thereupon identified as the repository of the full content of the theory. The Universalist deprives coalescence of its ability to distinguish between more and less physical examples of possibility most broadly construed. Enfranchising coalescence as a legitimate sort of content specification, we understand the content of quantum theories in a way that not only pays due heed to their formal structures, but also captures some of the texture of scientific practice.

To praise the coalescence account on that score is to assume that it is a virtue of an interpretation to capture the texture of scientific practice. Alas, explanations envisioned in the thermodynamic limit are susceptible to challenge on the ground that they are *obviously false*. For the phenomena to be explained (phase transitions) occur in finite systems (our coffee cups) which the thermodynamic limit models—falsely—as infinite. No explanation proceeding from such patently false premises, the challenge runs, is legitimate. The next two chapters develop a two-pronged response to this challenge.

13

Interlude
Symmetry Breaking in QSM

13.1 Introduction: Coalesced Structure

Chapter 12's arguments against extremism—which I'll collect under the heading of "the Coalesced Structure Argument"—exploit the presence in QM_∞ of "coalesced" structures such as phase observables or W^* dynamics. The features of a coalesced structure X crucial to the arguments against extremism include the following:

a. X has explanatory oomph: e.g. it characterizes targets of explanation or serves as their agents.

The global magnetization **m** of a ferromagnet is, for instance, the observable in whose terms phase coexistence is *characterized*; and the W^* dynamics of the Emch–Knops model is a dynamics with respect to which the collection of KMS states is structured in a way that accounts for phase coexistence.

b. X is without counterpart in the abstract C^* algebra $\mathfrak{A}$ fixing the canonical type of the theory to be interpreted.

The global magnetization **m** isn't an element of the canonical algebra $\mathfrak{A}$ for the ferromagnet, nor is the Emch-Knops dynamics implementable as an automorphism of that algebra.

c. The explanations involving X presuppose X to apply to a set of states larger than the folium of any single factor representation of $\mathfrak{A}$.

Sticking with the ferromagnet, interesting phase structure occurs only when disjoint factor states on its canonical algebra $\mathfrak{A}$ are in the domain of the global magnetization **m**; disjoint factor states sustain the W^* dynamics of the Emch–Knops model.

d. There exist states in the algebraic sense on $\mathfrak{A}$ to which X does not apply.

Still sticking with the ferromagnet, neither the global polarization **m** nor the Emch–Knops dynamics is well-defined on every state on its canonical algebra $\mathfrak{A}$.

Supposing that to support the explanations cited in (a) is to contour the possibility space of the theory in such a way that X has the generality requisite for participation in explanation-supporting laws, the Coalesced Structures Argument used features (b), (c), and (d) respectively to show that Algebraic Imperialism, Hilbert Space Conservatism, and Universalism are each incapable of such support. One weakness of that argument is the nature of its motivating examples, drawn from the thermodynamic limit of QSM. The thermodynamic limit puts the ∞ in QM_∞ by way of a patent idealization, the idealization that a ferromagnet or a superconductor or a cup of tea is infinite in extent. This leaves room to worry that coalesced structures crucial to the arguments against extremism are artifacts of this idealization (see Callender 2001 for the worry developed with respect to other foundational issues). The aim of this chapter and the next is to articulate and evaluate a strategy for assuaging this worry. QFT incorporates infinitely many degrees of freedom honestly, or at least through idealizations less patent than those committed by the thermodynamic limit.[1] A quantum field theoretic analog of coalesced structures would not be so easy to dismiss as bereft of theoretical interest. The strategy is to look for such an analog.

This chapter is a warm-up for the search. It undertakes to recast the case for explanatorily relevant coalesced structure in the thermodynamic limit of QSM with *broken symmetry* at center stage. It seeks applications of QSM where the physics of broken symmetry demand not only a wider range of states than the Hilbert Space Conservative can countenance but also nomically significant observables that neither the Algebraic Imperialist nor the untempered Universalist can recognize as such. Set at the thermodynamic limit, such a case for the physical significance of coalesced structure remains prone to dismissal on the grounds of idealization. Still, it promises to guide the search for coalesced structure in QFT, where broken symmetries are supposed to abound, and to matter.

The plan of this chapter is as follows. Section 13.2 discusses symmetries in physics, with a view toward debunking the claim that that they can be identified only having drawn a sharp distinction between laws and initial conditions—the very distinction non-extremist approaches to interpretation (called "coalescence accounts" by §12.4) would blur. Sections 13.3 and 13.4 offer two accounts, and two illustrations, of broken symmetry in QM_∞. Section 13.5 musters those illustrations to argue that coalesced structures emerge from accounts of broken symmetry

13.2 Symmetry and the Sharp Distinction

This section acknowledges, and attempts to rebut, a case that the very idea of a symmetry is at odds with the picture of coalesced content I claim that QM_∞ invites.

---

[1] Wallace (2006), for one, denies that QFT comes by its infinities honestly. Underlying the denial is the use of cutoffs in what §5.1 calls "physicist's QFT" calculations. The mathematically explicit free QFTs on which I focus here don't furnish similar grounds for denial. Of course, because they don't furnish many predictions either, they're not the sort of QFTs Wallace is interested in.

John Earman offers a framework for characterizing the symmetries of a physical theory.

A practitioner of mathematical physics is concerned with a certain mathematical structure and an associated set M of models with this structure. The sought after laws L of physics pick out a distinguished sub-class of models $M_L := mod(L) \subset M$, the models satisfying the laws L (or in more colorful, if misleading, language, the models that "obey" the laws L). Abstractly, a symmetry operation is a map $S : M \to M$. S is a symmetry of the laws L just in case it preserves M_L. (Earman 2004b, 4)

The framework suggests an account of symmetry in mathematical physics.

Definition 13.1 (symmetry of a theory). $S : M \to M$ is a **symmetry** of T iff $m \in M_L \Leftrightarrow S(m) \in M_L$.

This amounts to the requirement that the class of models satisfying T's laws be invariant under the action of the symmetry. The definition generalizes straightforwardly to symmetry *groups*. Unfortunately, Def. 13.1 captures at most a necessary condition, not a definition, of symmetry. A Scholium sketches why.

Scholium: shortfalls of the folk definition. Although Def. 13.1 enjoys something of a folk status in the philosophical literature, it furnishes at most a necessary condition for symmetry.[2] There are several ways to see this. First notice that *any* permutation S of the class M_L of models counts as a symmetry in Def. 13.1's sense. But for any pair of models in M_L, there's a permutation of M_L taking one to the other. So it follows from the folk definition that any model is related to any other by a symmetry. But we're wont to think of symmetries as made of sterner stuff: as connecting (not just any old models but) models that "have the same physics." Consider, for example, the n-body problem of Newtonian physics. According to Def. 13.1, its symmetries include diffeomorphisms of the manifold that is its state space, which diffeomorphisms form an infinite-dimensional group. But it's both more orthodox and more interesting to identify the symmetries of the Newtonian n-body problem with the (much smaller) Galilean group.

Footnote 8 of Chapter 12 suggested that Def. 13.1 be supplemented to require a putative symmetry S to preserve interesting structure—for instance, the symplectic structure that turns the phase space of a classical theory into a symplectic manifold. But finite dimensional symplectic manifolds admit infinite dimensional symplectic-structure-preserving symmetry groups. So even strengthened by this requirement to preserve interesting structure, the folk definition would saddle us with much larger symmetry groups than we'd like for theories such as the Newtonian n-body problem.

---

[2] This scholium owes much of its substance to Gordon Belot. For details and elaboration, see Giulini (2006, §2); Belot (2008, §2).

If we focus attention on the tractable case of theories whose models assume the form of solutions to differential equations, there is an established account of symmetry that fares better than Def. 13.1 and its modifications. On that account, a symmetry of a differential equation is a diffeomorphism from that equation's solution space to itself, which diffeomorphism arises from a "suitably local" transformation of the equation's dependent and independent variables (see Olver 1993, §5.1, for details; Belot 2008, §2, for a sketch). Unfortunately, the relevant notion of "suitably local" is nightmarish. Typically, spacetime transformations are symmetries in the well-developed official sense just in case they permute the solution space and qualify as a symmetry in the folk sense. This coincidence might explain the ascendancy of the folk notion.

Notice that the well-developed official sense of symmetry can leave us at sea when we come to quantum theories, which we sometimes obtain by quantizing theories to which the official sense applies directly and sometimes obtain by other means. Later in this section I will suggest that symmetries of quantum theories are identified on a variety of *ad hoc* bases. ♠

Although imperfect, Def. 13.1 has been influential, and represents a way of thinking about symmetry according to which the coalescence approach, championed here, is nonsense. The balance of this section explicates, and tries to respond to, this charge.

It is immediately apparent that to apply Definition 13.1, we need a handle on what the collection of law-abiding structures M_L is. A claim dear to Wigner and endorsed by Earman is that the only way to get this handle is to pull asunder what the coalescence approach would blur together. For Earman, "The relevant senses of symmetry and invariance *presuppose* a distinction between what holds as a consequence of the laws of physics and what is compatible with but does not follow directly from those laws" (2004a, 1229). Wigner and his co-authors are more expansive:

[Symmetry principles] are possible only because our knowledge of the physical world has been divided into two categories: initial conditions and laws of nature. The state of the world is described by initial conditions. These are complicated and no accurate regularity has been discovered in them. In a sense, the physicist is not interested in the initial conditions, but leaves their study to the astronomer, geologist, geographer, etc. . . . The invariance principles apply only to the second category of our knowledge of nature, the so-called laws of nature. . . . invariance principles can be formulated only if one admits the existence of two types of information which correspond in present-day physics to initial conditions and laws of nature. It would be very difficult to find a meaning for invariance principles if the two categories of our knowledge of the physical world could no longer be sharply separated. (Wigner, Houtappel, and van Dam 1965, 596)

According to the *Doctrine of the Sharp Distinction*, a theory's laws cull its models M_L from the collection M of models with a theory-appropriate structure. The exigencies of actual applications—"what is compatible with but does not follow directly from [the] laws"—merely label elements of M_L, in the manner of addresses. As such, they are individually of interest only to geographers and their ilk. Collectively, the models they label make up M_L, whose invariance under transformations anoints those transformations as symmetries. The collective M_L and the status of some transformations of M as symmetries "hold as a consequence of the laws," and merit the attention of physicists.

M_L otherwise described is the set of worlds possible according to a theory T, the set of worlds it is the task of an interpretation of T to characterize. Proponents of the Doctrine of the Sharp Distinction would presumably impose upon an interpretation the criterion of adequacy that the worlds it deems possible according to T are just the worlds T's laws allow—the criterion, that is, that the worlds possible according to an interpretation of T coincide with M_L. Such coincidence underlies the claim, forwarded in Chapter 1, that an interpretation of T is an explication of "lawfulness according to T". Viewed in light of the coincidence, the Doctrine of the Sharp Distinction expresses what we've been calling the ideal of pristine interpretation: the ideal that the content of a theory be contoured *without* recourse to geographical considerations.

Running roughshod over the Sharp Distinction, the coalescence approach allows "geographical" considerations arising from a particular application of a theory to configure the set of theoretical possibilities framing that application. According to the coalescence approach, there needn't be any such ungeographical thing as M_L period. Instead, there may be many application a-adapted $M_L(a)$'s, each constituted by appeal to geographical considerations. In this case, there is no Sharp Distinction and no way to follow the Earman and Wigner recipe for identifying symmetries. A defense of the coalescence approach, especially one appealing to the notion of symmetry, should deflect Earman and Wigner's contention that the Sharp Distinction is a presupposition of the very notion of symmetry.

The deflection I attempt in this section will begin by announcing a superficial puzzle the Doctrine of the Sharp Distinction raises about the role symmetry considerations play in interpreting physical theories. The point of the Doctrine of the Sharp Distinction is that we need a pre-existing fix on M_L in order to identify T's symmetries. But a fix on M_L *is* an interpretation of the theory. And if we have to have an interpretation of T in place in order to identify T's symmetries, symmetry considerations arise too late to inform the project of interpretation. This is especially puzzling because symmetry considerations have figured centrally in foundational debates from Parmenides onward.

The puzzle isn't very deep. One way to get out of it is to notice that standard interpretive practice is far subtler than the pseudo-puzzle's depiction of it allows. Consider, for instance, some methods by which those interested in the foundations of theories of QM$_\infty$ have identified their symmetries. For these purposes, let us adopt the working assumption that the set M of models with theory-appropriate structure for a quantum theory T is provided by the collection of states (in the algebraic sense) on a C^* algebra $\mathfrak{A}$. An interpretation will single out a (possibly improper) subset M_L of M as "physical." The pseudo-puzzle arising from the Doctrine of the Sharp Distinction contends that only having appealed to an interpretation of the theory and/or the theory's laws to distinguish some M_L of models of the theory, can we identify the symmetries of the theory. But in practice we identify *prima facie* symmetries in some of the following ways.

1. First, directly. Given a one-parameter family $\alpha(t)$ of *dynamical automorphisms* of $\mathfrak{A}$, we identify as a dynamical symmetry any automorphism σ of $\mathfrak{A}$ that commutes with $\alpha(t)$. We also *stipulate*, say by announcing an axiom to that effect, that certain automorphisms (e.g. those implementing Poincaré transformations) are symmetries. These identifications and stipulations proceed within the framework of the set M of models with theory-appropriate structure, and so prior to distinguishing a privileged subset $M_L \subset M$.
2. Second, by inheritance. The algebra $\mathfrak{A}$ for a theory of QM_∞ can sometimes be obtained by applying the Hamiltonian quantization recipe to a classical Lagrangian theory. Chapter 2 developed the example of the Weyl algebra $\mathfrak{A}$ expressing the quantization of the classical Klein–Gordon theory. Associating the classical theory with a symplectic vector space $(\mathcal{S}, \Omega)$, $\mathfrak{A}$ is the norm closure of a representation of the Weyl relations over $(\mathcal{S}, \Omega)$. It follows that symplectic symmetries of the classical theory—maps $s : \mathcal{S} \to \mathcal{S}$ preserving the symplectic product Ω^3—"lift" to automorphisms of $\mathfrak{A}$. (See §9.5's discussion of time-like isometries for an illustration of such lifting.) Arageorgis, Earman, and Ruetsche (2002) discuss symmetries obtained in this way for quantized Klein–Gordon theory set in a globally hyperbolic spacetime. Where Σ_1 and Σ_2 are Cauchy surfaces and ι an identity map between them, $\alpha^\iota_{\Sigma_1, \Sigma_2}$ is the automorphism of $\mathfrak{A}$ corresponding to the symmetry of the classical solution space that maps a solution with initial data ϕ on Σ_1 to the solution whose initial data on Σ_2 is the first solution's Σ_1 initial data, transferred to Σ_2 by the identity map. These symmetries are identified not by appeal to an interpretation of the *quantum* theory, but by appeal to the classical theory it quantizes.
3. Third, by extrapolation from symmetries of finite subsystems/dynamics. An example is the flip-flop symmetry of the Emch–Knops model (see §12.3.2). Because the finite-volume dynamics $\alpha_V(t)$ needn't norm-converge to a dynamical automorphism group $\alpha(t)$ on $\mathfrak{A}$, this method for identifying symmetries doesn't reduce to the first method. The extrapolation method can operate even where the presupposition of the first method, that there is an explicit C^* dynamics, fails. Once again, no interpretation of the theory obtained at the thermodynamic limit is required to identify these symmetries: they are identified by examining finite-volume theories.

The point of this enumeration is that each of these methods can proceed in total or partial ignorance of the contents of M_L, or its exact boundaries. Each, that is, can

---

[3] See the Scholium on pp. 293–4 an unattractive consequence of identifying these with full-blown symmetries.

proceed without a completed interpretation of T, understood as a pristine account of what possibilities its laws allow, in place. That the identification of *prima facie* symmetries can occur in advance of a specification of the collection of worlds possible according to a theory is what enables symmetry considerations to inform interpretive projects.

However an interpretation picks out a subset M_L of M as physical, the following suggests itself as a reasonable criterion to impose upon the selection:

Reasonable Criterion. If $S : M \to M$ is a symmetry of T, then S ought to leave M_L invariant.

Compare the Reasonable Criterion to the folkloric *definition* of symmetry (Def. 13.1). The latter holds M_L's invariance under S to be *constitutive* of S's status as a symmetry. Thus the definition assumes what interpretive practice rebuts: that it's *settled* what the contents of M_L are. The Reasonable Criterion, by contrast, addresses *proposals* for configuring M_L, and subjects them to the tribunal of *prima facie* symmetries. One aim of interpretation is to bring these into a desirable equilibrium, an equilibrium the Reasonable Criterion aims to characterize.

What gives the Reasonable Criterion traction in the context of QM_∞ is the failure of proposals for selecting M_L to march in lockstep with symmetries identified through the methods above. Consider:

a. The Hadamard criterion for identifying M_L runs afoul of symmetries $\alpha^t_{\Sigma_1,\Sigma_2}$ identified by inheritance. Some of those symmetries, for instance those corresponding to "nocturnal doubling," can map a Hadamard state ω on the quasilocal algebra $\mathfrak{A}$ to a state $\omega \circ \alpha^t_{\Sigma_1,\Sigma_2}$ which is not Hadamard.[4] M_L as characterized by the Hadamard condition is not invariant under symmetries as identified by inheritance.

b. Hilbert Space Conservatism for the infinite spin chain runs afoul of symmetries identified by extrapolation. A Hilbert Space Conservative who identifies the space M_L of possible states of an infinite spin chain with those in the folium of a translationally invariant ground state ω thereby admits that M_L is not invariant under the extrapolated rotational symmetry of the theory. As §3.3.1 demonstrated, if α is a rotation that flip-flops the prevailing Pauli spin of electrons in the privileged ground state, α is not unitarily implementable on ω's GNS representation, and $\omega \circ \alpha$ lies outside the folium M_L of physical states.

c. Similar troubles arise for the Hilbert Space Conservative from QFT. Consider the mass 0 free Klein–Gordon field, and suppose the Conservative identifies the folium of a Poincaré-invariant ground state of that field as M_L. As the next chapter elaborates, that folium is not closed under the action of symmetry transformations inherited from the classical theory.

---

[4] See Arageorgis et al. (2002). Recall that $\omega \circ \alpha$ on $\mathfrak{A}$ is defined as follows: $\omega \circ \alpha(A) = \omega(\alpha(A))$ for each $A \in \mathfrak{A}$.

d. The Algebraic Imperialist, who contends that every state on the quasilocal algebra $\mathfrak{A}$ for the infinite spin chain is physically possible, runs afoul of time-translation symmetries identified by extrapolation from local dynamics. Recall the example of §12.3.2, in which those dynamics involve long-range forces, and fail to converge in the thermodynamic limit to an automorphism of $\mathfrak{A}$. The local dynamics do, however, converge *weakly* in the GNS representations π_ω of *certain physically significant states*. Such convergence defines a W^* dynamics for the von Neumann algebra $\pi_\omega(\mathfrak{A})''$, which defines in turn a time-translation symmetry map on ω's folium. This symmetry map does not extend to the full set of algebraic states: it fails to satisfy a precondition of being a symmetry, in Def. 13.1's sense.

It seems that we must react in one of three ways to such mismatches between interpretive strategies for characterizing M_L and methods of *prima facie* symmetry identification. The first is to reject the interpretive strategy, and aim to develop in its stead a strategy for characterizing M_L that eventuates in a set of physical models that is invariant under the *prima facie* symmetries. The second is to retain the interpretive strategy but deny that the *prima facie* symmetries are genuine symmetries. Demoted, these pretender symmetries are no longer fodder for the Reasonable Criterion. The third reaction is to reject the Reasonable Criterion.

To decide ahead of time to react in the first way is to pledge allegiance to the *prima facie* symmetries, and hence reject any criterion of physical reasonableness that issues in an M_L not invariant under those symmetries. Blind adherence to this policy could cause us to fall short of whatever explanatory desiderata motivated the criteria of physical reasonableness to begin with. Example (a) above illustrates this pitfall.

To decide ahead of time to react in the second way is to pledge allegiance to a set M_L of physical states, then define symmetries *only* as those transformations that leave M_L invariant. Blind adherence to this strategy could cost us some beloved symmetries—even spacetime symmetries—and undermine heuristics invoking those symmetries. Examples (b)–(d) above illustrate this pitfall.

To react in the third way—to reject the Reasonable Criterion—would be unreasonable. Particularly when the symmetries in question are taken to connect physically equivalent states, the third reaction countenances symmetry-connected states that are so physically inequivalent that only one of them is physical.

But perhaps we can render this consequence palatable by blurring the Sharp Distinction between "what holds as a consequence of the laws" and "what is compatible with but does not follow directly from the laws." Characterizing the latter is the task of interpretation, pristinely pursued. It's the *output* of a pristine interpretive project. The realpolitik of interpretation allows considerations which vary from application to application to be part of the *input* of interpretation. The coalescence approach understands interpretation as a function from a set of situations to which a theory might apply to collections of worlds possible according to the theory—collections which sustain the explanatory (etc.) aspirations appropriate to those settings, e.g. by underwriting the definition of salient observables or relevant alternatives. *It could be that this function is a constant*—that situation-dependent particularities never contour interpretation. The ideal of pristine interpretation presupposes that it *must* be this way. But if we blur the Sharp Distinction, we can open interpretive options such as the following: we can reject the reasonable criterion *on grounds of equivocation*. Here I'll sketch how.

In characterizing the coalescence approach to interpretation, §12.4 distinguished between what it called a theory's "primordial" possibility space and its coalesced possibility spaces. The full set of algebraic states on $\mathfrak{A}$ are a good candidate for the primordial possibility space of a theory of canonical type $\mathfrak{A}$; examples of its coalesced possibility spaces might be Hadamard states on $\mathfrak{A}$ (coalesced in the service of semi-classical quantum gravity); states in the folium of a Killing particle representation (coalesced in the service of a project of particulate explanation); or states in the folium of an infrared coherent state on $\mathfrak{A}$ (coalesced in the service of solving QED's "infrared problem"). Examples of coalesced possibility spaces drawn from the thermodynamic limit include states in the domain of a phase observable (coalesced in the service of bringing critical phenomena within the amibit of renormalization group theory) or states on which W^* dynamics are well-defined (coalesced in the service of understanding a system's time evolution and/or phase structure). The coalescence account does not require a theory's primordial possibility space to coincide with its coalesced possibility spaces, nor does it require a theory's coalesced possibility spaces to coincide with one another. In the presence of mismatches, a distinction can be drawn between *prima facie* symmetries of the primordial possibility space—exemplified by automorphisms of a Weyl algebra $\mathfrak{A}$ "lifted" from symplectic symmetries of the underlying classical theory—and *prima facie* symmetries of the coalesced possibility spaces—exemplified by the time-translation symmetries of a W^* dynamics. While it may be reasonable to require a coalesced possibility space to be invariant under a coalesced symmetry, and it may be reasonable to require a primordial possibility space to be invariant under primordial symmetries, it isn't reasonable to require a coalesced possibility space to be invariant under primordial symmetries or vice versa. But by failing to allow for the difference between primordial and coalesced possibility spaces, the Reasonable Criterion requires exactly this: applied to example (a), it faults a coalesced possibility space for lacking invariance under a primordial symmetry; applied to example (d), it does the reverse. The exponent of the coalescence approach has another option: taking the Reasonable Criterion to equivocate between "primordial" and "coalesced," she can declare it inapplicable to examples (a) and (d), and regard the circumstances described in those examples as tolerable.

To render such an attitude unrepulsive, I must produce situations from the physical sciences clamoring for it. Although I think some of the examples summoned above count, that has not been my aim here. I have tried rather to contend that sense can be made of the symmetries of a theory of QM_∞ even in the absence of an antecedent commitment to the doctrine of the Sharp Distinction or the ideal of pristine interpretation. I have tried as well to indicate some interpretive maneuvers made available by the rejection of that doctrine and that ideal.

13.3 Broken symmetry in QM$_\infty$

This section presents two quantum theoretic accounts of broken symmetry, and two examples of quantum systems illustrating both those accounts. For the sake of the exposition, we'll associate symmetries of a theory of canonical type $\mathfrak{A}$ with automorphisms $\alpha : \mathfrak{A} \to \mathfrak{A}$ of that algebra.[5] The hope is to discern coalesced structures, fueling arguments against extremism, in QM$_\infty$ treatments of broken symmetry.

13.3.1 The individual sense

What I'll call the "individual sense" of broken symmetry is something a state on $\mathfrak{A}$ can achieve on its lonesome.

Definition 13.2 (individual BS). A symmetry α broken in the state ω on $\mathfrak{A}$ just in case α is not unitarily implementable on ω's GNS representation.

This "individual sense" of broken symmetry is the one most typically encountered in expositions both philosophical (Earman 2004a; 2004b; 2004c) and physical (Streater 1965; Aitchison 1982).

We can see immediately that broken symmetry does not arise in ordinary QM. There $\mathfrak{A} = \mathfrak{B}(\mathcal{H})$ for a separable $\mathcal{H}$. Because every automorphism of $\mathfrak{B}(\mathcal{H})$ is implemented by a unitary $U \in \mathfrak{B}(\mathcal{H})$, no ordinary QM density operator state on $\mathfrak{B}(\mathcal{H})$ breaks *any* symmetry. Extraordinary QM—quantum theories whose canonical algebras admit unitarily inequivalent representations—is what makes broken symmetry possible (see Liu 2003).

It will come as no surprise that the infinite spin chain provides an illustration of broken symmetry in the individual sense. Equip the spin chain with the familiar Heisenberg exchange Hamiltonian:

$$H = -\sum_{\langle i,j \rangle} J\sigma_i \cdot \sigma_j \qquad (13.1)$$

Recall that $\langle i, j \rangle$ denotes summation over nearest neighbors, σ_k is the three-component Pauli spin associated with the lattice site k, and $J > 0$ is a coupling constant.

These dynamics have a "flip-flop" symmetry. The algebra of observables pertaining to the chain is the C^* algebra $\mathfrak{A}$ generated by the Pauli spins for each site k. The flip-flop operation is the automorphism α_F of $\mathfrak{A}$ obtained by replacing $\sigma_k(z)$ with $-\sigma_k(z)$ for each k. This flip-flop symmetry is broken in both the ground state ϕ^+ in which all spins are aligned along the positive z axis and in the ground state ϕ^- in

---

[5] The so-called Jordan automorphisms of $\mathfrak{A}$, which preserve the algebraic structure of $\mathfrak{A}$'s self-adjoint part and which include C^*-automorphisms as a special case, are another candidate. See Emch (1972, 152–9).

which all spins are aligned along the negative z axis: $\phi^{\pm}$ and $\phi^{\pm} \circ \alpha_F$ are disjoint. This is because $\phi^{\pm} \circ \alpha_F = \phi^{\mp}$, and ϕ^+ and ϕ^- are disjoint.

13.3.2 The decompositional sense

The Heisenberg exchange Hamiltonian (13.1) has symmetries beyond the flip-flop one. For instance, it's invariant under a one-parameter group $\alpha(\theta)$ of rotational symmetries. The Heisenberg exchange Hamiltonian also has translationally invariant ground states in addition to $\phi^{\pm}$. Where n is an axis in space, ϕ^n, a state in which each site's Pauli spin points along the n-axis, is a ground state. In fact, every translationally invariant ground state is of this form. Just as with ϕ^+, ϕ^-, and the flip-flop symmetry, distinct translationally invariant ground states are disjoint, and each translationally invariant ground state is connected to every other by the group $\alpha(\theta)$.

We typically apply the concept of broken symmetry to a state ω with some antecedent claim on our attention: it's a ground, vacuum, or equilibrium state, say. Also ω is typically a member of a *collection* of similarly significant symmetry breaking states *connected to one another by the symmetry*. Contending that, "The essence of SSB in general for any system is *the existence of solutions, each of which has less symmetry than the symmetry of the problem, while the symmetry of the problem acts transitively on the set of all solutions*" (Liu and Emch 2005, 143), Liu and Emch fault the account of broken symmetry in the individual sense (which they attribute to Earman) for obscuring this essence. They develop an account of broken symmetry that foregrounds the obscured feature.

I don't want to enter the fray about which account of broken symmetry is better. I do want to use the Emch and Liu characterization of symmetry-breaking equilibrium states to illuminate connections between symmetry breaking and phase structure, with a view toward eventually reposing the arguments of the last chapter in terms of broken symmetry, and so in terms amenable to a QFT instantiation, supposing QFTs make use of broken symmetries.

Emch and Liu's analysis of broken symmetry brings a dynamical automorphism group τ_t into play alongside the automorphism α implementing the broken symmetry.

Definition 13.3 (BS: decompositional). A symmetry $\alpha : \mathfrak{A} \to \mathfrak{A}$ is spontaneously broken in an α-invariant state ϕ on $\mathfrak{A}$ that is KMS with respect to the automorphism group τ_t just in case:

  i. ϕ is not extremal KMS; and
  ii. Some of ϕ's extremal KMS components are not α-invariant.

Notice that condition (ii), coupled with ϕ's α-invariance, imply that α acts on some non-α-invariant extremal KMS component of ϕ to yield a distinct non-α-invariant extremal KMS component of ϕ. Thus does the decompositional account articulate the striking feature of broken symmetry: that different states of broken symmetry are connected by the symmetry they break.

Connection to symmetry breaking in the individual sense is mediated by the following facts. Let ϕ be a (not necessarily KMS) state on a C^* algebra $\mathfrak{A}$, and let $\mathfrak{Z}_\phi$ be the center of its affiliated von Neuman algebra.[6,g] Assume for the sake of simplicity that the identity operator I_Z for $\mathfrak{Z}_\phi$ has a discrete spectrum. This is tantamount to assuming that $\mathfrak{Z}_\phi$ is Type I. Because $\mathfrak{Z}_\phi$ is abelian, I_Z can be resolved into eigenprojections P_k that themselves are spectral projections for each self-adjoint element of $\mathfrak{Z}_\phi$. These P_k are *central projections* for the von Neuman algebra $\mathfrak{M}_\phi$. We can use them and the state ϕ to define a set of subalgebras of $\mathfrak{M}_\phi$ as follows:

$$\mathfrak{M}_k = P_k \mathfrak{M}_\phi P_k \tag{13.2}$$

$\mathfrak{M}_k$ is just the subalgebra of $\mathfrak{M}_\phi$ whose range coincides with P_k's.

We can also use the central projections and the state ϕ to define a set of states ϕ_k on $\mathfrak{M}_\phi$ as follows:

$$\phi_k(A) := \frac{\phi(P_k A P_k)}{\phi(P_k)} \text{ for all } A \in \mathfrak{M} \text{ s.t. } \phi(A) \neq 0 \tag{13.3}$$

According to the central decomposition theorem,[7]

i. $\mathfrak{M} = \oplus_k \mathfrak{M}_k$;
ii. $\phi = \sum_k \lambda_k \phi_k$, where $\lambda_k = \phi(P_k)$.

(ii) gives the unique *central decomposition* of ϕ. The states ϕ_k appearing in it are factor states. In the case that I_Z has a continuous spectrum, the sums in (i) and (ii) will be replaced by integrals complimented by an appropriate measure theory.

If ϕ is invariant under a symmetry α, we can say more:

iii. Either $\phi_k \circ \alpha = \phi_k$ or $\phi_k \circ \alpha = \phi_j, k \neq j$.

If ϕ happens to be a τ_t-KMS state, its central decomposition has a very special feature:

iv. If ϕ is a τ_t-KMS state, the states ϕ_i are extremal τ_t-KMS states.

Thus given a τ_t-KMS state ϕ with a central decomposition $\phi = \sum_k \lambda_k \phi_k$, we can draw two significant conclusions. First, the decomposing states ϕ_i share the distinguishing physical feature of ϕ: they're all τ_t-KMS. Second, the decomposing states ϕ_i and ϕ_j are disjoint if $i \neq j$.

Any state ϕ satisfying the conditions for broken symmetry in Emch and Liu's decompositional sense is an α-invariant τ_t-KMS state. So of such a state we can say all of the above. Such states ϕ occur in the thermodynamic limit of QSM. They are, for instance, the non-extremal KMS states implicated in Chapter 12's account of phase structure. Because distinct extremal KMS states are disjoint, (iii) entails that each state

---

[6] That is, $\mathfrak{Z}_\phi = \mathfrak{M}'_\phi \cap \mathfrak{M}_\phi$ where $\mathfrak{M}_\phi := \pi_\phi(\mathfrak{A})''$ (see §4.7 for a review).
[7] See Emch (1972, 213) for a discussion and a generalization to the case that I_Z has a continuous spectrum.

ϕ_i appearing in the central decomposition of an α-invariant non-extremal KMS state ϕ is a state that breaks the symmetry α in the individual sense: ϕ and $\phi \circ \alpha$ are disjoint.

In the course of their analysis, Emch and Liu introduce and deploy the notion of a *marker* of broken symmetry.[8] A marker is an observable that takes a different value on different components of the central decomposition of a symmetry breaking state:

Definition 13.4 (marker of broken symmetry). Where ϕ breaks the symmetry α in the decompositional sense, a marker D of this broken symmetry is an element of $\mathfrak{Z}_\phi$ such that for some $i \neq j$, $\phi_i(D) \neq \phi_j(D)$.

What Chapter 12 termed a *phase observable*—a macroscopic observable that takes different, dispersion-free values in different extremal KMS states—serves admirably as a marker for ϕ's breaking of symmetry in the decompositional sense. Indeed, as the discussion of the Heisenberg ferromagnet's broken flip-flop symmetry might lead us to expect, global observables, understood as infinite volume limits of spatial averages, make excellent markers. In general, a marker D of broken symmetry is not an element of $\mathfrak{A}$. Because distinct elements of the central decomposition are disjoint, $\mathfrak{M}_\phi = \pi_\phi(\mathfrak{A})'' = \oplus \pi_{\phi_k}(\mathfrak{A})'' = \oplus \mathfrak{M}_k$ (Fact 10.1). As an element of $\mathfrak{Z}_\phi$, then, a marker D will take the form $\oplus d_k I_k$, where I_k is the identity element of $\mathfrak{M}_k = \pi_{\phi_k}(\mathfrak{A})''$. Thus for each ϕ_i in the central decomposition, D will be dispersion-free on ϕ_i's folium. Supposing those foliums to be weakly equivalent, the requirement that for some $i \neq j$, $\phi_i(D) \neq \phi_j(D)$ and Fell's theorem imply D lies outside $\mathfrak{A}$.

Part of the point of embarking on an analysis of *broken* symmetry from the perspective of the α-invariant state ϕ is to constitute an observable that has each of the symmetry-breaking (in the individual sense) states ϕ_i in its domain, and that distinguishes abruptly between those states. That is, the decompositional account illuminates what's physically at stake between different states of broken symmetry in the individual sense. The marker D effecting the distinction isn't an element of $\mathfrak{A}$.

13.3.3 Comparing the accounts

We've already seen that whenever a KMS state ϕ breaks a symmetry α in the decompositional sense, its central decomposition features extremal KMS states ϕ_i that break α in the individual sense. The converse is not the case. The decompositional sense of broken symmetry applies only to KMS states; the individual sense also applies to states that purchase their physical significance by other coin. QFT vacuum or ground states are prominent among these. Of course, ground states typically *are* KMS states at inverse temperature 0. But because such states are pure, they provide their own central decomposition, which precludes them from breaking symmetry in the

---

[8] Their terminology is "witness." "Marker" is used here because Ch. 8 has already appropriated "witness" for another purpose.

decompositional sense. We will see that they're perfectly capable of breaking symmetry in the individual sense.

So the scope of the individual sense of broken symmetry is wider than the scope of the decompositional sense, and a symmetry broken in the decompositional sense implies a symmetry broken in the individual sense. I take these as grounds for peaceful coexistence. Perhaps we can build on these grounds. Perhaps we can develop a decompositional account of broken symmetry that applies whenever the individual account applies. A shot at such an account follows, with a statement of its inadequacies hard on its heels.

Where x is some sort of physical distinction—e.g. being a ground, vacuum, or KMS state—we might try to say:

Inadequate Generalization of Def. 13.3. An α-invariant x-state ϕ breaks the symmetry α if ϕ can be decomposed in terms of x-states ϕ_i which break the symmetry in the individual sense.

The idea can be illustrated by means of the flip-flop symmetry of the Heisenberg ferromagnet. The state $\phi = \frac{1}{\sqrt{2}}(\phi^+ + \phi^-)$ is a ground state of the Hamiltonian invariant under that symmetry; it's decomposed in terms of ground states ϕ^+ and ϕ^- on which the flip-flop symmetry fails to be implemented unitarily, and so states that each breaks the flip-flop symmetry in the individual sense.

This illustration notwithstanding, Inadequate Generalization's inadequacies are readily apparent. What makes the generalization a generalization is the lifting of the restrictions to KMS states and *central* (and therefore unique) decompositions. But these restrictions colluded significantly: the central decomposition of KMS states is into *other* KMS states, and is also always available. When physically interesting features other than "KMS" are substituted for x, there is no known decomposition preserving those features (Liu and Emch 2005, 146). The original decompositional account not only identifies the states to which it applies—non-extremal KMS states—but gives a recipe for decomposing them in terms of states of broken symmetry in the individual sense: find their central decomposition! The Inadequate Generalization does neither.

A non-extremal KMS state is a KMS state ϕ whose affiliated von Neumann algebra has a non-trivial center $\mathfrak{Z}_\phi$. This clears the way for markers of broken symmetry in the original decompositional sense: there will be elements of that non-trivial center, we know from Chapter 12, that take different, dispersion-free values in different extremal KMS states. The inadequately generalized decompositional account is much less amenable to markers. Even if it were supplemented to require α-invariant ϕ be such that $\mathfrak{Z}_\phi$ was non-trivial, the supplementation neither implies that ϕ can be decomposed into states of individual broken symmetry nor that macroscopic observables mark the differences between the decomposing states.

13.4 The decompositional account illustrated

13.4.1 The ferromagnet

The last section opened with an example of an infinite spin chain whose symmetry-breaking (in the individual sense) states were ground states of the Heisenberg exchange Hamiltonian. To make contact with the Emch–Liu analysis, we're going to consider an infinite spin chain whose states of broken symmetry are KMS states. We'll modify the Hamiltonian slightly, to model a *long-range force* through which every spin couples to every other through its z component. (Because only the z-component is involved, this is an Ising, as opposed to a Heisenberg, model. Like the Emch–Knops model of §12.3.2, it involves averages over the whole system, so it's still a mean field model.) Thus in a finite segment V of the chain, the Hamiltonian takes the form:

$$H_V = -\sum_{k \in V}\sum_{j \neq k \in V} J_V \sigma_k(z)\sigma_j(z) \tag{13.4}$$

$J_V > 0$ is a coupling constant encouraging spins in the segment to line up. Let $\phi_{V\beta}$ be the Gibbs (and hence KMS) state with respect to unitary dynamics $U_V(t)$ generated by H_V at inverse temperature β. Schrödinger dynamics generated by such a local Hamiltonian have a "flip-flop" symmetry. Indeed $\phi_{V\beta}$ is invariant under the action of the flip-flop automorphism α_F.

The thermodynamic limit of this model is well-behaved, or well-*enough* behaved. In the limit as $V \to \infty$, the unitaries $U_V(t)$ implementing the dynamics (13.4) on the local algebras *do not* converge to a norm limit on the quasilocal C^* algebra $\mathfrak{A}$ for the infinite spin chain. But the KMS states $\phi_{V\beta}$ *do* have an infinite volume limit, call it ϕ_β. And ϕ_β determines a representation of $\mathfrak{A}$ in which the local dynamics have a *weak* limit as $V \to \infty$. Let $\tau_{\phi,\beta}$ denote this weak limit. (NB $\tau_{\phi,\beta}$ is an element of the concrete von Neumann algebra $\pi_{\phi,\beta}(\mathfrak{A})''$, and $\tau_{\phi,\beta} : \pi_{\phi,\beta}(\mathfrak{A})'' \to \pi_{\phi,\beta}(\mathfrak{A})''$ is an automorphism of that algebra. $\tau_{\phi,\beta}$ is not an element of $\pi_{\phi,\beta}(\mathfrak{A})$, and does not define an automorphism of the C^* algebra $\mathfrak{A}$.) The thermodynamic limit of the model (13.4) is well-enough behaved because ϕ_β's unique extension to $\pi_{\phi,\beta}(\mathfrak{A})''$ (which unique extension it has, because it's $\pi_{\phi,\beta}$-normal) is a β KMS state with respect to the W^* dynamics $\tau_{\phi,\beta}$.

Another $V \to \infty$ limit that is well-defined in the weak topology of ϕ_β's GNS representation is the (by now familiar) spatial average defining each component of the net magnetization of the spin chain:

$$\mathbf{m}_z = \lim_{N \to \infty} \frac{1}{2N+1} \sum_{k=-N}^{N} \sigma_k(z) \tag{13.5}$$

(and analogously for $\mathbf{m}_x$ and $\mathbf{m}_y$). Each component of the the net magnetization resides in the center $\mathfrak{Z}_{\phi,\beta}$ of the von Neumann algebra $\pi_{\phi,\beta}(\mathfrak{A})''$ affiliated with ϕ_β. ϕ_β breaks the flip-flop symmetry α, in the decompositional sense, with $\mathbf{m}_z$ as its marker.

To see this, we must invoke a fact (see Liu and Emch 2005).

Fact 13.5. ϕ_β is *extremal* $\tau_{\phi,\beta}$ KMS iff $\beta \leq \frac{1}{J}$, in which case $\phi(\mathbf{m}_x) = \phi(\mathbf{m}_y) = 0$ and $\phi(\mathbf{m}_z)$ satisfies the consistency equation:

$$\phi(\mathbf{m}_z) = tanh(\beta J \phi(\mathbf{m}_z)) \tag{13.6}$$

When $\beta > \frac{1}{J}$, ϕ is not extremal KMS and two values of $\phi(\mathbf{m}_z)$ solve (13.6). Call these values m_z^+ and m_z^-. These values correspond respectively to states ϕ^{m+} and ϕ^{m-} on $\pi_{\phi,\beta}(\mathfrak{A})''$. Each of these states is extremal KMS with respect to the dynamics $\tau_{\phi,\beta}$. Together these states give the central decomposition of ϕ_β, each appearing with a coefficient of $\frac{1}{2}$, a weighting necessary to preserve ϕ_β's α-invariance. Thus the demands of symmetry-breaking in the decompositional sense are met. What's more, the polarization $\mathbf{m}_z$ marks this symmetry breaking: we've already seen that $\mathbf{m}_z \in \mathfrak{Z}_{\phi,\beta}$, and the foregoing implies that different elements of ϕ_β's central decomposition assign $\mathbf{m}_z$ different expectation values. Identifying $\phi^{m\pm}$ as pure thermodynamic phases breaking the flip-flop symmetry, we reveal $\mathbf{m}_z$ as an observable that marks the physical difference between them.

13.4.2 Superconductivity

Without getting too far into the details of the Bardeen–Cooper–Schrieffer model of superconductivity (for which see Emch 2007, 1122–5), we will remark its analogies to the ferromagnet. The BCS model is a historically significant example because it helped inspire the electroweak theory, whose broken symmetry the next chapter examines.

So-called "Cooper pairs," pairs of electrons that bond to one another at low temperatures, play a starring role in the BCS model. Their alliance, in the pop version of the model, explains their oblivion to resistance at low temperatures. Possessed of integral spin, a bound pair of electrons is a boson, and so immune to the Pauli exclusion principle. With that principle suspended, the Cooper pairs in a superconductor have the capacity to all occupy the same state simultaneously. What induces them to exercise this capacity is the presence of a discrete *energy gap* Δ between their ground and their first excited state. Discouraged from excitations by this energy gap, they all settle into the same ground state—forming, in essence, a Bose–Einstein condensate. When this happens, so do the peculiarities of superconductivity. Without other relatively excited electrons to kick them around any more, the Cooper pairs encounter 0 electrical resistance. Another empirical signature of superconductivity is the Meissner effect, in which a superconductor below its critical temperature achieves *perfect diamagnetism*, a condition in which no magnetic field infiltrates the superconducting material. Non-zero magnetic fields do persist in a thin layer at the surface of a superconductor below critical temperature; the *penetration depth* gives the thickness of this magnetizable skin.

13.4 THE DECOMPOSITIONAL ACCOUNT ILLUSTRATED

The energy gap Δ is temperature-dependent. The onset of superconductivity that occurs when $\Delta(T) \neq 0$ is a phase transition, and the structure of $\Delta(T)$'s temperature dependence near this critical point is a critical phenomenon exhibiting universality.

The BCS model is set in three-dimensional Euclidean space. A finite-volume Hamiltonian is introduced with a term describing the free energy of electrons and a term describing the interaction energy of so-called "Cooper pairs." We will focus on this second term. Let $a_s(p)$, $a_s^\dagger(p)$ be creation and annihilation operators for single electrons, where s gives the spin and p the momentum of the electron in question. A creation operator $b^\dagger(p)$ for a Cooper pair is a product of two $a_s^\dagger(p)$ with opposite spin and momentum. For a system confined to a cube of length L, the Hamiltonian looks like:

$$H_L = \text{free energy} + \sum_{p,q} b^\dagger(p) v_{pq} b(q) \tag{13.7}$$

where v_{pq} is a constant describing the interaction energy. The algebra $\mathfrak{A}_L$ for the system is generated by subjecting $a_s(p)$, $a_s^\dagger(p)$ to CARs. Its L-dependence comes from requiring $a_s(p)$ to create excitations that fit in an L-sized box (Emch and Liu 2002, 458, elaborates).

Next, a mean field approximation is performed. The second term in (13.7), a term summing over interactions between every possible pair of Cooper pairs, is replaced by a term in which each Cooper pair interacts with an average, taken with respect to a canonical equilibrium state ρ at inverse temperature β, over every other Cooper pair:

$$H_{L,\rho} = \text{free energy} + \sum_{p} b^\dagger(p) \Delta_\rho(p) \tag{13.8}$$

where:

$$\Delta_\rho(p) = \Sigma_q v_{pq} \langle b(q) \rangle_\rho \tag{13.9}$$

Don't fret that we incorporate ρ without specifying the dynamics with respect to which ρ is an equilbirium state. ρ will be constrained by a self-consistency equation for Δ_ρ that appears in the next paragraph.

Our next move is to perform a Bogoliubov transformation[9] on the $a_s(p)$s and $a_s^\dagger(p)$'s to obtain creation and annihilation operators $\gamma_{s,\rho}^\dagger(p)$, $\gamma_{s,\rho}(p)$ that diagonalize $H_{L,\rho}$:

$$H_{L,\rho} = \sum_{p,s} E_\rho(p) \gamma_{s,\rho}^\dagger(p) \gamma_{s,\rho}(p) \tag{13.10}$$

Thanks to our extensive training on models like the simple harmonic oscillator, we immediately recognize the r.h.s. of (13.10) as the Hamiltonian for a system of *free*

---

[9] Etymological note: these are so-called because Bogoliubov (1958), which showed that the BCS model (13.7) was exactly soluble in the $L \to \infty$ limit, put them to such famous use.

"quasiparticles" created and annihilated by $\gamma^\dagger_{s,\rho}(p)$, $\gamma_{s,\rho}(p)$. The energy $E(\rho)(p)$ for the particle created by $\gamma^\dagger_{s,\rho}(p)$ involves a term $\Delta_\rho(p)$ that's the *energy gap* whose non-zero value would induce superconductivity.

All of the foregoing implies a β-dependent self-consistency equation for $\Delta_\rho(p)$. Picking a solution of this equation defines a dynamics (13.8) and thence an equilibrium state with respect to that dynamics. For temperatures above some critical temperature T_c, the only solution to this self-consistency equation is $\Delta_\rho(p) = 0$. But *below the critical temperature, non-zero solutions occur*, engendering the qualitative expectation of superconducting phenomena. More quantitative expectations, having to do with (among other things) the temperature-dependence of the specific heat of the superconductor and its penetration depth, are fueled by numerical calculations, and upheld by experiment.

There is, however, a lingering embarrassment. It is that the original Hamiltonian (13.7) is invariant under the transformation:

$$\alpha(\theta) : a_s(p) \to e^{i\theta} a_s(p) \qquad (13.11)$$

These $\alpha(\theta)$ constitute a one-parameter group of symmetries of the theory. However, thanks to details about the Bogoliubov transformations that I've suppressed, the quasiparticle Hamiltonian (13.10) lacks $\alpha(\theta)$-invariance.

This embarrassment can be relieved by taking the foregoing finite-volume mean field approximation to the thermodynamic limit, reached by letting the length L of the box confining the system go to ∞. In this limit, the model is exactly soluble (Bogoliubov 1958; Haag 1962 gives an algebraic treatment), and the mean field approximation is not only well-defined but also well-behaved. The quasiparticle creation and annihilation operators no longer provide a mere approximation of the original Hamiltonian (13.7). They diagonalize it *exactly*.

Let the C^* algebra $\mathfrak{A}$ be the uniform closure of $\cup_{L \in \mathbb{R}} \mathfrak{A}_L$, the union, over growing box lengths L, of the algebras for systems confined to boxes of length L. $\mathfrak{A}$ is the quasilocal algebra the thermodynamic limit assigns the superconductor. As with the Heisenberg ferromagnet, the $L \to \infty$ limit of the mean field dynamics (13.10) is well-defined, not as a norm limit on $\mathfrak{A}$, but as a weak limit in the representation π_ρ of $\mathfrak{A}$ determined by a state obtained as the infinite volume limit of finite-volume equilbrium states ρ. (I'll continue to use ρ to denote the infinite-volume equilibrium state on $\mathfrak{A}$.) The key to π_ρ's capacity to sustain the infinite volume limit of (13.10) in the form of W^* dynamics for $\pi_\rho(\mathfrak{A})''$ is the fact that the energy gap observables $\Delta_V(p)$ the model associates with local regions converge in π_ρ's weak topology as $V \to \infty$. Their limit $\Delta_\rho(p)$ lies in the center of $\pi_\rho(\mathfrak{A})''$.

Now suppose ρ is an extremal KMS state. It follows that π_ρ is a factor representation. Lying in the center of $\pi_\rho(\mathfrak{A})''$, $\Delta_\rho(p)$ must be a multiple of the identity. And this is enough to indicate that the transformation (13.11) isn't unitarily implementable on π_ρ. For under the automorphism $\alpha(\theta)$, $\Delta_\rho(p)$ *changes*!

$$\alpha(\theta) : \Delta_\rho(p) \to e^{2i\theta} \Delta_\rho(p) \tag{13.12}$$

If $\alpha(\theta)$ were implemented unitarily on π_ρ, it would take $\Delta_\rho(p)$, a scalar multiple of the identity, to itself. But it doesn't. So $\alpha(\theta)$ isn't implemented unitarily on the factor state ρ, which is ergo disjoint from the factor state $\rho \circ \alpha(\theta) := \rho_\theta$.

For each θ, the state ρ_θ is an extremal KMS state breaking the symmetry (13.11). We can build a *non-extremal* KMS state ϕ by taking a direct integral of these. Sweeping many formalities under the rug, we write $\phi = \int \mu(\theta)\rho_\theta d\theta$, where $\mu(\theta)$ is an integrable measure. Now, ϕ satisfies the decompositional conditions for symmetry breaking. ϕ is $\alpha(\theta)$-invariant. Its GNS representation π_ϕ is a direct integral of factor representations π_{ρ_θ} corresponding to pure thermodynamic phases of the superconductor. Each breaks—and is related to every other by—the symmetry $\alpha(\theta)$. The "marker" of the difference between different pure phases is the energy gap $\Delta_\phi(p)$, which resides in the center of $\pi_\phi(\mathfrak{A})''$. $\Delta_\phi(p)$ is dispersion-free in each pure phase, and transforms non-trivially under the action of the symmetry $\alpha(\theta)$ (see Eq. (13.12)).

To make the analogies with the simpler case of the ferromagnet explicit:

	Ferromagnet	BCS
Symmetry	Flip-flop	$\alpha(\theta)$
Marker	Magnetization $\mathbf{M}_z$	Energy gap $\Delta_\phi(p)$
Pure phases	ϕ^{m+}, ϕ^{m-}	ρ_θ
Invariant state	$\frac{1}{2}(\phi^{m+} + \phi^{m-})$	$\phi = \int \mu(\theta)\rho_\theta d\theta$

At least one disanalogy should be noted. In the case of the ferromagnet, different values of the marker have an empirical signature: which way iron filings move in the presence of the ferromagnet. By contrast, in the BCS case, the markers $\Delta_\phi(p)$ seem to have been erased. That is, empirical data (from the low-temperature superconductors we can actually model) accord with the expectation values the non-extremal, $\alpha(\theta)$-invariant state ϕ assigns $\Delta_\phi(p)$, not with expectation values assigned by any of the continuously many symmetry-defying pure phases. Perhaps this is reassuring: $\alpha(\theta)$ implements transformations which are supposed to make no physical difference. It would be alarming if observations could indicate we were in one, rather than another, gauge-connected state. And we shouldn't lose sight of the empirical differences the energy gap Δ *does* track: the difference between the normal ($\Delta = 0$) and the superconducting ($\Delta \neq 0$) phases of the material.

13.5 Coalesced structures in broken symmetry

With these examples in hand, we turn next to the question of whether QSM's treatments of broken symmetry rely on coalesced structures that might fund arguments against extremism. Such arguments against extremism suppose that in order to have the generality that qualifies them for participation in laws, and so law-mediated explanations, a structure must be well-defined on every state. Coalesced structures are structures whose nomic significance (so understood) requires configuring the state space in a way no extremist can. A coalesced observable, for instance, applies to disjoint factor states, but lacks counterparts either in the canonical algebra $\mathfrak{A}$ or in $\mathfrak{A}$'s universal enveloping von Neumann algebra. Neither the Conservative nor the Imperialist nor the Universalist can attribute to such an observable the generality that secures its involvement in laws.

The discussion of broken symmetry furnishes several candidates for coalesced structure. Let the C^* algebra $\mathfrak{A}$ be the canonical algebra for a theory in which a symmetry $\alpha : \mathfrak{A} \to \mathfrak{A}$ is broken, let ϕ be a state that breaks the symmetry *in the decompositional sense*, and let $\{\phi_i\}$ be ϕ's extremal components, each of which breaks the symmetry α in the individual sense. Because distinct elements of $\{\phi_i\}$ are disjoint factor states, the Hilbert Space Conservative, claiming that physically possible states are confined to the folium of a factor representation of $\mathfrak{A}$, isn't even entitled to this *framework* for discussing symmetry breaking, much less structures coalesced within the framework. I take this to establish that the Conservative cannot make sense of the physics of broken symmetry in QSM.

To establish counterpart claims for the Imperialist and the Universalist requires a bit more work. A law-involved magnitude M whose domain is a proper subset of $S_{\mathfrak{A}}$ will do the trick. As Chapter 12 argued, no such magnitude appears in either extremist's catalog of physically significant observables. Markers of broken symmetry are magnitudes of this sort. Consider $\mathbf{m}_z$, the z-component of the polarization of the infinite spin chain. $\mathbf{m}_z$'s temperature-dependence near the critical temperature satisfies relations constituting universal behavior. Its value also explains the behavior of iron filings in the vicinity of spin chain. Participation in such explanatory relations secures $\mathbf{m}_z$'s nomic significance. $\mathbf{m}_z$ is limited in domain because it's an $L \to \infty$ spatial average that is well-defined only on representations of $\mathfrak{A}$ with respect to whose weak topologies the limits defining the average converge. Such representations correspond to a proper subset of the full set of states on $\mathfrak{A}$. *Mutatis mutandis* for the energy gap observable Δ in the BCS model.

The illustrations furnish other examples of coalesced structure, in the form of W^* dynamics. Neither model manifests the persistence and change through time constitutive of a natural system except in the form of W^* automorphisms coalesced as weak (but not norm) infinite volume limits of finite volume dynamics. Only a proper subset of states in the algebraic sense on $\mathfrak{A}$ instantiates the dynamic aspect of a theory of canonical type $\mathfrak{A}$ subject to W^* dynamics. Such dynamics are coalesced structures whose nomic import neither the Imperialist nor the Universalist can grasp.

A final example of a possible coalesced structure is more problematic. Let $\alpha(t)$ be a one-parameter family of symmetries. When ϕ breaks the symmetry $\alpha(t)$ in the decompositional sense, $\alpha(t)$ is strongly unitarily implementable on ϕ (because ϕ is α-invariant and thanks to Fact 5.3) but not on any of ϕ's extremal components ϕ_i that break α in the individual sense (because that's what it is to break α in the individual sense). Let $U(t)$ be the strongly continuous unitary group that implements $\alpha(t)$ in ϕ's GNS representation. Stone's theorem ensures that $U(t)$ has an an infinitesimal generator, call it G. $U(t)$ and G appear to be credible candidates for nomically significant coalesced structure.

But their appearance is marred by the following reflections. Suppose the broken symmetry $\alpha(t)$ has a marker, call it D. By definition, D resides in the center $\mathfrak{Z}_\phi$ of the von Neumann algebra $\pi_\phi(\mathfrak{A})''$ affiliated with ϕ's GNS representation. A natural way to attribute physical significance to coalesced structures involved in the decompositional account of broken symmetry is to identify the algebra of observables with $\pi_\phi(\mathfrak{A})''$, which contains not only markers of the broken symmetry but agents of the system's W^* dynamics. However *if we take $\pi_\phi(\mathfrak{A})''$ to be the algebra of observables pertaining to the system, the generator G of the symmetry transformations is not an observable.* Here's why. Let ϕ_i and ϕ_j be distinct states that break symmetry in the individual sense. To mark this broken symmetry, D must be such that $\langle\phi_i|D|\phi_i\rangle \neq \langle\phi_j|D|\phi_j\rangle$. To satisfy the demands of the decompositional account, there must be some $U(t)$ that maps ϕ_i to ϕ_j. Thus:

$$\langle\phi_i|D|\phi_i\rangle \neq \langle\phi_i|U^{-1}(t)DU(t)|\phi_i\rangle$$

But if $U(t)$ is an element of $\pi_\phi(\mathfrak{A})''$, $U(t)$ commutes with everything in $\pi_\phi(\mathfrak{A})''$'s center, including D. So:

$$\langle\phi_j|D|\phi_j\rangle = \langle\phi_i|U^{-1}(t)DU(t)|\phi_i\rangle = \langle\phi_i|U^{-1}(t)U(t)D|\phi_i\rangle$$

—which contradicts the condition just announced. Therefore $U(t)$ is not a member of $\pi_\phi(\mathfrak{A})''$. But if G were a member of $\pi_\phi(\mathfrak{A})''$, so would be $U(t)$, because the algebra is weakly closed. So G isn't a member of $\pi_\phi(\mathfrak{A})''$ either. If $\pi_\phi(\mathfrak{A})''$ is the algebra of physical observables, G isn't a physical observable. This disqualifies G from the status of coalesced structure. Of course, G can be restored to such status by loosening our criteria for admission to the collection of observables.

The broken symmetry examples of coalesced structure underwriting arguments against extremism are highly parasitic on Chapter 12's examples, framed in terms of phase structure. In particular, any observable marking the difference between different states of broken symmetry is a phase observable distinguishing abruptly between distinct pure thermodynamic phases. And the symmetry-breaking version of the arguments, like the original version, adverts to the thermodynamic limit, riddled by idealizations. The foundational significance of such settings has been questioned by (among others) Earman (2004c) and Callender (2001). So it might seem that we're back where we started at the beginning of this chapter: we have an argument against extremisms that appeals to suspect idealizations. We can, however, claim to have advanced this far. We now have a template for an argument against extremisms cast in terms of broken symmetry. Given the putative importance of broken symmetry in QFT, this suggests the strategy of looking for realizers of the template in that putatively fundamental setting. The next chapter pursues this strategy.

14

Broken Symmetry and Physicists' QFT

14.1 Introduction

Chapters 12 and 13 sought to extract a case—a case I called "the Coalesced Structures Argument"—against "extremist" interpretations of QM_∞ from accounts of phase structure and broken symmetry encountered at the thermodynamic limit of QSM. The crux of the case was the centrality to those accounts of *coalesced structures*, which structures the argument claimed to be features of the systems of interest whose apparent nomic significance no extremist can accommodate. Macroscopic observables "marking" the physical differences between different phases or different states of broken symmetry provide an example of coalesced structure, as do representation-dependent dynamics.

The thermodynamic limit *idealizes* manifestly finite systems, such as steaming teacups, as infinite in volume and infinitely populous. The availability of unitarily inequivalent representations of the physics of infinite systems is the pivot on which the Coalesced Structures Argument turns. Turning as it does on an idealization, the argument runs afoul of a stricture John Earman announces in the course of a discussion of Curie's principle:

[**Sound Principle**:] While idealizations are useful and, perhaps, even essential to progress in physics, a sound principle of interpretation would seem to be that no effect can be counted as a genuine effect if it disappears when the idealizations are removed. (Earman 2004c, 191)

When the idealizations required to reach the thermodynamic limit are removed, so too are the unitarily inequivalent representations on which the Coalesced Structures Argument pivots. That argument violates Earman's "sound principle of interpretation."

This chapter considers two attempts to exculpate the Coalesced Structures Argument. The first would parry the idealization charge by embracing Earman's Sound Principle but developing a quantum *field* theoretic instantiation of the Coalesced Structures Argument. Such an argument would exploit the presence in QFT of accounts deploying coalesced structures analogous to the coalesced structures encountered in the

14.1 INTRODUCTION 313

thermodynamic limit of QSM. The systems treated by QFT are *fields*. Because fields have infinitely many degrees of freedom, the unitarily inequivalent representations that would appear in a quantum field theoretic realization of the Coalesced Structures Argument *aren't* artifacts of idealization.[1] Thus the first attempted exculpation would place that argument beyond the reproach of Earman's sound principle.

Sustenance for this attempted exculpation may be found in textbook and popular accounts of the physics of the QFTs making up the Standard Model of High Energy Physics (HEP), accounts which portray broken symmetry as central to those theories, and make extensive analogical use of a notion of "phase" borrowed from the thermodynamic limit. Aitchison writes in his exemplary 1982 *Informal Introduction to Gauge Field Theories*:

Now the ground state of a complicated system... may well have unsuspected properties—which may, indeed, be very hard to prove as following from the Lagrangian. But we can postulate (even if we cannot yet prove) properties of the quantum field theory vacuum state $|0\rangle$ which are analogous to those of the ground states of many physically interesting many-body systems—such as ferromagnets, superfluids and superconductors, to name a few with which we shall actually be concerned. (1982, 75)

Later, having distinguished states of broken from states of unbroken symmetry by appeal to the kind of quanta those states contain, he continues:

One may speculate that we are *literally* talking about a phase transition picture, exactly as in the superfluid or ferromagnet case. Possibly, in the very early stages of the universe, the temperature was above the critical (transition) value, and so we might imagine the manifest symmetry phase as occurring then, the present lop-sided state of affairs having arisen as the universe cooled. (Aitchison 1982, 87)

Apparently drawing abrupt physical distinctions between different states of broken symmetry, conceptualized as different phases, HEP promises to furnish material for an idealization-free Coalesced Structures Argument. Sections 14.2–14.5 examine this promise, focusing in particular on two phenomena whose heuristic presentation is strongly suggestive of coalesced structure: Goldstone bosons and the Higgs mechanism. The former are sometimes described as quanta of a field identified as the order parameter that distinguishes between states of broken and unbroken symmetry—just as the global polarization does for the ferromagnet. If this order parameter has the character of a "phase observable," Goldstone bosons are candidates for coalesced structure.

Standard presentations of the Higgs mechanism associate different states of broken symmetry with different numbers of massive and massless particle types:

for the broken components of the symmetry... one has massive vector particles; for the other remaining scalar fields one again has massive scalar particles; and the only massless particles

---

[1] Setting aside the issue that, if spacetime is discrete at the Planck scale, the continuous spacetime of QFTs *is* an idealization.

which remain correspond to the unbroken components of the vector field... These numbers, of course, depend on the particular representation you choose... and it is perfectly possible, if you want to, to make this last number zero and have no massless particles at all. (Kibble 1967, 291)

These purported physical differences between different states of broken symmetry are also candidates for coalesced structures.

Limitations of space and expertise mean that I concentrate here on pedagogical presentations of HEP—indeed, on the particular strand of those presentations which emphasizes heuristics drawn from the thermodynamic limit. *Spoiler alert!* Section 14.5 will acknowledge that the dialectical strategy of relocating the Coalesced Structures Argument to the QFT context fails, simply because broken symmetry in HEP, *as this chapter treats it*, lacks models sufficiently explicit to undergird the Coalesced Structures Argument. This very circumstance suggests another strategy for responding to the idealization complaint, one which does not rely on the success of the relocation strategy. Section 14.6 develops this second response by making a case that Earman's sound principle is not so sound.

Before allowing the reader to proceed, I should emphasize that I don't take myself to have attained any sort of stable understanding, however inadequate, of the theoretical arena of HEP. I am venturing into it not to announce theses I take to be airtight, but so that I might gesture in the general direction of interesting foundational questions lurking there. My hope is that by so gesturing I can induce those better able to formulate and address the questions to do just that.

14.2 Broken symmetry in QFT

In textbook presentations of HEP, broken symmetry is often depicted as making a difference to what masses particles have, and even to how many different species of particles there are:

The ϕ field associated with this peculiar [because broken] symmetry is *massless*... The essential point, however, is that the massless ϕ field becomes combined with the massless gauge field to form a massive vector field... This is no isolated freak phenomenon, but possibly the simplest example illustrating the *generation of mass for vector particles is a spontaneously broken gauge theory*. (Aitchison 1982, 70, italics in original)

The picture promises to sustain a Coalesced Structures Argument. If different states of broken symmetry are unitarily inequivalent but nevertheless physically significant, broken symmetry in QFT stymies Hilbert Space Conservatism; if the physical differences between different states of broken symmetry resist assimilation to different expectation values assigned an element of an abstract algebra $\mathfrak{A}$ or its universal enveloping von Neumann algebra, Algebraic Imperialism and Universalism are stymied as well. The aim of the following sections is to probe discussions of broken symmetry in HEP, in search of fodder for an idealization-free Coalesced Structures Argument.

14.2.1 Overview

To get at the notion of broken symmetry applicable to QFT, we need to recall that QFTs typically arise as *quantizations* of classical Lagrangian field theories. In this context, the standard notion of symmetry is that of a *Lagrangian symmetry*, a transformation that leaves the Lagrangian $\mathcal{L}$ invariant. The *quantum* theoretic question of broken symmetry is the question of how symmetries of the Lagrangian of the classical theory to be quantized carry over to the ensuing QFT. Streater puts it this way: "The term 'spontaneous breakdown of symmetry'... has come to mean a field theory whose Lagrangian is invariant under a certain transformation of the fields, whereas there exist solutions, i.e. realizations of the algebra of operators, that do not possess the symmetry as a unitary transformation" (Streater 1965, 510). We can elaborate. Focus for starters on the straightforward case of classical Klein–Gordon theory, whose Lagrangian determines a symplectic vector space $(\mathcal{S}, \Omega)$ of classical solutions. To quantize this theory is to find a representation of the Weyl relations over $(\mathcal{S}, \Omega)$, i.e. a map from solutions $\phi \in \mathcal{S}$ to unitary operators $W(\phi)$ acting on a Hilbert space $\mathcal{H}$ to satisfy the Weyl relations. These operators $W(\phi)$ are generators of the Weyl algebra $\mathfrak{A}$.

Any symmetry $A: \phi \to \phi'$ of the classical solution space should lift to an automorphism group $\alpha: \mathfrak{A} \to \mathfrak{A}$ of the Weyl algebra via:

$$\alpha(W(\phi)) = W(A(\phi)) \tag{14.1}$$

Take the automorphism α to be the quantum realization of the classical symmetry. Now, given a concrete Hilbert space quantization of Klein–Gordon theory in the form of a Hilbert space representation $(\pi, \mathcal{H})$ of the Weyl algebra $\mathfrak{A}$, *it is a further question whether α is unitarily implementable in $(\pi, \mathcal{H})$*. Streater takes the symmetry to be *broken* in representations where it fails to be unitarily implemented. (Other physicists call such symmetries *hidden*.) As the discussion of § 13.3.3 would lead us to expect, the sense of broken symmetry at issue in QFT is broken symmetry in the individual sense (Def. 13.2). Indeed, symmetry-invariant states that break symmetry in Emch and Liu's collective sense are not prominent in what follows.

Symmetries of different sorts will feature in the saga of the electroweak theory, regaled in § 14.4. The following scholia introduce them (and may be skipped by those who have already made their acquaintance).

Scholium: internal vs. spacetime symmetries. Solutions of the equations of motion determined by a Lagrangian $\mathcal{L}(\phi_i)$ will assign to each point x of a spacetime $\mathcal{M}$ values for the field variables $\phi_i(x)$. Call the space of such solution $\mathcal{S}_\mathcal{L}$. A **spacetime** symmetry of this Lagrangian theory is just what it sounds like: a transformation that acts indirectly on the dependent field variables by acting directly on the independent variable x, a transformation under which relevant structures (such as $\mathcal{L}(\phi_i)$ or $\mathcal{S}_\mathcal{L}$)[2] are invariant. Diffeomorphisms of the differential manifold on which the

---

[2] See the Scholium starting on p. 293 for a cautionary note about identifying these structures.

metric and stress-energy tensors are defined are spacetime symmetries of GTR: if $(\mathcal{M}, g_{ab}, T_{ab})$ solves the Einstein field equations, so does its "drag-along" under a diffeomorphism $\mathcal{M} \to \mathcal{M}$. An **internal** symmetry, by contrast, is a transformation that acts directly on degrees of freedom of the dependent field variable $\phi_i(x)$, a transformation under which relevant structures are invariant. The homely phase transformation mapping a solution $\phi(x)$ of the Klein–Gordon equation to the solution $e^{i\theta}\phi(x)$ is an example. ♠

Scholium: The symmetry groups *U(1)* and *SU(2)*. $U(1)$ is the group of 1×1 Unitary matrices—i.e. complex numbers of unit modulus—with a group law given by matrix multiplication. A theory of ordinary QM is invariant under $U(1)$ in the following sense. $U(1)$ has an extremely simple representation on the theory's Hilbert space $\mathcal{H}$: the operator that multiplies an arbitrary vector by $e^{i\theta}$ represents the element $e^{i\theta}$ of $U(1)$. The space of solutions to Schrödinger's equation is invariant under the action of this internal symmetry group.

$SU(2)$ is the group of 2×2 Unitary matrices that are Special because their determinants are 1. As before, matrix multiplication gives the group law. That old warhorse, the Hilbert space $\mathbb{C}^2$ of pairs of complex numbers, furnishes a representation of $SU(2)$ with which we are already acquainted. $SU(2)$ is three-dimensional: each element is a linear combination of three "basis" elements. Represented on $\mathbb{C}^2$, these basis elements are three unitary operators whose generators are proportional to the Pauli matrices. (The constant of proportionality, $\frac{1}{2}$, is needed to ensure that the unitaries generated have determinant 1.) Call these generators τ_1, τ_2, and τ_3.

Satisfying (up to constants of proportionality) the Pauli relations, the generators τ_a fail to commute: thus $SU(2)$, considered as a symmetry group, is *non-abelian*. The unitaries $e^{i\alpha_1 \tau_1}, e^{i\alpha_2 \tau_2}, e^{i\alpha_3 \tau_3}$ which are the basis of the representation on $\mathbb{C}^2$ of $SU(2)$ can be thought of as implementing "rotations" between different, but unitarily equivalent, representations of the Pauli relations. For instance, the effect of acting on a trio $2\tau_1, 2\tau_2$, and $2\tau_3$ satisfying the Pauli relations with the unitary $e^{i\alpha_1 \tau_1}$ is to rotate the eigenvectors of $2\tau_2$, and $2\tau_3$ through a "phase angle" α_1. One obtains another representation of the Pauli relations, manifestly unitarily equivalent to the original. ♠

In the theory of weak interactions, $SU(2)$ is an internal symmetry. The internal degree of freedom exhibiting $SU(2)$ symmetry in weak interactions is known as *isospin*. In the lingo, the $SU(2)$ symmetry reflects a freedom to choose a trio of "isospin axes."

Scholium: Global and local symmetries. In QFT, a *global symmetry* $\phi_i(x) \to \phi'_i(x)$ of the field is one implemented by an operator that's independent of x. *Local symmetries*, by contrast, are implemented on the Hilbert space of fields $\phi_i(x)$ by operators that do depend on x. Thus, for instance, to make the global internal $U(1)$ symmetry $\phi(x) \to e^{i\theta}\phi(x)$ local, one allows the phase θ to vary with x: $\phi(x) \to e^{\theta(x)}\phi(x)$. ♠

14.2.2 An example: the massless Klein–Gordon field

The mass 0 Klein–Gordon field on Minkowski spacetime may give the simplest example of symmetry breaking in QFT. For $m = 0$, the classical Klein–Gordon equation reads:

$$\Box \phi(x) = \partial_\mu \phi(x) \partial^\mu \phi(x) = 0 \tag{14.2}$$

where x is a point in Minkowski spacetime and we have invoked the Einstein summation convention. It's clear from Eq. (14.2) that if $\phi(x)$ is a solution, so is $\phi(x) + \chi$ where $\chi \in \mathbb{R}$ is a constant function. Thus we have a one (real) parameter group of transformations $A_\chi : \phi(x) \to \phi(x) + \chi$ that leaves the space of solutions to (14.2) invariant. (The group law is $A_\chi A_{\chi'} = A_{\chi + \chi'}$.) The group A_χ is a group of symmetries of the classical theory.

As Noether's first theorem famously has it, where there's a symmetry there's a conservation law (see Goldstein, Poole, and Safko 2002, 589–98 for an elaboration; Brading and Brown 2003 and Earman 2004c are discussions aimed at philosophers). More carefully stated, the theorem concerns theories that may be obtained via variational principles from a Lagrangian $\mathcal{L}(\phi_i)$, which is a function of fields $\phi_i(x)$ (and their derivatives) defined on a spacetime we'll take to be four-dimensional. If $\mathcal{L}$ is invariant (up to a four-divergence that makes no difference to the Euler–Lagrange equations determined by $\mathcal{L}$) under a *continuous* family of transformations A_χ of the fields ϕ_i, then there is a four-vector current j_μ which is conserved in the sense that $\partial_\mu j^\mu = 0$. The spatial integral of j_μ's time (here, $\mu = 0$) component:

$$Q = \int j^0(x) d^3x \tag{14.3}$$

is a conserved (i.e. time-independent) charge. Q generates the symmetry, in the sense that the Poisson bracket $\{Q, \phi_i\} = \frac{\partial A_\chi \phi_i}{\partial \chi}$ gives the infinitesimal variation of the field under the symmetry transformation A_χ.

A_χ is a one-parameter Lie group. In more generality, Noether's first theorem states that an action obtained by integrating a Lagrangian is invariant under an n-parameter Lie group of continuous symmetries if and only if there are n conservation laws of the form $\partial_\mu j^\mu = 0$.

For illustrations of Noether's first theorem, we need look no further than a Lagrangian that's invariant under continuous families of rotations and spatial translations—say the Lagrangian for n interacting point masses in classical mechanics. Such a Lagrangian describes a system in which linear and angular momentum are conserved; these are the Noether charges corresponding to the continuous translation and rotation symmetries. Each spatial component of linear momentum generates translations along the corresponding spatial axis, and so on, and so on.

We have so far been considering symmetries of *classical* Klein-Gordon theory. What about its quantization? The connection Noether's theorem forges between symmetries and conservation laws conditions the expectation that quantized Klein–Gordon theory exhibits analogous symmetries. Aitchison gives voice to the expectation:

> Suppose that a given Lagrangian is invariant under some one-parameter continuous internal symmetry with a conserved Noether current j^μ, $\partial_\mu j^\mu = 0$. The associated charge is the Hermitian operator $Q = \int j^0 d^3x$, and $\dot{Q} = 0$. We have hitherto assumed (though the nature of the assumption has been emphasized) that the transformations of this $U(1)$ group are representable on the Hilbert space $\mathcal{H}$ of physical states by unitary operators $U(\lambda) = e^{i\lambda Q}$ for arbitrary λ, with the vacuum invariant under U, so that $Q|0\rangle = 0$. (Aitchison 1982, 71)[3]

We can resolve Aitchison's expectation into two parts. First, a one real parameter internal symmetry (such as A_χ) of the classical theory is implemented on its quantized counterpart's Hilbert space of states by a continuous unitary group $U(\chi) = e^{i\chi R}$ whose self-adjoint generator is the quantum analog of the classically conserved charge associated with the symmetry. A tradition consolidated and championed by Wigner, of identifying symmetry transformations of quantum theories with [anti]unitary operators on the Hilbert spaces framing those theories, motivates this component of the analogy. Second, the vacuum state $|0\rangle$ of the quantized theory is an eigenstate of R with eigenvalue 0.

To motivate the second component of the analogy, recall that an internal symmetry transformation acts, not on the spacetime over which fields are defined, but directly on field degrees of freedom themselves. We'd therefore expect such symmetries to commute with spacetime symmetries. It follows that the Poincaré-invariant vacuum state $|0\rangle$—which is supposed to be the unique Poincaré-invariant state in its Hilbert space representation[4]—is invariant under U_χ:

Argument Sketch 14.1. The axioms require the group Λ of Poincaré transformations to be implemented unitarily on the vacuum representation by a group U_λ, $\lambda \in \Lambda$ of unitary operators. $|0\rangle$ is Poincaré-invariant, so for any $\lambda \in \Lambda$, $U_\lambda|0\rangle = |0\rangle$. Thus $U_\chi U_\lambda|0\rangle = U_\chi|0\rangle = U_\lambda U_\chi|0\rangle$ (because U_λ and U_χ commute). Thus $U_\chi|0\rangle$ is invariant under every Poincaré transformation, hence—because $|0\rangle$ is the *unique* Poincaré-invariant vacuum state—equivalent to $|0\rangle$. ♠

Since $U(\chi) = e^{i\chi R}$, $|0\rangle$'s invariance under U_χ implies that it's an eigenvalue 0 eigenstate of the charge R assumed to generate the quantum symmetry transformation.

The expectations just announced have non-trivial presuppositions. One is that the charge associated with a classical symmetry has a quantum counterpart: e.g. that $Q = \int j^0 d^3x$ is a well-defined Hermitian operator on the space of quantum states. But Q so defined is an infinite volume ($R \to \infty$) limit of spatial integrals $Q_R = \int_{|x|<R} j^0 d^3x$ giving the charge contained in a sphere of radius R. Infinite volume limits are the

---

[3] Peskin and Schroeder (1995, 308–10) offer a partial justification of the analogy.

[4] For an account of the relations between the supposition and the axioms of AQFT, see Halvorson and Müger (2007, 743).

sorts of quantity, we've learned from the infinite spin chain and other examples, whose existence can be representation-dependent. Given a one-parameter family A_χ of Noether symmetries of classical Klein–Gordon theory, we are warranted in assuming that the Weyl algebra $\mathfrak{A}$ for that theory's quantization enjoys a corresponding family α_χ of automorphisms. But it is a further and distinct step to suppose those automorphisms to be unitarily implemented in a representation π of $\mathfrak{A}$, by a family of unitaries generated by some quantum counterpart to the classical Noether charge.

The quantized mass 0 Klein–Gordon field upsets these further suppositions. A one parameter family A_χ of classical symmetries isn't implemented unitarily on a Poincaré-invariant vacuum representation:

Argument Sketch 14.2. (following Aitchison 1982, 69). Suppose that the classical symmetry A_χ corresponds to a transformation of the *quantum* field operators $\hat{\phi}(x) \to \hat{\phi}(x) + \chi$ which is implemented unitarily on a Poincaré-invariant vacuum representation[5]: $U_\chi \hat{\phi}(x) U_\chi^{-1} = \hat{\phi}(x) + \chi$ for U_χ. Where $|0\rangle$ is the vacuum state of that representation, we know from Argument Sketch 14.1 that

$$U_\chi |0\rangle = |0\rangle \tag{14.4}$$

But this can't be. We are assuming that the vacuum is a Fock space vacuum. $\mathcal{H}$, the single-particle Hilbert space over which the Fock space in question is a Fock space, is spanned by vectors of the form $a_k^\dagger |0\rangle$. Because $\langle 0|a_k^\dagger|0\rangle = \langle 0|a_k|0\rangle = 0$, the vacuum is orthogonal to such vectors. It follows that the vacuum is orthogonal to every other Fock space vector, and so orthogonal to the state $\hat{\phi}(x)|0\rangle$. This implies that $\langle 0|\hat{\phi}(x)|0\rangle = 0$, which implies that:

$$\langle 0|\hat{\phi}(x) + \chi|0\rangle = \chi \tag{14.5}$$

But from (14.4):

$$\langle 0|\hat{\phi}(x) + \chi|0\rangle = \langle 0|U_\chi \hat{\phi}(x) U_\chi^{-1}|0\rangle = \langle 0|\hat{\phi}(x)|0\rangle = 0 \tag{14.6}$$

This is a contradiction. Thus the symmetry fails to be unitarily implementable. ♠

We can use the apparatus of abstract algebras to restate these results about the mass 0 Klein–Gordon field. The symmetries A_χ lift to an automorphism group α_χ of the Weyl algebra $\mathfrak{A}$ encapsulating the quantization of Klein–Gordon theory. The Poincaré group Λ is represented on $\mathfrak{A}$ by the automorphism group α_λ, $\lambda \in \Lambda$. The axioms posit an α_λ-invariant vacuum state ω. But if ω is α_λ-invariant, so is $\omega_\chi := \omega \circ \alpha_\chi$, since α_χ and α_λ commute. (This follows from Argument Sketch 14.1 above.) If α_χ isn't unitarily implementable on ω (which is what Argument Sketch 14.2 showed), ω and ω_χ are

---

[5] For the time being, Hilbert space operators designated by the same letters as classical magnitudes will sport hats, to avert ambiguity.

unitarily inequivalent Poincaré invariant states.[6] This manifests one familiar signature of broken symmetry: degenerate vacuum (or ground) states, connected to one another by the symmetry in question.

14.3 Goldstone bosons

This section considers the contention that broken symmetries in QFT are heralded by the appearance of Goldstone bosons. Goldstone bosons may aid and abet a field theoretic Coalesced Structures Argument because they're sometimes said to correspond to oscillations in a symmetry-breaking order parameter. In the thermodynamic limit, such order parameters, exemplified by the global magnetization of a ferromagnet, can be coalesced structures.

Goldstone bosons sometimes play the role of shadowy, unbidden, and menacing figures in the saga of the electroweak theory. Like almost all good sagas, this one does not take the form of a deductively valid argument. Peopled by heuristics, analogies, and inspired guesses, the saga of the electroweak theory is propelled more often by rules of thumb than by theorems. We'll review some of these for the light they shed on Goldstone bosons and other candidates for coalesced structure in QFT. Here's the first:

Rule of Thumb 14.3. "The hallmark of a hidden symmetry situation is the existence of some field with *non-vanishing vacuum expectation value*" (Aitchison 1982, 73).

A standard motivation for this rule of thumb invokes $\hat{Q}$, the quantum observable posited to represent the charge associated with a global, internal, continuous symmetry (Aitchison 1982, 73; Kibble 1967, 280). $[\hat{Q}, \hat{\phi}(y)]$ gives the infinitesimal variation of the field operator $\hat{\phi}(y)$ under the symmetry transformation. In the case of a "manifest" (i.e. unitarily implemented) symmetry, $\hat{Q}|0\rangle = 0$, and $\langle 0|[\hat{Q}, \hat{\phi}(y)]|0\rangle = 0$. Thus if $\langle 0|[\hat{Q}, \hat{\phi}(y)]|0\rangle$ does not vanish, the symmetry is broken. If $\langle 0|[\hat{Q}, \hat{\phi}(y)]|0\rangle$ does not vanish, there exists a field operator $\hat{\phi}' := [\hat{Q}, \hat{\phi}(y)]$ whose vacuum expectation value is non-zero. Thus, the rule of thumb.

The motivation for the rule of thumb is not a proof. For one thing, it supposes that the charge Q generating a broken symmetry is well-defined on the symmetry-breaking vacuum representation. But we've been put on notice that Q needn't be. Indeed, a result due to Fabri and Picasso suggests that Q *isn't* well-defined on a symmetry-breaking representation. In the general case of a classical Lagrangian $\mathcal{L}$ invariant under a continuous family A_χ of global internal symmetries associated with the charge Q, and a representation of the corresponding quantization on a Hilbert space $\mathcal{H}$ with Poincaré-invariant vacuum state $|0\rangle$, Fabri and Picasso (1966) show that either:

---

[6] Nevertheless, representations π_ω and π_{ω_χ} are "relativistically equivalent"—their respective representations of the Lorentz group Λ are unitarily equivalent (Streater 1965).

(a) The symmetry is implemented in $|0\rangle$'s representation by a unitary group $U_\chi = e^{iQ\lambda}$, where $Q|0\rangle = 0$; or

(b) $Q|0\rangle$ is not an element of $\mathcal{H}$.

In case (b), the case physicists call "hidden" or "spontaneously broken" symmetry (Aitchison 1982, 71), Q threatens not to be a well-defined operator on the space $\mathcal{H}$ of states supposed to be physical. There Q takes the physical state *primus inter pares*—the vacuum state $|0\rangle$—outside of $\mathcal{H}$. Q's failure to be a bounded operator on $\mathcal{H}$ is often excused by noting that really only expressions such as $\langle 0|[\hat{Q}, \hat{\phi}(y)]|0\rangle$ are encountered in use, and these can be well-defined even if Q isn't.

Goldstone's theorem concerns a QFT satisfying standard axioms including translation covariance, positivity of energy, and microcausality. It states that if such a QFT breaks a continuous symmetry which is "locally generated," then massless bosons (in the sense of states of the QFT whose energy goes to zero as their three-momentum does) exist.[7] This statement of the theorem is patently vague, and spelling out criteria for local symmetry generation turns out to be a delicate business (see Strocchi 2005, 161–76). But the symmetries to have in mind are Noether symmetries, generated by charges obtained by integrating local current densities.

We've already seen an illustration of Goldstone's theorem in the broken symmetry α_χ of massless Klein–Gordon field, a broken symmetry evidenced, in accord with the first rule of thumb, by the non-vanishing vacuum expectation value $\langle 0|[\hat{Q}, \hat{\phi}(y)]|0\rangle$. Here, Aitchison observes, "there is a massless particle present—the ϕ-quantum itself!" (1982, 74).

To articulate the character of Goldstone bosons, consider an exactly soluble model of a symmetry breaking system: our old friend, the Heisenberg ferromagnet. Model the magnet as an infinite chain of spin $\frac{1}{2}$ systems interacting with their immediate neighbors via the Heisenberg exchange Hamiltonian:

$$H = - \sum_{<i,j>} J\sigma_i \sigma_j \quad (14.7)$$

where J is a coupling constant and $<i,j>$ denotes summation over nearest neighbors. Ground states of this manifestly rotationally symmetric Hamiltonian are states where all the spins are aligned. There are as many ground states as there are distinct axes in space along which the spins might be aligned. Each such ground state determines a representation of the CARs for the infinite spin chain unitarily inequivalent to every other.

Heuristically, each ground state of the continuously many unitarily inequivalent representations of the infinite spin chain can be obtained from every other by the symmetry transformation of rotation: to get the ground state in which all spins are

---

[7] See Aitchison (1982, 73–9), for a presentation of the theorem in the context of conventional field theory; Kastler, Robinson and Swieca (1966) for an old-fashioned algebraic treatment; and Buchholz, Longo, and Roberts (1992) for a newfangled algebraic treatment exploiting DHR selection theory.

aligned along the x axis, rotate the ground state in which all the spins are aligned along the z axis through an angle of $90°$ in the x-z plane. The degeneracy, and unitary inequivalence, of distinct ground states testifies that the evident rotational symmetry of the Hamiltonian is broken. A final symptom of symmetry breaking is the presence of a "field"—the global polarization $\mathbf{m} = w\text{-}lim_{n \to \infty} \frac{1}{2n+1} \Sigma^n_{k=-n} \sigma_k$—with a distinct non-vanishing expectation value in each of the degenerate ground states. In the parlance of Emch and Liu, $\mathbf{m}$ is a *marker* of the broken symmetry.

The Heisenberg ferromagnet breaks the continuous rotational symmetry of its dynamics. A non-relativistic version of Goldstone's theorem applies,[8] and we expect massless bosons. Where are they? The intuition pumping answer is that they're "spin waves", in which the spins σ_i associated with each site rotate, with characteristic wavelength λ, as you move along the chain. As $\lambda \to \infty$, a spin wave in the vacuum of one representation approaches the vacuum of another representation. Because both are ground states, this suggests that the energy required to create a spin wave of characteristic wavelength λ goes to 0 as λ goes to ∞. De Broglie taught us that the wave momentum k is inversely proportional to the wavelength λ. So as λ approaches ∞, k approaches 0. The presence of states whose energy goes to zero as their momentum does is the signature of a Goldstone boson. These heuristic considerations expose $\lambda \to \infty$ spin waves as Goldstone bosons.

Aitchison summarizes a key lesson of the example:

Rule of Thumb 14.4. "The Goldstone (massless) excitations are *space-time-dependent oscillations in the order parameter (field) whose non-vanishing expectation value in the global theory defines the hidden symmetry*" (Aitchison 1982, 77).

In the example of the ferromagnet, that order parameter is the polarization $\mathbf{m}$, also the marker of broken symmetry. There, the analysis of Goldstone bosons implicates exactly the sort of observable the Coalesced Structures Argument uses as ammunition against both the Imperialist and the Universalist: a nomically significant observable whose domain is a proper subset of the full set of algebraic states.

An eponymous and field theoretic illustration of Goldstone's theorem will pave the way for our discussion of the Higgs mechanism. In the Goldstone model (see Aitchison and Hey 2004, §17.5), a complex scalar field $\phi = \frac{1}{\sqrt{2}}(\phi_1 - i\phi_2)$ is supposed to obey the classical Lagrangian:

$$\mathcal{L} = \partial_\mu \phi^* \partial^\mu \phi - V(\phi) \qquad (14.8)$$

where:

$$V(\phi) = \frac{1}{4}\lambda(\phi^*\phi)^2 - \mu^2 \phi^*\phi \qquad (14.9)$$

---

[8] See Guralnik, Hagen, and Kibble (1964, 630 ff.). The absence of long-range forces plays the role in the non-relativistic version that Microcausality plays in the relativistic version.

14.3 GOLDSTONE BOSONS

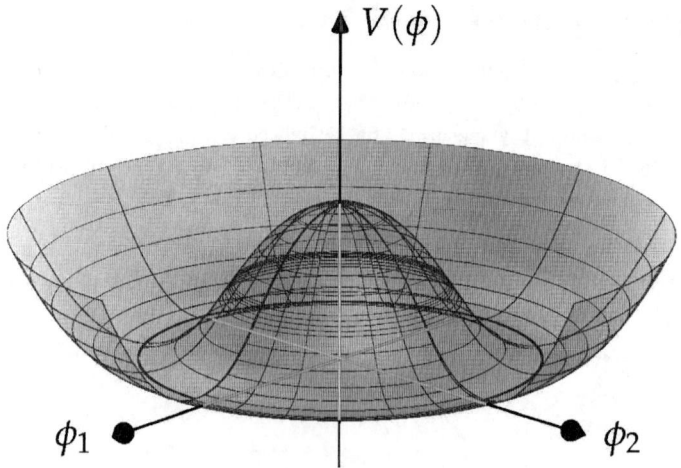

Figure 14.1. A Mexican hat potential

This Lagrangian enjoys a global $U(1)$ symmetry: it is invariant under transformations $\phi \to e^{i\alpha}\phi$ (which are just rotations of the axes used to coordinate ϕ_1 and ϕ_2). When $\mu^2 > 0$, $V(\phi)$ describes the notorious classical "Mexican hat potential" (see Figure 14.1). Because $V(\phi) = 0$ when $\phi^*\phi = \frac{2\mu^2}{\lambda}$, the rotationally symmetric Lagrangian admits a continuum of *asymmetric* ground states (ϕ_1, ϕ_2) lying along the circle $\phi_1^2 + \phi_2^2 = 2|\phi|^2 = \frac{4\mu^2}{\lambda} := f^2$ at which $V(\phi)$ is minimized.

The classical ground states are degenerate, and asymmetric. The quantum theory should reflect this. It will, if we start with a "semi-classical" approximation: that a vacuum state $|0\rangle$ of the QFT assigns the quantum field amplitude $\hat{\phi}$ an expectation value very near the value characterizing the ground states of the classical Lagrangian. So let's suppose that $|\langle 0|\hat{\phi}|0\rangle| = \frac{|f|}{\sqrt{2}}$. According to Rule of Thumb 14.3, this non-zero vacuum expectation value signals a broken symmetry. Re-expressing the complex field ϕ by mean of a polar decomposition in terms of a real amplitude ρ and a complex phase θ:

$$\phi = \frac{\rho}{\sqrt{2}} e^{-i\frac{\theta}{dim}} \quad (14.10)$$

(where *dim* designates factors included to make the units come out nice in the end) suggests that we use the complex phase θ to distinguish distinct ground states $|0\rangle_\theta$ of the QFT: supposing $_0\langle 0|\hat{\phi}|0\rangle_0 = \frac{f}{\sqrt{2}}$, $_\theta\langle 0|\hat{\phi}|0\rangle_\theta = e^{-i\theta}\frac{f}{\sqrt{2}}$. (Notice that for both vacua, $|\langle\hat{\phi}\rangle| = \frac{|f|}{\sqrt{2}}$, as per the semiclassical approximation.) This suggests that the symmetry being broken is the global $U(1)$ symmetry relating distinct vacuum states to one another—a symmetry whose breakage should be accompanied by massless Goldstone bosons.

At this point in his exposition, Aitchison interjects frankly:

Before proceeding further it is as well to realise that this argument for the nature of the vacuum may be fine for classical fields, but its status is not so clear in the true quantum field theory case. A rigorous proof that the Lagrangian [(14.8)] has indeed this [$|0\rangle_0$] as the vacuum state, and that [$_0\langle 0|\hat{\phi}|0\rangle_0 \neq 0$] is consistent, seems not to be available. We shall accept it as an *assumption*. (1982, 85)

Undaunted, we proceed. The quanta of our QFT will take the form of small oscillations around a vacuum state. In light of the decomposition (14.10) and the semiclassical approximation $\langle 0|\hat{\phi}|0\rangle = e^{-i\theta}\frac{f}{\sqrt{2}}$, we express the field of which these quanta are quanta as follows:

$$\hat{\phi} = \frac{1}{\sqrt{2}}(f + \hat{h})e^{-i\frac{\hat{\theta}}{dim}} \qquad (14.11)$$

In keeping with the spirit of quantization, we have promoted the classical variables ρ and θ to operators. Now we rewrite the Lagrangian (14.8) in terms of the quantum field operators we're approximating, to first order, by (14.11). The point is to examine the result for pieces we recognize as describing free particles of antecedently familiar types. We obtain:

$$\mathcal{L} = \frac{1}{2}\partial_\mu \hat{h} \partial^\mu \hat{h} - \mu^2 \hat{h}^2 + \frac{1}{2}\partial_\mu \hat{\theta} \partial^\mu \hat{\theta} - \mu^4/\lambda \qquad (14.12)$$

which is immediately recognizable to those who have been properly brought up as describing two species of particles, one (the quanta of the field $\hat{h}$) of mass $\sqrt{2}\mu$, the other (the quanta of the field $\hat{\theta}$) massless. In the latter, *we have found our Goldstone bosons*. Beyond signaling the presence of Goldstone bosons, the symmetry-breaking vacuum expectation value has further empirical significance: it contributes to the probability amplitude (Aitchison and Hey 2004b, 215) for a transition between the vacuum state and a state with one Goldstone boson.

The ferromagnet's spin waves are not the only Goldstone bosons populating serious physical theories. Other examples are Landau phonons in superfluids, phonon excitations in crystals, and the ("anomalously light") pions associated with the broken chiral symmetry of QCD—a broken symmetry Nambu introduced on the basis of an explicit analogy with the BCS model of superconductivity prosecuted at thermodynamic limit. Such field theoretic Goldstone bosons are candidates for coalesced structures. The next section considers another area of QFT where coalesced structures, in the forms of robust physical differences between different states of broken symmetry, appear to figure prominently.

14.4 The Higgs mechanism

The Higgs mechanism, in which a broken gauge symmetry enables the masses of various particles to agreeably sort themselves out, attains its full glory in the context of the electroweak theory. This theory exhibits a "non-abelian" $SU(2) \times U(1)$ symmetry. Section 14.4.2 will illustrate the Higgs mechanism by means of a theory with the simpler $U(1)$ symmetry. A very cheap tour of the electroweak theory precedes the illustration.

14.4.1 The saga of the electroweak theory

The unification of the electromagnetic and weak interactions is made possible by the mechanism of spontaneous symmetry breaking. (Strocchi 2008, 193)

What follows is a highly schematic and impressionistic reconstruction of pedagogical presentations of the electroweak theory. We start with a Lagrangian $\mathcal{L}_0$ invariant under the group $SU(2) \times U(1)$ of global symmetry transformations. This is a composition of the symmetry group $U(1)$ of electromagnetic interactions and the symmetry group $SU(2)$ of weak interactions. (Weak interactions have a hand in beta decay and radioactivity; I'll have only a very little to say about their habits here. For these, consult Peskin and Schroeder 1995, 700–19.)

Next, we apply the *gauge principle*, which has proven a powerful heuristic for describing interactions in HEP (see Aitchison 1982, 28–30, for a discussion). The gauge principle directs us, when confronted with a Lagrangian exhibiting a global symmetry, to posit a corresponding *local* symmetry. For example, given a Lagrangian invariant under *global* phase transformations $\phi(x) \to e^{i\theta}\phi(x)$, we construct the corresponding *local* symmetry by allowing the phase angle to be x-dependent: $\phi(x) \to e^{i\theta(x)}\phi(x)$. The global symmetry having been converted to a local one, the gauge principle exhorts us to find a new Langrangian invariant under the local symmetry. To build such a Lagrangian, we augment the original Lagrangian with terms involving new (vector-valued) fields which transform suitably under the local (a.k.a. "gauge") symmetry transformation. These new fields are *gauge* fields; the revamped Lagrangian typically includes terms coupling them to other fields. Thus following the gauge principle we develop models of not only new fields but also their forms of interaction with fields we already knew about. It is part of the physicist's catechism that *the gauge fields obtained by following this prescription are massless*. Otherwise, terms involving the mass of the field would have to appear in the Lagrangian, and these terms would break the very symmetries the gauge fields were introduced to restore (see Aitchison 1982, 29).

A bevy of philosophical questions, largely ignored here, surround gauge theories, both classical and quantum. See Healey (2007) for a taste of these. It is a commonplace for physicists and philosophers alike that gauge transformations are unphysical: situations connected by a gauge transformation are widely held to be physically equivalent (see Earman 2004a; 2004b; 2004c; Witten 1991, 1149; t'Hooft 2007, §5.3). But there are dissenters (Giulini 1994; Balachandran 1994, §4; Bais 2004, §5.1), who will have a return engagement later in this chapter.

Applying the gauge principle to electroweak theory, we convert the global $SU(2) \times U(1)$ symmetry to a local one, and demand a Lagrangian invariant under these local symmetries. To satisfy this requirement, we add terms to the Langrangian $\mathcal{L}_0$ to obtain a Lagrangian $\mathcal{L}_g$. These new terms involve vector fields that transform under the local symmetries so as to effect the desired invariance of $\mathcal{L}_g$. To sustain local $SU(2)$ invariance, we add three gauge fields $W_\mu^a, a \in \{1, 2, 3\}$ to the Lagrangian; to sustain local $U(1)$ invariance, we add one gauge field B_μ.

Another piece of the physicist's catechism is that the force associated with a global symmetry that, made local, induces the introduction of a gauge field is the force mediated by that gauge field. Thus the three gauge fields W_μ^a mediate the weak force and the gauge field B_μ mediates the electromagnetic force. The particle associated with a gauge field is a *gauge boson*; it's said to carry the corresponding force.

But there's a problem with this picture: gauge fields introduced by following the gauge principle are invariably massless, but the mediators of the weak force ought to be massive, because it's a short-range force. This follows from yet another bit of the physicist's catechism: if the gauge boson mediating a force is massive, the "time-energy uncertainty principle" implies that its mean lifetime is limited, so that it can range only so far. Gauge bosons are massless; vector bosons mediating the short-range weak force, evidently, are not. Should we aim, then, for a non-gauge theory of weak interactions? Not, it seems, if we want that theory to exhibit the virtue of renormalizability. The nature of this virtue deserves much more discussion than it receives here,[9] where we will take it to be the virtue a theory exhibits when calculations involving it, and finitely many (typically experimentally determined) input parameters, can be made to yield tractably finite predictions. (State of the art techniques for taming QFT infinities and divergences are drawn from renormalization group theory, originally developed for critical phenomena.) It is believed that the renormalizability of the electroweak theory hinges crucially on possession by that theory of gauge symmetry (see Aitchison and Hey 2004, 251–5). The quandary is that we want the bosons mediating the electroweak force to be gauge bosons, for the sake of renormalizability, but we also want them to be massive, for the sake of curtailing the range of that force.

Weinberg and Salam independently struck upon the same solution to this problem. The solution hinges on another rule of thumb.

Rule of Thumb 14.5. "Spontaneous symmetry breaking can, under certain circumstances, give mass to a massless vector field" (Kaku 1993, 322).

Section 14.4.2's sketch of the Higgs mechanism will illustrate this rule of thumb, which suggests that the bosons conveying the weak force can be bulked up in representations where symmetries are broken. However, this solution to the problem of the massless vector bosons introduces further problems of its own. First, we don't want *all* the gauge bosons of the electroweak theory to be massive, because the photon is among them,

---

[9] For this, see e.g. Teller (1995, ch. 7); Folland (2008, ch. 7); Zee (2010, pt. III).

and it's massless. The solution to this problem, hinted at in the digression below, is to pick a vacuum (representatation) with residual symmetry.

Scholium: Residual symmetry. A $U(1)$ symmetry demands absolute compliance. Either it's unitarily implemented on the Hilbert space of quantum states or it isn't. More nuanced symmetries, such as $SU(2)$, are another matter. Possessed of three generators, $SU(2)$ extends a quantum theory the option of breaking some, but not all, of the symmetries composing it. For a simple example, recall the representation of $SU(2)$ on the Hilbert space $\mathbb{C}^2$, which we'll take to be equipped with a representation of the Pauli relations for a single spin system. $SU(2)$'s three generators are a trio of operators proportional to the Pauli matrices $\sigma_i, i \in \{x, y, z\}$. We can regard a unit vector $\psi \in \mathbb{C}^2$ as picking out an axis in $\mathbb{R}^3$: the axis along which ψ is a $+\frac{1}{2}$ eigenvector of the associated component of spin. The $SU(2)$ generator proportional to σ_x generates a unitary family implementing rotations around the x-axis, and *mutatis mutandis* for y and z. Now, the $\sigma(z)$ eigenvector $\begin{pmatrix} 0 \\ 1 \end{pmatrix}$ "breaks" $SU(2)$'s full rotational symmetry, by picking out as a preferred direction in space the positive z-axis, along which the net spin of the system is oriented. But $\begin{pmatrix} 0 \\ 1 \end{pmatrix}$ *exhibits* some rotational symmetries: it is invariant under rotations in the xy plane. Just so, in cases of sophisticated symmetries, a QFT can break them part way, in the sense that some, but not all, of their generators give rise to families of transformations that are unitarily implemented on the Hilbert space of states. ♠

A more dire problem with the scheme to endow vector bosons with mass by breaking symmetry is that the execution of the scheme could bring Goldstone's theorem into play: if symmetries of the wrong sort are broken, massless Goldstone bosons are introduced. The physicist's catechism holds massless particles ought to be easy to detect, since no mc^2 term appears in the energy required to create them. But we haven't seen any massless particles not otherwise accounted for. So the prediction of novel massless particles would be a problem for the electroweak theory.

The solution to this more dire problem is a choice of representation. We exploit gauge freedom to set fields that might otherwise correspond to Goldstone bosons to 0 everywhere, thus neutralizing their threat.

After some artful neglect of higher-order terms, the electroweak saga leaves us with a Lagrangian involving three massive fields—W_ν^1, W_ν^2, and Z_μ—corresponding to the trio of weak bosons, and one massless field A_ν, corresponding to the photon. The Z_μ and A_ν are each mixtures of gauge fields implicated in the preservation of local $SU(2)$ and $U(1)$ symmetries; this is taken to go some way toward unifying the electromagnetic and weak forces (see Maudlin 1996 for a discussion). The Lagrangian emerging from the electroweak saga also features another massive scalar field σ. This Higgs field is often said to be what's left of the field postulated to break the global $SU(2) \times U(1)$ symmetry after the Goldstone bosons associated with that broken symmetry have been gauged away.

14.4.2 The abelian Higgs model

In the Higgs mechanism, the gauge bosons of the electroweak theory acquire mass through a broken symmetry accompanied, *not* by massless Goldstone bosons, but by a new scalar field. This section illustrates the maneuver for the simpler case of a theory with a $U(1)$ symmetry. Be warned that my exposition is crude, and sets to 1 a number of constants inserted to make the outcome palatable to physicists.[10]

The theory in question is obtained by applying the gauge prinicple to the Lagrangian of §14.3's Goldstone model. That Lagrangian is invariant under global $U(1)$ transformations $\phi(x) \to e^{-i\chi}\phi(x)$, so following the gauge principle, we demand *local* $U(1)$-invariance, that is, invariance under transformations:

$$\phi(x) \to e^{-i\chi(x)}\phi(x) \qquad (14.13)$$

To construct a Lagrangian invariant under these transformations, we need to add to it terms involving a vector field A_μ transforming as:

$$A_\mu(x) \to A_\mu(x) + \partial_\mu \chi(x) \qquad (14.14)$$

when $\phi(x)$ transforms as (14.13) (compare Aitchison 1982, §5.3–5.5). These local transformations are gauge transformations. The gauge-invariant Lagrangian we obtain is:

$$\mathcal{L} = -\frac{1}{4}(F_{\mu\nu})^2 + |D_\mu \phi|^2 - V(\phi) \qquad (14.15)$$

where:

$$F_{\mu\nu} := \partial_\mu A_\nu - \partial_\nu A_\mu \qquad (14.16)$$

and:

$$D_\mu := \partial_\mu + iA_\mu \qquad (14.17)$$

The Lagrangian (14.15) describes a complex scalar field coupled to itself as well as to an electromagnetic field.

As before, we are assuming that the potential $V(\phi)$ assumes the notorious "Mexican hat" form, so that a semi-classical approximation warrants us in concluding that the quantum field $\hat{\phi}$ has a continuum of non-zero vacuum expectation values, corresponding to the continuum of degenerate minima of the classical potential. One global $U(1)$ symmetry breaking vacuum state would be $|0\rangle$ such that $\langle 0|\hat{\phi}|0\rangle = \rho_{vac}$, a real number that minimizes the classical potential.

As with the Goldstone model, we investigate the quantization of the Lagrangian (14.15) by "expanding around the vacuum": treating the quantum field as a first-order

---

[10] For more thorough treatment, see Aitchison (1982, §6.9); Peskin and Schroeder (1995, §20.1); Kaku (1993, §10.2); Zee (2010, IV.6); Folland (2008, §9.3).

perturbative expansion around a vacuum state described by the semi-classical approximation. In terms of the decomposition of the complex field by means of a real amplitude and a complex phase, we get $\hat{\phi}(x) = (\rho_{vac}(x) + \hat{\sigma}(x))e^{-i\hat{\theta}(x)}$, where $\rho_{vac}(x)$, the classical field amplitude, is a constant; $\hat{\sigma}(x)$ is a small quantum perturbation in the field amplitude, and $\hat{\theta}(x)$ is the phase.

The Goldstone model eventually related this phase to the field whose quanta were the Goldstone bosons associated with the breaking of that model's global $U(1)$ symmetry. But (let us suppose) we don't want these Goldstone bosons in our own eventual theory. We are, after all, developing it as a simplified model of the electroweak theory, which has too many massless and not enough massive bosons to start with!

We know how the phase and amplitude of the field $\phi(x)$ behave under the *gauge transformation* $\phi(x) \rightarrow e^{-i\chi(x)}\phi(x)$. The amplitude is unchanged; the phase transforms as:

$$\theta(x) \rightarrow \theta(x) + \chi(x) \qquad (14.18)$$

Viewing Eq. (14.18) through the lens of gauge freedom, we see immediately how to neutralize the threat of Goldstone bosons. Simply *fix* the gauge by setting:

$$\chi(x) = -\theta(x) \qquad (14.19)$$

at each $x \in \mathcal{M}$. Considering the situation from the perspective of this *unitary gauge* essentially does away with the complex part of $\phi(x)$, and so sets the field $\theta(x)$ whose quanta (in the Goldstone model) are Goldstone bosons, to 0 everywhere. The corresponding gauge transformation for the A_μ fields is:

$$A_\mu(x) \rightarrow A_\mu(x) - \partial_\mu \theta_x := W_\mu(x) \qquad (14.20)$$

Now we rewrite the Lagrangian (14.15) in terms of the perturbative expansion $\hat{\phi}(x) = (\rho_{vac}(x) + \hat{\sigma}(x))e^{-i\hat{\theta}(x)}$ in the unitary gauge that suppresses ϕ's θ dependence, putting hats hither and yon to emphasize its association with a quantum field. The rewritten Lagranian includes terms which are quadratic in the interesting fields $\hat{W}_\mu$ and $\hat{\sigma}$:

$$\frac{1}{4}\tilde{F}^2_{\mu\nu} + \frac{1}{2}\rho_{vac}\hat{W}^2_\mu + \frac{1}{2}(\partial_\mu(\hat{\sigma}))^2 - \frac{1}{2}\frac{\partial^2 V(\rho_{vac})}{\partial t^2}\hat{\sigma}^2 \qquad (14.21)$$

Here $\tilde{F}_{\mu\nu}$ is defined on the model of (14.16), with W_μ playing the role of A_μ.

These terms are immediately recognizable to those properly brought up as describing two species of particles: a vector boson, the quantum of the field W_μ, of mass $\frac{\rho_{vac}}{\sqrt{2}}$, and a scalar field σ of mass $\sqrt{\frac{\partial^2 V(\rho_{vac})}{\partial t^2}}$. Regarding W_μ's mass spectrum in light of (14.20), we are tempted to exclaim: "the massless gauge field [A_μ] has 'swallowed' the Goldstone field [θ]…to make the massive vector field [W_μ]" (Aitchison and Hey 2004, 261). That is, we have equipped the gauge boson, better known as the photon, with a mass that depends on the vacuum expectation value ρ_{vac}. (Historically, the Lagrangian (14.15) was introduced to model superconductivity, where the association of a mass with the photon helps account for the Meissner effect (see §13.4.2), in which a magnetic field pierces a superconductor only to a finite penetration depth. In the presence of symmetry-breaking state of the superconductor, the bosons mediating the magnetic force acquire mass and therefore enjoy only a limited range.) We have also introduced a new massive scalar field σ. Its quanta are *Higgs bosons*.

The foregoing maneuver is known as the *Higgs mechanism*. Generalized to the non-abelian case (see Peskin and Schroeder 1995, 692 ff., for an account), it endows the gauge bosons of the electroweak theory with masses while shielding that theory from the heartbreak of Goldstone bosons. It accomplishes this by the cunning choice of the unitary gauge (14.19).

There are also gauges in which the electroweak theory is demonstrably renormalizable. These gauges are *different* gauges from the unitary gauge in which the number and masses of the predicted particle species work out. Aitchison comments, "We rely on the gauge invariance to let us have our cake (unitarity) and eat it (renormalizability) i.e. one gauge exhibits one desirable property, and another the other, but by gauge-invariance it is all the same theory. Naturally this is not a proof" (1982, 99).

This attitude poses a threat to field theoretic realizations of the Coalesced Structures Argument, if any there be. The threat is that if you embrace renormalizability while chanting the mantra of gauge equivalence, you can't very well contend, in support of a Coalesced Structures Argument, that breaking gauge symmetry makes a physically significant difference, ineffable to extremists. While suggesting that the mantra of gauge equivalence is, in this context, optional, the next section gives a dim view of the prospects for a QFT realization of the Coalesced Structures Argument.

14.5 Coalesced structures in QFT?

14.5.1 Promissory notes

Recall our aim: to silence the "idealization complaint" about the Coalesced Structures Argument by re-posing that argument in a setting all hands would (modulo cavils about discreteness at the Planck scale) admit requires infinitely many degrees of freedom to model: the setting of quantum fields. The Coalesced Structures Argument hinges on physically respectable explanations that configure the content of a quantum theory in a way none of our extremists can. This section asks: do Goldstone bosons or the saga of the electroweak theory furnish us with quantum field theoretic counterparts of such explanations?

Fanning expectations of a positive answer are two considerations. The first is that putatively disjoint states appear in the saga of the electroweak theory. Let the automorphism α be a symmetry of a C^* algebra $\mathfrak{A}$. If a state ω on $\mathfrak{A}$ breaks the symmetry α, then so does the state $\omega \circ \alpha$, which is unitarily inequivalent to ω. That is, *different states of broken symmetry are disjoint* (provided they're pure). And different

14.5 COALESCED STRUCTURES IN QFT? 331

states of broken symmetry populate the saga of the electroweak theory: the vacuum state in the unitary gauge and the vacuum states of gauges in which the theory is demonstrably renormalizable are examples. Each breaks the gauge symmetry of the theory. The Hilbert Space Conservative can't regard both as physically possible.

The second consideration is that the saga of the electroweak theory *appears* to draw physical distinctions between different states of broken symmetry, distinctions that could be understood in terms of observables available to neither the Algebraic Imperialist nor the Universalist. The abelian Higgs model of §14.4.2 illustates this: the mass of the vector boson W_μ depends on the symmetry-breaking vacuum expectation value ρ_{vac}. Thus the saga of the electroweak theory posits a general physical difference between states that break $SU(2) \times U(1)$ symmetry and states that don't: in the former, but not the latter, gauge bosons associated with the symmetry have mass. The saga of the electroweak theory also posits several sorts of specific physical differences between different states of broken symmetry: *which* gauge bosons acquire mass depends on what residual symmetries the representation displays. More salient to our purposes, *the massive gauge bosons mark the distinction between unitarily inequivalent symmetry breaking representations*. In representations that don't break symmetry, gauge bosons are massless; in representations breaking symmetry, some gauge bosons are massive. This suggests that the mass of gauge bosons could serve as a marker, in something like the Emch–Liu sense, of broken symmetry, and thereby as fodder for the Coalesced Structures Argument.

The Higgs mechanism is the second chapter in which broken symmetry appears to assume physical significance. There it is a local (gauge) $SU(2) \times U(1)$ symmetry that is broken, and in a very specific way. It is broken by the gauge fixing that spirits Goldstone bosons away. What distinguishes the unitary gauge—the state of broken symmetry suiting the Higgs mechanism—from other states of broken gauge symmetry, as well as from states of unbroken gauge symmetry, then, is this: the Higgs state is free of unwanted massless particles and graced by an additional massive scalar particle, the Higgs boson. Again, this suggests that magnitudes characterizing the number of different types of particle at large in a representation, and their masses, could serve as markers of broken symmetry, and thereby as fodder for a field theoretic Coalesced Structures Argument. Goldstone bosons, at least as realized by the pions connected to the broken chiral symmetry of QED, are also fodder, if we are warranted in regarding them as oscillations in a coalesced symmetry-breaking order parameter.

There are two major impediments to developing these suggestions. First, the explicit and precise framework of operator algebras casts the mould for the Coalesced Structures Argument. The examples of broken symmetry in HEP chronicled here (at least *as* chronicled here) defy formulation in these terms. Commutation relations, the algebras they generate, and their representations have been notably absent from the accounts here forwarded of Goldstone bosons and the Higgs mechanism. Those accounts are just too loose for the apparatus of the Coalesced Structures Argument to get any purchase on them. Second, the Coalesced Structures Argument requires the explanations plying inequivalent representations to be physically respectable. But some commentators have cast aspersions on some of the explanatory work broken symmetry is meant to do on behalf of the electroweak theory (Earman 2004b; Healey 2007). The balance of this section acknowledges the force of each of these grounds for skepticism. The next section summons reasons to take the Coalesced Structures Argument seriously anyway.

14.5.2 Unphased

HEP is sometimes presented as though positing physical differences between different states of broken symmetry. But that supports a quantum field theoretic Coalesced Structures Argument only if the physical differences are expressed in terms analogous to those in which the Coalesced Structures Argument, as well as the interpretive positions it targets, is conducted. *But the electroweak theory is formulated in no such framework.* Its Lagrangian is a springboard for daunting feats of physics derring-do, ranging from applications of the gauge principle, to renormalization in the t'Hooft gauge, to describing the decay of the μ meson. Its Lagrangian is *not* an easy ingredient to use in a simple Hamiltonian quantization recipe. (Difficulties attendant upon the quantization of gauge theories are partially to blame.) Applying that recipe to the Klein–Gordon Lagrangian, we obtain the Weyl algebra as the canonical algebra for quantized Klein–Gordon theory. There is no analogous critter $\mathfrak{A}_{EW}$ for a quantum theory based on the electroweak Lagrangian. It follows that questions about the unitary implementability in significant states on that algebra of significant automorphisms of that algebra—questions about broken symmetry and the unitary equivalence or lack thereof of physically salient representations—are *ill-posed*. We can make *guesses* about how the putative phenomena collected under the heading of the saga of the electroweak theory would be embedded in an explicit operator-theoretic realization of that theory. But because we lack the operator-theoretic realization, they would only be guesses.

Let us make them anyway. In particular, let us attempt to characterize an operator-theoretic realization of the electroweak theory that would sustain a Coalesced Structures Argument. Imagine on behalf of the electroweak theory a canonical algebra $\mathfrak{A}_{EW}$ on which are defined automorphisms α_λ and α_χ implementing (respectively) non-gauge and gauge symmetries. Take the electroweak theory to traffic in states ω on $\mathfrak{A}_{EW}$ in which these automorphisms are not unitarily implementable, signaling the breakage of the corresponding symmetries. Putative markers of the broken global $U(1) \times SU(2)$ symmetry are the masses of the weak vector bosons. A putative marker of the broken gauge symmetry is that in the unitary gauge, but not in others, Goldstone bosons are suppressed and the Higgs field is a scalar. Let ω_H be a state of broken gauge symmetry implementing the Higgs mechanism. *Suppose* that a marker of the broken gauge symmetry exists as an element M_γ of the von Neumann algebra $\pi_{\omega_H}(\mathfrak{A})''$ affiliated with ω_H's GNS representation but missing from the image under π_{ω_H} of the canonical algebra $\mathfrak{A}_{EW}$ of the electroweak theory. Suppose also that a marker of the broken global symmetry exists as an element M_χ with analogous features. Call this *the Naive Story*.

Given this collection of extraordinary guesses about the sort of algebraic formulation the electroweak theory would admit, if it admitted an algebraic formulation, there is some hope that a QFT-adapted Coalesced Structures Argument would go through. But so shot through with wishful thinking and self-serving speculation is the Naive Story that few would accept it without complaint. For one thing, the guesses are guided by the very examples whose susceptibility to the idealization complaint prompted the present investigation of symmetry breaking in HEP. Although this is standard operating procedure within HEP, in the present dialectical context—responding to the idealization complaint by relocating the Coalesced Structures Argument to QFT—it is self-defeating. For another, the Naive Story is only imperfectly analogous to the QSM base case. For the ferromagnet, it is a property that *already* distinguishes between degenerate *classical* ground states—the direction of the net polarization—that serves as a marker of broken symmetry.[11] In the Naive Story, the putative markers to broken symmetry lack classical precursors. So in addition to being highly speculative, the Naive Story has a highly surprising ending.

Work done on classical gauge theories suggests another route to a field theoretic Coalesced Structures Argument.[12] Considering classical gauge theories, Giulini (1994) contends that a charge associated with global gauge transformations at infinity generates a transformation which connects different ground states to one another. His somewhat heterodox view is that the charge is a physical quantity, and the different ground states are physically distinct. Modeling our interpretation of the Higgs mechanism on this classical precedent, we'd describe the different symmetry-breaking vacua as physically distinct, perhaps even physically distinct in a way no extremist could reconstruct. Call this the Controversial Story.

A serious problem with both the Naive and the Controversial Stories would be that they make a fundamental mistake about the nature of gauge freedom. Among philosophers of physics it is catechismic that "the objective facts about a possible world (as described by a theory) are exactly those that can be stated in gauge-invariant terms." It follows that *genuine* "observables [are] gauge independent quantities of the theory" (Earman 2004a, 1233–4). I'll call this the Principle of Gauge Equivalence. It has implications for how to implement the gauge automorphisms (Earman's θs, below) of a theory whose genuine observables lie in an algebra $\mathfrak{A}$:

if the root notion of a gauge transformation is that of a transformation that connects different descriptions of the same physical state, it follows that whereas nature can break a non-gauge symmetry θ by choosing a non-θ symmetric state, She cannot break a gauge symmetry in the same fashion. We, not Nature, break the gauge symmetry by choosing a particular gauge

---

[11] The degenerate classical ground states are those of the Ising model, which assigns ± to each site on a lattice, and subjects those classical spins to a scalar form of the Heisenberg model interaction. Its degenerate ground states are the states where every spin points up and every spin points down.

[12] Thanks to Gordon Belot for bringing these approaches to my attention.

condition. Consequently the automorphism θ induced by the gauge symmetry must be construed as acting on a field algebra $\mathfrak{F} \supset \mathfrak{A}$ that is larger than the algebra $\mathfrak{A}$ of genuine, gauge independent observables since otherwise $\omega \circ \theta$ and ω would be genuinely different states. (Earman 2004c, 186)

Thinking of the gauge automorphism θ as acting on the big field algebra $\mathfrak{F}$, Earman assumes that the "algebra $\mathfrak{A}$ of genuine, gauge independent observables" is point-wise invariant under θ. It follows that states on $\mathfrak{A}$ are also θ-invariant. And (because if ω is α-invariant, then α is unitarily implementable on ω) this implies that $\mathfrak{A}$ *admits no state breaking gauge symmetry*. But physical differences between different states breaking gauge symmetry are central characters in both the Naive and the Controversial Stories. Hence both stories fundamentally mistake the nature of gauge. Insofar as the QFT version of the Coalesced Structures Argument endorses either of these stories, it too fundamentally mistakes the nature of gauge.

Earman thinks the Naive Story has company in mistaking the nature of gauge. In the conventional presentation of Higgs mechanism, the Goldstone bosons associated with the breaking of global $U(1) \times SU(2)$ acquire mass by settling into the unitary gauge and "eating" part of the gauge field. To Earman, the gauge field is "descriptive fluff", and "The popular slogan can be counterbalanced by the cautionary slogan that neither mass not any other genuine attribute can be gained by eating descriptive fluff" (2004, 189–90).

I believe that there is some room to resist Earman's skepticism about broken gauge symmetry and the Higgs mechanism. The gauge symmetries at issue in the Principle of Gauge Equivalence are symmetries of a Lagrangian that circumscribes the theory to be quantized. Although I have not made much of it here, focusing instead on canonical Hamiltonian quantization, strategies for quantization are legion and still under construction, particularly when it comes to theories whose gauge freedom stymies a rote application of canonical strategies. "Quantization" isn't, at least not in the present state of play of physics, a particularly well-understood or univocal process. And this undermines confident inferences from "θ is a gauge symmetry of $\mathcal{L}$" to "every state of the theory obtained by *quantizing* $\mathcal{L}$ is θ-invariant." We don't know enough about how the quantization of gauge theories goes to be sure.

One way to secure the inference would be to make it a *desideratum* for a quantization of a theory defined by a Lagrangian $\mathcal{L}$ that it be one whose states are invariant under gauge symmetries of $\mathcal{L}$. But given phenomena like the Bohm–Aharonov effect (whose conventional interpretation, much resisted by philosophers, is that a quantity that classically is "mere gauge" becomes physically significant in a quasi-quantum setting), this could be an ill-judged desideratum.

Earman has quite explicitly in mind the program of Dirac quantization which (to oversimplify[13]) subjects a classical gauge theory to the usual quantization strategy, then

---

[13] See Healey 2007 (173–4, appendix C) for an introduction aimed at philosophers; Henneaux and Teitelboim (1992) for a treatise.

imposes conditions which require states of the quantum theory to be gauge-invariant. A quantization obtained by following this program *would* satisfy the desideratum and frustrate the translation of the Coalesced Structures Argument into quantum field theoretic terms. *But when it comes to the theories under consideration, the program has not been successfully carried out.* Imagining that it might be, Earman constructs a trilemma for the conventional saga of the electroweak theory whose most prominent horns are that either there were no Goldstone bosons to begin with or gauge-fixing doesn't suppress them. He concludes:[14]

While there are too many what-ifs in this exercise to allow any firm conclusions to be drawn, it does suffice to plant the susupicion that when the veil of gauge is lifted, what is revealed is that the Higgs mechanism has worked its magic of suppressing zero mass modes and giving particles their masses by quashing spontaneous symmetry breaking. However, confirming the suspicion or putting it to rest require detailed calculations, not philosophizing. (Earman 2004c, 191)

But this misrepresents the situation. It is not as if the program of Dirac quantization of the electroweak Lagrangian would be successfully completed if only the relevant mathematical physicists could afford more supercomputing time. That is, the impediments are not merely "calculational." They're conceptual as well. Given a Lagrangian with gauge redundancy, the program of Dirac quantization quantizes, then eliminates the redundancy. Another principled approach to quantizing a gauge theory is to first eliminate redundant variables, then quantize. The question of whether Reduction and Quantization "commute" is not only open but also difficult to formulate in general (see Landsman 2004 for a recent discussion). If they don't, in cases where they don't, one should ask whether the result of the Dirac route is the "correct" quantization. And even if Reduction and Quantization do commute, there is the awkward circumstance that practicing physicists charged with quantizing gauge theories generally use the technique of BRST quantization, which (instead of reducing) begins by adding even more fields to the Lagrangian (see Nemenschansky, Preitschopf, and Weinstein 1988).

By predicating his trilemma for conventional glosses of the Higgs mechanism on an executed and accepted Dirac quantization of gauge field theory, Earman is making guesses about the future of physics, including guesses about the contents of the toolbox physicists will use to construct future physical theories. That's not only his prerogative but also his job. Still they're *only guesses*. A future physics wherein a mathematically rigorous and precise realization of interacting quantum field theories attributes them a structure sustaining the Coalesced Structures Argument is also possible. I have not argued that it is likely, and I'm not sure that it is. But I think that noticing the incomplete state of physics is the key to responding to the idealization complaint *on its own ground*. Which I will do in the next, and concluding, section.

---

[14] See Healey (2007, 172 ff.) for further reasons to think that asserting a physical difference between the unitary and other gauges "is confused."

14.6 A Sounder Principle?

The principle behind the idealization complaint bears requoting.

[**Sound Principle**:] While idealizations are useful and, perhaps, even essential to progress in physics, a sound principle of interpretation would seem to be that no effect can be counted as a genuine physical effect if it disappears when the idealizations are removed. (Earman 2004, 191)

The principle threaten to undermine the original Coalesced Structures Argument because the "effects" crucial to that argument—the availability of unitarily inequivalent representations of the physics of (say) a ferromagnet—disappear once the $n \to \infty$ and $V \to \infty$ idealizations of the thermodynamic limit are removed.

Against this threat, one might contend that the interpretive conclusions the Coalesced Structures Argument extracts from the treatment of phase transitions at the thermodynamic limit depend, not on features of the model (e.g. infinite volume) that are false idealizations, but on features of the model (e.g. the degeneracy of equilibrium states at phase transition temperatures) that apparently get things right. We—as philosophers of physics concerned about the relevance of interpretive projects—might hope that accounts of the content of quantum theory adequate to these features could serve as guides to the form this future theory might take. In view of this possibility, there is a point to using the Coalesced Structures Argument to interpret QM_∞.

To accommodate that point, I advocate revising Earman's sound principle.

[**Sounder Principle**:] No effect predicted by a non-final theory can be counted as a genuine physical effect if it disappears (and stays disappeared) from that theory's successors.

While (I would contend) sounder than Earman's principle, the revised principle has the pragmatic shortcoming that we can't apply it until we know what (all) the successors to our present theories are. Unpractical as the principle is, it carries a healthy moral: the scientific image is still under construction, and *guesses* about the future course of that construction can inform interpretations of its present condition.

To illustrate the role guesses play in interpretive disputes, consider Craig Callender (2001), who has criticized attempts to make interpretive hay out of the thermodynamic limit. Although he doesn't target the Coalesced Structures Argument explicitly, it is easy to imagine how a skeptic of Callender's stripe would. Rejecting the thermodynamic model framing the Coalesced Structures Argument, this skeptic would maintain instead that there is no single temperature at which different phases coexist. The discontinuities in thermodynamic functionals which accompany interesting phase structure will be absent from a full and proper theory of finite statistical mechanical systems, a theory which offers a differently structured account of critical phenomena. In that full and proper theory "the characteristic discontinuities and singularities accompanying phase transitions appear rounded and smeared; accordingly there are different criteria for phase transitions" (Callender 2001; see also Liu 2001).

14.6 A SOUNDER PRINCIPLE? 337

Such a skeptic could be understood as embracing the Sounder Principle, but supposing the $n \to \infty$ and $V \to \infty$ idealizations of the thermodynamic limit fail the criterion it announces. This skeptic guesses that future physics will have no need of the representational capacities our present models attain only by means of these idealizations. Conjoining the Sounder Principle to the guess, the skeptic concludes that those whose interpretive stances rest on features peculiar to the thermodynamic limit are (in Callender's words) "taking thermodynamics too seriously."

Those Callender targets (implicitly and explicitly) have responded (implicitly and explicitly) that the "idealizations" instrumental to taking the thermodynamic limit are "essential." Here is a physicist, responding implicitly:

the empirical fact that systems of different microscopic constitution exhibit similar macroscopic behavior provides grounds for suspecting that macrophysics is governed by very general features of the quantum properties of many-particle systems. Accordingly, it appears natural to pursue an approach to the theory of its emergence from quantum mechanics that is centred on macroscopic observables and certain general features of quantum structures, independently of microscopic details.... [Operator algebra theory] provides the natural framework for the generic model of macroscopic systems, idealised as infinitely extended assemblies of particles of finite mean number density. This idealisation is essential for the precise specification of different levels of macroscopicality and for the sharp mathematical characterization of quintessentially macroscopic properties of matter, such as phase transitions. (Sewell 2002, ix)

And here a philosopher of physics, responding explicitly: "Despite the fact that real systems are finite, our understanding of them and their behavior *requires*, in a very strong sense, the idealization of infinite system and the thermodynamic limit" (Batterman 2005, 231; emphasis mine).

One way to read this "Doctrine of Essential Idealization" is as a claim about the future of physics. But it can also be read as a claim about the present of physics. The readings needn't be in competition. Singly and together, the readings lend interpretive weight to the Coalesced Structures Argument.

Read as a claim about the future of physics, the Doctrine of Essential Idealization can be understood to embrace the Sounder Principle, but to make a guess different from Callender's skeptic's guess about the form future theories will take. On this reading, the Doctrine holds that the idealizations made to reach the thermodynamic limit are required to make available structures that will persist in, and underlie the empirical success of, future theories. Structures of exactly this sort are structures the Stronger Principle licenses us to lend interpretive weight. They can include the structures on which the Coalesced Structures Argument pivots.

Which camp—Earman's and Callender's curmudgeons, or adherents to this reading of the Doctrine of Essential Idealization—is correct? The revised sound principle of interpretation won't tell us. But it enables us to regard the camps as engaged in speculations about the future of physics. To so speculate is the prerogative of philosophy. Such speculations can be judged as more or less plausible, fruitful, coherent, or engaging. But they can't be judged, right now, as true or false. Whereas Earman's Sound Principle, by covert appeal to a guess about the future of physics, condemned the camp harboring the Coalesced Structures Argument to irrelevance, the revised sound principle condemns neither camp. Forwarding the revised sound principle keeps the Coalesced Structures Argument in play, at the same time as it allows how much company it has in the field of interpretation.

Sewell and Batterman are in that company. The passages quoted above suggest an alternative reading of the Doctrine of Essential Idealization.[15] On this reading, what makes some idealizations essential is the role features arising from those idealizations play in *present* physics. For Sewell, the idealization of the thermodynamic limit "is essential for the ... sharp mathematical characterization of quintessentially macroscopic properties of matter, such as phase transitions" (2002, ix). What marks the transition between the paramagnetic and ferromagnetic phases of a ferromagnet is the presence of a non-zero spontaneous magnetization. As Chapter 12 related, short of the thermodynamic limit, quantum models of the ferromagnet lack the capacity to represent pure thermodynamic phases associated with different magnetizations. They lack the capacity, partially constitutive of their success as present physics, to represent phase transitions. The idealization of the thermodynamic limit is essential to *that* success of present physics.

Batterman emphasizes other empirical and theoretical successes to which idealizations such as the thermodynamic limit are essential (see e.g. 2002; 2005; 2009). Phase diagrams such as Figure 12.1 chronicle the dependence of the magnetization of a ferromagnet as a function of its temperature and external magnetization. Essentially the same phase diagram applies to ferromagnets of different microphysical constitutions. It applies as well to fluids exhibiting "one phase-two phase" phase transitions (see Batterman 2002, 37–8 or Liu 2001), with pressure playing the role of external magnetization and fluid density "marking" the phase transition. The sense in which these disparate materials realize "the same" phase structure is captured by dimensionless parameters called "critical exponents," which describe how interesting bulk properties of a material—its spontaneous magnetization or density, say—vary as a function of temperature, as the critical temperature at which phase transitions occur is approached. The same critical exponents describe the phase behavior of ferromagnets and fluids.

This tendency of diverse media to instantiate patterns captured by the same critical exponents is known as "universality." The phenomenon of universality confronts us, Batterman urges, with an explanatory demand: *why* should similar patterns admit such diverse realizations? In statistical and condensed matter physics, the techniques of Renormalization Group theory rise to this demand (see Batterman 2002; 2009). The details of this story are fascinating and rich, but we needn't go into them to

---

[15] Although inspired by and obligated to the work of Sewell and Batterman, the sketch of that reading I offer here is just that: sketchy. Dramatically less developed than the positions of those authors, it shouldn't be mistaken for a representation of those positions.

extract from the story a reading of the Doctrine of Essential Idealization. On that reading, without the idealizations committed to reach the thermodynamic limit, we lack rigorous mathematical models of macroproperties like magnetization, and the relations into which those properties fall—including the relations constituting critical phenomena and exhibiting universality. Lacking models of critical phenomena in individual systems, we also lack a collection of models featuring the same critical behavior: we lack any *systematic* theoretical purchase on universality. Lacking this purchase, we cancel the explanatory agenda of explaining universality. Canceling that agenda, we do away with Renormalization Group theory, an approach whose explanatory *bona fides* come from advancing that agenda. In short, $n \to \infty$ and $V \to \infty$ idealizations of the thermodynamic limit are essential for modeling the full range of behavior that falls under the ambit of enormously fruitful Renormalization Group approaches to critical phenomena and universality. We denature a vast and powerful swathe of present scientific theorizing if we forsake the idealizations. In this sense, the very idealizations underlying the Coalesced Structures Argument are essential to the *present* of physics.

The Doctrine of Essential Idealization, cast as a claim about the present of physics, needn't be in tension with that Doctrine, cast as a claim about the future of physics. Indeed, the different readings can stand to one another in a relation of mutual support, if what "essential" idealizations contribute to the present of physics is also what's exploited to reach its future (see e.g. Batterman 2005, 431 ff.). A possible illustration of such mutual reinforcement is the following: Renormalization Group techniques devised for QSM have found application to physicists' QFTs. In QSM, those techniques help explain the insensitivity of critical phenomena to the detailed microphysics of the material prone to them. In physicists' QFT, Renormalization Group techniques help explain how the detailed high-energy physics of exact (but unknown) QFTs could have renormalizable effective theories as their low energy limits. Thus, Renormalization Group techniques, validated by taking thermodynamics quite seriously, are instrumental in identifying plausible future directions for QFTs. Theoretical features made available by idealizations are likely to persist in future theories when those features function as guides for theory development.

Their promise to guide future physics is only a plausibility consideration in favor of guessing that features made available by idealization will persist. It's a consideration (not a proof) in favor of guessing with the wingnuts and against the curmudgeons that the Sounder Principle will eventually license us in resting interpretive weight on the thermodynamic limit. It's a consideration (not a proof) in favor of taking the Coalesced Structures Argument seriously. It functions as such in the context of a commitment to interpret theories in a way that enables them both to discharge their explanatory duties and to serve as breeding grounds for successor theories. The roots and entailments of such a commitment, as well as the nature of its entanglement in the debate over scientific realism, are a focus of the next and final chapter. That chapter attempts to bring the discussions of QM_∞ mounted so far to bear on questions dear to the generalist.

15

Re: Interpreting Physical Theories

> To understand a scientific theory, we need to see how the world could possibly be the way that the theory says it is. An interpretation tells us that. The answer is not unique, because the question is not the sort of question to call for a unique answer.
>
> (van Fraassen 1991, 336–7)

Chapter 1 conjured the image of "generalist" philosophers of science clashing like Olympian gods over such epic questions as whether our best scientific theories are true, while far beneath their lofty heights, "particularist" philosophers of physics toil like badgers on such matters as the proper understanding of Haag's theorem. Ruing the suggestion that generalist inquiries could and should proceed in isolation from particularist ones, that chapter promised a particularist excursion through QM_∞ which would reveal lessons generalists would be well-advised to take to heart. This concluding chapter undertakes to characterize some of those lessons. They concern the related issues of scientific realism, physical law, and the nature of theoretical virtue.

15.1 An anatomy of scandal

Consider the realist who believes our successful scientific theories are (approximately) true on the grounds that those theories have withstood decades of scrutiny; predicted a wide range of phenomena, including unexpected phenomena, to a high degree of accuracy; underwritten advances both technological and conceptual; contributed efficiently to diverse regions of scientific practice; and in general exhibited all or most of the tangled panoply of virtues realists are wont to take as constitutive of good science. I would contend that, over the course of human history, the theory which most thoroughly realizes these virtues is QM.[1] Yet, after nearly a century of foundational

---

[1] QM notably lacks one virtue that might appear on the list: answering to whatever yen for explanation predisposes people to be realists in the first place. Many of the contributions to Cushing and McMullin (1989) explore this theme.

scrutiny, it remains utterly unclear what someone who believed QM would believe. Roger Jones puts it well: "Physicists don't know what deep explanatory structure of the microworld to be realists about" (1991, 191). This, I think, is a scandal. But it's a scandal whose origins reward investigation. The present section endeavors to dig up possible roots of the scandal. Following sections argue that scandal-producing conditions are rampant in QM_∞.

Laboring in abstraction, generalists have constructed a schematic argument for scientific realism, which I will call "the Miracles Argument." The *locus classicus* is Putnam's "The Nature of Mathematical Truth":

[T]he positive argument for scientific realism is that it is the only philosophy that doesn't make the success of science a miracle. That terms in mature scientific theories typically refer..., that the theories accepted in a mature science are typically approximately true, that the same term can refer to the same thing even when it occurs in different theories—these statements are viewed not as necessary truths but as part of the only scientific explanation of the success of science. (Putnam 1975, 73)

Here's one way to reconstruct the argument:

The Miracles Argument

1. Theory T is successful.
2. T's truth is the best explanation of this success.
∴ T is true.

The scandal arises when this Miracles Argument is de-schematized, by setting T equal to QM. Then its first premise is extremely well supported. But those who think they know how to believe the conclusion subscribe to interpretations of QM that mortify most realists. The Many Worlds Interpretation commits zany metaphysics; the Bohm theory implies the mere instrumental adequacy of a theory (the special theory of relativity) most realists are happier believing than they are believing QM; the Bare Theory implies we're almost always wrong about almost everything; and so on. Opining that "God is subtle but not malicious," Einstein probably meant that quantum theory was too deranged to be true of a universe subject to *benign* divine oversight. Many realists of my acquaintance join Einstein in hoping that QM turns out to be false.

Considered through the mists of lofty abstraction, the Miracles Argument has real intuitive pull. But when it's brought down to earth and concretized as an argument for realism about a particular, extraordinarily successful empirical theory, the content of its conclusion, as well as its persuasive force, is unclear. What's going on?

The first and most obvious moral to draw from the predicament is that particularist philosophers, wrestling with interpretations of particular physical theories, give content to positions in the realism debate, even positions that aren't realist. What a realist believes when she believes a theory T is an interpretation of T; the contents of various antirealisms about T can be spelt out in terms of attitudes toward the manifold of interpretations T admits.[2] This obvious moral, as it stands, is consistent with the two-levels picture, according to which generalist debates may as well proceed in isolation from particularist concerns. The force of the moral, on the two-level picture, is that once the generalists have decided what attitudes we ought to have toward empirically successful theories in general, particularists will supply the interpretations of specific successful theories towards which we're meant to strike those attitudes.

But it is possible to draw from the scandal morals which are more leveling, morals which call the two-level picture into question. Here are some examples:

[Moral?: the disunity of virtues] The nature of the virtues constituting T's success, invoked in Premise 1 of the Miracles Argument, *isn't a matter for generalists to decide.*

According to this moral, in the Miracles Argument it's not just the "T" that's schematic but also the "successful." Prediction, explanation, strength, systematization, novel extendability, and so on, may be virtues of scientific theories, but they aren't virtues that admit a uniform explication whose details are independent of the circumstances activating those virtues. The disunity of virtue could derail a deschematized Miracles Argument because a virtue like "explanatory success" might look a lot more truth-conducive when considered as though unified and in abstraction than it looks in particular instances where actual theories express it.

[Moral?: the disintegration of virtue.] There may be no single interpretation of a particular theory T under which all the virtues cited in the first premise of the Miracles Argument accrue to it.

The moral evokes the possibility that a successful scientific theory T *underdetermines its own interpretation*, and does so in such a way that different, and often incompatible, interpretations account for different parts of T's total virtuosity. What derails the concretized Miracles Argument in this case is that no single interpretation possesses the totality of virtues cited in premise (1) as reasons to believe (given premise (2)) that T is true. Since to accept the conclusion of the concretized Miracles Argument is to believe *an interpretation of T*, the concretized Miracles Argument commits a fallacy of many terms.

Chapter 1 suggested that one motivation for cleaving to the ideal of pristine interpretation is to prevent the disintegration of virtue. If we're vigilant against the contamination of our interpretations by "adulterating" factors keyed to particular applications of the theory we're interpreting, we foreclose the possibility that the interpretation our theory receives depends on its circumstances of application—and with it the possibility

---

[2] Here's a spelling out I like: "There is more than one tenable interpretation of quantum mechanics. It is true that each adds something to the theory. These additions are 'empirically superfluous.' ... Quantum theory is well understood. How can I say this when there are several mutually contradictory, perhaps equally plausible, interpretations in the field? Well, I assert it exactly because we do have interpretations that have so far stood the test of debate" (van Fraassen 1991, 336).

that as the empirical virtues a theory exhibits shift from application to application, the interpretation that secures those virtues on behalf of the theory shifts as well.

A variety of empirical success that most impresses many realists is the one for which Whewell coined the lovely term "consilience." A theory achieves consilience when it applies successfully to settings and problems distinct from those it was devised to address. Such novel successes strengthen the case for realism about a theory enjoying them only if it is *under the same interpretation* that the theory enjoys both its original successes and its novel ones. Here again, the ideal of pristine interpretation, the extremist commitment to interpret a theory in the same way no matter what conditions befall it, reinforces realism. So too does an extremist commitment to adopt a uniform interpretive strategy with respect to families of similar theories, so that the success of any of them redounds to the credit of all.

But commitment to pristine interpretation entails a risk: interpretations that abjure adulteration may thereby fail to make sense of the full (diverse, messy, conditioned?) range of empirical accomplishments that made us interested in the theory to begin with.

Ordinary QM can be understood to exemplify a moral even more dramatic:

[Moral?: the Decoupling of Truth and Virtue] *T* may have interpretations which *undermine* the Miracles Argument.

The infamous quantum measurement problem is a consequence of embracing the orthodox eigenstate–eigenvalue link along with the hypothesis that among the time evolutions described by the Schrödinger equation are those undertaken by composite object + apparatus systems in the course of measurement interactions. For it follows from these assumptions (along with further, apparently unobjectionable ones, about what kinds of interactions count as measurements[3]) that a measurement interaction will almost always leave its participating apparatus in a state that's *not* an eigenstate of the designated pointer observable, and so not a state, according to the eigenstate–eigenvalue link, in which the pointer observable possesses a determinate value. Bluntly put, it follows that most measurements don't have outcomes. But measurements have to have outcomes, if QM is to be empirically adequate. For those outcomes are the phenomena QM secures its empirical adequacy by saving. Without phenomena to save, QM isn't empirically adequate.

The orthodox response to this measurement problem invokes what textbooks tend to call "the Postulate of Measurement Collapse"—the supposition that during the course of a measurement interaction, unitary Schrödinger evolution is interrupted by a measurement collapse. Schrödinger evolution is deterministic, continuous, and reversible; measurement collapse is a discontinuous, irreversible, indeterministic change of the state of the object system to one of the eigenstates of the observable measured. This makes measurement collapse a miracle in Hume's sense, a violation of

---

[3] For more on these, see the exemplary Busch, Lahti, and Mittelstaedt (1996).

the law of nature encapsulated by the Schrödinger equation. Now recall the Miracles Argument, whose crux is that the success of false theories would be miraculous. We've just seen that if collapse-free QM, as interpreted by the eigenstate–eigenvalue link, is true, it takes a miracle for measurements to have outcomes *period* (much less ones that accord with quantum statistical algorithm). That interpretation severs the entailment, presupposed by the Miracles Argument (as well as by many of its critics), between a theory's truth and its empirical adequacy. If QM as interpreted by the eigenstate–eigenvalue link is true, then it's not empirically adequate.

In an effort to restore the connection between a theory's truth and its empirical adequacy, we might try interpreting QM by means of the eigenstate–eigenvalue link *plus* the postulate of measurement collapse. Alas, the theory so interpreted is either inconsistent (because it contains a pair of dynamical postulates that contradict one another) or ill-formed (because it lacks a criterion clearly demarcating situations governed by Schrödinger evolution from those stricken by measurement collapse). Either way, it lacks prominent virtues constitutive of the sort of empirical success evoked in Premise 1 of the Miracles Argument. Either way, we are back in the ambit of the second moral: that only under different and competing interpretations does a theory exhibit different virtues inclining us to belief. And, I suggest, similar fates plausibly await other candidate interpretations of QM: each of them strips the theory of some of the virtues realists would regard as truth-conducive. This is exactly why debate over how to interpret the theory continues nearly a century after its inception.

15.2 Realism sophisticated

Section 15.3 will argue that theories of QM_∞ underdetermine their own interpretation, and do so in such a way that different interpretations bear different parts of their overall explanatory burden. This is the disintegration of virtue. Section 15.3 will also argue that concrete empirical virtues exhibited by theories of QM_∞ under particular interpretations can look less cohesive and truth-conducive than their schematic counterparts. This is the disunity of virtue. Both the disunity and the disintegration of virtue challenge the Miracles Argument for scientific realism sketched in §15.1. But few would consider the realist position presented in that section to be state of the art. Thus this section discusses more sophisticated scientific realisms, under the headings of "global realism" and "Structural Realism." The discussions are cursory, and fail to characterize these more sophisticated positions in great detail or with real precision. Their aim is to suggest that the phenomena of virtue disunity and disintegration also attenuate the persuasive force of arguments promising to support these more sophisticated realisms.

15.2.1 *The global argument*

Sometimes "local" and "global" versions of the Miracles Argument are distinguished (see e.g. Ladyman 2002, 210–19). The argument of §15.1 is "local;" it presents the empirical success of a *particular* scientific theory as grounds for a realist attitude toward *that* theory. By contrast, the "global" version of the Miracles Argument, of which Richard Boyd is a leading advocate, takes the overall success of the many-splendored juggernaut that is "the mature sciences" as grounds for a realist attitude toward those sciences as a corporate body.[4]

For Boyd, scientific realism is "an empirical hypothesis which is justified because it provides the best scientific explanation for various facts about the ways in which scientific methods are epistemically successful" (1985, 3). Boyd calls "instrumentally reliable" those theories which "make approximately true predictions about observable phenomena" (Boyd 1985, 4). He thinks there are instrumentally reliable theories. But—in contrast, perhaps, to the "local" realist of the previous section—he does not regard the mere existence of these theories as the central reason to espouse scientific realism. Boyd lends more significance to the fact that the mature sciences frame and condition a *methodology* that is instrumentally reliable insofar as it furnishes "a reliable guide to the acceptance of theories which are themselves instrumentally reliable" (Boyd 1985, 4). Scientists devise theories and protocols for their assessment; having carried out these protocols, they reach judgments about the instrumental reliability of the theories in question, which judgments are borne out by the deliverances of subsequent science.

For Boyd, the crucial observation is that this instrumentally reliable method for assessing the instrumental reliability of theories *relies on a background constituted by theories making up the mature sciences.* This theoretical background guides judgments about which experiments are well-designed, which data patterns are significant, which conjectures are sufficiently plausible to reward investigation, and so on. Suppose that the startling efficiency of the prevailing methodology of the mature sciences demands explanation. Global realism about the mature sciences meets the demand. Prevailing methodology, a methodology informed through and through by those sciences, is instrumentally reliable because the claims of those sciences are approximately true, their theoretical terms refer, and so on. In short, the method works because the theories guiding the method, and guiding it in myriad interlocking ways, are right.

Boyd's position is, of course, immensely more subtle and extensive than the one just sketched. Still, the position just sketched constitutes a sophistication of the two-step local realism presented in §15.1, and we can articulate on behalf of this more sophisticated global realism a more sophisticated account of the empirical virtues realists regard as evidence of truth. The virtues contributing to the instrumental reliability of the mature sciences include the capacity of theories collectively composing the mature sciences to serve both as springboards for further theory development and as frameworks for understanding theoretical and experimental interventions. In virtue of the first capacity, the methodology of the mature sciences generates hypotheses

---

[4] I am grateful to Kevin Coffey for both writings and conversations about Boyd's brand of realism.

whose instrumental reliability it is able to assess in virtue of the second capacity. These are exactly the capacities best explained, according to the global argument, by the hypothesis of scientific realism.

If the outcome of this global argument is to be something more specific, and controversial, than a posture of goodwill toward the mature sciences and their capacity for self-improvement, that outcome must incorporate some realist attitude toward *present* mature sciences. The content of a scientific realism supported by the global argument would presumably be furnished by something like an interpretation of the conjunction of the mature sciences (if a consistent one can be found), or by a conjunction of interpretations of the mature sciences (if a consistent one can be found).

Whatever the details of that content, the global argument for scientific realism is just as susceptible to the disintegration of virtue critique as is the simple-minded local realist argument laid out in §15.1. The myriad successes constituting the instrumental reliability of the mature sciences function in Boyd's argument as reasons to accept the hypothesis of scientific realism. But if there is no "interpretation of mature science" which equips it with all those myriad successes, a gulf opens up between what we're being asked to believe and the reasons we're being given to believe it. Put in more concrete terms, if QFT's contribution to the instrumental reliability of the mature sciences is so structured that different, and rival, interpretations of QFT are required to make sense of that contribution, then it's not at all clear *what* belief in QFT is licensed by the instrumental reliability of the mature sciences.

I am supposing this to be a problem because I am supposing that to adopt the realist hypothesis with respect to our present mature sciences is to believe (among other things) that QFT is approximately true. If the latter supposition is false, I suspect that "realism about the mature sciences" is a realism deferred until such time as the regnant mature sciences grow out of the interpretive and theoretical awkwardnesses besetting our present mature sciences. Addressing this deference, §15.6 argues that while it may not be incoherent, neither is it compulsory.

15.2.2 Structural Realism

Another realist sophistication is Structural Realism, injected into contemporary debates by John Worrall, who attributes the basic position to Poincaré. Worrall finds it peculiar that the two most intuitively compelling arguments encountered in the debate over scientific realism support competing positions: the Miracles Argument, already outlined, is an argument *for* scientific realism; the Pessimistic Meta-induction—the argument that the history of science, as a history of discredited theories, constitutes forceful inductive evidence that our present theories will also be rejected as false—is an argument *against* scientific realism. Worrall markets Structural Realism as "the best of both worlds": a position *both* the Pessimistic Meta-induction and the Miracles Argument support.

The heart of Worrall's account is a move to reduce the inductive base invoked by the Pessimistic Meta-induction to past scientific theories which achieved consilience by enjoying novel or surprising successes beyond the domain of phenomena those theories were introduced to save. Regarding the theories in this reduced inductive base, Worrall observes that often some aspect of a "discredited" theory lives on in its successors. Take for instance Fresnel's wave theory of light, which reposes in the graveyard of rejected science, to which it was dispatched by the success of Maxwell theory. Nevertheless:

> There was an important element of continuity in the shift from Fresnel to Maxwell—and this was much more than a simple question of carrying over the successful empirical content into the new theory. At the same time it was rather less than a carrying over of the full theoretical content or full theoretical mechanism (even in "approximate" form)... There was continuity or accumulation in the shift, but the continuity is one of form or structure, not of content. (Worrall 1989, 117)

The Miracles Argument tells us that our successful scientific theories are getting something right. The Pessimistic Meta-induction alerts us that they're not getting everything right. The Structural Realist takes both lessons to heart by espousing a realism about the "form or structure" of our best theories, a realism which explains their empirical success; these forms or structures inoculate themselves against pessimistic inductions by "carrying over" to successor theories.

In the literature subsequent to Worrall, Structural Realism bifurcates into "epistemic" and "ontic" varieties. Roughly, the former holds that our best reasons for belief in scientific theories are reasons for belief in their structures; the latter is a thesis about physical reality to the effect that what there is is structural. So deep and roiling are these waters that I dare not wade far into them. But I will offer some observations from the safety of shore.

The broad-brush portrait of Structural Realism is undeniably attractive. But it doesn't tell us how to bring the position into sharp focus. Operating at the level of schematic abstraction, consider a successful scientific theory T. What I'll call the Content Question for Structural Realism is the question of what a Structural Realist believes about T. Structural Realism is a genuine position in the debate over scientific realism only if the Content Question has an informative answer. (I suspect that there are versions of ontic Structural Realism ("the world is structure") that demur on the Content Question. I don't mean to charge those versions with triviality. I mean rather to classify them as metaphysical theses, rather than positions in the scientific realism debate.)

I've already said that the Structural Realist about T believes in T's structure, but what this comes to depends on how two further questions are answered. First, what is a theory's structure? Second, what is it to believe in that structure? I'll focus here on the former. A patently inadequate metaphor for the "structure" on which Worrall keys is the "theoretical DNA" that survives through successive generations. The metaphor suggests the following answer to the Content question: "the Structural Realist believes in those structural elements of our current best theories which will be preserved in (all?)

their successors." Bracketing questions about how to understand the word "structure," this is a position consistent with Chapter 14's "Sounder Principle" of interpretation, which enjoined us to regard as significant only those features of physical theories that all future physics will continue to regard as significant. But like the Sounder Principle, this answer to the Content Question has a pronounced pragmatic shortcoming. Regarding what good Structural Realists are to believe *now*, it gives us no guidance.

An informative answer to the Content Question won't defer to future theories. One strategy for developing an informative answer will rest on principles that, given a successful theory T, identify "the structure of T." This strategy faces several challenges, rooted in the same possibilities animating the morals §15.1 draws about the crude local Miracles Argument. T's empirical successes are presumably myriad: otherwise the question of realism about T lacks urgency. Also presumably, the Structural Realist will take some variation of the Miracles Argument to warrant belief that T gets the structure of physical reality approximately right on the grounds that getting the structure of empirical reality approximately right explains T's empirical success. But it could be that different, and rival, takes on T's structure explain different instances or types of T's empirical success. It could also be that no single take on T's structure explains enough of T's empirical success for a structuralized Miracles Argument to be persuasive. The challenge to the Structural Realist is to find a principle for identifying T's structure that recoups "enough" of T's empirical success.

Now it might seem that the Structural Realist has a neat and promising solution to the challenges just posed. Those challenges evoke the possibility that a theory enjoys different empirical success under different (and rival) formulations and/or interpretations. The neat Structural Realist rejoinder is that such apparently rival interpretations are in fact "representations of the same structure" (Ladyman 2002, 420). The idea is that Structural Realism averts worries about the disintegration of virtue by identifying a common (structural) content shared by all virtue-supporting interpretations of the theory.

Unfortunately, it's hard to see what the neat answer comes to for theories of QM_∞. Candidates for the structure of a theory of QM_∞ include abstract C^* algebras, concrete von Neumann algebras, and instances of the latter which stand to instances of the former as universal enveloping von Neumann algebras. Cast in terms of Structural Realism, what I've tried to argue in the foregoing chapters (and what I'll try to underscore in the next section) is this: while structures of each of these types underlie *some* of the empirical successes of theories of QM_∞, the collection of significant successes enjoyed by theories of QM_∞ defy attribution to a single type of structure. Empirical successes mediated by the particle notion, and explanations relating the energy of quantum field states to that of their classical predecessors, require the structure of a privileged irreducible Fock space representation; such a structure stymies the physics of broken symmetry and of equilibrium at non-zero tempeatures. An abstract algebraic structure broad enough to accommodate a full slate of possible dynamics for QFT in curved spacetime is too broad to sustain mean field dynamics of the sort so successful in the thermodynamic limit. The heuristically powerful insistence on Poincaré symmetry disables a full-blooded explanation of quantum coherence. And so on.

Chapter 14 gave a reason to take coalesced structures seriously, notwithstanding the role idealizations play in constituting those structures. The reason was the possibility that those structures might persist in future theories, a possibility lent some plausibility by the role played by renormalization group techniques, devised to accommodate those structures, in shaping physicists' QFTs. This case for the possible *interpretive significance* of coalesced structures is different from, and weaker than, a case for realism (structural or otherwise) about theories in which those structures appear. Plausibly significant interpretations of a theory T are those that enable T to discharge its duties and exercise its virtues *as a theory of physics*. I have been urging that arguments for realism about T fall flat if it is only under different, and competing, interpretations that T discharges different, and significant, duties, and exhibits different, and significant, virtues. Thus a case that an interpretation of T is significant isn't on its own an argument that one should be a realist about T by believing that interpretation.

15.3 QM_∞ and the morals

We are in a position to give a partial account of what it takes for a theory T to exhibit the virtues animating these more sophisticated realisms—virtues those realisms regard as grounds for belief in the mature sciences or their constituents. T is virtuous if it exhibits explanatory reach, particularly an explanatory reach extending to novel and surprising phenomena. T is virtuous if it stands in relations of intercalation and continuity with present and future theories. So standing constitutes T's contribution to the instrumental reliability of the mature sciences. And T's continuities with future science supply that aspect of T about which epistemic Structural Realists are realists. Let us add these virtues to those celebrated by the simple-minded local realist, paradigmatically, T's predictive and explanatory success. In a line, the Miracles Argument sophisticated is that T's virtuosity warrants belief in T, because T's approximate truth is the best explanation of its virtuosity.

Section §15.1 insinuated that such arguments can become less compelling when de-schematized: when actual scientific theories are cast in the role of T. Once the notion of empirical virtue is concretized and made applicable to a particular, successful theory, virtues that appear unproblematically truth-conducive when considered in the abstract become complicated in a way that attenuates the Miracles Argument and its kin. This section chronicles the attenuations visited upon such arguments by features of QM_∞ discussed here.

"Fecundity"—the capacity to breed successful successor theories—is a virtue. Indeed, it's a virtue central to the sophistications (global and structural) of realism related in §§15.2.1 and 15.2.2. Boyd takes the wherewithal to successfully and efficiently *extend* science to be of the first importance. The "structure" about which (some) Structural Realists would be realists is what's preserved over theory change. But de-schematized, "fecundity" fractures into a variety of traits that needn't be co-instantiable. Take QFT in curved spacetime (see Chapter 10, especially §10.3). To attribute to this theory the virtue of fecundity is to equip it with content making it a fruitful breeding ground for quantum gravity. But different, and competing, interpretations, enable the theory to exercise this virtue. Insofar as semi-classical quantum gravity is cultivated for hints about future quantum theories of gravity, QFT states must be Hadamard (i.e. such as to yield an expectation value to the stress-energy observable). Thus interpretations limiting physically significant states to those satisfying the Hadamard condition inculcate the virtue of fecundity. But insofar as the Hawking effect is pressed into service as the closest thing there is to "data" constraining quantum gravity, non-Hadamard states must be allowed to participate, for they are among the states which articulate the phenomenology of Hawking radiation. In this respect, interpretations limiting physically significant states to those satisfying the Hadamard condition *inhibit* the virtue of fecundity. Fecundity is a virtue QFT in curved spacetime exhibits under different, and rival, interpretations.

Explanatory success is a virtue. Explanations framed by fundamental particle notions are supported by restricting physically significant states of the quantum field to the folium of the irreducible Fock space representation circumscribing the particle notion in question. But such an interpretive strategy leaves no conceptual space for different states of broken symmetry, and so hamstrings explanations (e.g. of phase transitions and critical phenomena) that might operate in that space. There is no single interpretive approach to QM_∞ that recoups, on behalf of the set of theories of QM_∞, the totality of their explanatory accomplishments. Hence, there is no single interpretation of QM_∞, belief in which is supported by that totality of explanatory success.

Chapter 12 argued that there are *individual* explanations offered by theories of QM_∞ no extremist approach to interpreting QM_∞ could secure. In particular, no extremist interpretation of QM_∞ could single-handedly enjoy the success of explaining phase structure. The explanation considered rested upon *disjoint* states *distinguished by* a macroscopic observable marking the difference between different phases. The Algebraic Imperialist and Universalist posit enough states to frame the explanation—but don't recognize the distinguishing observable as an observable. With a suitable choice of privileged folium, the Hilbert Space Conservative supports the distinguishing observable central to the explanation—but she doesn't recognize the disjoint states between which the observable distinguishes. The virtue of underwriting the explanation of phase structure is not a virtue any single pristine interpretation of QM_∞ exhibits on its own. No single pristine interpretation supports the whole explanation.

The foregoing serve as examples of individual virtues, regarded as truth-conducive by realists, breaking up on the interpretive shoals of theories of QM_∞—examples of what §15.1 labeled "the disunity of virtue." QM_∞ also illustrates what that section called "the disintegration of virtue"—the tendency of sets of virtues, regarded by realists as collectively truth-conducive, to dissociate themselves, and accrue to a theory only under different and rival interpretations. Here are a few examples.

Consistency with other well-established theories, and capacity to contribute to instrumentally reliable methodology are virtues—virtues the global realist position in particular supposes to cleave together. Let's suppose STR is a well-established theory, consistency with which is a virtue for theories of QM_∞. Interpretations of those theories confining physical possibilities to representations satisfying Poincaré covariance exhibit this virtue (see Ch. 5). But if the Standard Model is reckoned to be a well-established theory, consistency with which is a virtue for theories of QM_∞, the plot thickens. Although the Standard Model is itself consistent with STR, interpretations of QM_∞ fostering assessments of the Standard Model's instrumental reliability may not be. A small part of that assessment relies on models of "soft photons" involved in scattering interactions confirming the Standard Model. But soft photon modeling is accomplished by coherent state representations violating Poincaré symmetry (see §10.4.5). Interpretations which make sense of a small part of free QFT's contribution to the instrumental reliability of the mature sciences fail what seemed to be the bedrock desideratum of consistency.

This example of virtue disintegration might prompt some to exclaim: "So much the worse for soft photons!" But even supposing we reinforce the virtue of consistency with STR by restricting physically admissible QFT states to those in the folia of Poincaré-invariant representations, the disintegration of virtue threatens. For explanatory power is another virtue. And, as §10.4.5 recounted, certain explanations of the phenomenon of coherence, a phenomenon exhibited by many states passing our test of consistency, make use of a phase observable. According to this explanation, the coherence of coherent states derives from their status as eigenstates of the phase operator: this status insulates them from phase fluctuations which would undo the correlations (constitutive of coherence) between values of the electric field at different spacetime points. This explanation succeeds only if there is such a thing as a phase observable. And there is, but only in representations violating the requirement of Poincaré symmetry, instituted to preserve the virtue of consistency with STR. That virtue and the virtue of explaining the phenomenon of optical coherence disintegrate: no interpretation underwriting one underwrites the other.

15.4 Law and possibility

I have suggested that, when it comes to theories of QM_∞, phenomena of virtue disunity and disintegration are rampant. This undermines arguments for scientific realism deschematized to be arguments for realism about theories of QM_∞. But the phenomena also have repercussions for how we think about physical law and physical possibility, repercussions which are the subject of this section.

In slogan form, my contention is that, when it comes to QM_∞ at least, there is a pragmatic dimension to theory articulation. What set of possible worlds we associate with a theory of QM_∞ can depend on what we'd like to do with that theory: what explanations, involving which magnitudes and guided by what laws, we aspire to; what phenomenological models we need to construct; what projects of theory development we'd like to sponsor.

To endorse such "pragmatization" of theoretical content might seem flaky. But it is kin to a hard-headed empiricist position John Earman articulates in his 1986 *Primer on Determinism*. The setting is the "best systems" account of physical law, the account Earman reckons to have the best chance of passing what he terms "The empiricist loyalty test" (1986, 85) that any two possible worlds that agree on all occurrent facts also agree on laws.[5] As formulated by David Lewis, the best systems account holds that "a contingent generalization is a law of nature if and only if it appears as theorem (or axiom) in each of the true deductive systems [i.e. deductively closed, axiomatizable sets of true sentences] that achieves a best combination of simplicity and strength" (1973, 73). Much more should be said about what constitute "strength," "simplicity," and their best balance. To make the point that's initially relevant here, I will suppose for now that these commodities are sufficiently well understood.

Let us say that world w is *nomically accessible* from a world w' if and only if all the laws of w' are true of w. It follows quite plausibly from the best systems account that "nomic accessibility" isn't a symmetric relation. For a vivid illustration of why, let world W be a simplified solar system, best systematized by Newton's laws, including the Law of Universal Gravitation. Let world W' be a world consisting of a single spherically symmetric mass. Now W's laws are true of W', making W' nomically accessible from W. But W's laws aren't axioms in the simplest strongest deductive system with respect to the totality of W' facts: deductive systems omitting W laws uninstantiated in W' will be just as strong but much simpler. Supposing that the best system for W' includes "For all x, x just sits there," the laws of W' aren't going to be true of W, which includes bodies that don't just sit there. So even though W' is nomically accessible from W, W isn't nomically accessible from W'. This is the asymmetry.

Here is Earman's commentary on it:

For the empiricist, there are no irreducible modal facts. A world W is a world of non-modal facts. Uniquely associated with each such world is a set L_W of non-modal propositions true in W—the laws of W. To mark off the elements of this set we may prefix "it is physically necessary that," but that prefix is merely an honorific. Accessibility is a defined relation, not a metaphysical given: world W' is nomically accessible from W iff the L_W are all true in W'. There are then myriad subsets of physically possible worlds, each radiating outward from a logically possible world. No one of these collections is more powerful or potent than any other. (Earman 1986, 99)

[5] See van Fraassen (1989) and Earman (1993) for variant expressions of loyal empiricism.

Stripped of empiricist sermonizing, the upshot is that physical possibility is *naturalized* (cf. Earman 1995, 176–9), insofar as what's possible depends on *how the world is*: even *within* the space of worlds nomologically accessible from our own, what the laws are at each world varies with the facts of that world. Consider, by contrast, a view according to which the worlds nomologically accessible from our world are exactly the worlds which share the laws of our world (see e.g. Carroll 1994). This view leaves no scope for the laws of a nomologically accessible world to change with its facts.

Earman's picture has the feature that to each world w, there corresponds *one* circle of worlds nomically accessible *from it*. These are just the worlds of which w's laws are true (although they needn't be worlds for which w's laws are laws). In order to draw this picture, Earman needs to waive concerns about problematic ambiguities in the notion of best system.

Now, the coalescence account of physical content naturalizes physical possibility, for instance, when it suggests that local dynamical facts contour the space of possibilities accessible at the thermodynamic limit, shrinking it from the full set of states on the relevant canonical algebra $\mathfrak{A}$ to only those states on which W^* dynamics are well-defined. *Different* local dynamical facts can call for *different* shrinkages. What is the case shapes what might be.

But the coalescence account can also *pragmatize* possibility. The account allows that from the actual world W there may emanate many different sets of physically possible worlds, each indexed to a sense of "physically possible" *native to W*. But multiple, different, even rival senses of "physically possible" are native to W because W natives encounter a variety of scientific challenges that require different contourings of the space of physical possibility. The coalescence account allows that for the sake of modeling critical phenomena, the space of states possible for a system subject to such phenomena shrink from the full set of states on the relevant canonical algebra $\mathfrak{A}$ to only those states on which "phase" observables, abrupt changes in whose values mark such phenomena, are well-defined. Different contexts may call for different shrinkages. The coalescence account allows that for the sake of pursuing semi-classical quantum gravity, the space of possible configurations of a QFT on curved spacetime shrink from the full set of states on the relevant quasilocal algebra to only those states satisfying the Hadamard condition. The coalescence allows projects of axiomatic theory development which require representations to be Poincaré-invariant to be physical. But it also allows non-Poincaré-invariant infrared coherent states to guide the interpretation of scattering experiments.

When physical possibility is naturalized, sets of physically possible worlds are indexed with respect to the "anchor" world whose facts shape that set. When physical possibility is pragmatized, sets of possible worlds are indexed (or indexed *as well*) to circumstances of application within the "anchor" world. When physical possibility is pragmatized, there's a single way the world is, but (as gleaned by physics) there isn't a single set of ways it might be. Instead, there are many sets of ways it might be. Taking these sets seriously is part and parcel of commitment to a physical theory: to call something an "electron" is to take on commitments regarding how it would behave in a variety of circumstances, and to offer explanations, evaluations, and further theory constructions in ways constrained by those commitments. But we take on different, internally cohesive collections of such commitments in different circumstances. Provided we can keep the circumstances calling for commitments straight, this doesn't make us incoherent. It makes us resourceful. And that suggests another way to conceive of empirical virtue.

15.5 Virtue reconceived

I don't think it's too controversial that theories have a modal dimension. What is controversial is how to analyze that dimension: as metaphysically basic, merely defined, or somewhere (if there is anywhere) in between. I am contending that "the" modal dimension of a theory is really manifold: at least some successful scientific theories, theories of QM_∞, are best understood not as determining a single set of possible worlds, sharply defined by principles of extremist interpretation, but as encompassing many such sets, at least some of which correspond to adulterated interpretations of limited but undeniable use. Here I want to suggest that this is no accident. I claim that it's a central theoretical virtue—it better enables a theory to discharge the many obligations it faces—to foster manifold spaces of physical possibility, spaces that are naturalized and pragmatized. Again, this isn't inconsistent. It's not saying theories contradict themselves with respect to how the world is. It's saying good theories are good theories because of a certain inbuilt flexibility about how the world might be.

The venerable realist thinks theories work so well because they're right: successful theories latch on to an unambiguous way the world is. I think there is an unambiguous way the world is. But I don't think latching on to it is the key to theoretical success. Unambiguously what it is, the world engages a physical theory in a variety of ways. It enhances a theory's capacity to emerge victorious from these engagements to embed its picture of how the world is in a variety of accounts of how the world might be. The account a theory offers of how the world might be is a window to its laws; they judge which logically possible worlds are also physically possible. A theory indulging in a variety of accounts of how the world might be is not a theory to which we can assign a unique and unambiguous set of laws. Even if laws are "best picture," because bestness is indexed to contexts, the right instrument for seeing the best picture is a kaleidoscope.

Theories explain, predict, model, and spawn new theories. These are all virtues. But what underlies or secures them isn't, or needn't be, truth. It can be ambiguity. This isn't to say anything goes. The ambiguity in question is severely *tempered* by our criteria for individuating theories, in the form of rooted physical relationships central to those theories. These criteria won't tolerate unlimited variation in the interpretations a theory receives. Those interpretations must be consistent with our sense of *what that theory is*. But that sense rarely issues in a unique interpretation, and when it admits many interpretations, the manifold of interpretations functions as a theoretical resource.

This reconception of theoretical virtue extends a metaphor notorious from van Fraassen's *Scientific Image*:

> I claim that the success of current scientific theories is no miracle. It is not even surprising to the scientific (Darwinist) mind. For any scientific theory is born into a life of fierce competition, a jungle red in tooth and claw. Only the successful theories survive—the ones which *in fact* latched on to actual regularities in nature. (1980, 40)

According to van Fraassen's metaphor, our best current theories abound in empirical adequacy because only abundantly empirically adequate theories are likely to survive the peculiar selection pressures of modern science. Extending this evolutionary metaphor, I take fitness—the trait or traits of a theory that promotes its survival—to consist not only in empirical adequacy, narrowly understood as saving the phenomena, but as empirical success more broadly construed. A theory that underdetermines its own interpretation is like a healthy breeding population: it has a shot at enough diversity to (under some interpretation or another) meet the variety of demands its scientific environment places on it. Like survival, "empirical success" is a convoluted, chancy, and conditioned thing. Like genetic diversity, possibility pragmatized situates its possessor to respond successfully to the changing circumstances on which its survival depends.

15.6 Fundamental physics

Realists might move to dismiss the morals suggested in this chapter on several grounds. The morals are inspired by theories of QM_∞, and indeed by some of their less central applications. One grounds for dismissal attests, "Present quantum theory is bound to be abandoned in favor of some theory of quantum gravity that has a straightforward interpretation palatable to the realist." The testimony expresses a concrete version of the hope that "fundamental" physical theory will be free of those features of our current best theories which are problematic for realism. A variation on these grounds objects that many of the applications I've discussed aren't "fundamental." We should interpret theories only in light of their "fundamental" applications. In sum, the morals drawn are to be dismissed because their origins are dubious: they come from physics that isn't fundamental.

I should confess that I myself am not in the grip of the idea of Fundamental Physics. Why believe that there is any such thing? Why expect that there will ever be a *theory*—as opposed to a "way the world is"—fundamental in the relevant sense (whatever that sense is exactly)? What are philosophers of physics supposed to talk about in the meantime? But I will set these idiosyncratic hangups to one side.

More to the point, I prefer to understand issues of scientific realism to animate our engagement with current science. According to the Fundamentalist dismissal, the only theory about which the question of realism arises in earnest is Fundamental Physics. The central difference between me and Fundamentalists lies not only in the attitudes we take to the theories ready to hand but also in the attitude we take toward a theory decidedly not ready to hand (and possibly inevitably present at hand): Fundamental Physics. Even if the very idea of Fundamental Physics is coherent, we have little inductive evidence that the physical theories humans develop and deploy will ever conform to the ideal of Fundamental Physics. I think this partly because scientific practice selects for theories whose reserves of ambiguity are empirical resources, and I suppose Fundamental Physics, whatever it is exactly, to harbor no reserves of ambiguity. So while I concede to Fundamentalists the possibility of Fundamental Physics, I think they owe me an argument making that possibility relevant to the attitude we ought to have about the theories we do have, and the theories we'll have after them, and the theories we'll have after *those* theories, and so on.

I haven't offered any proofs that future, even near future, science *won't* realize the possibility of Fundamental Physics. So I don't consider Fundamentalism irrational. It just takes faith in physics of a different order than the faith I've got.

References

Accardi, L., and C. Cecchini (1982), "Conditional expectations in von Neumann algebras and a theorem of Takesaki," *Journal of Functional Analysis* 45: 245–73.

Acerbi, F., G. Morchio, and F. Strocchi (1993a), "Infrared singular fields and non-regular representations of canonical commutation relation algebras," *Journal of Mathematical Physics* 34: 899–914.

—— —— —— (1993b), "Nonregular representations of CCR algebras and algebraic fermion bosonization," *Reports on Mathematical Physics* 33: 8–19.

Aitchison, Ian (1982), *An Informal Introduction to Gauge Field Theories* (Cambridge: Cambridge University Press).

—— and Anthony J. G. Hey (2003), *Gauge Theories in Particle Physics: A Practical Introduction*, vol. I (New York: Taylor & Francis).

—— —— (2004), *Gauge Theories in Particle Physics: A Practical Introduction*, vol. II (New York: Taylor & Francis).

Albert, David (1994), *Quantum Mechanics and Experience* (Cambridge, MA: Harvard University Press).

—— (2000), *Time and Chance* (Cambridge, MA: Harvard University Press).

Arageorgis, Aristidis (1995), "Fields, particles, and curvature: foundations and philosophical aspects of Quantum Field Theory in curved space-time", Ph.D. dissertation, University of Pittsburgh.

—— John Earman, and Laura Ruetsche (2002), "Weyling the time away: the non-unitary implementability of quantum field dynamics on curved spacetime and the algebraic approach to QFT," *Studies in the History and Philosophy of Modern Physics* 33: 151–84.

—— —— —— (2003), "Fulling Non-uniqueness, Rindler Quanta, and the Unruh Effect: a primer on some aspects of quantum field theory," *Philosophy of Science* 70: 164–202.

Araki, Huzihiro (1964), "Type of von Neumann algebra associated with free field," *Progress in Theoretical Physics* 32: 956–65.

—— (1969), "Gibbs states of a one dimensional quantum lattice," *Communications in Mathematical Physics* 14: 120–57.

—— (1999), *Mathematical Theory of Quantum Fields* (Oxford: Oxford University Press).

—— and E. J. Woods (1963), "Representations of the canonical commutation relations describing a nonrelativistic infinite free Bose gas," *Journal of Mathematical Physics* 4: 637–62.

Aristotle [1996], *Physics*, ed. D. Bostock and trans. R. Waterfield (Oxford: Oxford University Press).

Ashtekar, Abhay (2007), "What did we learn from quantum gravity?" in T. M. Nieuwenhuizen (ed.), *Beyond The Quantum* (Hackensack, NJ: World Scientific), 20–34.

—— S. Fairhurst, and J. L. Willis (2003), "Quantum gravity, shadow states, and quantum mechanics," *Classical and Quantum Gravity* 20: 1031–61.

—— J. Lewandowski, and H. Sahlmann (2003), "Polymer and Fock representations for a scalar field," *Classical and Quantum Gravity* 20: L11.

Ashtekar, Abhay, T. Pawlowski, and P. Singh (2006), "Quantum nature of the big bang," *Physical Review Letters 96*: 141–301.

—— and Anne Magnon (1975), "Quantum fields in curved spacetime," *Proceedings of the Royal Society (London) A346*: 375–94.

Baez, John, Irving Segal, and Z. Zhou (1992), *Introduction to Algebraic and Constructive Quantum Field Theory* (Princeton, NJ: Princeton University Press).

Bais, F. A. (2004), "To be or not to be? Magnetic monopoles in non-Abelian gauge theories," LANL archive hep-th/0407197.

Baker, David J. (2009), "Against field interpretations of quantum field theory," *British Journal for the Philosophy of Science 60*: 585–609.

—— and Hans Halvorson (2010), "Antimatter," *British Journal for the Philosophy of Science 61*: 93–121.

Balachandran, A. P. (1994), "Gauge symmetries, topology and quantization," *AIP Conference Proceedings 317*: 1–81.

Barrett, Jeff (1999), *The Quantum Mechanics of Minds and Worlds* (Oxford: Oxford University Press).

—— (2002), "The measurement problem in relativistic quantum field theory," in M. Kuhlman, H. Lyre, and A. Wayne (eds.), *Ontological Aspects of Quantum Field Theory* (Singapore: World Scientific), 165–80.

Batterman, Robert (2002), *The Devil in the Details: Asymptotic Reasoning in Explanation, Reduction, and Emergence* (Oxford: Oxford University Press).

—— (2005), "Critical phenomena and breaking drops: infinite idealizations in physics," *Studies in History and Philosophy of Modern Physics 36*: 225–44.

—— (2009), "Idealization and modelling," *Synthese 169*: 427–46.

Bäuerle, G. G. A., and A. J. Koning (1988), "Transient phenomena in the Unruh effect," *Physica A 152*: 189–98.

Beaume, R., J. Manuceau, A. Pellet, and M. Sirand (1974), "Translation invariant states in quantum mechanics," *Communications in Mathematical Physics 38*: 29–45.

Bell, John L., and M. Machover (1977), *A Course in Mathematical Logic* (Amsterdam: North-Holland).

Bell, John S. (1987), *The Speakable and Unspeakable in Quantum Mechanics* (Cambridge: Cambridge University Press).

Belot, Gordon (1998), "Understanding electromagnetism," *British Journal for the Philosophy of Science 49*: 531–55.

—— (2007), "The representation of time and change in classical mechanics," in Butterfield and Earman (2007a), 133–227.

—— (2008), "Background independence" (unpublished MS).

Beltrametti, E., and G. Casinelli (1981), *The Logic of Quantum Mechanics* (Reading, MA: Addison-Wesley).

Benatti, F. (1993), *Deterministic Chaos in Infinite Quantum Systems.* (Berlin: Springer).

Birrell, N. D., and P. C. W. Davies (1982), *Quantum Fields on Curved Space* (Cambridge: Cambridge University Press).

Bogoliubov, N. N. (1958), "A new method in the theory of Superconductivity, I, " *Soviet Physics JETP 34*: 41–6.

Borchers, Hans-Jürgen (2000), "Modular groups in quantum field theory," *Quantum Field Theory (Tegernsee, 1998)* (Berlin: Springer), 26–42.

Borek, R. (1985), "Representations of the current algebra of a charged massless Dirac field," *Journal of Mathematical Physics 26*: 339–44.

Boyd, Richard (1985), "*Lex Orandi est Lex Credendi*," in P. M. Churchland and C. A. Hooker (eds.), *Images of Science* (Chicago: University of Chicago Press), 3–34.

Brading, Katherine, and Harvey Brown (2003), "Symmetries and Noether's Theorems," in Katherine Brading and Elena Castellani (eds.), *Symmetries in Physics: Philosophical Reflections* (Cambridge: Cambridge University Press), 89–109.

Bratteli, Ola, and Derek W. Robinson (1987[1971]), *Operator Algebras and Quantum Statistical Mechanics I*, 2nd edn (Berlin: Springer).

——— (1997[1981]), *Operator Algebras and Quantum Statistical Mechanics II*, 2nd edn (Berlin: Springer).

Brighouse, Carolyn (1995), "Spacetime and holes," in David Hull, Mickey Forbes, and Richard Burian (eds.), *PSA: Proceedings of the Biennial Meeting of the Philosophy of Science Association, 1994, vol. 1: Contributed Papers* (East Lansing, MI: Philosophy of Science Association), 117–25.

Brunetti, Romeo, and Giuseppe Ruzzi (2007),"Superselection sectors and general covariance, I," *Communications in Mathematical Physics 270*: 69–108.

Bryant, Robert (1995), "An introduction to Lie groups and symplectic geometry," in Daniel Freed and Karen Uhlenbeck (eds.), *Geometry and Quantum Field Theory* (Providence, RI: American Mathematical Society), 13–181.

Bub, J. (1977), "Von Neumann's Projection Postulate as a probability conditionalization rule," *Journal of Philosophical Logic 6*: 381–90.

——— (1997), *Interpreting the Quantum World* (Cambridge: Cambridge University Press).

Buchholz, Detlev, C. D'Antoni, and K. Fredenhagen (1987), "The universal structure of local algebras," *Communications in Mathematical Physics 111*: 123–35.

——— and Sergio Doplicher (1984), "Exotic infrared representations of interacting systems," *Annales de l'Institut Henri Poincaré. Physique Théorique 32*: 175–84.

——— ——— Robert Longo, and John E. Roberts (1992), "A new look at Goldstone's theorem," *Reviews in Mathematical Physics 4 (special issue)*: 49–83.

Busch, Paul, Pekka Lahti, and Paul Mittelstaedt (1996), *The Quantum Theory of Measurement* (Berlin: Springer).

Butterfield, Jeremy (1995), "Vacuum correlations and outcome independence in algebraic quantum field theory," *Annals of the New York Academy of Sciences 755*: 768–85.

——— (2007), "On symplectic reduction in classical mechanics," in Butterfield and Earman (2007a), 1–131.

Butterfield, Jeremy, and John Earman (eds.) (2007a), *Handbook of Philosophy of Science: Philosophy of Physics*, vol. 1 (Amsterdam: Elsevier).

——— ——— (eds.) (2007b), *Handbook of Philosophy of Science: Philosophy of Physics*, vol. 2 (Amsterdam: Elsevier).

Callender, C. (2001), "Taking thermodynamics too seriously," *Studies in the History and Philosophy of Modern Physics 32*: 539–53.

Camp, Joseph (2002), *Confusion: A Study in the Theory of Knowledge* (Cambridge, MA: Harvard University Press).

Candelas, P. (1980), "Vacuum polarization in Schwarzschild spacetime," *Physical Review D 21*: 2185–202.

Carroll, John (1994), *Laws of Nature* (Cambridge University Press).

Carroll, John (ed.) (2004), *Readings on Laws of Nature* (Pittsburgh, PA: University of Pittsburgh Press).

——(2007), "Nailed to Hume's cross?" in T. Sider and J. Hawthorne (eds.), *Contemporary Debates in Metaphysics* (Oxford: Oxford University Press), 67–81.

Castellani, Elena (ed.) (1998), *Interpreting Bodies: Classical and Quantum Objects in Modern Physics* (Princeton, NJ: Princeton University Press).

Cavallaro, S., G. Morchio, and F. Strocchi (1999), "A generalization of the Stone–von Neumann theorem to non-regular representations," *Letters in Mathematical Physics 47*: 307–20.

Chaiken, J. (1967), "Finite-particle representations and states of the canonical commutation relations," *Annals of Physics 42*: 23–80.

Chmielowski, P. (1994), "States of a scalar field on spacetimes with two isometries with timelike orbits," *Classical and Quantum Gravity 11*: 41–56.

Clifton, Robert (2000), "The modal interpretation of algebraic quantum field theory," *Physics Letters A 271*: 167–77.

——et al. (1998), "Super entangled states," *Physical Review A 58*: 135–45.

——and Hans Halvorson (2001a), "Entanglement and open systems in algebraic quantum field theory," *Studies in the History and Philosophy of Modern Physics 32*: 1–31.

————(2001b), "Are Rindler quanta real? Inequivalent particle concepts in quantum field theory," *British Journal for the Philosophy of Science 52*: 417–70.

————(2002), "Reconsidering Bohr's reply to EPR," in T. Placek and J. Butterfield (eds.), *Non-locality and Modality* (Dordrecht: Kluwer), 3–18.

Connes, Alain, and Carlo Rovelli (1994), "Von Neumann algebra automorphisms and time-thermodynamics relation in generally covariant quantum theories," *Classical Quantum Gravity 11*: 2899–917.

Cushing, James, and Ernan McMullin (eds.) (1989), *Philosophical Consequences of Quantum Theory: Reflections on Bell's Theorem* (Notre Dame: University of Notre Dame Press).

Davies, P. C. W. (1984), "Particles do not exist," in S. M. Christensen (ed.), *Quantum Theory of Gravity* (Bristol: Hilger), 66–77.

De Bièvre, Stephan (2007), "Where's that quantum?" in *Contributions in Mathematical Physics* (New Delhi: Hindustan Book Agency), 123–46.

DeWitt, Bryce S. (1975), "Quantum field theory on curved spacetime," *Physics Reports 19*: 295–357.

Dimock, J. (1980), "Algebras of local observables on a manifold," *Communications in Mathematical Physics 77*: 219–28.

Dirac, P. A. M. (1930a), "Quelques problèmes de mécanique quantique," *Annales de l'Institut Henri Poincaré 1*: 357–400.

——(1930b), *The Principles of Quantum Mechanics* (Oxford: Clarendon).

Doplicher, Sergio (1992), "Progress and problems in algebraic quantum field theory," *Ideas and Methods in Quantum and Statistical Physics* (Cambridge: Cambridge University Press), 390–404.

——F. Figliolini, and D. Guido (1984), "Infrared representations of free boson fields," *Annales de l'Institut Henri Poincaré. Physique Théorique 41*: 49–62.

——Rudolf Haag, and John E. Roberts (1971), "Local observables and particle statistics, I," *Communications in Mathematical Physics 23*: 199–230.

————(1974), "Local observables and particle statistics, II," *Communications in Mathematical Physics 35*: 49–85.

Dürr, Detlef, et al. (2006), "Topological factors derived from Bohmian mechanics," *Journal Annales Henri Poincaré* 7: 791–807.

Earman, John (1986), *A Primer on Determinism* (Boston, MA: Reidel).

—— (1989), *World Enough and Space-Time: Absolute Versus Relational Theories of Space and Time* (Cambridge, MA: MIT Press).

—— (1993), "In defense of laws: reflections on Bas van Fraassen's *Laws and Symmetry*," *Philosophy and Phenomenological Research* 53: 413–19.

—— (1995), *Bangs, Crunches, Whimpers, and Shrieks: Singularities and Acausalities in Relativistic Spacetimes* (Oxford: Oxford University Press).

—— (2004a), "Laws, symmetry, and symmetry breaking: invariance, conservation principles, and objectivity," *Philosophy of Science* 71: 1227–41.

—— (2004b), "Laws, symmetry, and symmetry breaking: invariance, conservation principles, and objectivity," philsci-archive.pitt.edu/archive/00000878/.

—— (2004c), "Curie's Principle and spontaneous symmetry breaking," *International Studies in the Philosophy of Science* 18: 173–98.

—— (2008), "Superselection rules for philosophers," *Erkenntnis* 69: 377–414.

—— and Doreen Fraser (2006), "Haag's Theorem and its implications for the foundations of quantum field theory," *Erkenntnis* 64: 305–44.

—— and John Norton (1987), "What price spacetime substantivalism: the hole story," *British Journal for the Philosophy of Science* 38: 515–25.

—— and Laura Ruetsche (2005), "Relativistic invariance and modal interpretations," *Philosophy of Science* 72: 557–83.

Emch, Gérard (1972), *Algebraic Methods in Statistical Mechanics and Quantum Field Theory* (New York: Wiley).

—— (1984), *Mathematical and Conceptual Foundations of Twentieth Century Physics* (Amsterdam: North-Holland).

Emch, Gérard (1997), "Beyond irreducibility and back," *Reports on Mathematical Physics* 40: 187–93.

—— (2007), "Quantum statistical physics," in Butterfield and Earman (2007b), 1075–1182.

—— and H. J. F. Knops (1970), "Pure thermodynamic phases as extremal KMS states," *Journal of Mathematical Physics* 11: 3008–18.

—— and Chuang Liu (2002), *The Logic of Thermostatistical Physics* (Berlin: Springer).

Everett, Hugh (1957), " 'Relative state' formulation of quantum mechanics," *Reviews of Modern Physics* 29: 454–62.

Fabri, E., and L. E. Picasso (1966), "Quantum field theory and approximate symmetries," *Physical Review Letters* 16: 408–10.

Farmelo, Graham (2009), *The Strangest Man: The Hidden Life of Paul Dirac* (London: Faber & Faber).

Fine, Arthur (1986), *The Shaky Game: Einstein and the Quantum Theory* (Chicago: University of Chicago Press).

Florig, Martin, and Stephen J. Summers (2000), "Further representations of the canonical commutation relations," *Proceedings of the London Mathematical Society* 80: 451–90.

Folland, G. (1989), *Harmonic Analysis in Phase Space* (Princeton, NJ: Princeton University Press).

—— (2008), *Quantum Field Theory: A Tourist Guide for Mathematicians* (Providence, RI: American Mathematical Society).

Fraser, Doreen (2009), "Quantum field theory: underdetermination, inconsistency, and idealization," *Philosophy of Science 76*: 536–67.

Friedman, Michael (1983), *Foundations of Space-Time Theories: Relativistic Physics and Philosophy of Science* (Princeton, NJ: Princeton University Press).

Friedrichs, K. O. (1953), *Mathematical Aspects of the Quantum Theory of Fields* (New York: Interscience).

Frisch, M. (2005), *Inconsistency, Asymmetry, and Non-Locality: A Philosophical Investigation of Classical Electrodynamics* (Oxford: Oxford University Press).

Fulling, Stephen A. (1972), "Scalar quantum field in a closed universe of constant curvature," Ph.D. dissertation, Princeton University.

—— (1989), *Aspects of Quantum Field Theory in Curved Space-Time* (Cambridge: Cambridge University Press).

Gårding, L., and A. Wightman (1954), "Representations of the commutation relations," *Proceedings of the National Academy of the Sciences USA 10*: 622–6.

Gel'fand, I. M., and N. Vilenkin (1964), *Generalized Functions*, vol. IV (New York: Academic Press).

Geroch, Robert (1978), *General Relativity from A to B* (Chicago: University of Chicago Press).

Giulini, Domenico (1994), "Asymptotic symmetry groups of long-range gauge configurations," gr-qc/9410042.

—— (2006), "Algebraic and geometric structures of special relativity," LANL preprint 0602018.

Glauber, Roy (1963), "Coherent and incoherent states of the radiation field," *Physical Review 131*: 2766–88.

Glimm, James (1969), "The foundations of quantum field theory," *Advances in Mathematics 3*: 101–25.

—— and Arthur Jaffe (1972), "The $\lambda\phi_2^4$ quantum field theory without cutoffs IV: perturbations of the Hamiltonian," *Journal of Mathematical Physics 13*: 1568–84.

Glymour, C. (1970), "Theoretical realism and theoretical equivalence," in R. Buck and R. Cohen (eds.), *Boston Studies in Philosophy of Science VII* (Dordrecht: Reidel), 275–88.

Goldstein, Herbert, Charles Poole, and John Safko (2002), *Classical Mechanics*, 3rd edn. (San Francisco: Addison-Wesley).

Gotay, M. J. (2000), "Obstructions to quantization," in *Mechanics: From Theory to Computation* (New York: Springer), 171–216.

Gottfried, Kurt, and Tung-Mow Yan (2003), *Quantum Mechanics: Fundamentals* (New York: Springer).

Grib, A. A., S. G. Mamayev, and V. M. Mostepanenko (1976), "Particle creation from vacuum in homogeneous isotropic models of the universe," *General Relativity and Gravitation 7*: 535–47.

Guralnik, G. S., C. R. Hagen, and T. W. B. Kibble (1964), "Global conservation laws and massless particles," *Physical Review Letters 13*: 585–7.

Haag, Rudolf (1955), "On quantum field theories," *Det Kongelige Danske Videnskabernes Selskab Matematisk-Fysiske Meddelelser 29*: 1–37.

—— (1962), "The mathematical structure of the Bardeen–Cooper–Schrieffer model," *Nuovo Cimento 25*: 287–99.

—— (1992), *Local Quantum Physics* (New York: Springer).

—— and Daniel Kastler (1964), "An algebraic approach to quantum field theory," *Journal of Mathematical Physics 5*: 848–61.

Hajicek, P. (1976), "On quantum field theory in curved space-time," *Nuovo Cimento 33B*: 597–612.

Hall, D., and A. S. Wightman (1957), "A theorem on invariant analytic functions with applications to relativistic quantum field theory," *Det Kongelige Danske Videnskabernes Selskab Matematisk-Fysiske Meddelelser 31*: 1–47.

Halvorson, Hans (2001), "On the nature of continuous physical quantities in classical and quantum mechanics," *Journal of Philosophical Logic 30*: 27–50.

—— (2004), "Complementarity of representations in quantum mechanics," *Studies in History and Philosophy of Modern Physics 35*: 45–56.

—— and Rob Clifton (1999), "Maximal beable subalgebras of quantum mechanics," *International Journal of Theoretical Physics 38*: 2441–84.

—— —— (2000), "Generic Bell correlations between arbitrary local algebras in quantum field theory," *Journal of Mathematical Physics 41*: 1711–17.

—— —— (2002), "No place for particles in relativistic quantum theories," *Philosophy of Science 69*: 1–28.

—— and Michael Müger (2007), "Algebraic quantum field theory," in Butterfield and Earman (2007a), 731–922.

Healey, Richard (2007), *Gauging What's Real: The Conceptual Foundations of Contemporary Gauge Theories* (Oxford: Oxford University Press).

Helfer, A. D. (1996), "The stress-energy tensor," *Classical and Quantum Gravity 13*: L129–34.

Hempel, C. (1965), "The theoretician's dilemma," in *Aspects of Scientific Explanation* (New York: Free Press), 173–226

Henneaux, M,. and C. Teitelboim (1992), *Quantization of Gauge Systems* (Princeton, NJ: Princeton University Press).

Hepp, Klaus (1963), "On the asymptotic condition in a local relativistic quantum field theory," *Acta Physica Austriaca 17*: 85–95.

—— (1972), "Quantum theory of measurement and macroscopic observables," *Helvetica Physica Acta 45*: 237–45.

Hollands, Stefan, and Robert M. Wald (2010), "Axiomatic quantum field theory in curved spacetime," *Communications in Mathematical Physics 293*: 85–125.

Honegger, Reinhard, and Alfred Rieckers (1990), "The general form of non-Fock coherent boson states," *Publications of the Research Institute for Mathematical Science 26*: 397–417.

Huggett, Nick (1995), "What are quanta and why does it matter?" in David Hull, Mickey Forbes, and Richard Burian (eds.), *PSA: Proceedings of the Biennial Meeting of the Philosophy of Science Association, 1994*, vol. 2 (East Lansing, MI: Philosophy of Science Association), 69–78.

—— (2000), "Philosophical foundations of quantum field theory," *British Journal for the Philosophy of Science 51*: 617–37.

Hughes, R. I. G. (1989), *The Structure and Interpretation of Quantum Mechanics* (Cambridge, MA: Harvard University Press).

Husain, V., and O. Winkler (2004), "Singularity resolution in quantum gravity," *Physical Review D 69*: 084016.

Isham, C. J. (1983), "Topological and global aspects of quantum theory," in B. S. DeWitt and R. Stora (eds.), *Les Houches, Session XL: Relativité, Groupes et Topologie II* (Amsterdam: Elsevier).

Jammer, Max (1966), *The Conceptual Development of Quantum Mechanics* (New York: McGraw-Hill).

Jones, Roger (1991), "Realism about what?" *Philosophy of Science* 58: 185–202.

Jordan, P., and E. P. Wigner (1928), "Über das Paulische Aequivalenzverbot," *Zeitschrift für Physik* 47: 631–51.

Jørgensen, P. E. T., and R. T. Moore (1984), *Operator Commutation Relations* (Dordrecht: Reidel).

Kadison, Richard (1965), "Transformations of states in operator theory and dynamics," *Topology* 3, Suppl. 2: 177–98.

Kadison, R. V., and J. R. Ringrose (1997a[1983]), *Fundamentals of the Theory of Operator Algebras*, vol. 1 (New York: Academic Press).

—————— (1997b[1986]), *Fundamentals of the Theory of Operator Algebras*, vol. 2 (New York: Academic Press).

Kaku, M. (1993), *Quantum Field Theory: A Modern Introduction* (Oxford: Oxford University Press).

Kastler, D. (1964), "A C^*-algebra approach to field theory," *Theory and Application of Analysis in Function Space* (Cambridge, MA: MIT. Press), 179–91.

—————— D. Robinson, and A. Swieca (1966), "Conserved currents and associated symmetries: Goldstone's theorem," *Communications in Mathematical Physics* 2: 108–20.

Kay, Bernard S. (1985), "The double wedge algebra for quantum fields on Schwarzschild and Minkowski spacetimes," *Communications in Mathematical Physics* 100: 57–81.

—————— (1988), "Quantum field theory in curved spacetime," in *Differential Geometrical Methods in Theoretical Physics (Como, 1987): NATO Advanced Sciences Institute Series C Mathematical and Physical Sciences* 250: 373–93.

—————— and Robert M. Wald (1991), "Theorems on the uniqueness and thermal properties of stationary, nonsingular, quasifree states on spacetimes with a bifurcate Killing horizon," *Physics Reports* 207: 49–136.

Kibble, T. W. B. (1967), "Symmetry breaking in non-abelian gauge theories," *Physical Review* 155: 1554–61.

Kitajima, Yuichiro (2004), "A remark on the modal interpretation of algebraic quantum field theory," *Physics Letters A331*: 181–6.

Klauder, J. R., and B. S. Skagerstram (1985), *Coherent States, Applications in Physics and Mathematical Physics* (Singapore: World Scientific).

—————— and E. C. G. Sudarshan (1968), *Fundamentals of Quantum Optics* (New York: Benjamin).

Kronz, Frederick M., and Tracy A. Lupher (2005), "Unitarily inequivalent representations in algebraic quantum theory," *International Journal of Theoretical Physics* 44: 1239–58.

Kubo, R. (1957), "Statistical mechanics of irreversible processes," *Journal of the Physical Society of Japan* 12: 570–86.

Ladyman, James (2002), *Understanding Philosophy of Science* (New York: Routledge).

Landsman, N. P. (1990), "C^*-algebraic quantization and the origin of topological quantum effects," *Letters in Mathematical Physics* 20: 11–18.

—————— (1998), *Mathematical Topics Between Classical and Quantum Mechanics* (New York: Springer).

—————— (2004), "Quantum mechanics and representation theory: the new synthesis," *Acta Applicandae Mathematicae* 81: 167–89.

—————— (2006), "Lie groupoids and Lie algebroids in physics and non-commutative geometry," *Journal of Geometry and Physics* 56: 24–54.

—— (2007), "Between classical and quantum," in Butterfield and Earman (2007a), 417–553.
Letaw, J. R., and J. D. Pfautsch (1981), "Quantized scalar field in the stationary coordinate systems of flat spacetime," *Physical Review D* 24: 1491–8.
—— —— (1982), "The stationary coordinate systems in flat spacetime," *Journal of Mathematical Physics* 23: 425–31.
Lévy-Leblond, Jean-Marc (1976), "Who is afraid of nonhermitian operators? A quantum description of angle and phase," *Annals of Physics* 101: 319–41.
Lewis, David (1973), *Counterfactuals* (Cambridge, MA: Harvard University Press).
—— (1979), "Counterfactual dependence and time's arrow," *Noûs* 13 (Special Issue on Counterfactuals and Laws): 455–76.
Liu, Chuang (2001), "Infinite systems in SM explanations: thermodynamic limit, renormalization (semi-) groups, and irreversibility," *Philosophy of Science* 68: S325–44.
—— (2003), "Classical spontaneous symmetry breaking," *Philosophy of Science* 70: 1219–32.
—— and Gérard Emch (2005), "Explaining quantum spontaneous symmetry breaking," *Studies in the History and Philosophy of Modern Physics* 36: 137–63.
Mackey, George (1998), "The relationship between classical mechanics and quantum mechanics," in L. Coburn and M. Reiffel (eds.), *Perspectives on Quantization* (Providence, RI: American Mathematical Society), 91–109.
Malament, David (1984), "'Time travel' in the Godel universe," in *PSA: Proceedings of the Biennial Meeting of the Philosophy of Science Association, 1984, Volume Two: Symposia and Invited Papers*, 91–100.
—— (1996), "In defense of dogma," in Robert Clifton (ed.), *Perspectives on Quantum Reality: Non-Relativistic, Relativistic, and Field-Theoretic* (Dordrecht: Kluwer), 1–10.
Manchak, John (2008), untitled MS.
Marsden, J. E., and T. S. Ratiu (1994), *Introduction to Mechanics and Symmetry* (Berlin: Springer).
Martin, P. C., and J. Schwinger (1959), "Theory of many particle systems," *Physical Review* 115: 1342–73.
Maudlin, Tim (1994), *Quantum Non-locality and Relativity* (Oxford: Oxford University Press).
—— (1996), "On the unification of physics," *Journal of Philosophy* 93: 129–44.
Mermin, David (1993), "Hidden variables and the two theorems of John Bell," *Reviews of Modern Physics* 65: 803–15.
Messiah, Albert (1961), *Quantum Mechanics*, trans. G. M. Temmer (New York: Interscience).
Modesto, L. (2006), "Loop quantum black hole," *Classical and Quantum Gravity* 23: 5587–601.
Moore, Walter (1989), *Schrödinger: Life and Thought* (Cambridge: Cambridge University Press).
Morandi, G. (1992), *The Role of Topology in Classical and Quantum Physics* (Berlin: Springer).
Morchio, G., and F. Strocchi (1985), "Spontaneous symmetry breaking and energy gap generated by variables at infinity," *Communications in Mathematical Physics* 99: 153–75.
—— (1986), "Infrared problem, Higgs phenomenon and long range interactions," in *Fundamental Problems of Gauge Field Theory (Erice, 1985): NATO Advanced Sciences Institute Series B Physics* 141: 301–44.
Morgan, M. S., and M. Morrison (eds.) (1999), *Models as Mediators* (Cambridge: Cambridge University Press).
Murray, F. J. and J. von Neumann (1936), "On rings of operators," *Annals of Mathematics* 37: 116–229.

Nemeschansky, Dennis, C. Preitschopf, and Marvin Weinstein (1988), "A BRST primer," *Annals of Physics 183*: 226–68.
Norton, John (2008), "The dome: an unexpectedly simple failure of determinism," *Philosophy of Science 75*: 786–98.
Ojima, Izumi (2003), "A unified scheme for generalized sectors based on selection criteria: order parameters of symmetries and of thermality and physical meanings of adjunctions," *Open Systems and Information Dynamics 10*: 235–79.
Olver, Peter (1993), *Applications of Lie Groups to Differential Equations* (Berlin: Springer).
Parker, Leonard (1969), "Quantized fields and particle creation in expanding universes, I," *Physical Review 183*: 1057–68.
—— (1971), "Quantized fields and particle creation in expanding universes, II,' *Physical Review D3*: 346–56.
Peskin, Michael, and Daniel Schroeder (1995), *An Introduction to Quantum Field Theory* (Reading, MA: Addison-Wesley).
Pickering, Andrew (1995), *The Mangle of Practice: Time, Agency, and Science* (Chicago: University of Chicago Press).
—— (2010), *The Cybernetic Brain* (Chicago: University of Chicago Press).
Primas, H. (1983), *Chemistry, Quantum Mechanics, and Reductionism* (New York: Springer).
Provost, J. P., F. Rocca, and G. Vallee (1975), "Coherent states, phase states, and condensed states," *Annals of Physics 94*: 307–19.
Prugovečki, Edouard (1971), *Quantum Mechanics in Hilbert Space* (New York: Dover).
Przibaum, K. (ed.) (1986), *Letters on Wave Mechanics*, trans. M. Klein (New York: Philosophical Library).
Psillos, Stathis (1999), *Scientific Realism: How Science Tracks Truth* (New York: Routledge).
Putnam, Hilary (1975), *Mathematics, Matter, and Method*, vol. 1 (Cambridge: Cambridge University Press).
—— (1989), "Equivalence," in *Realism and Reason: Philosophical Papers*, vol. 3 (Cambridge: Cambridge University Press), 26–45.
Rawls, John (1974), "The independence of moral theory," *Proceedings and Addresses of the American Philosophical Assocation 48*: 5–22.
Rédei, Miklós (1998), *Quantum Logic in Algebraic Approach* (Dordrecht: Kluwer).
—— (2010), "Operational independence and operational separability in algebraic quantum mechanics," *Foundations of Physics*: 0015–9018.
—— and Stephen Summers (2007), "Quantum probability theory," *Studies in the History and Philosophy of Modern Physics 38*: 390–417.
—— —— (2010), "When are quantum systems operationally independent?" *International Journal of Theoretical Physics*: 0020–7748.
—— and Giovanni Valente (2010), "How local are local operations in quantum field theory?" *Studies in the History and Philosophy of Modern Physics 41*: 346–53.
Redhead, Michael (1983), "Quantum Field Theory for Philosophers," in P. Asquith and T. Nickes (eds.), *PSA: Proceedings of the Biennial Meeting of the Philosophy of Science Association, 1982*, vol. 2: 57–99.
—— (1986), "Relativity and quantum mechanics: conflict or peaceful coexistence?" *Annals of the New York Academy of Sciences 480*: 14–20.
—— (1989), *Incompleteness, Nonlocality, and Realism* (Oxford: Oxford University Press).

—— (1995a), "More ado about nothing," *Foundations of Physics* 25: 123–37.

—— (1995b), "The vacuum in relativistic quantum field theory," in David Hull, Mickey Forbes, and Richard Burian (eds.), *PSA: Proceedings of the Biennial Meeting of the Philosophy of Science Association*, vol. 2 (East Lansing, MI: Philosophy of Science Association), 77–87.

Reed, Michael, and Barry Simon (1980a), *Methods of Modern Mathematical Physics, I: Functional Analysis* (New York: Academic Press).

—— —— (1980b), *Methods of Modern Mathematical Physics, III: Scattering Theory* (New York: Academic Press).

Reents, Georg, and Stephen Summers (1994), "Beyond coherent states: higher order representations," in *On Klauder's Path: A Field Trip* (River Edge, NJ: World Scientific), 179–88.

Rivasseau, V. (2003), "An introduction to renormalization," in B. Duplantier and V. Rivasseau (eds.), *Poincaré Seminar 2002: Vacuum Energy—Renormalization* (Basel: Birhauser), 139–77.

Roberts, J. E., and G. Roepstorff (1969), "Some basic concepts of algebraic quantum theory," *Communications in Mathematical Physics* 11: 321–38.

Robinson, Derek (1966), "Algebraic aspects of relativistic quantum field theory," in M. Chretien and S. Deser (eds.), *Axiomatic Field Theory* (New York: Gordon & Breach).

—— (1968), "Statistical mechanics of quantum spin systems, II," *Communications in Mathematical Physics* 7: 337–48.

Roepstorff, G. (1970), "Coherent photon states and spectral condition," *Communications in Mathematical Physics* 19: 301–14.

Rosenberg, Jonathan (2004), "A selective history of the Stone–von Neumann Theorem," in *Operator Algebras, Quantization, and Noncommutative geometry* (Providence, RI: American Mathematical Society), 331–54.

Ruelle, D. (1969), *Statistical Mechanics* (New York: Benjamin).

Ruetsche, Laura (2003), "Modal semantics, modal dynamics, and the problem of state preparation," *International Studies in the Philosophy of Science* 17: 25–41.

—— (2004), "Intrinsically mixed states: an appreciation," *Studies in the History and Philosophy of Modern Physics* 35: 221–39.

—— and John Earman (2010), "Probabilities in quantum field theory and quantum statistical mechanics," in Claus Beisbart and Stephan Hartmann (eds.), *Probabilities in Physics* (Oxford: Oxford University Press).

Sakai, S. (1971), C^*-*algebras and* W^*-*algebras* (New York: Springer).

Schmüdgeon, K. (1990), *Unbounded Operator Algebras and Representation Theory* (Basel: Birkhauser).

Schrödinger, Erwin (1926), "Der stetige Übergang von der Mikro- zur Makromechanik," *Naturwissenschaften* 14: 664–6.

Segal, Irving E. (1959), "The mathematical meaning of operationalism in quantum mechanics," in L. Henkin, P. Suppes, and A. Tarski (eds.), *Studies in Logic and the Foundations of Mathematics* (Amsterdam: North-Holland), 341–52.

—— (1967), "Representations of the canonical commutation relations," in *Cargèse Lectures in Theoretical Physics: Application of Mathematics to Problems in Theoretical Physics* (New York: Gordon & Breach), 107–70.

Sellars, Wilfrid (1963), "Philosophy and the scientific image of man," in *Science, Perception, and Reality* (New York: Humanities Press).

Sewell, Geoffrey (1973), "States and dynamics of infinitely extended physical systems," *Communications in Mathematical Physics 33*: 43–51.

—— (1986), *Quantum Theory of Collective Phenomena* (Oxford: Oxford University Press).

Sewell, Geoffrey (2002), *Quantum Mechanics and its Emergent Metaphysics* (Princeton, NJ: Princeton University Press).

Shankar, Ramamurti (1994), *Principles of Quantum Mechanics* (New York: Plenum).

Simon, B. (1972), "Topics in functional analysis," in R. F. Streater (ed.), *Mathematics of Contemporary Physics* (London: Academic Press).

Sklar, Lawrence (1993), *Physics and Chance: Philosophical Issues in the Foundations of Statistical Mechanics* (Cambridge: Cambridge University Press).

—— (2002), *Theory and Truth* (Oxford: Oxford University Press).

Streater, R. F. (1965), "Spontaneous breakdown of symmetry in axiomatic theory," *Proceedings of the Royal Society, Series A 287*: 510–18.

—— (1967), "The Heisenberg ferromagnet as a quantum field theory," *Communications in Mathematical Physics 6*: 233–47.

—— and A. S. Wightman (1964), *PCT, Spin and Statistics, and All That* (Reading, MA: Addison-Wesley).

Strocchi, F. (1988), "Long range dynamics and spontaneous symmetry breaking in many-body systems," in *Fractals, Quasicrystals, Chaos, Knots and Algebraic Quantum Mechanics (Maratea, 1987): NATO Advanced Sciences Institute Series C Mathematical and Physical Sciences 235*: 265–85.

—— (2005), *Symmetry Breaking* (Berlin: Springer).

—— (2008), *Elements of Quantum Mechanics of Infinite Systems* (Philadelphia: World Scientific).

Summers, Stephen (1990), "On the independence of local algebras in quantum field theory," *Reviews in Mathematical Physics 2*: 201–47.

—— (1999), "On the Stone–von Neumann uniqueness theorem and its ramifications," in *John von Neumann and the Foundations of Quantum Physics* (Budapest, 1999), 135–152, *Vienna Circle Institute Yearbook, 8* (Dordrecht: Kluwer).

—— and R. Werner (1987), "Maximal violation of Bell's inequalities is generic in quantum field theory," *Communications in Mathematical Physics 110*: 247–50.

Sunder, V. S. (1987), *An Invitation to von Neumann Algebras* (New York: Springer).

Suppe, F. (ed.) (1977), *The Structure of Scientific Theories* (Urbana: University of Illinois Press).

Takagi, S. (1986), "Vacuum noise and stress induced by uniform acceleration," *Progress of Theoretical Physics*, Supplement 88: 1–142.

Takesaki, Masamachi (1994), "Twenty-five years in the theory of Type III von Neumann algebras," in R. S. Doran (ed.), *C*-algebras, 1943–1993: A Fifty Year Celebration* (Ann Arbor, MI: American Mathematical Society).

Takesaki, Masamachi (2002), *Theory of Operator Algebra I* (Berlin: Springer).

—— (2003), *Theory of Operator Algebra II* (Berlin: Springer).

Teller, Paul (1979), "Quantum mechanics and the nature of continuous physical quantities," *Journal of Philosophy 7*: 345–61.

—— (1995), *An Interpretive Introduction to Quantum Field Theory* (Princeton, NJ: Princeton University Press).

Thirring, W., and N. Narnhofer (1992), "Covariant QED without indefinite metric," *Reviews of Mathematical Physics, Special Issue*: 197–211.

t'Hooft, G. (2007), "The conceptual basis of quantum field theory," in Butterfield and Earman (2007a), 661–729.
Tipler, Frank (1969), *Modern Physics* (Berkeley: University of California Press).
Uffink, Jos (2007), "Compendium of the foundations of classical statistical physics," in Butterfield and Earman (2007b), 923–1074.
Unruh, W. H. (1976), "Notes on black hole evaporation," *Physical Review D 14*: 870–92.
—— (1990), "Particles and fields," in J. Audretsch and V. de Sabbata (eds.), *Quantum Mechanics in Curved Spacetime* (New York: Plenum Press).
—— and R. M. Wald (1984), "What happens when an accelerating observer detects a Rindler particle?" *Physical Review D 29*: 1047–56.
Updike, John (1963), *From Telephone Poles and Other Poems* (New York: Knopf).
van Fraassen, Bastian C. (1980), *The Scientific Image* (Oxford: Oxford University Press).
—— (1989), *Laws and Symmetry* (Oxford: Oxford University Press).
—— (1991), *Quantum Mechanics: An Empiricist View* (Oxford: Oxford University Press).
Wald, Robert M. (1984), *General Relativity* (Chicago: University of Chicago Press).
—— (1994), *Quantum Field Theory in Curved Spacetime and Black Hole Thermodynamics* (Chicago: University of Chicago Press).
—— (1995), "Quantum field theory in curved spacetime," gr-qc/9509057.
—— (1999). "Gravitation, thermodynamics, and quantum theory," *Classical and Quantum Gravity 16*: A177–90.
Wallace, David (2003), "Everettian rationality: defending Deutsch's approach to probability in the Everett interpretation," *Studies in History and Philosophy of Modern Physics 34*: 415–39.
—— (2006), "In defense of naiveté: the conceptual status of lagrangian quantum field theory," *Synthese 151*: 33–80.
Wightman, A. S., and S. S. Schweber (1955), "Configuration space methods in relativistic quantum field theory," *Physical Review 98*: 812–37.
Wigner, Eugene (1959), *Group Theory and its Application to Quantum Mechanics of Atomic Spectra* (New York: Academic Press).
—— R. M. F. Houtappel, and H. van Dam (1965), "The conceptual basis and use of the geometric invariance principles," *Reviews of Modern Physics 37*: 595–632.
Witten, Edward (1999), ... in P. Deligne et al. (eds.), *Quantum Fields and Strings: A Course for Mathematicians*, vol. 2 (Ann Arbor, MI: American Mathematical Society).
Woodhouse, Nicholas (1992), *Geometric Quantization* (Oxford: Oxford University Press).
Worrall, John (1989), "Structural realism: the best of both worlds," *Dialectica* 43: 99–124.
Zee, A. (2010), *Quantum Field Theory in a Nutshell*, 2nd edn. (Princeton, NJ: Princeton University Press).
Zhu, Kehe (1993), *An Introduction to Operator Algebras* (Boca Raton, FL: CRC Press).

Index

Normally, only the pages on which technical terms are defined or introduced (indicated by italics) are indexed.

φ_2^4 theory 103, 109, 255–6

Abelian 23
Abelian Higgs model 328–30
Accardi, L. and Cecchini, C. 112
Acerbi, F. 53
Aitchison, Ian 300, 313–14, 318–22, 324–6, 328–30
Albert, David 263
Algebra 73–4
    Abelian 74
    Local 104
    Generator of 75
    Quasi-local 105
    Unital 74
See also */C* algebra, CAR algebra, von Neumman algebra, Weyl algebra
Algebraic Imperialism 16, 119, 132–43, 145–6, 261, 281, 290, 292, 298, 310, 314, 322, 331, 350
Algebraic Quantum Theory 15, 73–101, 148–68
Algebraically complete set of operators 200-1
Anticommutator 60
Arageorgis, Aristidis 16, 70, 115, 132, 203, 216–20, 241, 235, 296–7
Araki, Huzihiro 104, 106, 166, 274, 278, 281, 286
ARIS principle 286–87, 289
Aristotle 139, 140, 141, 143, 146, 286
Ashtekar, Abhay 53, 205, 207, 232
Asymptotic completeness 254
Asymptotically abelian action 266–7
Axioms for QFT/QSM 102–16
    Covariance 105
    Independence 114
    Isotony 105
    Microcausality 106, 109–13
    Primitive Causality 107
    Primitivity 107
    Vacuum 108
    Weak Additivity 108

Baez, John 117, 197
Bais, F. A. 325
Baker, David 129, 220, 226, 240
Balachandran, A. P. 325

Banach space 76
Barrett, Jeff 181
Batterman, Robert 263, 289, 337–9
Bäuerle, G. G. A. and Koning, A. J. 217
BCS model. See Superconductivity
Beable 16, 169–70, 173–89
Bead on a circle 54, 57–9
Beaume, R. 350
Bell, John L. and Machover, M. 157, 178
Bell, John S. 174–5
Bell Inequalities 106, 109, 112–14
Belot, Gordon 25, 34, 57, 59, 293, 294, 333
Beltramettie E. and Casinelli, G. 180
Belushi, John 3
Bennati, F. 266, 267
Big Picture principle 237–9
Birrell, N. D. 211, 217
Black hole 67, 238–9
Bogoliubov, N. N. 308
Bogoliubov transformations 211, 307–8
Bohm–Aharonov effect 58, 334
Bohm theory 10, 29, 58, 169, 180, 185–6, 280, 341
Boolean (two-valued) homomorphism *178*
Borchers, Hans 162
Borek, R. 167
Born Rule 7, 111, 169, 180–1
Bose gas 161, 167, 274, 284
Bose–Einstein condensate 268, 284, 287, 306
Bose–Einstein statistics 130
Boson 53, 60–1, 130 *See also* Klein–Gordon field.
Boyd, Richard 345–6, 349
Bracket 32
Brading, Katherine 317
Bratteli, Ola 27, 40, 73, 77–8, 83, 86–91, 93–5, 98, 117, 122, 138, 155, 158, 161–2, 165, 167, 210, 211, 279, 281
Brighouse, Caro 12
Broken symmetry 17, 132, *126*, 270–2, 291–339
    Decompositional sense of *301*–4, 305–9
    Individual sense of *300*–1, 303–4
    Marker of *303*–4, 305, 309–11, 322, 331–3
Brown, Harvey 317
Brunetti, Romeo and Ruzzi, Giuseppe 131
Bryant, Robert 32, 53

Bub, Jeffrey 170, 176, 179, 247, 249–50
Buchholz, Detlev 166, 167, 321
Busch, Paul, Lahti, Pekka and Mittelstaedt, Paul 343
Butterfield, Jeremy 34

/C Algebra 74–5/76–7
C* dynamical system 115
C* statistical dynamical system 266
Calibration condition 217–18
Callender, Craig 292, 311, 336–7
Camp, Joseph 103
Candelas, P. 238
Carroll, John 9, 353
Castellani, Elana 193
Chmielowski, P. 236
Cauchy surface 107
CAR Algebra 60–1, 83–4, 95, 116, 150, 159
Canonical Anticommutation Relations (CARs) 20
   as constitutive of quantum theory 7, 14, 25, 119–22
Canonical Commutation Relations (CCRs) 14, 20
   as constitutive of quantum theory 7, 14, 25, 119–22
   Heisenberg form of 36
   Weyl form of 37–9
canonical coordinates 31
center. *See von* Neumann algebra, center of
Central Decomposition Theorem 302–4
characteristic function 81
Circular Canonical Commutation Relations (CCCRs) 58–9
Classical mechanics 30–5,
Clifton, Rob 26, 46, 114, 134, 143, 156, 158, 170, 174–6, 182, 186–8, 193, 211–12, 222–3, 283
Coalesced Structures Argument 17–18, 291–2, 309–11, 312–14, 320, 322, 330–9
Coalesence Approach 147, 289–92, 349, 353
   And broken symmetry 289–300, 309–11
Commutant of an algebra 86
Commutator 60
Complementarity 48
Complete. *See norm*, completeness with respect to
Complex structure 208
Configuration space 30
   Non-vanilla 33, 53–9
conjugate momentum 30–1
Connes, Alain 152, 153, 164
consistency, intertheoretic 10, 13, 109, 341
Content. *See also* interpretation
Coalesced vs primordial 288–90

Convergence. *See also* Topologies
   Point-wise 78
   Uniform 78
Convex set 40–1
Cosmological particle creation 207, 247, 249, 257–60
Creation and annihilation operators 60–1, 68–72, 209–10. *See also* Fock space
Critical exponent/phenomenon/temperature 17, 269, 272, 289–90, 299, 306–10, 313, 326, 336–9, 353
Curie's principle 312
Curved spacetime 35, 104, 124, 131, 193, 197, 206, 233, 237–8, 247, 350, 353
Cushing, James and McMullin, Ernan 106
Cyclic vector 92

Van Dam, H. 294
D'Antoni, C. and Fredenhagen, K. 166
Davies, P. C. W. 211, 216–17
(DE) Dynamical equivalence 29
De Biévre, Stephan 193
Dense set 92
Density operator 21
Determinism 12, 107, 119, 343
DeWitt, Bryce 216–18
DHR selection theory 126–32, 321
Dimension function 154
Dimock, J. 104–8
Dirac, Paul 15, 200–201
Dirac quantization 334
Disjoint states/representation 96–7
Doctrine of Essential Idealization 337–9
Domain of dependence 107
Doplicher, Sergio 127–9, 167
Dürr, Detleff 58–9
Dynamics 8

Earman, John 3, 4, 40, 115, 164, 173, 187, 214, 219, 231, 251, 279, 293–6, 300–1, 311–14, 317, 325, 332–8, 352–3
Einstein, Albert 19, 106, 130, 167, 342
Einstein field equations 7, 8, 12, 66, 82, 235, 257, 289, 316
Einstein summation convention 65
Electrodynamics 10
electromagnetic field 217, 239, 244, 249
electroweak theory 325–8
Emch, Gérard 20, 50, 52, 60–1, 115, 117, 145, 150, 159–61, 167, 263, 265–8, 270, 273, 276, 279–82, 284–7, 291, 296, 300–7, 315, 322, 331
Emch-Knops model 281, 285–7, 291, 296, 305
Empirical success 1, 5, 13, 18, 103, 132, 174, 337–9, 340–56
Energy principle 232–4, 237

Entanglement 39, 109–114. *See also* (non)-locality
Equal time commutation relations (ETCCRs) *195*, 251–2
Equilibrium. *See* KMS condition/state.
Ergodicity 167, 263–8. *See also* state, ergodic.
Ergodic system *267*
Everett, Hugh 3
Explanatory capacity 4–5, 7–8, 11, 17, 28–9, 125–6, 130–1, 139, 142, 238–9, 246, 247–9, 261–8, 291–2, 298–9, 330–2, 338–9, 340–56
Extendability principle *236–9*
Extremist interpretations as caricatures 119. *See also* Interpretation

Fabri, E. 362
Fairhurst, S. and Willis, J. L. 53
Farmelo, Graham 22
Fecundity 34, 289, 339, 349–50
Fell's theorem 134–6, 222, 225–6, 303
Fermi gas 167
Fermi–Dirac statistics 130, 218
Fermion 53, 60–1, 130, 193
Ferromagnetism 262, 268–72, 284–7, 291–3, 303–6, 308–9, 313, 321–2, 338
Feynmann diagram 17, 247–53, 255–6
Field interpetation of QFT 203, 220
Field operators *209*
Figliolini, F. 167
Fine, Arthur 3, 226
Florig, Martin 208, 242–3
Folium *96*
Folland, G. 51, 56, 326, 328
Fock condition 199
Fock space 9, *68–70*, 195–9, 206–11
frequency-splitting heuristic 68, 195–9
Friedman, Michael 207
Friedrichs, K. O. 147
Frisch, Matthias 10
Fulling, Stephen 35–6, 71, 119–20, 131, 204, 211–13, 222, 235, 238, 257–8
"fundamental" in the physicist's vs. the metaphysician's sense 200
Fundamental observables 31–2, 53, 213–14, 223–5, 228
   And algebraic completeness 200–4
fundamental set of observables *202*
Future physics 12, 18, 238, 335–9, 348–9, 355–6
Van Fraassen, Bas 2, 6–9, 121, 171, 340, 342, 352, 355
Fraser, Doreen 103, 251
Frisch, Mathias 10

Gårding, L. 70, 254
Gauge principle *325*

Gauge theory 25, 30, 31, 103, 312–36
Gel'fand, I. M. 92
Generalists and particularists 1–5, 13, 18, 340–3
generalized coordinates *30*
General Theory of Relativity (GTR) 7, 8, 10, 66–7, 124, 193
Geroch, Robert 67
Giulini, Domenico 293, 325, 333
Glauber, Roy 239, 244
Gleason's theorem 21, 89, 90, 95, 121
   Generalized 95
Glimm, James 109, 255–6
Glymour, Clark 26, 45, 145
Geography 3–4, 294–5
Goldstein, H. 317
Goldstone boson 313, 320–4, 327, 328–30, 331–2, 334–5
Goldstone model 320–4, 328–30
Goldstone's theorem 321–4, 328
Gotay, M. J. 51, 57
Gottfried, Kurt 240, 241
Grib, A. A. 247, 257
Group *23*
Guido, D. 167
Guralnik, G. S. 322

Haag–Kastler axioms 104–9
Haag, Rudolf 104, 107, 109–12, 117–19, 127–8, 133–7, 252, 308
Haag–Ruelle scattering theory 250, 253–5
Haag's theorem 117, 193, 248, 250–6, 340
Hadamard condition 124, 221–2, 235–9, 289, 297, 299, 350
Hadamard principle *235–9*, 260
Haeccity 193
Hacijek, P. 212
Hall, D. 117, 252
Halvorson, Hans 16, 48, 50, 63, 91, 103–4, 112, 120, 127–30, 134, 143, 151, 156, 158, 164–6, 182–4, 188, 193, 211–12, 222–3, 318
Hamilton's equations *31–3*
Harmony principle 123, 139, 141–6, 223, 227
Hawking radiation 216, 238–9, 350
Healey, Richard 25, 31, 59, 325, 332, 334–5
Heisenberg, Werner 19
Heisenberg group 41, 54–7
Heisenberg model of ferromagnetism 278, 281, 284–6, 300–5, 308, 321–2, 333
Heisenberg picture *23*, 115, 160–1
Heisenberg relations *196*
Heisenberg Uncertainty Principle 50, 239
Hempel 11, 247–8, 255–7, 259, 281
Henneaux, M. and Teitelboim, C. 31, 334
Hepp, Klaus 255, 273
Hey, Anthony 322, 324, 326, 329

Higgs mechanism 313, 322, 324–6
High Energy Physics (HEP) 13, 191–2, 312–35. *See also* Standard Model, Physicists' QFT.
Hilbert space 21
 Direct sum of *39*
 Invariant subspace of *39*
 Non-separable 48–9, 63
 Separable *21*, 48, 63
Hilbert Space Conservatism 16–7, 119, 122–6, 127, 130, 133, 142–5, 193–5, 232–4, 259, 261, 278–9, 281–90, 292, 297, 310, 331
 Privileging strategies for 123–5
Hilbert space representation *14*
Honneger, Reinhard and Rieckers, A. 245
Van Hove obstruction 14, 117
Houtappel, R. M. F. 294
Huggett, Nick 194, 220
Hughes, R. I. G. 20, 22, 40
Husain, V. and Winkler, O. 53

Idealization 17, 102, 311–14, 330, 333, 335–9, 349
Ideal of Pristine Interpretation 3–4, 8–10, 12–13, 15–17, 118–19, 121–3, 125, 132, 146–8, 170, 174–6, 188–9, 233–4, 237–8, 246, 262, 295, 297–300, 343, 350
ineffability 183–4
infinitesimal generator *20*
infinite spin chain 59–65, 80–1, 83, 95, 115–16, 126, 135, 140, 150, 152, 159, 272–8, 281, 284–7, 300–1, 305–6, 310, 321–2
infrared model/representation 131, 235, 244–6, 299, 353
infrared problem 245–6, 299
Initial/boundary conditions 2–3, 9, 119, 292, 194
Instance principle *141*
Instantiation 139–41, 185–6
Instrumental reliability 345–6, 349–53
Interpretation 1–18, 103–4, 113–14, 117–47, 169–88, 193–5, 340–55
 Adulterated 4–5, 13–14, 16–17, 119, 123, 125, 146–7, 189, 232–4, 237–9, 246, 260, 262, 342–3
 Already partial 3, 7–8, 36
 Criteria of adequacy for 10–12, 247–8, 255–7, 259, 281
 Extremist 4, 16–18, 119, 122, 234, 261–2, 281–90, 291–3, 309–11, 312, 330, 333, 343, 350, 354
 Particle. *See* particle interpretation
 Pristine. *See* Ideal of Pristine interpretation
 Semantic phase of 8–10, 12, 121, 169–89

Structure-specifying phase of *8*, 12, 121, 126, 189
And physical law 2–9, 13, 118–19, 139, 142–3, 295–8, 340, 351–3
And scientific realism 2, 4–5, 12–13, 15, 340–56
Isham, Chris 57–8
Isometry. *See* symmetry, spacetime
Isomorphism 77
Inner automorphism *166*

Jack and Jill 190–1, 204–5, 214–16, 218–20, 221–4, 234–7, 260
Jaffe, Arthur 109, 256
Jammer, Max 19
Jones, Roger 341
Jordan–Wigner Theorem *62*
Jørgensen, P. E. T. 52

Kadison, Richard 24, 27, 73, 75, 77, 81, 82, 86, 89, 94, 145, 147, 150–7, 163, 165, 227
Kaku, M. 68, 326, 328
Kay, Bernard 104, 124, 210, 215, 231–3, 236–7, 247, 260
Kastler, D. 104, 107, 109, 118, 132–6, 145, 152, 321
(ke) weak kinematic equivalence 25–9, 97–8
(KE) kinematic equivalence 26–9, 98
(KE*) kinematic* equivalence 28–9
Kibble, T. W. B. 314, 320
Killing field *207*
Killing time *207*
Kinematics *8*, 14
Kinematic pair *23*
Kitajima, Yuichiro 187
Klauder, J. R. 239, 241, 245
Klein–Gordon equation *67*
Klein–Gordon field 35, 65–71, 115, 124, 131, 134, 166–7, 190–3, 185–99, 203–17, 234–8, 242–5, 296–7, 315–21
 Massless 217, 242–3, 317–21
KMS criterion for equilibrium 125, 159–61. *See also* State, KMS.
 Connection to modular theory 162–6
Knops, H. F. J. 281, 285–7, 291, 296, 301, 304–6
Kronz, Frederick 145, 283
Kubo, R. 160

Ladyman, James 344, 348
Landsman, N. P. 30, 42, 59, 93, 335
Lattice theory *177–9*
 Boolean lattice 157, 178–9
 Projection lattice 88, 157, 177–8
Laws of nature/physics 2–4, 6–9, 13, 26, 52–3, 102, 118–19, 139–46, 173,

185–6, 263, 289–90, 294–8, 309–10, 351–54
  Best systems account of 352–3
Lebesgue measure 23
Letaw, J. R. and Pfautsch, J. D. 216, 218, 224
Lévy Leblond, Jean-Marc 57
Lewandowski, J. and Sahlmann, H. 53
Lewis, David 6, 28, 352
Lie Algebra 32, 54–8
Lie Group 54
Lightcone structure 66–7
Liu, Chuang 160, 161, 270, 300–4, 307, 315, 332, 336, 338
Longo, Robert 321
Loop Quantum Gravity 53
Lüders Rule 96, *170–2*
Lupher, Tracy 145, 228, 283

Mackey, George 41, 56
Mackey's imprimitivity theorem 41
Magnetization observable. *See* polarization observable.
Magnon, Anne 199, 205, 207, 232
Malament, David 52, 67, 193
Manchak, John 52
Marsden, J. E. 30, 31
Mary and Martha 205, 217
Martin, P. C. 160
Maudlin, Tim 112, 113, 327
Maximal abelian subalgebra *179*
Maximal beable algebra *176*
Maximal beable approach (MBA) 16, 169–70, 173–89, 201
Maximal beable recipe *179*
Mean field models *285*, 307–8
Measurement problem 11, 170, 343
Mechanics,
  Hamiltonian 8–9, 30–3
  Lagrangian 30
  Matrix 19–20, 22
  Wave 20–2
Mermin, David 141
Messiah, Albert 22
"metaphysical" principles of interpretation 3–4, 119, 122, 169, 174–7, 188–9, 204, 237, 260, 288, 347
mexican hat potential *323*, 328
Miracles Argument 13, 340–56
Mixed strategies of interpretation 143–4
Mixed thermodyamic phases 273, 279–80
Modal interpretation 170, 175, 180. *See also* no-collapse interpretation
  Of QFT 187–8
Modesto, L. 53
Modular theory 162–6
  Modular automorphism group *164*

Modular invariants *165*
Modular operator *162*
Modular spectrum *165*–6
Moore, T. 52
Moore, Walter 19
Morandi, G. 33
Morchio, G 48, 53, 245, 285
Morgan, Mary 10
Murray, F. J. 149
(*) morphism 77
Morrison, Margaret 10

Naive Story *332*–5
Nambu, Yoichiro 324
von Neumann Algebra *86–8*
  Affiliated with a representation 87
  Atomless 151, 155–9, 166–8, 169–89
  Center of 99
  Central summands of *154*
  Factor 99
  Finite 153
  Normal state of 95–6
  Semifinite 153
  Type Theory for 149–55
  Type I 148, *150*
  Type II *152*
  Type III *152*
  Universal enveloping *145*, 261, 283, 310, 314, 348
  Weight on *163*
Von Neumann, John 19, 20, 24
Newtonian theory 10, 30, 31, 270, 293, 252
Noether's theorem 317–19
No-collapse interpretations 169, 170–1, 180, 280, 244. *See also* Modal interpretation.
(Non)-locality 106, 109–14
norm *75–6*
  completeness with respect to 76
normal conception of states *21*, 61, 90, 138
normal ordering *197*
Norton, John 4, 66, 287
n point function *210*
number operator 44–5, 69–70

observable 8
  at ∞ 276
  macroscopic 273–7, 281, 282, 285–6, 288, 303–4, 312, 337–8
  spatially averaged *273*
occupation number formalism 70, 202
one particle structure *231*
Ojima, Izumi 127, 131, 132
Olver, Peter 294
operationalism 97, 104–5, 134–9, 175, 216–18, 288
order parameter 269, 286, 313, 320, 322, 331

Ordinary QM 14–6, 20–5, 48, 89–90, 92–3, 99–100, 118–23, 148–9, 169–70, 180–1, 262–72
   OQM *21*–5, 36, 71, 118
   OQS *21*–5, 36, 71, 118
   OQD *21*–5, 36, 71, 118
Ordinary QSM *263*–72
Orthogonal complement *85*

Parker, Leonard 247, 257–8
Parmenides 295
"Particle behind the moon" argument 135–6
Particle interpretation 16–7, 124, 190–60
   Fundamental 191–206, 209–10, 219–20
   Phenomenological 17, 247–60
   And Fock space structure 69–70, 195–9, 206–13
Particle notions,
   Extended 222–9
   Incommensurable 191, 205–16, 219–20, 222, 237, 248–9, 251–3, 257–60
   Killing vs μ-born 229–32
   Operationalized 216–18
   Universalized 226–9
Particle types vs particle varieties 191–2
Particularist. *See* generalist and particularist.
Pauli exclusion principle 306
Pauli relations *59–60*
   And CARs 60
Pauli spins *59–60*
Pauli, Wolfgang 19
Pawlowski, T. and Singh, P. 53
Peskin, Michael 186, 197, 245, 249, 318, 325, 328, 330
Pessimistic Metainduction 13, 346–9
Phase Argument 218, 281–4, 286
Phase diagram 270, 289, 338
Phase observable *282*–3, 299, 303, 311, 313, 351
Phase structure xii, 17, 146, 160, 272–90, 291, 301–2, 311, 338, 350
Phase transition 270, 282, 289, 307, 313, 336, 338
Philosophy of physics 2–3, 6, 8, 10, 130, 333, 336, 337, 340, 355
Physical equivalence 14–5, 24–9, 45, 85–6, 97–8, 122–3, 134–9, 330
   Content coincidence criterion of 25
Physical possibility. 2–9, 15–16, 18, 25, 72, 141–2, 146–9, 236–7, 286–90, 292, 299, 351–5. *See also* Laws of nature/physics.
   Naturalized 352–3
   pragmatized 353–5
   Unimodal 3–4, 10, 17
Physicists' QFTs 17, 103, 256, 312–36
Picasso, L. E. 362

Pickering, Andrew 217
Planck scale 313, 330
Planck's constant *20*
Plato 1, 73
Poincaré, Henri 346
Poincaré group *66*
Poisson Bracket *32*
   ... goes to commutator rule 37, 51, 54, 58
Polar decomposition 162–3
polarization observable *63*–5
Poole, Charles 317
Powers, R. T. 153
Preparation problem 169, 170–3
Primas, Hans 114, 146
Projection operator(s),
   Abelian *150*
   Central *302*
   Equivalent *149*
   Infinite *149*
   Minimal *150*
   Range of *149*
   Subprojection *149*
   Weaker than *149*
Projective unitary representation 56
Provost, J. P., Rocca, F., and Vallee, G. 245
Prugoveki, Edouard 41, 46–8, 200
Przibaum, K. 19
Psillos, Stathis 13
Pure thermodynamic phase *272–5*
Putnam, Hilary 29, 341

Quanta concept 194
Quantization,
   Hamiltonian 35–7
   Ideologies of 51–3, 334–5
Quantum Field Theory (QFT) 2, 11–2, 14, 16–8, 65–71, 102–14, 127–32, 166–7, 186–8, 208–60, 312–39.
Quantum gravity 11–2, 53, 125, 193, 235, 238–9, 350, 353, 355
QM$_\infty$ 2
   Individuating theories of 119–20
Quantum probability, interpretation of 170–89
Quasi-equivalence 87

Ratiu, T. S. 30, 31
Rawls, John. 1–2, 5
Real inner product μ 210
Reasonable Criterion (for interpreting symmetry) 297–300
Redei, Miklós 11–2, 152
Redhead, Michael 4, 109, 113, 145, 175, 193–4, 204
Reed, Michael 23, 254
Reents, George 208, 243
Reeh–Schlieder theorem 113–14
Relative state formulation 3, 169, 180–1

Renormalization 104, 235, 326, 330–32
Renormalization Group 269, 289, 299, 338–39, 349
Representation. *See also* state
  Coherent 242–6
  Cyclic 92
  Factor 99
  Faithful 84
  GNS 92
  (ir)reducible 39, 93
  irreducible Fock space 199
  momentum 50
  non-degenerate 94
  non-regular 46–3
  Of a C^* algebra 82–5
  Of canonical relations 36
  position 46–3
  positive energy 108
  Rindler 131, 215–16
  Universal 145–6, 261, 279, 282–3, 288
Representational Realism 130
Ringrose, J. R. 24, 27, 73, 75, 77, 81, 82, 86, 89, 94, 145, 150–7, 163, 165, 227
Rivasseau, V. 102
Roberts, John E. 97, 127, 129, 321
Robinson, Derek 27, 40, 73, 77–8, 83, 86–91, 93–5, 98, 117, 122, 123, 133–4, 138, 155, 158, 161–2, 165, 167, 210, 211, 279, 281, 284
Roepstorff, G. 97, 109–20
Rosenberg. Johnathan 42
Rovelli, Carlo 164
Ruelle, D. 158, 273

Safko, John 317
Sakai, S. 122, 155
Scholia
  "Bra-ket" notation 22
  Cauchy surfaces 107
  Direct Sum of Hilbert spaces 39
  Folk definition of "symmetry" 293
  Internal and spacetime symmetries 315–16
  Measurable functions 23
  Notational variants on Klein–Gordon equation 67–8
  Residual symmetry 327
  Some common Hilbert spaces 23
  Static and stationary spacetimes 207
  The symmetry groups $U(1)$ and $SU(2)$
Schrödinger equation 23, 185–6, 343–4
Schrödinger, Erwin 19, 20, 256
Schrödinger picture 23
Schrödinger representation 37–45, 49–50, 54, 55, 57
Schmüdgeon, K. 52
Schroeder, Dan 186, 197, 245, 249, 318, 325, 328, 330

Schwartz space 52
Schweber, S. S. 117
Schwinger, J. 160
Scientific realism 2, 4–5, 12–13, 147, 340–56
  Global 344–6, 349–51
  Structural 346–51
    The Content Question for 347–9
Segal, Irving 39, 50, 53, 70, 92, 102, 106, 117, 132, 137–8, 142, 144, 197
SEGOP principle 138–39, 141–3, 175
Sellars, Wilfrid 7
"Semantic view" of theories 5–6
Semiclassical quantum gravity 11–12, 125, 235, 238–9, 289, 299, 350, 353, 355
Sewell, Geoff 61, 63–5, 114, 143, 161, 244, 273, 276–7, 285–86, 337–38
Shankar, Ramamurti 22
Sharp Distinction, Doctrine of 292–9
Simon, B. 23, 41, 63, 254
Simple harmonic oscillator 42
  Quantization of 42–5, 239–42, 307
  QFT analogs 68–70, 195–8
Simplex 280
single particle Hilbert space 68
Skagerstram, B. S. 245
Sklar, Larry 263–4
smeared field operator 196
Sound Principle of interpretation 312–14, 336–8
Sounder Principle of interpretation 336–9, 348
spacetime
  crash course 65–7
  expanding 256–9
  globally hyperbolic 107
  metric 65–6
  Minkowski 65–7
  Rindler 214–15
  Static 207
  Stationary 207
  Schwartzschild 238
  time orientable 104
Spectral Theory 46–8
Spectrum condition 108
"Standard Account" of theoretical content 5–10, 24, 72, 146
Standard Model 18, 59, 191–3, 247, 256, 312–39
state space 8
state 25, 89
  coherent 239–46
  "converging to λ" 91, 157–8, 183–5
  countably additive 21, 80
  cyclic 69. *See also* cyclic vector.
  Dispersion-free 175–7
  Ergodic 165, 187–8
  Factor 99

state (cont.)
  Ground 167
  Hadamard 235
  Intrinsically mixed 158–9, 166–7, 184
  Invariant 106
  KMS 125, 159–61, 218–19, 261–309
  linear 21
  left kernel of 156
  mixed 21, 89–90
  normal 90, 95
  positive 21
  pure 21, 89–90
  quasi-free 210, 230–7, 243
  regular ground 231
  support projection of 156
  tracial 163
  Type III factor 158
  vacuum state 44, 69
  vector 22
  ultraweakly continuous 94
  uniformly/ultraweakly/and weakly continuous 94, 137–8
Stone, Marshall 24
Stone–von Neumann theorem 15, 24, 29, 30–44, 46
Stone's theorem 24
Streater, R. 60, 284, 300, 315, 320
Stress energy tensor 8
Strocchi, G 48, 53, 199–200, 245–6, 285, 321, 325
(strong) unitary implementability of an automorphism 106
Subrepresentation 39, 84–5
Substantivalism 7, 12
Sudarshan, E. C. G. 239, 241
Summers, Stephen 37, 49, 111, 114, 136, 152, 208, 242–3
Sunder, V. S. 149, 152–4, 162–4, 166
Superconductivity 59, 167, 268, 284, 287, 306–10, 324, 330
Superselection 40, 93–4, 135–6
  And DHR selection theory 127–132
Suppe, Fred 2, 8, 9
symmetry,
  broken. See Broken Symmetry
  folk definition of 293
  gauge 127–32, 309, 324–36
  hidden 315. See also broken symmetry
  internal vs spacetime 315–16
  Lagrangian 315
  Noether 319, 321
  Poincaré 66
  Prima facie 295–9
  residual 327
  spacetime (isometry) 66
  symplectic 55
  unitarily implementable 126

symplectic product 33
symplectic manifold 34
symplectic vector space 33–5

Takagi, S. 218
Takesaki, Masamachi 145, 152–3, 156, 162–4, 167
Tamir, Mike 77
(thermodynamic limit of) Quantum Statistical Mechanics 2, 11, 114–16, 125, 142, 158–62, 167, 186, 261–312, 339
Teller, Paul 44, 46, 51, 70, 103, 193–4, 198, 202, 213, 244, 326
Theory of canonical type X 120
Thirring, W. and Narnhofer, N. 53
t'Hooft, G. 103, 325, 332
Tipler, Frank 249
Topologies
  Norm operator. See Uniform operator.
  Strong operator. 78
  Ultraweak operator 26–7, 79
  Uniform operator 78
  Weak operator 79
Transition probability 27–8, 45, 97, 223

Uffink, Jos 263–4
Ultrafilter extension theorem 147
Unitary equivalence
  Of ordinary quantum theories, 27
  of representations 85
  Illustrated 44–5, 62–4
  As a criterion for physical equivalence 14–5, 24–30
  And Algebraic Imperialism 133
Unitary inequivalence, Examples of 46–72, 117, 203–4, 211–14, 258, 322
Universalism 119, 145–6, 261, 282–4, 286–7, 290, 292, 310, 314, 322, 331, 350
Tempered 220
Unruh, Bill 190, 238
Unruh effect 16, 131, 190–1, 204–6, 218–19, 221–2, 225–6, 260
Unruh vacuum 238–9
Updike, John 190

Valente, Giovanni 112
Virtue, theoretical 3, 4–5, 13, 17–8, 245, 326, 340–56. See also individual virtues, e.g., empirical success, explanatory capacity, fecundity.
  Decoupling of Truth and 343–4
  Disintegration of 14, 342–56
  Disunity of 342–56

W* algebra. See von Neumann algebra
W* argument 281, 284–7, 288

Wald, Robert 12, 31, 33, 35, 65, 68, 104, 120, 124, 134, 196–9, 204, 206–7, 210–11, 216, 218, 222, 225, 231–3, 235, 237, 260
Wallace, David 103, 181, 292
Weak equivalence *133*–8, 142, 225–6
Werner, R. 62, 114
Weyl Algebra *115, 118*
Weyl, Hermann 55
Weyl relations/Weyl form of the CCRs *37*–9
Wightman, A. 60, 70, 117, 252, 254
Wigner, Eugene 3, 28, 62, 191, 270, 294–5, 318
Wigner's Theorem 55, 270, 318
Witness for a state *171*–3, 188, 303
Witten, Ed 325
Wittgenstein, Ludwig 6
Woodhouse, Nicholas 68
Worrall, John 346–7

Yan, Tung-Maw 240, 241

Zee, A. 269, 326, 328
Zhou, Z. 177, 197
Zhu, Kehe 162
Zorn's lemma 155, 157, 182